Span, Multiparameter Equations of State

Springer
*Berlin
Heidelberg
New York
Barcelona
Hong Kong
London
Milan
Paris
Singapore
Tokyo*

R. Span

Multiparameter Equations of State

An Accurate Source
of Thermodynamic Property Data

With 151 Figures and 36 Tables

 Springer

Dr.-Ing. Roland Span
Ruhr-Universität-Bochum
Lehrstuhl für Thermodynamik

44780 Bochum / Germany

e-mail: R.Span@thermo.ruhr-uni-bochum.de

ISBN 3-540-67311-3 Springer-Verlag Berlin Heidelberg New York

CIP data applied for

Die Deutsche Bibliothek - CIP-Einheitsaufnahme
Span, Roland: Multiparameter equations of state : an accurate source of thermodynamic property data / R. Span. - Berlin ; Heidelberg ; New York ; Barcelona ; Hong Kong ; London ; Milan ; Paris ; Singapore ; Tokyo : Springer, 2000
ISBN 3-540-67311-3

This work is subject to copyright. All rights are reserved, whether the whole or part of the material is concerned, specifically the rights of translation, reprinting, reuse of illustrations, recitation, broadcasting, reproduction on microfilm or in other ways, and storage in data banks. Duplication of this publication or parts thereof is permitted only under the provisions of the German Copyright Law of September 9, 1965, in its current version, and permission for use must always be obtained from Springer-Verlag. Violations are liable for prosecution under German Copyright Law.

Springer-Verlag is a company in the BertelsmannSpringer publishing group
© Springer-Verlag Berlin Heidelberg 2000
Printed in Germany

The use of general descriptive names, registered names, trademarks, etc. in this publication does not imply, even in the absence of a specific statement, that such names are exempt from the relevant protective laws and regulations and therefore free for general use.

Typesetting: Camera-ready by author
Cover layout: Medio, Berlin
SPIN: 10742955   Printed on acid-free paper   62 / 3020 hu - 5 4 3 2 1 0

Preface

As a basis for printed property charts and tables, empirical multiparameter equations of state have been the most important source of accurate thermodynamic property data for more than 30 years now. However, due to increasing demands on the accuracy of thermodynamic property data in computerised calculations as well as the availability of appropriate software tools, and the ever increasing computer power, such formulations are nowadays becoming a valuable tool for everyday work. This development has substantially increased the number of scientists, engineers, and students who are working with empirical multiparameter equations of state, and it continues to do so.

Nevertheless, common knowledge on this kind of thermodynamic property models and on the ongoing progress in this scientific discipline is still very limited. Multiparameter equations of state do not belong to the topics which are taught intensively in thermodynamic courses in engineering and natural sciences and the books and articles where they are published mainly deal with the thermodynamic properties of certain substances rather than with the theoretical background of the used equations of state. In contrast to this, my concern mainly was to give a survey of the theoretical background of multiparameter equations of state both with regard to their application and their development. This book is written for users of multiparameter equations of state who need some background information for their own application developments as well as for scientists who are active in the development of such equations or who want to become involved in it. As the only current and comprehensive work of this kind, it is intended to improve common knowledge on multiparameter equations, to fuel their application, and to attract more scientists to the work on such thermodynamic property models.

For my own part, I have been involved in the development of multiparameter equations of state since I joined the research group of Wolfgang Wagner at the Ruhr-Universität Bochum as a student almost 15 years ago. After all this time, I am firstly and above all grateful for all the support I have received from Prof. Wagner. He is not only an outstanding expert in this field, but also an extraordinary personality who has over the years slowly turned from a professor into a colleague – it has always been fun to work in his group. Many of the assessments given in this book result from intensive discussions with him and the main reason why this book is authored only by me and not by *Span and Wagner* is that it is based on the habilitation thesis which I submitted last year. Secondly, I am grateful to V. Marx who introduced me to the work on multiparameter equations of state years ago. I wish I had as much time for my students as he had for me. Besides this, I would like to thank all those who later worked with me on this topic, namely

A. Pruß, Ch. Tegeler, A. Kruse, J. Smukala, M. Kubessa, R. Klimeck, M. Künne, H.-J. Collmann, C. Bonsen, D. Bücker, and M. Waschke. In very different ways, they have all contributed to the scientific foundation of this book. However, work on multiparameter equations of state is always closely related to experimental work as well. In place of all the experimentalists who have helped me to understand the possibilities and the limitations of experimental work, I would like to thank R. Kleinrahm, N. Kurzeja, R. Gilgen, and M. Trusler for many fruitful discussions. And finally, I am grateful to E. W. Lemmon not only for the inspiring discussions which we have had over the course of several co-operative projects but most of all for turning the manuscript of this book into readable English.

Bochum, February 2000   Roland Span

Contents

Nomenclature XI

1 Introduction 1

2 History and Potentials – an Overview 5
   2.1 A Brief History 5
   2.2 Publications on Thermodynamic Reference Data 11
   2.3 Current Work and Future Challenges.. 12

3 Using Multiparameter Equations of State for Pure Substances 15
   3.1 Different Kinds of Multiparameter Equations of State 15
        3.1.1 Equations in Terms of the Helmholtz Energy 16
             3.1.1.1 Correlations for the Helmholtz Energy of the
                         Ideal Gas 18
             3.1.1.2 Common Formulations for the Residual Helmholtz
                         Energy 23
        3.1.2 Equations in Terms of Pressure 25
             3.1.2.1 Integration of Pressure Explicit Equations 26
        3.1.3 Special Functional Forms... 28
             3.1.3.1 Hard Sphere Terms 29
             3.1.3.2 Critical Region Terms.. 31
        3.1.4 Simple Equations for Gas Phase Calibrations............................ 35
   3.2 Calculating Thermodynamic Properties from the Helmholtz Energy...... 35
        3.2.1 Properties in the Homogeneous Region 36
        3.2.2 Properties of the Vapour-Liquid Equilibrium Phases 42
        3.2.3 Properties in the Vapour-Liquid Two Phase Region................... 45
   3.3 Iterative Procedures... 46
        3.3.1 Calculations Based on Temperature and Pressure 48
        3.3.2 Calculations Based on Pressure and Density 51
        3.3.3 Calculations Based on Pressure and Enthalpy 52
        3.3.4 Calculations Based on Pressure and Entropy.............................. 54

      3.3.5   Calculation of Phase Equilibria 54
              3.3.5.1   Multiple Maxwell Loops 56

4. Setting Up Multiparameter Equations of State for Pure Substances 61

4.1   Linear and Nonlinear Fitting 62
4.2   Describing the Helmholtz Energy of the Ideal Gas 66
4.3   Describing the Residual Helmholtz Energy 74
      4.3.1   The Multiproperty Approach 74
      4.3.2   Defining Residua for Linear Algorithms 77
      4.3.3   Defining Residua for Nonlinear Algorithms 80
      4.3.4   Assigning Weights to Experimental Data 85
4.4   Optimising the Functional Form 90
      4.4.1   Defining a Bank of Terms 91
              4.4.1.1   Hard Sphere Terms 93
      4.4.2   Setting Up a Regression Matrix 97
      4.4.3   The Stepwise Regression Analysis (SEEQ) 98
              4.4.3.1   Introduction of a Pairwise Exchange 103
      4.4.4   The Evolutionary Optimisation Method (EOM) 106
      4.4.5   The Optimisation Algorithm by Setzmann and Wagner (OPTIM) 107
              4.4.5.1   Adapting OPTIM to Equations of State 114
              4.4.5.2   Introduction of a Pairwise Exchange 116
      4.4.6   The Nonlinear Optimisation Algorithms by Tegeler et al. 117
              4.4.6.1   The Nonlinear Quality Criterion 118
              4.4.6.2   The Nonlinear Stepwise Regression Analysis (NLREG) 121
              4.4.6.3   The Nonlinear Optimisation Algorithm (NLOPT) 125
              4.4.6.4   Speeding Up Nonlinear Optimisation Algorithms 126
      4.4.7   Automated Optimisation Algorithms 128
              4.4.7.1   Optimisation of Simplified Equations for Mixtures with Constant Composition 129
      4.4.8   Independent Developments and Future Perspectives 131
4.5   Describing Properties in the Critical Region 133
      4.5.1   Predictions from Theory 134
              4.5.1.1   Theoretically Founded Equations of State 138
      4.5.2   Capabilities of Empirical Multiparameter Equations of State 141
      4.5.3   Setting Up Equations with Modified Gaussian Bell Shaped Terms 151
      4.5.4   Setting Up Equations with Nonanalytic Terms 152

        4.5.5   Semiempirical Approaches 156
                4.5.5.1   The Use of Switching Functions 157
                4.5.5.2   The Transformation Approach 158
                4.5.5.3   The Approach of Kiselev and Friend 159
    4.6   Consideration of the Extrapolation Behaviour 161
        4.6.1   Comparisons with Data Beyond the Range of Primary Data 163
        4.6.2   The Influence of the Functional Form 166
        4.6.3   The Representation of Ideal Curves 168

5 The Performance of Multiparameter Equations of State 173
    5.1   Comparisons with Thermal Properties 177
    5.2   Comparisons with Caloric Properties 191
    5.3   Properties at Vapour-Liquid Phase Equilibrium 206
    5.4   Some General Assessments 214

6 Generalised Functional Forms 219
    6.1   Simultaneous Optimisation of Functional Forms 223
    6.2   Simultaneously Optimised Equations for Technical Applications 227
        6.2.1   Accuracy Versus Numerical Stability – A Compromise 235
                6.2.1.1   An Investigation of Numerical Stability 237
                6.2.1.2   The Influence of Uncertain Critical Parameters 242
                6.2.1.3   Conclusions Regarding Required Data Sets 245
        6.2.2   Results for Non- and Weakly Polar Fluids 248
        6.2.3   Results for Polar Fluids 262
    6.3   Simultaneously Optimised Reference Equations of State 275

7 Generalised Equations of State 277
    7.1   BACKONE Equations of State 279
        7.1.1   Fitting BACKONE Equations of State to Data 281
        7.1.2   Some Results 285
    7.2   Generalised Empirical Equations of State 291
        7.2.1   The Approach by Platzer and Maurer 292
                7.2.1.1   Some Results 293
        7.2.2   The Approach by Span and Wagner 300
                7.2.2.1   Fitting the Substance Specific Parameters 303
                7.2.2.2   Results for Non- and Weakly Polar Fluids 306
                7.2.2.3   Numerical Stability 314

8 Describing Mixtures with Multiparameter Equations of State ... 319
8.1 Using Composition Dependent Sets of Coefficients ... 320
8.1.1 The AGA8-DC92 Equation of State for Natural Gases ... 324
8.2 Extended Corresponding States Approaches ... 327
8.2.1 Interpolation Between Reference Fluids ... 328
8.2.2 The Shape Factor Concept ... 329
8.3 Helmholtz Models with Departure Functions ... 332
8.3.1 Mixture Specific Departure Functions ... 335
8.3.2 Generalised Departure Functions ... 336

References ... 341

Index ... 363

Nomenclature

Symbols

a	Helmholtz energy (molar or specific)
	Adjustable parameter
	Process parameter
	Term or a combination of derivatives of a term
	Element of the matrix **A**
A	Adjustable parameter
	Term
	Rotational constant
AAD	Average absolute deviation
b	Exponent
	Adjustable parameter
	Universal parameter in scaling approaches
	Element of the regression matrix **B**
	Binary interaction parameter
	Density exponent (AGA8-DC92)
B	2nd virial coefficient
	Adjustable parameter
c	Heat capacity (molar or specific)
	Speed of light in vacuum
	Adjustable parameter
	Generalised coefficient
	Ancillary variable for matrix transformations
C	3rd virial coefficient
	Number of components
	Number of constraints
	Arbitrary amplitude
C^*	Composition dependent coefficient (AGA8-DC92)
d	Derivative
d	Parameter
	Density exponent
	Ancillary variable for matrix transformations
D	4th virial coefficient
	Discriminant
e	Parameter

XII Nomenclature

f	Switching function
	Parameter
	Ancillary variable for matrix transformations
	Equivalent substance reducing factor
	Composition dependent function
F	Parameter of the Fisher F statistic
	Scaling factor for generalised departure functions
g	Gibbs energy
	Generalised coefficient
Gi	Ginzburg number
h	Enthalpy (molar or specific)
	Planck constant
	Equivalent substance reducing factor
i	Serial number
I	Upper limit of a serial number
	Number of terms in the bank of terms
	Moment of inertia
j	Massieu function (molar or specific)
	Serial number
J	Process parameter
	Upper limit of a serial number
	Parameter of the bank of terms
k	Expansion coefficient
	Boltzmann constant
	Adjustable parameter
	Density exponent (AGA8-DC92)
K	Number of independent variables
	Parameter of the bank of terms
	Kernel term
	Mixture size parameter (AGA8-DC92)
L	Likelihood function
	Dimension of the regression matrix
m	Coefficient
	Adjustable parameter
	Serial number
M	Number of data points
n	Coefficient
	Adjustable parameter
	Number of possible exchanges
N	Number of adjustable parameters, terms
NA	Number of investigated additions, optimisation parameter
NC	Number of constraints, optimisation parameter
NE	Number of investigated exchanges, optimisation parameter

NG	Number of generations, optimisation parameter	
NM	Number of mutations, optimisation parameter	
NP	Number of parameter vectors, optimisation parameter	
NPE	Number of investigated pairwise exchanges, optimisation parameter	
NR	Number of regression runs, optimisation parameter	
p	Pressure Density exponent in the exponential function Element of the vector $\mathbf{P}$	
P	Number of properties	
PS	Number of preselected terms, optimisation parameter	
q	Ancillary parameter Element of the matrix $\mathbf{Q}$	
Q^*	Reduced quadrupole moment	
r	Ancillary parameter Parametric variable	
R	Gas constant (molar or specific) Kernel term parameter	
RES	Residuum	
RMS	Root mean square deviation	
s	Entropy (molar or specific)	
S	Sum of squares in the regression matrix Probability	
t	Parameter of the Student t statistic Temperature exponent	
T	Temperature	
u	Internal energy (molar or specific) Temperature exponent (AGA8-DC92)	
U	Potential energy barrier	
v	Volume (molar or specific) Element of vector $\mathbf{V}$	
w	Speed of sound Adjustable parameter	
x	Vapour fraction (quality) Independent variable (measured value) Element of the vector $\mathbf{x}$, molar concentration	
x^*	Liquid fraction	
X	Independent variable (exact value) An arbitrary property Substituted property	
y	Dependent variable (measured value) Any property	

Y	Dependent variable (exact value)
	Transformed compression factor
	Crossover function
	An arbitrary property
z	Any implicit property
	Standard normal value
	Normally distributed random number
Z	Compression factor
	An arbitrary property

Greek Symbols

α	Reduced Helmholtz energy (reduced Massieu function)
	Critical exponent
α_v	Volume expansivity
β	Adjustable parameter
	2nd acoustic virial coefficient
	Critical exponent
γ	Coefficient in the exponential function
	Adjustable parameter
	Precorrelation factor (speed of sound)
	Critical exponent
Γ	Gamma function
δ	Reduced density
	Throttling coefficient
	Critical exponent
∂	Partial derivative
Δ	Distance to the critical point
	Any difference
	Universal constant (kernel term)
$\Delta\alpha$	Departure function for the reduced Helmholtz energy
ε	Adjustable parameter
	Convergence criterion
	Excited state
	Precorrelation factor (isobaric heat capacity)
ζ	Empirical relation for any property
	Distance function
η	Adjustable parameter
θ	Reduced temperature
	Temperature-like parameter
	Reduced distance to the phase boundary
	Ancillary variable

	Parametric variable
	Shape factor
ϑ	Reduced temperature-like parameter
λ	Lagrangian multiplier
μ	Joule-Thomson coefficient
	Chemical potential
μ^*	Reduced dipole moment
ν	Wavenumber
	Degree of freedom
	Adjustable parameter
ξ	Packing fraction
	Physical relation for any property
π	Reduced pressure
	Geometrical constant ($\pi = 3.14159265$)
ρ	Density (molar or specific)
σ	Standard deviation
σ^2	Variance
τ	Inverse reduced temperature
$\hat{\tau}$	Reduced difference to T_c
	Transformed reduced temperature
φ	Anisotropy parameter
	Fugacity coefficient
	Angle
ϕ	Shape factor
χ	Polar factor
χ^2	Sum of squares
X^2	Sum of sum of squares
X^{*2}	Sum of reduced sum of squares
ψ	Exponential (damping) function
ω	Acentric factor

Vectors and Matrices

A	Matrix resulting from the system of normal equations
	Ancillary matrix
B	Regression matrix
C	Vector of constraints
n	Vector of coefficients
N	Vector of coefficients
P	Parameter vector
Q	Matrix resulting from the system of normal equations

XVI Nomenclature

V	Ancillary parameter vector
X	Vector of exact independent variables
x	Vector of measured independent variables
	Vector of compositions
λ'	Vector of Lagrangian multipliers

Superscripts

o	Ideal gas state
	Hypothetical ideal gas state at given values for T and ρ
I	First integration
II	Second integration
(0)	Isotropic reference fluid
(r)	Anisotropic reference fluid
'	Saturated liquid state
"	Saturated vapour state
*	Different set of parameters
	Ancillary vector or matrix
~	Reduced properties
a	Contribution from attractive dispersion
d	Contribution from dipolar interaction
e	Exponential term
E	Excess
f	Modifying the mixing rule for $f(\mathbf{x})$
h	Modifying the mixing rule for $h(\mathbf{x})$
p	Polynomial term
q	Contribution from quadrupolar interaction
r	Residual contribution
r*	Residual contribution in connection with hard sphere terms
T	Transposed
T	Modifying the mixing rule for $T_c(\mathbf{x})$
ρ	Modifying the mixing rule for $\rho_c(\mathbf{x})$

Subscripts

o	Substance specific reference
0	Reference state
	Fixed serial number
	Background
	Initial state

Nomenclature XVII

1,2	Different sets of parameters Fixed serial numbers
2	Property of a two phase system
68	Temperature according to IPTS-68
90	Temperature according to ITS-90
a	Acoustic
an	Analytical
AGA	Calculated from the AGA8-DC92 equation
B	Boyle curve
c	Critical parameter Integration constant Constraint
calc	Calculated
comb	Combined
cor	Correction
double	Pairwise exchange of terms
el	Electron
eq	Calculated from an equation of state
eq−1	Equation truncated by one term
est	Estimated
ev	Evaporation
exp	Experimental
expl	Explicit property
Exp	Exponential term
GBS	Modified gaussian bell shaped term
h	Hugoniot
i	Serial number
I	Ideal curve
irot	Internal rotation
j	Serial number
J	Joule inversion curve
JT	Joule-Thomson inversion curve
k	Serial number
l	Local
m	Serial number (for measured data)
M	Mixture
max	Maximum
min	Minimum
NA	Nonanalytic term
p	Isobaric

XVIII Nomenclature

PE	Planck-Einstein term
Pol	Polynomial term
r	Reducing parameter
rot	Rotational
s	At saturation
	Starting value
s	Isentropic
sc	Scaled
simp	Simplified
single	Exchange of single terms
t	Triple point parameter
T	Isothermal
trans	Translational
v	Isochoric
vib	Vibrational
VLE	Vapour-liquid phase equilibrium
δ	Partial derivative with respect to δ
τ	Partial derivative with respect to τ
σ	Along the saturated liquid line

1 Introduction

Knowledge of thermodynamic properties of involved fluids is one of the fundamentals of common calculations in engineering applications and in the natural sciences. However, the requirements with respect to both the accuracy and the availability of thermodynamic property data strongly depend on the application. Where only few properties are needed at single states, property tables are still a common source of information. Available tables range from very general compilations such as the VDI-Wärmeatlas (Schlünder et al., 1997) to compilations for a certain group of substances (see for instance Tillner-Roth et al. (1998) for refrigerants or Jacobsen et al. (1997) for cryogenic fluids) to internationally agreed reference tables for single substances (see for instance Wagner and de Reuck (1996) for methane or Wagner and Kruse (1998) for water). In practical use, such tables are being more and more replaced by convenient interactive programs which are executable on common personal computers (see for example Span et al., 1995a; Wagner et al., 2000b; Friend and Huber, 1994).

Where higher demands on the availability of thermodynamic property data are formulated, routines for the calculation of such data are included directly into user-defined programs or into commercial process simulation packages. The correlations which are used in such routines range from simple estimative approaches to highly accurate equations of state. Ideally, the decision for a certain property model should depend on requirements on accuracy which result from the considered application. However, formulations which are accurate enough to satisfy advanced demands on accuracy are available only for relatively few pure fluids and mixtures and most users are not aware of the capabilities and limitations of different kinds of property models. Thus, in practice the decision for a certain formulation depends more often on the availability of models and readily programmed algorithms.

This work focuses on the most accurate source of thermodynamic property data: on empirical *multiparameter equations of state*. Properly designed multiparameter equations of state are able to represent thermodynamic properties of a certain substance within the accuracy of the most accurate experimental data. In this way, the equation becomes *the reference* for the properties of the corresponding substance; nobody needs to directly refer to experimental data any longer. Such reference equations of state have been the basis for partly internationally agreed reference tables for more than 25 years now. As a result of the increasing computing power and the availability of suitable software solutions highly accurate equations of state are nowadays also being used in everyday applications.

However, the available multiparameter equations of state form a rather inhomogeneous group of thermodynamic property models. State-of-the-art reference equations of state are used simultaneously with obsolete formulations which are more than 25 years old. And in general, knowledge of this kind of equations of state is rather restricted, since they do not belong to the topics which are taught intensively in thermodynamic courses in engineering and natural sciences. As a consequence, restrictions which are valid for obsolete simple multiparameter equations of state are often attributed to state-of-the-art reference equations and, even worse, expectations on accuracy and reliability which would be justified for reference equations of state are attributed to obsolete formulations. The results are frustrating in the best case and sometimes even disastrous.

Even though the development of accurate multiparameter equations of state goes back a long time, the number of substances for which such equations are available is still quite limited. To establish increasingly complex "state-of-the-art" multiparameter equations of state requires a substancial amount of time, both to measure the underlying sets of accurate data and to establish the equation of state itself. Since the number of experts actively working in this field is rather small, accurate equations of state are still available only for a limited number of fluids. This problem has been addressed by very recent approaches which adapt "know-how" from the development of highly accurate reference equations of state to the development of new classes of multiparameter equations of state designed especially for use in advanced technical applications. Due to the set-up of these *simultaneously optimised* (see Sect. 6.2) and *generalised* (see Sect. 7.2.2) equations it becomes comparatively easy and less time-consuming to establish accurate equations of state. Based on these formulations, a fast increase of the group of substances for which accurate equations of state are available becomes possible. A comprehensive report on these activities is published here for the first time.

Finally, the performance of multiparameter equations of state has always been less convincing when dealing with fluid mixtures instead of pure fluids. Recent developments have shown that the related problems can be solved as well. Today, even mixtures with quite non-ideal mixing behaviour can be described accurately using empirical multiparameter approaches. Although this work concentrates mainly on pure fluids, developments regarding the description of mixtures will be summarised as well. In this way, the reader has a structured introduction to this topic and is supplied with a number of relevant references as a starting point for more detailed work on mixtures. To discuss the description of mixtures with multiparameter equations as detailed as the description of pure fluids would require a second volume and still it could not be complete since some of the most interesting aspects are just developing.

For reasons of homogeneity, equations of state are discussed only as formulations in form of the Helmholtz energy as a function of temperature and density here. Recent empirical equations of state are formulated in this form almost exclusively, but older equations which were usually formulated in form of the pressure or the compression factor are still frequently used. Such equations are considered in Sect. 3.1.2, where rules for their application within Helmholtz energy based algorithms are given. Multiparameter equations of state which are based on different combinations of variables, such as the internationally agreed industrial formulation for water, IAPWS-IF97 (see Wagner et al., 2000a), are not discussed explicitly. However, many of the procedures described in Chaps. 3 and 4 can be adapted to such equations as well.

This work is aimed mainly at three groups of readers. *Users of multiparameter equations of state* are supplied with detailed information on capabilities and restrictions of different kinds of formulations. Algorithms which are required to apply multiparameter equations are outlined in some detail to facilitate the development of user programs. For users who rely on readily developed program packages, the corresponding background information will increase the sensitivity for disadvantages of such programs and for risks implied by them. Readers who are interested only in the application of multiparameter equations of state are recommended to read Chaps. 3 and 5. Where simultaneously optimised equations of state, generalised equations of state, or mixture models are to be used, additional information is given in Sects. 6.2.2 and 6.2.3, in Chap. 7, and in Chap. 8, respectively. The general set-up of this work allows such sequential reading; where information from other sections is required to understand certain statements, explicit references are given.

The second group this work aims at are *scientists who are engaged in the field of thermodynamic properties* themselves. They should be encouraged to contribute to the subject of accurate empirical multiparameter equations of state by experimental work or by the development of own equations of state. The development of "state-of-the-art" equations of state for technical applications has been significantly simplified by recent results summarised in Chap. 6 and in Sect. 7.2.2. With regard to the development of highly accurate reference equations of state, both experimental and theoretical challenges are discussed e.g. in Sect. 3.3.5.1 and in different sections of Chaps. 4 and 5.

The third group this work aims at are *experts in the field of equations of state* who are looking either for a reference covering most of the fundamentals which are necessary to use and to develop empirical multiparameter equations of state or for an advanced textbook which can be used in high-level courses on equations of state or to introduce co-workers into the subject. A comparable compilation, including recent and even unpublished results, has not been available to date.

Chapter 2 of this work gives a brief summary of the historical development of empirical multiparameter equations of state. Although this information is of course needed neither for the application nor for the development of such equations, it is helpful for an understanding of the background and for a very first assessment of equations of state which are not explicitly discussed here. Topics of current work and future challenges are briefly summarised.

Chapter 3 summarises information which is required for using equations of state. The underlying thermodynamic relations are discussed as well as different set-ups of multiparameter equations of state and iterative algorithms which are needed for practical applications.

Chapter 4, the far most extensive chapter, describes the theoretical background which is required to establish state-of-the-art multiparameter equations of state.

The optimisation of functional forms is discussed in much detail, illustrating the development from the stepwise regression analysis by Wagner (1974) to recent nonlinear algorithms. This very detailed discussion seems justified, since increasingly sophisticated optimisation algorithms have been the foundation for many of the improvements which could be achieved during the last decades. The discussion on the description of properties in the critical region and at very high temperatures and pressures, see Sects. 4.5 and 4.6, is important for users of multiparameter equations of state as well, if they are interested in properties in these regions.

Chapter 5 summarises information on typical performances of different kinds of multiparameter equations of state. This chapter is helpful especially for unexperienced users, who need to decide whether a certain equation of state suits their needs or not. Although the given general assessments necessarily simplify a process which may become very difficult when carried out in detail, they can give a general feeling for the kind of equation which is required for a certain task.

Chapter 6 deals with a recent development, namely with equations of state with simultaneously optimised functional form. This development is expected to improve the situation with regard to technical applications, where reliable thermodynamic property data are required for a broad variety of fluids, while usually only few reliable experimental results are available for the corresponding substances. Since most of the results summarised in Chap. 6 have not yet been published, the parameter sets of the corresponding equations of state are given explicitly in this section.

Chapter 7 summarises information on sophisticated generalised equations of state which can be established for substances where even multiparameter equations of state with simultaneously optimised functional form fail due to very restricted data sets. A very recent, still unpublished approach which promises to increase the accuracy of such generalised equations of state significantly is discussed in detail in Sect. 7.2.2. Parameters of the corresponding equations of state are given in this section explicitly as well.

Finally, *Chapter 8* discusses the application of multiparameter equations of state to mixtures in the way explained above.

2 History and Potentials – an Overview

2.1 A Brief History

While the thermal equation of state of ideal gases was established by Boyle and Marriot already in the 17th century, properties of real fluids could not be described until late in the 19th century. The famous equation by van der Waals (1873) was the first formulation which described gaseous, liquid, and supercritical states as well as vapour-liquid phase equilibria qualitatively correctly. This equation was the origin of the so called "*cubic equations of state*" which are frequently used to date. The development of cubic equations of state has been discussed in other publications (see for instance Schreiner, 1986) and will not be discussed in detail here.

Almost thirty years later, Kammerlingh Onnes (1901) introduced the so called "*virial equation of state*",

$$\frac{p(T,\rho)}{\rho RT} = Z(T,\rho) = 1 + B(T)\rho + C(T)\rho^2 + D(T)\rho^3 + \ldots \qquad (2.1)$$

where p is the pressure, T the thermodynamic temperature, ρ the density, R the gas constant, and Z the compression factor. B, C, and D are the second, third, and fourth virial coefficient, respectively. The virial coefficients depend only on temperature and are defined as elements of an endless series. Although introduced as a purely empirical description, virial coefficients were later identified with the influence of interactions between pairs of molecules (B), triplets of molecules (C), and so on (see Lucas, 1991). Theoretically, Eq. (2.1) can be used to describe the properties of fluids with moderate accuracy within the whole range of fluid states (see for example Altunin and Gadetskii, 1971). However, such an approach requires empirical formulations for the temperature dependence of about 8 virial coefficients and thus a fairly large number of empirical parameters, which was not feasible early in the 20th century. Thus, the virial equation was usually truncated after the third virial coefficient and was used to describe only gaseous and gas-like supercritical states. These states could be described significantly more accurately than with cubic equations of state.

About 40 years later, Benedict et al. (1940) introduced an exponential function into a truncated virial expansion for the first time. In a generalised form, the resulting "*BWR-type equation of state*" can be written as

$$\frac{p(T,\rho)}{\rho RT} = Z(T,\rho) = 1 + \sum_{i=1}^{6} n_i T^{t_i} \rho^{d_i} + \sum_{i=7}^{8} n_i T^{t_i} \rho^{d_i} \exp\left(-(\rho/\rho_r)^2\right), \qquad (2.2)$$

where the reducing parameter ρ_r is roughly equal to the critical density. With 8 adjustable coefficients n_i, this formulation describes properties in the whole range of fluid states qualitatively correctly. An application of Eq. 2.2 to mixtures followed soon, see Benedict et al. (1942). With regard to the accuracy of calculated properties, Eq. 2.2 was clearly superior to cubic equations of state at least for pure fluids. BWR-type equations and simple modifications of Eq. 2.2 (see for example Strobridge, 1962) were used for the calculation of "accurate" properties for more than 30 years. In technical applications such equations of state can still be found in use today.[1]

In the late 1960's, reasonable computing power became available to scientists working in thermodynamics. This was the starting point of a dynamic development of empirical multiparameter equations of state. From a recent point of view, two slightly different trends can be distinguished:

On the one hand, a rather broad variety of substances was described with equations of state which can still be regarded as simple modifications of Eq. 2.2, such as the *"Starling-type equation of state"*, see Starling (1973), or the equation by Lee and Kessler (1975). Since they were designed mainly for applications in the chemical and petrochemical industry with their multitude of relevant pure substances and mixtures, mixing rules and generalised forms were important features of such equations right from the beginning. These formulations are still frequently used in technical applications.

On the other hand, more sophisticated modifications of Eq. 2.2 were developed to describe properties of selected fluids with significantly higher accuracy (see for instance Bender, 1970; Altunin and Gadetskii, 1971; Jacobsen and Stewart, 1973). An important precondition for the development of more accurate equations of state was the consideration of phase equilibrium data during the development of the equations. In this way, multiparameter equations of state became able to describe accurately not only properties at homogeneous states but also properties of the vapour-liquid phase equilibrium. Furthermore, an accurate description of the phase equilibrium is a precondition for an accurate prediction of caloric properties at liquid states, see Sect. 3.2.2. Corresponding algorithms were developed independently by Bender (1970) and Wagner (1970).[2] To fit coefficients of equations of state simultaneously to $p\rho T$ data and to phase equilibrium data was the first step into *"multiproperty fitting"* in a more general sense.

The most prominent equations of state of this period are the *"Bender-type equation of state"* by Bender (1970) and the *"MBWR-type[3] equation of state"* by Jacobsen and Stewart (1973). Both formulations were fitted to a considerable

---

[1] For an assessment from a recent point of view, see Chap. 5.

[2] At the same time, McCarty (1970) used the phase equilibrium condition $g'(T) = g''(T)$ to improve the representation of caloric properties in the liquid phase, but he did not consider the thermal equilibrium conditions simultaneously.

[3] Although all of these equations are modifications of Eq. 2.2, the equation by Jacobsen and Stewart (1973) became known as *"the modified BWR equation."*

number of fluids and in both cases equations for further substances were still published more than twenty years after the original equations had been released (see Polt and Maurer, 1992; Polt et al., 1992; Younglove and McLinden, 1994; Outcalt and McLinden, 1996).

With 19 substance dependent coefficients, Bender-type equations of state can be regarded as the most sophisticated *"technical equations of state"* today. They were applied to mixtures with some success (see for example Bender, 1973) and generalised forms became available as well (see Platzer and Maurer, 1989). Thus, Bender-type equations superseded less accurate modifications of Eq. 2.2 in several technical applications.

The MBWR-type equation by Jacobsen and Stewart (1973) can be regarded as the first successful example of a *"reference equation of state."* It described the experimental data which were available for nitrogen at that time basically within their experimental uncertainty and was used as basis of the IUPAC tables for nitrogen (Angus et al., 1979) and directly as a reference for thermodynamic properties of nitrogen in several scientific and technical applications.

In the late 70's, Ahrendts and Baehr (1979a/1981a, 1979b/1981b) presented theoretically profound articles on the multiproperty approach and on nonlinear algorithms for setting up equations of state (see Sect. 4.3). Although not all of the ideas presented were completely new and though the multiproperty approach was still limited to data for only a few properties, these articles can be regarded as milestones − except for the optimisation of functional forms, the theoretical basis for the development of equations of state was settled. A few years later, Angus (1983), the director of the IUPAC Thermodynamic Tables Project Centre, summarised the available knowledge on the development of empirical multiparameter equations of state in a CODATA Bulletin. In theoretical work on the development of science, scientific disciplines are often considered as beeing mature if monographs are available which summarise the recent knowledge on a high scientific level − in this sense, the development of empirical multiparameter equations of state became a mature scientific discipline in the early 80's.

However, new trends which resulted in substantial improvements of multiparameter equations had already begun. The stepwise regression analysis, which had been developed and applied to vapour pressure equations by Wagner (1974), was soon applied to equations of state as well (de Reuck and Armstrong, 1979). Up to that point, only the coefficients n_i in a multiparameter equation could be determined exactly. The functional form of the equation of state, thus the t_i and d_i in Eq. 2.2 as well as the number of terms of a certain kind, had to be established in a trial and error process based only on the experience of the correlator. With the stepwise regression analysis, these parameters could be determined mathematically exactly for the first time. From a general set-up, the *"bank of terms"*, which should contain all terms which are considered as suitable for the description of a given problem, the optimisation algorithm selects the combination of terms which yields the best results for the given problem.

Compared to the results found for vapour pressure equations of state by Wagner (1974), the first results found for equations of state were not very encouraging. The limited success can be attributed mostly to shortcomings of the stepwise re-

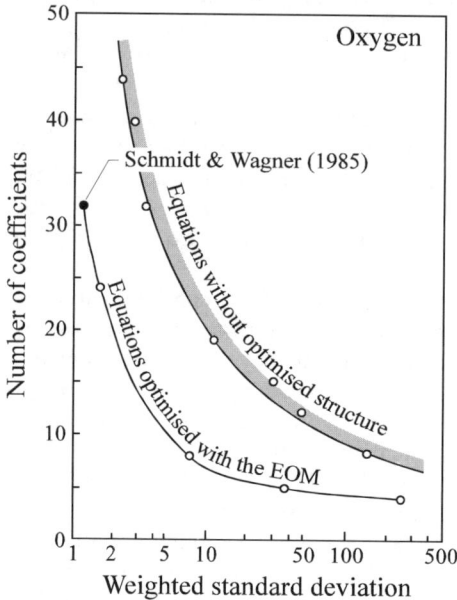

Fig. 2.1. Relations between the number of terms in different multiparameter equations of state and weighted standard deviations which result from fits to accurate experimental data for oxygen. The filled circle corresponds to the result obtained with the 32 term reference equation of state by Schmidt and Wagner (1985).

gression analysis: as a purely deterministic algorithm, the stepwise regression analysis cannot find global optima for complex optimisation problems. While banks of terms with about 20 terms are sufficient to optimise a vapour pressure equation, banks of terms with several hundred terms are considered as optimum for the development of equations of state, see Sect. 4.4.1. Optimisations based on such extensive set-ups are too complex for the simple regression analysis. Well aware of this limitation, de Reuck and Armstrong (1979) used a very small bank of terms with just 50 terms. In this way, and due to the used pressure explicit form (see below), the optimisation algorithm could not find really convincing results. More powerful non deterministic (Ewers and Wagner, 1982a/b; Setzmann and Wagner, 1989a) and recently even nonlinear optimisation algorithms (Tegeler et al., 1997) had to be developed to cope with the demands which resulted from the optimisation of increasingly sophisticated reference equations of state. This ongoing development is traced in Sect. 4.4 in detail.

With regard to the development of highly accurate multiparameter equations of state, the continuous improvement of optimisation algorithms was the most important methodical development during the last 20 years.[4] The success of optimisation algorithms has been illustrated by Schmidt and Wagner (1985) using a figure similar to Fig. 2.1. When plotting the number of coefficients of different equations of state versus the weighted standard deviation which results from a multiproperty fit to accurate experimental data, it becomes obvious that there is a direct relation between both properties. However, equations with a functional form which was selected by trial and error and equations with a functional form which was optimised using the "*evolutionary optimisation method*" (EOM) by Ewers and Wagner (1982a/b) yield significantly different results. In the range of typical reference equations of state, optimised functional forms were found to be superior by about a factor of 3. It also becomes obvious, that the advantage of optimised equations cannot be compensated by using equations of state which contain a larger number of terms; the use of optimisation algorithms resulted in a new class of reference equations of state with previously unattainable accuracy. However, the quantitative relations found for oxygen are arbitrary to a certain degree since they depend mainly on the quality of the data sets available for a certain substance. For substances with very restricted data sets, the advantage of optimised functional forms becomes small in terms of the weighted standard deviation. As will be shown in Chap. 6, the advantages of optimised equations of state depend mainly on their superior "*numerical stability*" for such substances. On the other hand, the same comparison would result in even more significant advantages of optimised functional forms when being carried out for well measured reference substances such as argon, nitrogen, or methane.

Simultaneously to the introduction of optimisation algorithms, a second development took place which had a major impact on the development of equations of state. Equations of state in terms of pressure, $p(T,\rho)$, or compression factor, $Z(T,\rho)$, are no "*fundamental equations of state*". Such equations need to be integrated to calculate caloric properties and they need to be supplemented by an independent formulation for the temperature dependence of the isobaric heat capacity, c_p^o, of the ideal gas. The corresponding fundamental equation is the one of the Helmholtz energy, $a = a(T,\rho)$. If a formulation for $a(T,\rho)$ is known, all thermodynamic properties can be calculated by combining derivatives of a with respect to T and ρ. This fact has often been quoted as the crucial advantage of equations of state in terms of the Helmholtz energy, which became common in the mid 80's.

However, this argument is only a formal one. In fact, equations of state in terms of the Helmholtz energy are always split up into one part which describes the residual part a^r (the equivalent to the pressure explicit equation of state) and a second part which describes the ideal gas contribution a^o (the equivalent to the formulation for c_p^o). Reasonable pressure explicit equations of state have always been formulated in such a way that an analytic integration is possible. As early as 1940, Benedict et al. formulated their equation both in terms of pressure and in terms of the residual Helmholtz energy, a^r. And a relation for a^o can easily be derived from an equation for c_p^o, see Sect. 3.1.1.1. Thus, there is no principle disadvantage in

---

[4] The impact of improved experimental techniques, such as single- and two-sinker densimeters and spherical resonators, was at least as important as the development of optimisation strategies. However, experimental techniques will not be discussed here.

using pressure explicit equations of state. The only, but decisive, advantage of formulations in terms of the Helmholtz energy is, that they are more flexible with regard to the used terms. The fact that each term in the equation needs to be integrated leads to severe restrictions especially for exponential terms in pressure explicit equations, see Sect. 3.1.2.1. More complex functional forms, as they are typical for highly accurate equations of state today (see Sect. 3.1.3.2), could never have been developed for pressure explicit equations of state.

The first equations of state which were formulated exclusively in form of the Helmholtz energy, the equations by Keenan et al. (1969) and Meyer-Pittroff and Grigull (1973), could have been formulated in terms of pressure as well. The first equation which used terms which could not have been used in a pressure explicit equation was the one by Pollak (1975) for water. However, the real breakthrough came when the flexibility of functional forms in terms of the Helmholtz energy was combined with the use of optimisation algorithms. Schmidt and Wagner (1985) published an equation of the general form

$$\frac{a^r(T,\rho)}{RT} = \alpha^r(\tau,\delta)$$

$$= \sum_{i=1}^{13} n_i \tau^{t_i} \delta^{d_i} + \sum_{i=14}^{24} n_i \tau^{t_i} \delta^{d_i} \exp(-\delta^2) + \sum_{i=25}^{32} n_i \tau^{t_i} \delta^{d_i} \exp(-\delta^4), \quad (2.3)$$

with $\tau = T_r/T$ and $\delta = \rho/\rho_r$; T_r and ρ_r correspond to the critical temperature and density, respectively. For the first time, this equation used exponential functions with higher density powers. Just one year later, scientists from the *Center for Applied Thermodynamic Studies* (CATS) of the university of Idaho published three equations of state (Jacobsen et al., 1986a/b; Jahangiri et al., 1986; Katti et al., 1986) which read

$$\frac{a^r(T,\rho)}{RT} = \alpha^r(\tau,\delta) = \sum_{i=1}^{I_{pol}} n_i \tau^{t_i} \delta^{d_i} + \sum_{j=1}^{6} \sum_{i=1}^{I_j} n_{j,i} \tau^{t_{j,i}} \delta^{d_{j,i}} \exp(-\delta^j). \quad (2.4)$$

For the first time, these equations used exponential terms with density powers ranging from 1 to 6. Neither the equation by Schmidt and Wagner (1985) nor the equations developed by the group around R. T Jacobsen could have been established in terms of pressure. These equations are the origin of modern fundamental equations of state which have become available since 1985, see Table 5.1.

During the 90's, two different developments could be observed. The working group *Annex 18* of the *International Energy Agency* (IEA) which was composed of various internationally well known experts pushed the development of equations of state for the new, non ozone depleting HFC and HCFC working fluids. International standards were established for the refrigerants R32, R123, R125, R134a, and R143a, see Table 5.1 for details. Although this work was very successful with regard to the development of standards needed by refrigeration industry, it did not result in new findings on reference equations of state for pure substances. None of the equations was established making use of functional forms or relevant techniques which have not been used at the time when Eqs. 2.3 and 2.4 were developed. However, interesting and rather accurate mixture approaches (for an overview see Lemmon, 1997) resulted from the work on alternative refrigerants, although no standards were defined for refrigerant mixtures.

During the same time, a new generation of highly accurate reference equations was developed. Characteristic features of the recent equations of state for methane (Setzmann and Wagner, 1991), carbon dioxide (Span and Wagner, 1996), water (Wagner and Pruß, 1997/2000), argon (Tegeler et al., 1997/1999), nitrogen (Span et al., 1998), and ethylene (Smukala et al., 1999) are a highly accurate representation of properties in the technically and scientifically most important regions, a previously unattained performance in the critical region, and a reasonable extrapolation behaviour up to extreme temperatures and pressures. These equations define a new standard with regard to the accuracy of reference equations of state, see also Chap. 5. The advantages achieved were closely linked to improved optimisation algorithms (see Sects. 4.4.5 and 4.4.6), to new functional forms which enabled an improved description of the critical region (see Sect. 4.5), and to an intense investigation of the extrapolation behaviour of empirical equations of state (see Sect. 4.6).

2.2 Publications on Thermodynamic Reference Data

Relevant Journals and series of monographs are always part of the history of a scientific discipline as well, at least if the results are closely linked to the development of some kind of standard. Thus, it is no surprise that typical references are difficult to specify for multiparameter equations of state which do not have reference character. Several equations were published in the *International Journal of Thermophysics* or in *Fluid Phase Equilibria*. However, depending on the considered substances and on the scientific background of the authors, several other Journals have published multiparameter equations of state as well. And some equations have only been published in monographs or reports, which are often difficult to access. For reference equations of state, the situation is different. Since the concept of reference equations of state became feasible in the early 70's, almost all of these equations have been published in one of three series.

Since 1973, the *Journal of Physical and Chemical Reference Data* which is maintained basically by the National Institute of Standards and Technology of the USA as part of the National Standard Reference Data System, has published more than 25 extensive articles on accurate multiparameter equations of state which were mostly accompanied by rather detailed property tables. In general, this journal yields the fastest access to reference equations of state since monographs are usually published with some delay, if they are published at all.

The *Thermodynamic Properties of ...* series of monographs was maintained by the National Standard Reference Data Service of the USSR; the monographs are available as English translations.[5] All of the equations of state presented in this

---

[5] *Thermodynamic Properties of ...* (1) Argon; Sychev et al. (1987a). (2) Nitrogen; Sychev et al. (1987b). (3) Methane; Sychev et al. (1987c). (4) Ethane; Sychev et al. (1987d). (5) Oxygen;

series were established by Russian scientists. In the 70's, these equations were of similar quality as those established in the western world. However, the more recent equations are no longer comparable to recent international standards. For applications where high accuracy is mandatory, these tables cannot be recommended.

The *International Thermodnamic Tables of the Fluid State* series[6] of monographs was maintained by the IUPAC Thermodynamic Tables Project Centre at the Imperial College of Science, Technology and Medicine in London. This series is based on equations of state which were regarded as the most accurate source of data for the corresponding substance, at least at the time that the tables were published. The results are internationally agreed and are frequently used where international standards are needed. However, some of the older tables have been superseded either by later tables for the same substance or by more accurate equations of state published in the Journal of Physical and Chemical Reference Data.[7]

Unfortunately, the IUPAC Thermodynamic Tables Project Centre was recently closed down due to funding problems. In this way, the only well established organisation which defined international standards for accurate thermodynamic properties and equations of state for various technically and scientifically important substances was abandoned.[8] The Journal of Physical and Chemical Reference Data, which is maintained by a national rather than an international standards organisation, has become the only source of thermodynamic reference data. In an increasingly international world with its need for international standards, the re-establishment of an internationally accepted organisation like the IUPAC Thermodynamic Tables Project Centre is an important organisational challenge.

2.3 Current Work and Future Challenges

However, there are a variety of scientific challenges as well. In more detail, the corresponding topics are discussed in their context in Chaps. 3–8; this section is intended only as a brief overview and maybe as an encouragement to deal more closely with one of the mentioned topics.

Sychev et al. (1987e). (6) Air; Sychev et al. (1987f). (7) Ethylene; Sychev et al. (1987g). (8) Freons, Part 1; Altunin et al. (1987a). (9) Freons, Part 2; Altunin et al. (1987b). (10) Neon, Argon, Krypton, and Xenon; Rabinovich et al. (1988). (11) Propane; Sychev et al. (1991). (12) Butane; Sychev et al. (1995).

[6] *International Thermodynamic Tables of the Fluid State* ... (1) Argon, 1971; Angus and Armstrong (1971). (2) Ethylene, 1972; Angus et al. (1974). (3) Carbon dioxide; Angus et al. (1976). (4) Helium; Angus and de Reuck (1977). (5) Methane; Angus et al. (1978). (6) Nitrogen; Angus et al. (1979). (7) Propene (Propylene); Angus et al. (1980). (8) Chlorine - tentative tables; Angus et al. (1985). (9) Oxygen; Wagner and de Reuck (1987). (10) Ethylene (Ethene); Jacobsen et al. (1988). (11) Fluorine; de Reuck (1990). (12) Methanol; de Reuck and Craven (1993). (13) Methane; Wagner and de Reuck (1996).

[7] A listing of recent reference equations of state is given in Table 5.1.

[8] The *International Association for the Properties of Water and Steam (IAPWS)* continues to work on international standards for water (see Wagner and Pruß, 1997/2000; Wagner and Kruse, 1998; Wagner et al., 2000a/b). However, the activities of the IAPWS are strictly restricted to water substance.

Today, the work on empirical multiparameter equations of state focuses mainly on three subjects:
- The development of *accurate reference equations of state*. The group of substances for which highly accurate reference equations of state are available is still too small both for technical and scientific needs (e.g., for calibrations or as a basis for accurate extended corresponding states approaches). Highly accurate equations for further substances will become available, but this development proceeds only slowly due to the enormous amount of work which has to be invested for each of these equations. A further improvement of the representation of caloric properties in the critical region based on functional forms which are still formulated in T and ρ will be one of the major theoretical challenges, see Sects. 4.5.4 and 4.5.5.3. Algorithms need to be developed which allow a reasonable description of the instable part of the two phase region without sacrificing accuracy at homogeneous states and on the phase boundary, see Sect. 3.3.5.1. Other necessary steps such as an improved description of caloric properties at liquid and liquid-like supercritical states and an improved description of properties at very high pressures and temperatures are first of all experimental challenges, see Sects. 5.1 and 5.2.
- The development of new *technical equations of state*. Very recently, a new generation of accurate but rather simple multiparameter equations of state has been developed; the corresponding results are summarised in Sects. 6.2 and 7.2.2 for the first time. To establish these equations, state-of-the-art techniques developed for reference equations of state were adapted to equations which are designed to satisfy the needs of advanced technical applications. The resulting equations should supersede obsolete equations of state which are often used in technical applications today and should be used to describe substances for which no reliable equations of state are available at all to date. To establish equations of this kind requires much less experimental and theoretical work than setting up reference equations of state. However, to describe a significant amount of the multitude of substances which are used in technical applications is a challenge which includes both theoretical and experimental aspects as well. Generalised functional forms need to be established for groups of substances which have not been considered yet, consistent databases need to be established for a variety of substances which are mostly not available in high purity, and basically known experimental techniques need to be adapted to less convenient fluids.
- The accurate *description of mixtures with models based on multiparameter equations of state* is the key to a broad variety of technical and scientific applications where accurate multiparameter equations of state are hardly used today. Recent projects on refrigerant mixtures and on natural gases show that selected, technically and scientifically relevant mixtures can be described accurately with models which are based on multiparameter equations of state, see Chap. 8. With the new generation of technical equations of state, accurate equations for the most relevant pure fluids will soon become available. Mixture models, which allow an accurate description of mixtures with only a few adjustable parameters, have been developed for rather simple systems and need to be estab-

lished for more complex systems as well. At the same time, experimental techniques need to be improved in particular for vapour-liquid equilibrium measurements in order to establish the experimental basis for more accurate thermodynamic property models.

All three topics share one common feature: fast success can only be expected in exemplary applications, which are of little practical use in most cases. To describe the variety of fluids and fluid mixtures which are relevant especially in technical applications requires long lasting efforts of more than just a few research groups. The accurate description of thermodynamic properties of fluids is a scientific discipline which has developed continuously during the last 30 years and which will continue to do so.

3 Using Multiparameter Equations of State for Pure Substances

During the last several years, accurate multiparameter equations of state have become a common data base for thermodynamic properties both in scientific and industrial applications. This development has been fueled not only by the general increase in the available computing power but also by the availability of corresponding software. Today, *interactive programs* are available which support common Microsoft Windows® data exchange formats both for numerical and graphical results. These programs are widely used as replacements for printed data tables and charts. When thermodynamic properties have to be included in process calculations directly, *Dynamic Link Libraries* (DLL's) are often used in combination with, e.g., Microsoft Excel® sheets. This technique enables calculations of thermodynamic properties within spreadsheets – calculating thermodynamic properties becomes almost as convenient as calculating trigonometric functions in this way. At the same time, the use of classical *source code libraries* for the calculation of thermodynamic properties, where the user has to specify his or her own main program and very often has to deal with problems depending on features of special compilers, is retrogressive and is becoming increasingly restricted to applications where the computational speed is a major criterion. Finally, there is still a small number of users who develop their own programs since they have special requirements either with regard to the code or with regard to the algorithms.

Thus, the enlargement of the group of scientists, engineers and students who are working with multiparameter equations of state advances with decreasing demands on the thermodynamic knowledge of the users. However, there are still tasks where the user has to understand the application of such equations in detail and without a basic understanding of the underlying algorithms even error-free software may produce misleading results. Thus, this chapter is intended to support users who must develop their own algorithms and to supply the necessary background information to users who rely on available software.

3.1 Different Kinds of Multiparameter Equations of State

Even though this book deals only with equations of state which use temperature and density as independent variables, this still covers a wide variety of formulations which are either explicit in pressure or Helmholtz energy and which range from simple virial equations to complex nonanalytic wide range equations of state. The fact that only derivatives of the Helmholtz energy with respect to temperature and density are needed to calculate thermodynamic properties makes both the algorithms for the calculation of properties and the development of equations more systematic; most of the very accurate Helmholtz equations available today could not have been developed in terms of pressure. Thus, the set of algorithms given here is based exclusively on Helmholtz equations.

Relations for the calculation of properties from pressure explicit equations can be found in textbooks on thermodynamics (see for instance Baehr, 1992) or more explicitly in specialised books like the ones of Rowlinson and Swinton (1982) or Bender (1973). Here, pressure explicit equations are discussed only briefly in Sect. 3.1.2. To enable their use in combination with the relations and algorithms presented in Sects. 3.2 and 3.3, rules for the conversion of pressure explicit equations of state into Helmholtz equations are given in Sect. 3.1.2.1.

3.1.1 Equations in Terms of the Helmholtz Energy

Helmholtz equations are generally formulated in reduced form as

$$\alpha(\tau,\delta) = \frac{a(T,\rho)}{RT} \quad \text{with} \quad \tau = \frac{T_r}{T} \quad \text{and} \quad \delta = \frac{\rho}{\rho_r}, \tag{3.1}$$

where a is the specific Helmholtz energy, R the gas constant, T the temperature, and ρ the density. Usually, but not necessarily, the critical parameters ρ_c and T_c are used for ρ_r and T_r to reduce the density and the inverse temperature.

According to the definitions given by *ISO* (ISO, 1992) and the *International Union of Pure and Applied Chemistry* (IUPAC, 1993/1996), $\alpha(\tau,\delta)$ does not correspond to a reduced specific *Helmholtz energy* which is defined by

$$a(T,v) = u - Ts, \tag{3.2}$$

where v is the specific volume, u the specific internal energy and s the specific entropy, but to a reduced specific *Massieu function* which is defined by

$$j(T,v) = s - \frac{u}{T} = -\frac{a(T,v)}{T}. \tag{3.3}$$

Although this is formally unquestionable since a is reduced with T and not with a constant T_r in Eq. 3.1, it is still international use to talk about equations explicit in the reduced Helmholtz energy or simply Helmholtz equations; to change this convention would cause unnecessary confusion.

In Eq. 3.1, T and T_r are thermodynamic temperatures in Kelvin, where the temperature scale must be considered for applications on a high level of accuracy. The formulations which are used today are usually based either on the *International Practical Temperature Scale of 1968* (*IPTS-68*) which was defined by Barber (1969) and Preston-Thomas (1976) or on the current *International Temperature Scale of 1990* (*ITS-90*) which was defined by Preston-Thomas (1990). When using equations which are valid on IPTS-68, recent values for the temperature T have to be converted to this temperature scale before using them in Eq. 3.1. Simple correlations for this conversion were given by Preston-Thomas (1990) and explained

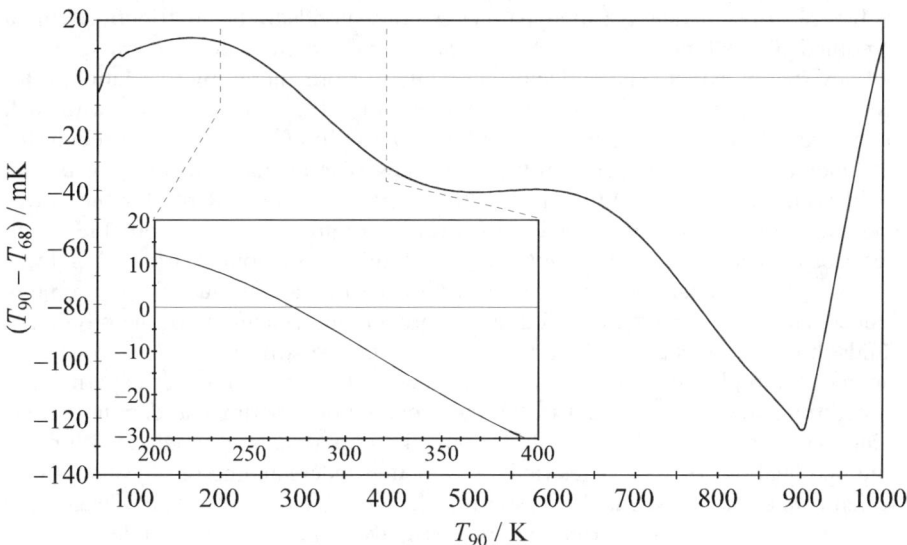

Fig. 3.1. Illustration of the temperature shift caused by conversions between the ITS-90 temperature scale and the IPTS-68 temperature scale.

later by Rusby (1990); Rusby et al. (1994) gave a revised relation for the conversion of temperatures above 903 K. To give an impression of whether this conversion is relevant for a certain application or not, Fig. 3.1 shows a plot of the temperature shift $(T_{90} - T_{68})$.

The units of R, ρ, and ρ_r determine whether Eq. 3.1 yields specific or molar results. When calculating on a molar basis the current, internationally accepted standard for the universal gas constant is

$$R = 8.314\,472 \text{ J mol}^{-1} \text{ K}^{-1} \tag{3.4}$$

with a standard uncertainty of $\pm 0.000\,015$ J mol^{-1} K^{-1} (Mohr and Taylor, 2000). This value is essentially based on the experimental work by Moldover et al. (1988). However, the value $R = 8.314\,510$ J mol^{-1} K^{-1} which was established as international standard by CODATA in 1986 (Cohen and Taylor, 1986) is still used in most equations of state. A comprehensive selection of molar masses which are needed for conversions is given by Coplen (1997); regular updates follow every two years. If values for the gas constant or the molar mass which were used when setting up the equation of state are given by the authors, these values have to be used even if they do not agree with recent standards.

When using Eq. 3.1, the Helmholtz energy is usually split into one part which describes the behaviour of the hypothetical ideal gas at given values of temperature and density and a second part which describes the residual behaviour of the real fluid. Thus, Eq. 3.1 becomes

$$\frac{a(T,\rho)}{RT} = \frac{a^\circ(T,\rho) + a^r(T,\rho)}{RT} = \alpha^\circ(\tau,\delta) + \alpha^r(\tau,\delta). \tag{3.5}$$

In terms of statistical thermodynamics a^o corresponds to the contribution from a hypothetical ideal gas where molecules can be considered as dot-like masses without any spatial extent. The molecules are able to store energy by translation, rotation and intramolecular effects such as vibration but they do not interact with each other, except by nonelastic impact. With this definition for a^o, a^r has to account for all kinds of interactions between the molecules. The available models for a^o can yield accurate results if sufficient spectroscopic data are available for the corresponding molecule (see Sect. 4.2). However, there are no models available which describe a^r accurately for the whole range of fluid states. Monte Carlo and molecular dynamics simulations are sometimes discussed as a way out of this problem. Indeed, simulations have increased our theoretical understanding of the physics of fluids, but even on state of the art computers they are still far too slow to be considered as a replacement for equations of state – they are "computer experiments". In addition, simulations do not match the accuracy of experimental data in regions which are experimentally accessible. Thus, the importance of simulations for setting up equations of state is restricted to a better understanding of physical effects which cannot be investigated by experimental means (see for instance Saager and Fischer, 1992) and to information on thermodynamic properties under extreme conditions which are not accessible for experiments (see for instance Wagner and de Reuck, 1996).

In general, accurate equations for a^r still depend on an empirical representation of experimental data (for details, see Chap. 4). Such data are very often scarce and inaccurate in the low density gas region where some experimental techniques fail completely. Splitting up Eq. 3.1 according to Eq. 3.5 avoids the disadvantages which would result from these experimental problems otherwise. If the contribution from a^r vanishes reasonably in the limit of low densities, the complete formulation meets the ideal gas limit at low densities. This limit is known from theoretical results and spectroscopic data and describes real fluids accurately for sufficiently small densities.

3.1.1.1 Correlations for the Helmholtz Energy of the Ideal Gas

Starting from the well known relation between Helmholtz energy, internal energy, and entropy, the Helmholtz energy of the ideal gas is

$$a^o(T,\rho) = u^o(T) - Ts^o(T,\rho). \tag{3.6}$$

When introducing the isochoric heat capacity of the ideal gas one obtains the expression

$$a^o(T,\rho) = u_0^o - Ts_0^o + \int_{T_0}^{T} c_v^o \, dT - T\int_{T_0}^{T} \frac{c_v^o}{T} \, dT + RT \ln\left(\frac{\rho}{\rho_0}\right). \tag{3.7}$$

For thermal properties, this relation can be evaluated without any information on the heat capacity. With $p = -(\partial a / \partial v)_T$, Eq. 3.7 results in the simple thermal equation of state for ideal gases,

3.1 Different Kinds of Multiparameter Equations of State

$$p^o = \rho^2 \left(\frac{\partial a^o}{\partial \rho}\right)_T = RT\rho. \tag{3.8}$$

But for any calculation of caloric properties, a correlation for the heat capacity of the ideal gas is required to obtain the temperature dependence of the Helmholtz energy of the ideal gas. Such correlations are usually formulated for the isobaric heat capacity using the general form

$$\frac{c_p^o}{R} = n_0^* + \sum_{i=1}^{I_{\text{Pol}}} n_i^* T^{t_i^*} + \sum_{k=1}^{K_{\text{PE}}} m_k \left(\frac{\theta_k}{T}\right)^2 \frac{e^{(\theta_k/T)}}{\left(e^{(\theta_k/T)} - 1\right)^2}. \tag{3.9}$$

For relatively narrow temperature ranges the isobaric heat capacity can be described by pure polynomial equations with sufficient accuracy. Thus in Eq. 3.9, K_{PE} becomes 0. This form of very simple formulations is common for special equations which are valid only for limited temperature ranges (see for instance Wagner and Span, 1993), for the description of substances which dissociate at fairly low temperatures such as the halogenated hydrocarbon refrigerants (see Younglove and McLinden, 1994; Tillner-Roth and Baehr, 1994; Baehr and Tillner-Roth, 1995; Outcalt and McLinden, 1996). Such equations may also be used for correlations for technical applications, if the speed of computations is regarded as more important than high accuracy of the formulation (see for instance McBride et al., 1993). Broad temperature ranges can be described with high accuracy with pure polynomial equations as well, but a large number of polynomial terms ($I_{\text{POL}} >$ 10) has to be used (see for instance Baehr and Diedrichsen, 1988). For such purposes, it is more convenient to introduce the so called "Planck-Einstein" functions which correspond to the second sum in Eq. 3.9. This functional form can be derived from theoretical results for the contribution of internal vibrations of the molecule (for a more detailed discussion, see Sect. 4.2) and enables an accurate description of broad temperature ranges with only a few terms and with improved extrapolation behaviour. Planck-Einstein terms can either be used in combination with polynomial terms (see for instance Jacobsen et al., 1986a) or without additional polynomial terms (see for instance Span and Wagner, 1996).

To derive a formulation for the reduced Helmholtz energy it is more convenient to rewrite Eq. 3.9 for the dimensionless isochoric heat capacity with reduced variables:

$$\frac{c_v^o(\tau)}{R} = n_0 + \sum_{i=1}^{I_{\text{Pol}}} n_i \tau^{t_i} + \sum_{k=1}^{K_{\text{PE}}} m_k (\vartheta_k \tau)^2 \frac{e^{\vartheta_k \tau}}{\left(e^{\vartheta_k \tau} - 1\right)^2} \tag{3.10}$$

where

$$\tau = T_r/T, \; n_0 = n_0^* - 1, \; n_i = n_i^* \cdot T_r^{t_i^*}, \; t_i = -t_i^* \text{ and } \vartheta_k = \theta_k/T_r.$$

Starting from Eq. 3.2, the internal energy becomes

$$u = a - T \cdot \left(\frac{\partial a}{\partial T}\right)_v = RT\alpha - T \cdot \left(\frac{\partial (RT\alpha)}{\partial T}\right)_v = -RT^2 \cdot \left(\frac{\partial \alpha}{\partial T}\right)_v \quad (3.11)$$

and for the isochoric heat capacity we find

$$c_v = \left(\frac{\partial u}{\partial T}\right)_v = -2RT \cdot \left(\frac{\partial \alpha}{\partial T}\right)_v - RT^2 \cdot \left(\frac{\partial^2 \alpha}{\partial T^2}\right)_v. \quad (3.12)$$

When replacing the derivatives with respect to T at constant v by derivatives with respect to τ at constant δ according to

$$\left(\frac{\partial \alpha}{\partial T}\right)_v = -\frac{\tau^2}{T_r}\left(\frac{\partial \alpha}{\partial \tau}\right)_\delta \quad \text{and} \quad \left(\frac{\partial^2 \alpha}{\partial T^2}\right)_v = 2\frac{\tau^3}{T_r^2}\left(\frac{\partial \alpha}{\partial \tau}\right)_\delta + \frac{\tau^4}{T_r^2}\left(\frac{\partial^2 \alpha}{\partial \tau^2}\right)_\delta \quad (3.13)$$

the reduced isochoric heat capacity becomes

$$\frac{c_v}{R} = -\tau^2 \cdot \left(\frac{\partial^2 \alpha}{\partial \tau^2}\right)_\delta = -\tau^2 \cdot \left(\left(\frac{\partial^2 \alpha^\circ}{\partial \tau^2}\right)_\delta + \left(\frac{\partial^2 \alpha^r}{\partial \tau^2}\right)_\delta\right). \quad (3.14)$$

The relation between the reduced Helmholtz energy of the ideal gas and the isochoric heat capacity is

$$\left(\frac{\partial^2 \alpha^\circ}{\partial \tau^2}\right)_\delta = -\frac{c_v^\circ}{R\tau^2}. \quad (3.15)$$

Thus, the Helmholtz energy of the ideal gas can be determined from a twofold integration of Eq. 3.10. Expanding the Planck-Einstein terms in Eq. 3.10 to higher orders by $e^{-2\vartheta_k \tau}$, this relation results in

$$\begin{aligned}\alpha^\circ &= \int\left(\int -\frac{c_v^\circ}{R\tau^2}\mathrm{d}\tau\right)\mathrm{d}\tau + \ln\left(\frac{\delta}{\delta_0}\right) \\ &= \int\left(\int\left(-\frac{n_0}{\tau^2} - \sum_{i=1}^{I_{\mathrm{Pol}}} n_i \tau^{t_i - 2} - \sum_{k=1}^{K_{\mathrm{PE}}} m_k \vartheta_k^2 \frac{e^{-\vartheta_k \tau}}{(1-e^{-\vartheta_k \tau})^2}\right)\mathrm{d}\tau\right)\mathrm{d}\tau + \ln\left(\frac{\delta}{\delta_0}\right),\end{aligned} \quad (3.16)$$

where the density dependence is considered independently according to Eq. 3.7. The first integration leads to

$$\alpha^\circ = c^1 + \int\left(\frac{n_0}{\tau} - \sum_{i=1}^{I_{\mathrm{Pol}}} \frac{n_i}{t_i - 1}\tau^{t_i - 1} + \sum_{k=1}^{K_{\mathrm{PE}}} m_k \vartheta_k \left(\frac{1}{1-e^{-\vartheta_k \tau}} - 1\right)\right)\mathrm{d}\tau + \ln\left(\frac{\delta}{\delta_0}\right) \quad (3.17)$$

where the constant factor -1 in the second sum is an integration constant to ensure that the contribution of these terms vanishes for small temperatures. Finally, the second integration leads to

3.1 Different Kinds of Multiparameter Equations of State

Table 3.1. Derivatives of Eq. 3.19 for the Helmholtz energy of the ideal gas

Derivative	Density dependence	Constants and temperature dependence
α^o	$\alpha^o = \ln(\delta)$	$+c^{II} + c^{I}\tau + c_0 \ln(\tau) + \sum_{i=1}^{I_{Pol}} c_i \tau^{t_i} + \sum_{k=1}^{K_{PE}} m_k \ln(1-e^{-\vartheta_k \tau})$
$\left(\dfrac{\partial \alpha^o}{\partial \delta}\right)_\tau$	$\alpha^o_\delta = \dfrac{1}{\delta}$	$+0$
$\left(\dfrac{\partial^2 \alpha^o}{\partial \delta^2}\right)_\tau$	$\alpha^o_{\delta\delta} = -\dfrac{1}{\delta^2}$	$+0$
$\left(\dfrac{\partial^2 \alpha^o}{\partial \delta \partial \tau}\right)$	$\alpha^o_{\delta\tau} = 0$	$+0$
$\left(\dfrac{\partial \alpha^o}{\partial \tau}\right)_\delta$	$\alpha^o_\tau = 0$	$+c^{I} + \dfrac{c_0}{\tau} + \sum_{i=1}^{I_{Pol}} c_i t_i \tau^{t_i-1} + \sum_{k=1}^{K_{PE}} m_k \vartheta_k \left(\dfrac{1}{1-e^{-\vartheta_k \tau}} - 1\right)$
$\left(\dfrac{\partial^2 \alpha^o}{\partial \tau^2}\right)_\delta$	$\alpha^o_{\tau\tau} = 0$	$-\dfrac{c_0}{\tau^2} + \sum_{i=1}^{I_{Pol}} c_i t_i (t_i-1) \tau^{t_i-2} - \sum_{k=1}^{K_{PE}} m_k \vartheta_k^2 \dfrac{e^{-\vartheta_k \tau}}{\left(1-e^{-\vartheta_k \tau}\right)^2}$

$$\alpha^o(\tau,\delta) = c^{II} + c^{I}\tau + n_0 \ln(\tau) - \sum_{i=1}^{I_{Pol}} \frac{n_i}{t_i(t_i-1)} \tau^{t_i} + \sum_{k=1}^{K_{PE}} m_k \ln(1-e^{-\vartheta_k \tau}) + \ln(\delta) \quad (3.18)$$

where the constant contribution from δ_0 is included in the second integration constant c^{II}.

Scientific articles on multiparameter equations of state usually contain either equations for the isobaric heat capacity of the ideal gas according to Eq. 3.9 or equations for the Helmholtz energy of the ideal gas according to

$$\alpha^o(\tau,\delta) = c^{II} + c^{I}\tau + c_0 \ln(\tau) + \sum_{i=1}^{I_{Pol}} c_i \tau^{t_i} + \sum_{k=1}^{K_{PE}} m_k \ln(1-e^{-\vartheta_k \tau}) + \ln(\delta). \quad (3.19)$$

Equation 3.19 is related to Eq. 3.9 by

$$c_0 = n_0^* - 1, \; c_i = -n_i^* / \left(t_i^*(t_i^*+1)\right) \cdot T_r^{t_i}, \; t_i = -t_i^* \text{ and } \vartheta_k = \theta_k / T_r;$$

the m_k remain unchanged. Table 3.1 summarises the derivatives of Eq. 3.19 which are required to calculate common thermodynamic properties (see Sect. 3.2).

The integration constants c^{I} and c^{II} are needed to define the zero states of all caloric properties which are connected directly to the Helmholtz energy or to its first derivative with respect to τ. For scientific use, the enthalpy and entropy of the ideal gas are commonly defined to become zero at a reference state, i.e., at

$T_0 = 298.15$ K and $p_0 = 0.101325$ MPa. For this condition and with $\tau_0 = T_r/T_0$ the integration constants are

$$c^{\mathrm{I}} = -\left(\frac{1+c_0}{\tau_0} + \sum_{i=1}^{I_{\mathrm{Pol}}} c_i t_i \tau_0^{t_i-1} + \sum_{k=1}^{K_{\mathrm{PE}}} m_k \vartheta_k \left(\frac{1}{1-e^{-\vartheta_k \tau_0}} - 1\right)\right) \quad (3.20)$$

and

$$c^{\mathrm{II}} = c_0(1 - \ln(\tau_0)) + \sum_{i=1}^{I_{\mathrm{Pol}}} c_i (t_i - 1) \tau_0^{t_i}$$

$$+ \sum_{k=1}^{K_{\mathrm{PE}}} m_k \vartheta_k \left(\tau_0 \left(\frac{1}{1-e^{-\vartheta_k \tau_0}} - 1\right) - \ln\left(1 - e^{-\vartheta_k \tau_0}\right)\right) - \ln(\delta_0). \quad (3.21)$$

The reduced density δ_0 corresponds to $\delta_0 = \rho_0/\rho_r$, where ρ_0 has to be calculated from T_0 and p_0 using the thermal equation of state of the ideal gas, Eq. 3.8.

For substances of special engineering interest, such as refrigerants or water, it is common to use points on the saturated liquid line to define the zero points either of enthalpy and entropy or of internal energy and entropy. In this case the constants c^{I} and c^{II} depend also on the arbitrary equation for the residual part of the Helmholtz energy and no closed expression in terms of the parameters introduced for the Helmholtz energy of the ideal gas can be given to define them.

In 1995, Jaeschke and Schley published accurate equations for the ideal gas isobaric heat capacity of 19 substances which use the functional form

$$\frac{c_p^o}{R} = n_0^* + m_1 \left(\frac{\theta_1/T}{\sinh(\theta_1/T)}\right)^2 + m_2 \left(\frac{\theta_2/T}{\cosh(\theta_2/T)}\right)^2 + m_3 \left(\frac{\theta_3/T}{\sinh(\theta_3/T)}\right)^2 + m_4 \left(\frac{\theta_4/T}{\cosh(\theta_4/T)}\right)^2$$

$$= n_0^* + \sum_{k=1,3} m_k \left(\frac{\vartheta_k \tau}{\sinh(\vartheta_k \tau)}\right)^2 + \sum_{k=2,4} m_k \left(\frac{\vartheta_k \tau}{\cosh(\vartheta_k \tau)}\right)^2; \quad (3.22)$$

see also Schley (1994). This slightly different functional form was originally proposed by Aly and Lee (1981) with two terms and later expanded to four terms by Savidge and Shen (1989). Although this kind of correlation is rarely found in other publications it became important as a result of the article of Jaeschke and Schley, since this article contains accurate equations for the heat capacity of higher hydrocarbons (up to n-decane) which are needed, among other applications, for the description of caloric properties of natural gases (see also Sect. 6.2.2). Table 3.2 summarises the required derivatives which result from the Helmholtz energy of the ideal gas when expressed in terms of Eq. 3.22; for c_0 and ϑ_k the relations given before hold.

When setting enthalpy and entropy as 0 for a reference point at T_0 and p_0 in the ideal gas state, the constants c^{I} and c^{II} become

$$c^{\mathrm{I}} = -\left(\frac{1+c_0}{\tau_0} + \sum_{k=1,3} m_k \frac{\vartheta_k}{\tanh(\vartheta_k \tau_0)} - \sum_{k=2,4} m_k \vartheta_k \tanh(\vartheta_k \tau_0)\right) \quad (3.23)$$

3.1 Different Kinds of Multiparameter Equations of State

Table 3.2. Derivatives of the Helmholtz energy of the ideal gas derived from the equations of Jaeschke and Schley (1995), Eq. 3.22

Derivative	Density dependence		Constants and temperature dependence
α^o	$\alpha^o =$	$\ln(\delta) + c^{\mathrm{II}} + c^{\mathrm{I}}\tau + c_0 \ln(\tau) + \sum_{k=1,3} m_k \ln\|\sinh(\vartheta_k \tau)\| - \sum_{k=2,4} m_k \ln\|\cosh(\vartheta_k \tau)\|$	
$\left(\dfrac{\partial \alpha^o}{\partial \delta}\right)_\tau$	$\alpha^o_\delta =$	$\dfrac{1}{\delta}$	$+ 0$
$\left(\dfrac{\partial^2 \alpha^o}{\partial \delta^2}\right)_\tau$	$\alpha^o_{\delta\delta} =$	$-\dfrac{1}{\delta^2}$	$+ 0$
$\left(\dfrac{\partial^2 \alpha^o}{\partial \delta \partial \tau}\right)$	$\alpha^o_{\delta\tau} =$	0	$+ 0$
$\left(\dfrac{\partial \alpha^o}{\partial \tau}\right)_\delta$	$\alpha^o_\tau =$	0	$+ c^{\mathrm{I}} + \dfrac{c_0}{\tau} + \sum_{k=1,3} m_k \dfrac{\vartheta_k}{\tanh(\vartheta_k \tau)} - \sum_{k=2,4} m_k \vartheta_k \tanh(\vartheta_k \tau)$
$\left(\dfrac{\partial^2 \alpha^o}{\partial \tau^2}\right)_\delta$	$\alpha^o_{\tau\tau} =$	0	$-\dfrac{c_0}{\tau^2} - \sum_{k=1,3} m_k \dfrac{\vartheta_k^2}{(\sinh(\vartheta_k \tau))^2} - \sum_{k=2,4} m_k \dfrac{\vartheta_k^2}{(\cosh(\vartheta_k \tau))^2}$

and

$$c^{\mathrm{II}} = c_0(1 - \ln(\tau_0)) + \sum_{k=1,3} m_k \left(\frac{\tau_0 \vartheta_k}{\tanh(\vartheta_k \tau_0)} - \ln|\sinh(\vartheta_k \tau_0)|\right)$$
$$- \sum_{k=2,4} m_k \left(\tau_0 \vartheta_k \tanh(\vartheta_k \tau) - \ln|\cosh(\vartheta_k \tau_0)|\right) - \ln(\delta_0) \quad (3.24)$$

for the ideal gas Helmholtz energies based on the equations of Jaeschke and Schley (1995).

3.1.1.2 Common Formulations for the Residual Helmholtz Energy

At first glance, many Helmholtz equations, especially those for substances which are mainly of engineering interest, still look very similar to the classic equation proposed by Benedict et al. (1940) – they are still composed of pure polynomial terms and of polynomial terms which are combined with an exponential function in density. With the inverse reduced temperature τ and the reduced density δ as independent variables these equations read

$$\alpha^{\mathrm{r}}(\tau, \delta) = \sum_{i=1}^{I_{\mathrm{Pol}}} n_i \tau^{t_i} \delta^{d_i} + \sum_{i=I_{\mathrm{Pol}}+1}^{I_{\mathrm{Pol}}+I_{\mathrm{Exp}}} n_i \tau^{t_i} \delta^{d_i} \exp(-\gamma_i \delta^{p_i}). \quad (3.25)$$

When using ρ_c to reduce the density, the γ_i are usually equal to 1. The density exponents d_i and p_i have to be positive integer values to make sure that α^r and all its derivatives vanish in the limit of zero density. For p_i values up to 8 have been used and even higher values are common for d_i. The t_i are mostly positive and can be real numbers, resulting in infinite contributions to the Helmholtz energy at temperatures near zero. Equations of state are usually valid only down to the triple point temperature T_t of the corresponding substance and even where lower temperature limits are defined, these are far from 0 Kelvin. Thus, the limiting behaviour for very low temperatures is arbitrary, but the user should be aware that extrapolations to temperatures below the given range of validity are usually more questionable than extrapolations to high temperatures. Details of the implications which result from the choice of the d_i, p_i and t_i are discussed in Sects. 4.4.1 and 4.6.2.

The main advantage of Eq. 3.25 is that the derivatives which are required to calculate thermodynamic properties can be given as analytic expressions for any

Table 3.3. Derivatives of Eq. 3.25 for the residual part of the Helmholtz energy, with respect to τ and δ

Derivative, abbreviation and formulation in τ and δ

$$\alpha^r = \sum_{i=1}^{I_{Pol}} n_i \, \delta^{d_i} \tau^{t_i} + \sum_{i=I_{Pol}+1}^{I_{Pol}+I_{Exp}} n_i \, \delta^{d_i} \tau^{t_i} \exp(-\gamma_i \delta^{p_i})$$

$$\left(\frac{\partial \alpha^r}{\partial \delta}\right)_\tau = \alpha^r_\delta = \sum_{i=1}^{I_{Pol}} n_i \, d_i \, \delta^{d_i-1} \tau^{t_i} + \sum_{i=I_{Pol}+1}^{I_{Pol}+I_{Exp}} n_i \, \delta^{d_i-1} (d_i - \gamma_i p_i \delta^{p_i}) \tau^{t_i} \exp(-\gamma_i \delta^{p_i})$$

$$\left(\frac{\partial^2 \alpha^r}{\partial \delta^2}\right)_\tau = \alpha^r_{\delta\delta} = \sum_{i=1}^{I_{Pol}} n_i \, d_i (d_i - 1) \delta^{d_i-2} \tau^{t_i} +$$

$$\sum_{i=I_{Pol}+1}^{I_{Pol}+I_{Exp}} n_i \, \delta^{d_i-2} \left((d_i - \gamma_i p_i \delta^{p_i})(d_i - 1 - \gamma_i p_i \delta^{p_i}) - \gamma_i^2 p_i^2 \delta^{p_i}\right) \tau^{t_i} \exp(-\gamma_i \delta^{p_i})$$

$$\left(\frac{\partial \alpha^r}{\partial \tau}\right)_\delta = \alpha^r_\tau = \sum_{i=1}^{I_{Pol}} n_i \, t_i \, \delta^{d_i} \tau^{t_i-1} + \sum_{i=I_{Pol}+1}^{I_{Pol}+I_{Exp}} n_i \, t_i \, \delta^{d_i} \tau^{t_i-1} \exp(-\gamma_i \delta^{p_i})$$

$$\left(\frac{\partial^2 \alpha^r}{\partial \tau^2}\right)_\delta = \alpha^r_{\tau\tau} = \sum_{i=1}^{I_{Pol}} n_i \, t_i (t_i-1) \delta^{d_i} \tau^{t_i-2} + \sum_{i=I_{Pol}+1}^{I_{Pol}+I_{Exp}} n_i \, t_i (t_i-1) \delta^{d_i} \tau^{t_i-2} \exp(-\gamma_i \delta^{p_i})$$

$$\left(\frac{\partial^2 \alpha^r}{\partial \delta \partial \tau}\right) = \alpha^r_{\delta\tau} = \sum_{i=1}^{I_{Pol}} n_i \, d_i \, t_i \, \delta^{d_i-1} \tau^{t_i-1} + \sum_{i=I_{Pol}+1}^{I_{Pol}+I_{Exp}} n_i \, t_i \, \delta^{d_i-1} (d_i - \gamma_i p_i \delta^{p_i}) \tau^{t_i-1} \exp(-\gamma_i \delta^{p_i})$$

$$\left(\frac{\partial^3 \alpha^r}{\partial \delta \partial \tau^2}\right) = \alpha^r_{\delta\tau\tau} = \sum_{i=1}^{I_{Pol}} n_i \, d_i \, t_i (t_i-1) \delta^{d_i-1} \tau^{t_i-2} +$$

$$\sum_{i=I_{Pol}+1}^{I_{Pol}+I_{Exp}} n_i \, t_i (t_i-1) \delta^{d_i-1} (d_i - \gamma_i p_i \delta^{p_i}) \tau^{t_i-2} \exp(-\gamma_i \delta^{p_i})$$

combination of d_i and p_i, while possible combinations are very restricted for pressure explicit equations (see Sect. 3.1.2.1). The required derivatives of Eq. 3.25 are given in Table 3.3 and the relations between these derivatives and common thermodynamic properties are discussed in Sect. 3.2.

More complex functional forms are common for semiempirical equations, where hard sphere terms are widely used, and for highly accurate reference equations for selected substances of special scientific or engineering interest. To improve the representation of properties in the extended critical region, special terms have been introduced in such reference equations. Both hard sphere and critical-region terms are discussed in Sect. 3.1.3.

3.1.2 Equations in Terms of Pressure

In addition to the Helmholtz energy equations discussed above, various pressure-explicit wide-range equations of state are in use today and new ones are still being developed especially for substances which are mainly of interest for engineering applications with moderate requirements regarding accuracy. Most of these equations can be written in reduced form in terms of the compression factor as

$$\frac{p}{\rho RT} = Z(\tau,\delta) = 1 + \sum_{i=1}^{I_{\text{Pol}}} n_i \tau^{t_i} \delta^{d_i} + \sum_{i=I_{\text{Pol}}+1}^{I_{\text{Pol}}+I_{\text{Exp}}} n_i \tau^{t_i} \delta^{d_i} \exp(-\gamma_i \delta^2) \quad (3.26)$$

and can be described as modifications of the *BWR equation* proposed by Benedict et al. (1940) which originally had a total of 8 terms. Well known simple modifications are, for instance, those of Starling (1973) with 11 terms or Strobridge (1962) with 15 terms. Today the more complex modifications of Bender (1970) with 13 polynomial and 6 exponential terms and of Jacobsen and Stewart (1973) with 19 polynomial and 13 exponential terms, referred to as *the MBWR equation* in scientific literature, are widely used. Combined with an equation for the isobaric heat capacity in the ideal gas state, c_p^o, these equations represent thermal and caloric data both in the homogeneous regions and on the liquid vapour phase boundary and are formally equivalent to simple Helmholtz equations. The most comprehensive listing of parameter sets for the Bender equation was given by Polt (1987). Examples for recent pressure explicit equations are the MBWR equations of Younglove and McLinden (1994) for refrigerant R123 and of Outcalt and McLinden (1996) for R152a; a new modification of the BWR equation with 18 terms was used by Piao and Noguchi (1998) for R125.

As already mentioned, no relations for the application of pressure explicit equations will be given here; for such information see for instance Rowlinson and Swinton (1982) or Bender (1973). Instead of this, rules for the conversion of pressure explicit equations to equations explicit in the Helmholtz energy will be given in the next section. Using these rules, pressure explicit equations can be used in combination with the relations described in Sect. 3.2. To maintain the necessary similarity with Eq. 3.25, we will describe the conversion of pressure explicit equations to Helmholtz energy equations based on the reduced form given in Eq. 3.26. However, different forms such as

$$\frac{p}{\rho RT} = Z(T,\rho) = 1 + \sum_{i=1}^{I_{\text{Pol}}} n_{1,i} T^{t_{1,i}} \rho^{d_{1,i}} + \sum_{i=I_{\text{Pol}}+1}^{I_{\text{Pol}}+I_{\text{Exp}}} n_{1,i} T^{t_{1,i}} \rho^{d_{1,i}} \exp\!\left(-\gamma_i \left(\frac{\rho}{\rho_{\text{r}}}\right)^{\!2}\right) \quad (3.27)$$

or

$$p = \rho RT + \sum_{i=1}^{I_{\text{Pol}}} n_{2,i} T^{t_{2,i}} \rho^{d_{2,i}} + \sum_{i=I_{\text{Pol}}+1}^{I_{\text{Pol}}+I_{\text{Exp}}} n_{2,i} T^{t_{2,i}} \rho^{d_{2,i}} \exp\!\left(-\gamma_i \left(\frac{\rho}{\rho_{\text{r}}}\right)^{\!2}\right) \quad (3.28)$$

are more common in scientific literature. The relations which are needed to transform Eq. 3.27 to Eq. 3.26 are

$$n_i = n_{1,i}\, \rho_{\text{r}}^{d_{1,i}}\, T_{\text{r}}^{t_{1,i}}, \quad t_i = -t_{1,i}, \quad d_i = d_{1,i}$$

and to transform Eq. 3.28 to Eq. 3.26 the relations

$$n_i = n_{2,i}\, \rho_{\text{r}}^{d_{2,i}-1}\, T_{\text{r}}^{t_{2,i}-1}\, R^{-1}, \quad t_i = -t_{2,i}+1, \quad d_i = d_{2,i}-1$$

hold. Where inverse temperatures are used in published equations, the sign of $t_{1,i}$ or $t_{2,i}$ has to be changed correspondingly.

3.1.2.1 Integration of Pressure Explicit Equations

The relation between pressure and Helmholtz energy is given by

$$p = -\left(\frac{\partial a}{\partial v}\right)_T = \rho^2 \left(\frac{\partial a}{\partial \rho}\right)_T. \quad (3.29)$$

Using reduced properties, Eq. 3.29 becomes

$$\frac{p}{\rho RT} = \rho \left(\frac{\partial(a/(RT))}{\partial \rho}\right)_T = \delta \left(\frac{\partial \alpha}{\partial \delta}\right)_\tau = \delta \left(\left(\frac{\partial \alpha^{\text{o}}}{\partial \delta}\right)_\tau + \left(\frac{\partial \alpha^{\text{r}}}{\partial \delta}\right)_\tau\right) = 1 + \delta \left(\frac{\partial \alpha^{\text{r}}}{\partial \delta}\right)_\tau. \quad (3.30)$$

Thus, the residual part of the reduced Helmholtz energy becomes

$$\alpha^{\text{r}}(\tau,\delta) = \int_0^\delta \frac{Z(\tau,\delta)-1}{\delta}\, d\delta. \quad (3.31)$$

Starting from Eq. 3.26, this integration results in the simple relation

$$\alpha^{\text{r}}_{\text{Pol}} = \sum_{i=1}^{I_{\text{Pol}}} \frac{n_i}{d_i}\, \tau^{t_i}\, \delta^{d_i} \quad (3.32)$$

for the polynomial terms. For the exponential terms, the substitution $X = \delta^2$ leads to

3.1 Different Kinds of Multiparameter Equations of State

$$\alpha_{\text{Exp}}^{\text{r}} = \int_0^{\delta} \sum_{i=I_{\text{Pol}}+1}^{I_{\text{Pol}}+I_{\text{Exp}}} n_i \tau^{t_i} \delta^{d_i-1} \exp(-\gamma_i \delta^2) \, d\delta$$

$$= \int_0^{\sqrt{X}} \sum_{i=I_{\text{Pol}}+1}^{I_{\text{Pol}}+I_{\text{Exp}}} n_i \tau^{t_i} X^{\frac{d_i-1}{2}} \exp(-\gamma_i X) \frac{d\delta}{dX} \, dX$$

$$= \sum_{i=I_{\text{Pol}}+1}^{I_{\text{Pol}}+I_{\text{Exp}}} n_i \tau^{t_i} \frac{1}{2} \int_0^{\sqrt{X}} X^{\frac{d_i-2}{2}} \exp(-\gamma_i X) \, dX. \tag{3.33}$$

Carrying out the integration and replacing X by δ again leads to the expression

$$\alpha_{\text{Exp}}^{\text{r}} = \sum_{i=I_{\text{Pol}}+1}^{I_{\text{Pol}}+I_{\text{Exp}}} -n_i \tau^{t_i} \exp(-\gamma_i \delta^2) \frac{1}{2} \sum_{k=1}^{\infty} \frac{\prod_{j=1}^{k}(d_i - 2(j-1))}{\gamma^k \, 2^{k-1} \, d_i} \delta^{d_i-2k} \Big|_0^{\delta}. \tag{3.34}$$

If d_i is an even number, all terms in the endless series become zero for $k > d_i/2$ and the integration can be carried out analytically.

When using integrated forms of pressure explicit equations with Helmholtz algorithms, the second sum in Eq. 3.34 leads to a significant increase of the number of terms, since every term of the second sum is combined with a different density exponent (d_i-2k), and has to be considered as a separate term in Eq. 3.25. The increase in the number of terms can be reduced if terms with the same density and temperature exponents are combined. The result of this rearrangement is shown in Table 3.4 for the exponential terms of the Bender equation and in Table 3.5 for the exponential terms of the MBWR equation. In Tables 3.4 and 3.5 the exponential terms with a density power of 0 result from Eq. 3.34 with $k = d_i/2$. To make sure

Table 3.4. Results of Eq. 3.34 for the exponential terms in the equation by Bender (1970)

	Bender according to Eq. 3.26			Bender integrated to Eq. 3.25			
i	d_i	t_i	γ_i	n_i	d_i	t_i	γ_i
14	2	3	γ	$n_{14}/2\gamma + n_{17}/2\gamma^2$	0	3	–[a]
15	2	4	γ	$n_{15}/2\gamma + n_{18}/2\gamma^2$	0	4	–[a]
16	2	5	γ	$n_{16}/2\gamma + n_{19}/2\gamma^2$	0	5	–[a]
17	4	3	γ	$-n_{14}/2\gamma - n_{17}/2\gamma^2$	0	3	γ
18	4	4	γ	$-n_{15}/2\gamma - n_{18}/2\gamma^2$	0	4	γ
19	4	5	γ	$-n_{16}/2\gamma - n_{19}/2\gamma^2$	0	5	γ
20				$-n_{17}/2\gamma$	2	3	γ
21				$-n_{18}/2\gamma$	2	4	γ
22				$-n_{19}/2\gamma$	2	5	γ

[a] additional terms which compensate for the zero density contribution of terms 17–19

3 Using Multiparameter Equations of State for Pure Substances

Table 3.5. Results of Eq. 3.34 for the exponential terms in the equation by Jacobsen and Stewart (1973)

	MBWR according to Eq. 3.26			MBWR integrated to Eq. 3.25			
i	d_i	t_i	γ_i	n_i	d_i	t_i	γ_i
20	2	3	γ	$\dfrac{n_{20}}{2\gamma}+\dfrac{n_{22}}{2\gamma^2}+\dfrac{n_{24}}{\gamma^3}+\dfrac{3n_{26}}{\gamma^4}+\dfrac{12n_{28}}{\gamma^5}+\dfrac{60n_{30}}{\gamma^6}$	0	3	–[a]
21	2	4	γ	$n_{21}/2\gamma + n_{25}/\gamma^3 + 12n_{29}/\gamma^5 + 60n_{31}/\gamma^6$	0	4	–[a]
22	4	3	γ	$n_{23}/2\gamma^2 + 3n_{27}/\gamma^4 + 60n_{32}/\gamma^6$	0	5	–[a]
23	4	5	γ	$-\dfrac{n_{20}}{2\gamma}-\dfrac{n_{22}}{2\gamma^2}-\dfrac{n_{24}}{\gamma^3}-\dfrac{3n_{26}}{\gamma^4}-\dfrac{12n_{28}}{\gamma^5}-\dfrac{60n_{30}}{\gamma^6}$	0	3	γ
24	6	3	γ	$-n_{21}/2\gamma - n_{25}/\gamma^3 - 12n_{29}/\gamma^5 - 60n_{31}/\gamma^6$	0	4	γ
25	6	4	γ	$-n_{23}/2\gamma^2 - 3n_{27}/\gamma^4 - 60n_{32}/\gamma^6$	0	5	γ
26	8	3	γ	$-\dfrac{n_{22}}{2\gamma}-\dfrac{n_{24}}{\gamma^2}-\dfrac{3n_{26}}{\gamma^3}-\dfrac{12n_{28}}{\gamma^4}-\dfrac{60n_{30}}{\gamma^5}$	2	3	γ
27	8	5	γ	$-n_{25}/\gamma^2 - 12n_{29}/\gamma^4 - 60n_{31}/\gamma^5$	2	4	γ
28	10	3	γ	$-n_{23}/2\gamma - 3n_{27}/\gamma^3 - 60n_{32}/\gamma^5$	2	5	γ
29	10	4	γ	$-n_{24}/2\gamma - 3n_{26}/2\gamma^2 - 6n_{28}/\gamma^3 - 30n_{30}/\gamma^4$	4	3	γ
30	12	3	γ	$-n_{25}/2\gamma - 6n_{29}/\gamma^3 - 30n_{31}/\gamma^4$	4	4	γ
31	12	4	γ	$-3n_{27}/2\gamma^2 - 30n_{32}/\gamma^4$	4	5	γ
32	12	5	γ	$-n_{26}/2\gamma - 2n_{28}/\gamma^2 - 10n_{30}/\gamma^3$	6	3	γ
33				$-2n_{29}/\gamma^2 - 10n_{31}/\gamma^3$	6	4	γ
34				$-n_{27}/2\gamma - 10n_{32}/\gamma^3$	6	5	γ
35				$-n_{28}/2\gamma - 5n_{30}/2\gamma^2$	8	3	γ
36				$-n_{29}/2\gamma - 5n_{31}/2\gamma^2$	8	4	γ
37				$-5n_{32}/2\gamma^2$	8	5	γ
38				$-n_{30}/2\gamma$	10	3	γ
39				$-n_{31}/2\gamma$	10	4	γ
40				$-n_{32}/2\gamma$	10	5	γ

[a] additional terms which compensate for the zero density contribution of terms 23–25

that the residual contribution to α vanishes in the low density limit, it is important to consider the lower limit of Eq. 3.34 too, which results in the additional polynomial terms with a density power of 0.

3.1.3 Special Functional Forms

Besides the simple polynomial and exponential terms in Eq. 3.25 two further groups of terms have become important for multiparameter equations of state – hard sphere terms which were originally introduced for semiempirical equations of state and special functional forms for an improved description of properties in the critical region which are typical for highly accurate equations of state for reference substances. Again the discussion of these functional forms focuses on the necessary information for users in this section; for more detailed information see Sects. 4.4.1.1 and 4.5.

3.1.3.1 Hard Sphere Terms

A logical extension of the concept behind Eq. 3.5 is, theoretically, the use of hard sphere terms. In Eq. 3.5 the Helmholtz energy is split into an ideal gas part and a residual part, since the ideal gas part is known accurately from statistical thermodynamics. In the same way, one may argue that more recent perturbation theories yield accurate results for hard body fluids as well, which are good approximations of the real fluid behaviour in the limit of very high densities. Following this argument, the Helmholtz energy can be written as

$$\frac{a(T,\rho)}{RT} = \frac{a^\circ(T,\rho) + a^h(T,\rho) + a^{r*}(T,\rho)}{RT} = \alpha^\circ(\tau,\delta) + \alpha^h(\tau,\delta) + \alpha^{r*}(\tau,\delta), \quad (3.35)$$

where the residual part α^{r*} is now responsible for mainly the attractive part of the molecular interaction, while the hard body part α^h is used as an approximation for the contribution from repulsive forces.

A well known early work on the effect of intermolecular repulsion is that of Carnahan and Starling (1969/1972) who wrote the compression factor as

$$Z = 1 + Z^h = \frac{1 + \xi + \xi^2 - \xi^3}{(1-\xi)^3} \quad (3.36)$$

where ξ is assumed to be a temperature independent packing fraction of spherical molecules. Later work has focused on effects which result from the anisotropy of molecules and on the influence of the temperature dependence of ξ (see for instance Chen and Kreglewski, 1977; Deiters, 1981a/b). At this point we want to focus on a recent functional form proposed by Saager et al. (1992) which was used in multiparameter equations for the Lennard-Jones fluid (Mecke et al., 1996) and with slightly modified elongation-dependent parameters for the two centre Lennard-Jones fluid (Mecke et al., 1997; Kriebel et al., 1998). However, the most important point is that this kind of term is used in the semiempirical BACKONE equation which will be discussed in more detail in Sect. 7.1.

Following Saager et al. (1992), the hard sphere contribution to α becomes

$$\alpha^h = (\varphi^2 - 1)\ln(1-\xi) + \frac{(\varphi^2 + 3\varphi)\xi - 3\varphi\xi^2}{(1-\xi)^2}. \quad (3.37)$$

φ is an anisotropy parameter which becomes 1 for spherical molecules; with $\varphi = 1$ Eq. 3.37 becomes equal to Eq. 3.36. The packing fraction is assumed to be temperature dependent according to

Table 3.6. Derivatives of the formulation proposed by Saager et al. (1992) for the hard sphere contribution to the Helmholtz energy

Derivatives of α^h with respect to τ and δ

$$\left(\frac{\partial \alpha^h}{\partial \delta}\right)_\tau = \left(\frac{d\alpha^h}{d\xi}\right)\left(\frac{\partial \xi}{\partial \delta}\right)_\tau \qquad \left(\frac{\partial^2 \alpha^h}{\partial \delta^2}\right)_\tau = \left(\frac{d^2\alpha^h}{d\xi^2}\right)\left(\frac{\partial \xi}{\partial \delta}\right)_\tau^2$$

$$\left(\frac{\partial \alpha^h}{\partial \tau}\right)_\delta = \left(\frac{d\alpha^h}{d\xi}\right)\left(\frac{\partial \xi}{\partial \tau}\right)_\delta \qquad \left(\frac{\partial^2 \alpha^h}{\partial \tau^2}\right)_\delta = \left(\frac{d^2\alpha^h}{d\xi^2}\right)\left(\frac{\partial \xi}{\partial \tau}\right)_\delta^2 + \left(\frac{d\alpha^h}{d\xi}\right)\left(\frac{\partial^2 \xi}{\partial \tau^2}\right)_\delta$$

$$\left(\frac{\partial^2 \alpha^h}{\partial \delta \partial \tau}\right) = \left(\frac{d^2\alpha^h}{d\xi^2}\right)\left(\frac{\partial \xi}{\partial \tau}\right)_\delta \left(\frac{\partial \xi}{\partial \delta}\right)_\tau + \left(\frac{d\alpha^h}{d\xi}\right)\left(\frac{\partial^2 \xi}{\partial \delta \partial \tau}\right)$$

$$\left(\frac{\partial^3 \alpha^h}{\partial \delta \partial \tau^2}\right) = \left(\frac{d^3\alpha^h}{d\xi^3}\right)\left(\frac{\partial \xi}{\partial \tau}\right)_\delta^2 \left(\frac{\partial \xi}{\partial \delta}\right)_\tau + \left(\frac{d^2\alpha^h}{d\xi^2}\right)\left[\left(\frac{\partial^2 \xi}{\partial \tau^2}\right)_\delta \left(\frac{\partial \xi}{\partial \delta}\right)_\tau + 2\left(\frac{\partial^2 \xi}{\partial \delta \partial \tau}\right)\left(\frac{\partial \xi}{\partial \tau}\right)_\delta\right] + \left(\frac{d\alpha^h}{d\xi}\right)\left(\frac{\partial^3 \xi}{\partial \delta \partial \tau^2}\right)$$

Derivatives of α^h with respect to ξ

$$\alpha^h = \left(\varphi^2 - 1\right)\ln(1-\xi) + \frac{\left(\varphi^2 + 3\varphi\right)\xi - 3\varphi\xi^2}{(1-\xi)^2} \qquad \left(\frac{d\alpha^h}{d\xi}\right) = -\frac{\varphi^2 - 1}{1-\xi} + \frac{\left(\varphi^2 + 3\varphi\right) + \left(\varphi^2 - 3\varphi\right)\xi}{(1-\xi)^3}$$

$$\left(\frac{d^2\alpha^h}{d\xi^2}\right) = -\frac{\varphi^2 - 1}{(1-\xi)^2} + \frac{3\left(\varphi^2 + 3\varphi\right) + \left(\varphi^2 - 3\varphi\right)(1+2\xi)}{(1-\xi)^4}$$

$$\left(\frac{d^3\alpha^h}{d\xi^3}\right) = -\frac{2\left(\varphi^2 - 1\right)}{(1-\xi)^3} + \frac{6\left[2\left(\varphi^2 + 3\varphi\right) + \left(\varphi^2 - 3\varphi\right)(1+\xi)\right]}{(1-\xi)^5}$$

Derivatives of ξ with respect to τ and δ

$$\xi = n\delta\left[a + (1-a)\tau^{-\gamma}\right]^{-1} \qquad \left(\frac{\partial \xi}{\partial \delta}\right)_\tau = n\left[a + (1-a)\tau^{-\gamma}\right]^{-1} \qquad \left(\frac{\partial^2 \xi}{\partial \delta^2}\right)_\tau = 0$$

$$\left(\frac{\partial \xi}{\partial \tau}\right)_\delta = n\delta \frac{(1-a)\gamma\,\tau^{-(\gamma+1)}}{\left(a + (1-a)\tau^{-\gamma}\right)^2} \qquad \left(\frac{\partial^2 \xi}{\partial \delta \partial \tau}\right) = n \frac{(1-a)\gamma\,\tau^{-(\gamma+1)}}{\left(a + (1-a)\tau^{-\gamma}\right)^2}$$

$$\left(\frac{\partial^2 \xi}{\partial \tau^2}\right)_\delta = -n\delta \frac{(1-a)\gamma\,\tau^{-(\gamma+2)}\left((\gamma+1)(a + (1-a)\tau^{-\gamma}) - 2\gamma(1-a)\tau^{-\gamma}\right)}{\left(a + (1-a)\tau^{-\gamma}\right)^3}$$

$$\left(\frac{\partial^3 \xi}{\partial \delta \partial \tau^2}\right) = -n \frac{(1-a)\gamma\,\tau^{-(\gamma+2)}\left((\gamma+1)(a + (1-a)\tau^{-\gamma}) - 2\gamma(1-a)\tau^{-\gamma}\right)}{\left(a + (1-a)\tau^{-\gamma}\right)^3}$$

$$\xi = n \left(\frac{\rho}{\rho_r} \right) \left[a + (1-a) \left(\frac{T}{T_r} \right)^{\gamma} \right]^{-1} = n\delta \left[a + (1-a)\tau^{-\gamma} \right]^{-1}. \qquad (3.38)$$

In Eq. 3.37 and Eq. 3.38, ρ_r, T_r and φ are adjustable parameters while

$$n = 0.1617, \quad a = 0.689, \quad \text{and} \quad \gamma = 0.3674$$

are constants. The temperature dependence in ξ is necessary to improve the representation of caloric properties, but it makes derivatives of Eq. 3.37 with respect to τ rather complex. The derivatives which are necessary to calculate common thermodynamic properties are summarised in Table 3.6. In Sect. 3.2, the calculation of thermodynamic properties will be explained based on the more common concept $\alpha = \alpha^o + \alpha^r$ again; when using hard sphere terms α^h has to be considered as part of $\alpha^r = \alpha^h + \alpha^{r*}$ to use the relations given in Sect. 3.2.

Until 1995 another semiempirical formulation with a hard sphere like contribution was important for multiparameter equations of state, namely the "base function" in the equation of Haar et al. (1982). This equation was accepted by the *International Association for the Properties of Water and Steam* (*IAPWS*) as standard for scientific and general use of water properties in 1984. In 1995 it was replaced by the equation of Pruß and Wagner (1995; see also Wagner and Pruß, 1997 / 2000) and thus it will not be discussed further at this point.

3.1.3.2 Critical-Region Terms

Equations of state which use only simple polynomial and exponential terms are able to describe the whole fluid region with highest accuracy, except in the critical region. To mimic the crossover from classical to nonclassical behaviour when approaching the critical point (for details on the underlying theory see Sect. 4.5.1 or Sengers and Levelt Sengers, 1986) terms are required which yield larger gradients in the derivatives of α, but which do not interfere with the classical behaviour outside of the critical region. Today, such terms are a characteristic feature of highly accurate reference equations for substances with experimentally well described critical regions. To set-up equations with special critical-region terms requires considerable "know-how" which is summarised in Sect. 4.5. However, only a few more complicated derivatives are required to use such terms.

Several functional forms have been proposed to improve the representation of properties in the critical region (see Haar et al., 1982; Huang et al., 1985; Pitzer and Schreiber, 1988; Saul and Wagner, 1989). Nevertheless, there are only two types of formulations which are typically used today and which fit into the set-up proposed here since they use temperature and density as independent variables.

In 1982, Haar et al. introduced a kind of two dimensional Gaussian bell shaped terms to improve the representation of properties in the critical region of water. Unfortunately, in the way they were formulated, these terms distorted the representation of properties on the phase boundary close to the critical temperature (see Saul and Wagner, 1989). To overcome this problem, Setzmann and Wagner (1991) reformulated the Gaussian bell shaped terms by introducing one additional parameter and changing values of other parameters by orders of magnitude. These modified Gaussian bell shaped terms resulted in a remarkable improvement of the representation both of thermal and caloric properties in the critical region and have been used in several reference quality equations of state since then.

Supplemented by modified Gaussian bell shaped terms, Eq. 3.25 becomes

$$\alpha^r(\tau,\delta) = \alpha^r_{\text{Pol}} + \alpha^r_{\text{Exp}} + \alpha^r_{\text{GBS}} = \sum n_i \tau^{t_i} \delta^{d_i} + \sum n_i \tau^{t_i} \delta^{d_i} \exp(-\gamma_i \delta^{p_i})$$
$$+ \sum n_i \tau^{t_i} \delta^{d_i} \exp\left(-\eta_i(\delta-\varepsilon_i)^2 - \beta_i(\tau-\gamma_i)^2\right). \qquad (3.39)$$

Usually, the number of Gaussian bell shaped terms in Eq. 3.39 varies from 3 to 5 while the total number of terms in highly accurate reference equations is usually in the order of 40. η_i, β_i, ε_i, and γ_i are additional internal parameters. In general these parameters are different from term to term, but ε_i has always been equal to 1 to date. Table 3.7 summarises the necessary derivatives of the Gaussian bell shaped terms.

Equations with Gaussian bell shaped terms are able to represent even the most accurate experimental data for thermodynamic properties in the critical region within their experimental uncertainty, except for isochoric heat capacities and speeds of sound very close to the critical point (within about $|T-T_c|/T_c < \approx 0.002$ and $|\rho-\rho_c|/\rho_c < \approx 0.25$), where these properties cannot be described by analytic

Table 3.7. Derivatives of the Gaussian bell shaped terms in Eq. 3.39

Derivative and formulation in τ and δ

$$\alpha^r_{\text{GBS},i} = n_i \delta^{d_i} \tau^{t_i} e^{-\eta_i(\delta-\varepsilon_i)^2 - \beta_i(\tau-\gamma_i)^2}$$

$$\left(\frac{\partial \alpha^r_{\text{GBS},i}}{\partial \delta}\right)_\tau = n_i \delta^{d_i} \tau^{t_i} e^{-\eta_i(\delta-\varepsilon_i)^2 - \beta_i(\tau-\gamma_i)^2} \left[\frac{d_i}{\delta} - 2\eta_i(\delta-\varepsilon_i)\right]$$

$$\left(\frac{\partial^2 \alpha^r_{\text{GBS},i}}{\partial \delta^2}\right)_\tau = n_i \delta^{d_i} \tau^{t_i} e^{-\eta_i(\delta-\varepsilon_i)^2 - \beta_i(\tau-\gamma_i)^2} \left[\left(\frac{d_i}{\delta} - 2\eta_i(\delta-\varepsilon_i)\right)^2 - \frac{d_i}{\delta^2} - 2\eta_i\right]$$

$$\left(\frac{\partial \alpha^r_{\text{GBS},i}}{\partial \tau}\right)_\delta = n_i \delta^{d_i} \tau^{t_i} e^{-\eta_i(\delta-\varepsilon_i)^2 - \beta_i(\tau-\gamma_i)^2} \left[\frac{t_i}{\tau} - 2\beta_i(\tau-\gamma_i)\right]$$

$$\left(\frac{\partial^2 \alpha^r_{\text{GBS},i}}{\partial \tau^2}\right)_\delta = n_i \delta^{d_i} \tau^{t_i} e^{-\eta_i(\delta-\varepsilon_i)^2 - \beta_i(\tau-\gamma_i)^2} \left[\left(\frac{t_i}{\tau} - 2\beta_i(\tau-\gamma_i)\right)^2 - \frac{t_i}{\tau^2} - 2\beta_i\right]$$

$$\frac{\partial^2 \alpha^r_{\text{GBS},i}}{\partial \delta \partial \tau} = n_i \delta^{d_i} \tau^{t_i} e^{-\eta_i(\delta-\varepsilon_i)^2 - \beta_i(\tau-\gamma_i)^2} \left[\frac{d_i}{\delta} - 2\eta_i(\delta-\varepsilon_i)\right]\left[\frac{t_i}{\tau} - 2\beta_i(\tau-\gamma_i)\right]$$

$$\frac{\partial^3 \alpha^r_{\text{GBS},i}}{\partial \delta \partial \tau^2} = n_i \delta^{d_i} \tau^{t_i} e^{-\eta_i(\delta-\varepsilon_i)^2 - \beta_i(\tau-\gamma_i)^2} \left[\left(\frac{t_i}{\tau} - 2\beta_i(\tau-\gamma_i)\right)^2 - \frac{t_i}{\tau^2} - 2\beta_i\right]\left[\frac{d_i}{\delta} - 2\eta_i(\delta-\varepsilon_i)\right]$$

Table 3.8. Derivatives of the nonanalytic terms in Eq. 3.40

Derivatives of α_{NA}

$$\alpha^r_{NA,i} = n_i \delta \Delta^{b_i} \psi \quad \text{with} \quad \Delta = \theta^2 + d_i\left[(\delta-1)^2\right]^{a_i}, \quad \theta = (1-\tau) + c_i\left[(\delta-1)^2\right]^{\frac{1}{2\beta_i}},$$

$$\text{and} \quad \psi = e^{-e_i(\delta-1)^2 - f_i(\tau-1)^2}$$

$$\left(\frac{\partial \alpha^r_{NA,i}}{\partial \delta}\right)_\tau = n_i \left[\Delta^{b_i}\left(\psi + \delta \frac{\partial \psi}{\partial \delta}\right) + \frac{\partial \Delta^{b_i}}{\partial \delta} \delta \psi\right]$$

$$\left(\frac{\partial^2 \alpha^r_{NA,i}}{\partial \delta^2}\right)_\tau = n_i \left[\Delta^{b_i}\left(2\frac{\partial \psi}{\partial \delta} + \delta \frac{\partial^2 \psi}{\partial \delta^2}\right) + 2\frac{\partial \Delta^{b_i}}{\partial \delta}\left(\psi + \delta \frac{\partial \psi}{\partial \delta}\right) + \frac{\partial^2 \Delta^{b_i}}{\partial \delta^2} \delta \psi\right]$$

$$\left(\frac{\partial \alpha^r_{NA,i}}{\partial \tau}\right)_\delta = n_i \delta \left[\frac{\partial \Delta^{b_i}}{\partial \tau} \psi + \Delta^{b_i} \frac{\partial \psi}{\partial \tau}\right]$$

$$\left(\frac{\partial^2 \alpha^r_{NA,i}}{\partial \tau^2}\right)_\delta = n_i \delta \left[\frac{\partial^2 \Delta^{b_i}}{\partial \tau^2} \psi + 2\frac{\partial \Delta^{b_i}}{\partial \tau}\frac{\partial \psi}{\partial \tau} + \Delta^{b_i}\frac{\partial^2 \psi}{\partial \tau^2}\right]$$

$$\left(\frac{\partial^2 \alpha^r_{NA,i}}{\partial \delta \partial \tau}\right) = n_i \left[\Delta^{b_i}\left(\frac{\partial \psi}{\partial \tau} + \delta \frac{\partial^2 \psi}{\partial \delta \partial \tau}\right) + \delta \frac{\partial \Delta^{b_i}}{\partial \delta}\frac{\partial \psi}{\partial \tau} + \frac{\partial \Delta^{b_i}}{\partial \tau}\left(\psi + \delta \frac{\partial \psi}{\partial \delta}\right) + \frac{\partial^2 \Delta^{b_i}}{\partial \delta \partial \tau} \delta \psi\right]$$

$$\left(\frac{\partial^3 \alpha^r_{NA,i}}{\partial \delta \partial \tau^2}\right) = n_i \left\{\left[\frac{\partial^2 \Delta^{b_i}}{\partial \tau^2} \psi + 2\frac{\partial \Delta^{b_i}}{\partial \tau}\frac{\partial \psi}{\partial \tau} + \Delta^{b_i}\frac{\partial^2 \psi}{\partial \tau^2}\right]\right.$$

$$\left.+\delta\left[\frac{\partial^3 \Delta^{b_i}}{\partial \delta \partial \tau^2}\psi + \frac{\partial^2 \Delta^{b_i}}{\partial \tau^2}\frac{\partial \psi}{\partial \delta} + 2\frac{\partial^2 \Delta^{b_i}}{\partial \delta \partial \tau}\frac{\partial \psi}{\partial \tau} + 2\frac{\partial \Delta^{b_i}}{\partial \tau}\frac{\partial^2 \psi}{\partial \delta \partial \tau} + \frac{\partial \Delta^{b_i}}{\partial \delta}\frac{\partial^2 \psi}{\partial \tau^2} + \Delta^{b_i}\frac{\partial^3 \psi}{\partial \delta \partial \tau^2}\right]\right\}$$

formulations. To overcome this remaining problem, Span (1993) proposed nonanalytic terms which enable an equation of state to represent the steep increase of the isochoric heat capacity and the steep decrease of the speed of sound when approaching the critical point. A slightly modified form of nonanalytic terms was used in the recent reference equations for carbon dioxide (Span and Wagner, 1996) and water (Pruß and Wagner, 1995; Wagner and Pruß, 1997 / 2000).

When using nonanalytic terms, the formulation for the reduced Helmholtz energy reads

$$\alpha^r(\tau,\delta) = \alpha^r_{Pol} + \alpha^r_{Exp} + \alpha^r_{GBS} + \alpha^r_{NA} = \sum n_i \tau^{t_i} \delta^{d_i} + \sum n_i \tau^{t_i} \delta^{d_i} \exp(-\gamma_i \delta^{p_i})$$

$$+ \sum n_i \tau^{t_i} \delta^{d_i} \exp\left(-\eta_i(\delta-\varepsilon_i)^2 - \beta_i(\tau-\gamma_i)^2\right)$$

$$+ \sum n_i \delta \Delta^{b_i} \exp\left(-e_i(\delta-1)^2 - f_i(\tau-1)^2\right) \tag{3.40}$$

$$\text{with} \quad \Delta = \left\{(1-\tau) + c_i\left[(\delta-1)^2\right]^{1/(2\beta_i)}\right\}^2 + d_i\left[(\delta-1)^2\right]^{a_i}.$$

Table 3.8. Derivatives of the nonanalytic terms in Eq. 3.40 – Continued

Derivatives of the distance function Δ^{b_i}

$$\frac{\partial \Delta^{b_i}}{\partial \delta} = b_i \, \Delta^{b_i-1} \frac{\partial \Delta}{\partial \delta}$$

$$\frac{\partial^2 \Delta^{b_i}}{\partial \delta^2} = b_i \left\{ \Delta^{b_i-1} \frac{\partial^2 \Delta}{\partial \delta^2} + (b_i-1) \Delta^{b_i-2} \left(\frac{\partial \Delta}{\partial \delta} \right)^2 \right\}$$

$$\frac{\partial \Delta^{b_i}}{\partial \tau} = -2\theta b_i \, \Delta^{b_i-1}$$

$$\frac{\partial^2 \Delta^{b_i}}{\partial \tau^2} = 2b_i \, \Delta^{b_i-1} + 4\theta^2 b_i (b_i-1) \Delta^{b_i-2}$$

$$\frac{\partial^2 \Delta^{b_i}}{\partial \delta \partial \tau} = -c_i b_i \frac{2}{\beta_i} \Delta^{b_i-1} (\delta-1) \left[(\delta-1)^2 \right]^{\frac{1}{2\beta_i}-1} - 2\theta b_i (b_i-1) \Delta^{b_i-2} \frac{\partial \Delta}{\partial \delta}$$

$$\frac{\partial^3 \Delta^{b_i}}{\partial \delta \partial \tau^2} = 2b_i (b_i-1) \Delta^{b_i-2} \left\{ \frac{\partial \Delta}{\partial \delta} \left(1 + 2\theta^2 (b_i-2) \Delta^{-1} \right) + \frac{4}{\beta_i} \theta c_i (\delta-1) \left[(\delta-1)^2 \right]^{\frac{1}{2\beta_i}-1} \right\}$$

$$\frac{\partial \Delta}{\partial \delta} = (\delta-1) \left\{ c_i \, \theta \, \frac{2}{\beta_i} \left[(\delta-1)^2 \right]^{\frac{1}{2\beta_i}-1} + 2 d_i a_i \left[(\delta-1)^2 \right]^{a_i-1} \right\}$$

$$\frac{\partial^2 \Delta}{\partial \delta^2} = \frac{1}{(\delta-1)} \frac{\partial \Delta}{\partial \delta} + (\delta-1)^2 \left\{ 4 d_i a_i (a_i-1) \left[(\delta-1)^2 \right]^{a_i-2} + 2 c_i^2 \left(\frac{1}{\beta_i} \right)^2 \left\{ \left[(\delta-1)^2 \right]^{\frac{1}{2\beta_i}-1} \right\}^2 + c_i \, \theta \, \frac{4}{\beta_i} \left(\frac{1}{2\beta_i} - 1 \right) \left[(\delta-1)^2 \right]^{\frac{1}{2\beta_i}-2} \right\}$$

Derivatives of the exponential function ψ

$$\frac{\partial \psi}{\partial \delta} = -2 e_i (\delta-1) \psi$$

$$\frac{\partial^2 \psi}{\partial \delta^2} = \left\{ 2 e_i (\delta-1)^2 - 1 \right\} 2 e_i \, \psi$$

$$\frac{\partial \psi}{\partial \tau} = -2 f_i (\tau-1) \psi$$

$$\frac{\partial^2 \psi}{\partial \tau^2} = \left\{ 2 f_i (\tau-1)^2 - 1 \right\} 2 f_i \, \psi$$

$$\frac{\partial^2 \psi}{\partial \delta \partial \tau} = 4 e_i f_i (\delta-1)(\tau-1) \psi$$

$$\frac{\partial^3 \psi}{\partial \, \partial \tau^2} = \left\{ 2 f_i (\tau-1)^2 - 1 \right\} (-1) 4 e_i f_i \, \psi$$

The number of nonanalytic terms in Eq. 3.40 varies between 2 and 3 and a_i, b_i, c_i, d_i, e_i, f_i, and β_i are additional internal parameters of the nonanalytic terms which are different from term to term. The necessary derivatives of the nonanalytic terms are given in Table 3.8.

It is obvious that the numerical expense is increased significantly by using nonanalytic terms, while their advantages are restricted to a narrow region around the critical point. Thus, these terms are only useful for reference equations describing substances for which the critical region is regarded as very important. Recent reference equations for nitrogen (Span et al., 1998b), argon (Tegeler et al., 1997/1999) and ethylene (Smukala et al., 1999) describe the critical region without nonanalytic terms again (see also Sects. 4.5.2 – 4.5.4).

3.1.4 Simple Equations for Gas Phase Calibrations

Accurate values for thermodynamic properties of reference fluids are needed for the technical calibration of measuring devices for gas flows or gas properties. Such applications are often restricted to moderate temperatures, while pressures can be as high as 10 to 30 MPa. For typical reference fluids like argon, nitrogen or methane this range corresponds to supercritical states with maximum densities just above the critical density. This range can be described accurately by simple polynomial equations in form of the Helmholtz energy which read

$$\alpha^r(\tau,\delta) = \sum_{i=1}^{I} n_i \tau^{t_i} \delta^{d_i} \;; \qquad (3.41)$$

the required simple derivatives are already given in Table 3.3. When written in terms of the compression factor, the close relation of Eq. 3.41 to the well known virial expansion becomes obvious. According to

$$Z(T,\rho)-1 = \delta\left(\frac{\partial \alpha^r}{\partial \delta}\right)_\tau = \sum_{i=1}^{I} \frac{n_i}{d_i} \tau^{t_i} \delta^{d_i} = \sum_{i=1}^{I} \frac{n_i \tau^{t_i}}{d_i \rho_r^{d_i}} \rho^{d_i} \qquad (3.42)$$

all terms with $d_i = 1$ contribute to the second virial coefficient B, all terms with $d_i = 2$ contribute to the third virial coefficient C, and so on.

Such simple but nevertheless very accurate equations of state have been published for example by Wagner and Span (1993) and Span et al. (1998b). Aside from the primary task of serving as an easy-to-access data source for gas phase calibrations, they are useful for tests of program packages or for educational applications.

3.2 Calculating Thermodynamic Properties from the Helmholtz Energy

It has been mentioned repeatedly that thermodynamic properties can be calculated from the Helmholtz energy and its derivatives since

$$da(T,v) = -s\,dT - p\,dv \qquad (3.43)$$

is one of the fundamental equations of thermodynamics which contain all the information necessary to describe a system holding a constant amount of substance. This general statement can be found in common textbooks on thermodynamics, but the relations which are necessary to calculate thermodynamic properties from Eq. 3.43 are usually not given. This section gives examples of the techniques which are used to establish such relations; comprehensive listings of results for common thermodynamic properties are given in tables again. To simplify calculations based on equations which are formulated in reduced properties, all relations are given for derivatives of α with respect to τ and δ.

3.2.1 Properties in the Homogeneous Region

Properties in the homogeneous region can easily be derived from the Helmholtz equation, if their definition implies only derivatives of $a(T,\rho)$ with respect to its independant variables T and ρ for constant values of the other independent variable. Examples were already given for the isochoric heat capacity c_v in Eqs. 3.11–3.14 and for the compression factor Z in Eqs. 3.29–3.30. A crucial point in deriving such relations is usually the step where derivatives of a with respect to T and ρ have to be replaced by derivatives of α with respect to τ and δ. This step can easily be managed using the relations

$$\rho \left(\frac{\partial (a/RT)}{\partial \rho} \right)_T = \delta \left(\frac{\partial \alpha}{\partial \delta} \right)_\tau, \tag{3.44}$$

$$\rho^2 \left(\frac{\partial^2 (a/RT)}{\partial \rho^2} \right)_T = \delta^2 \left(\frac{\partial^2 \alpha}{\partial \delta^2} \right)_\tau, \tag{3.45}$$

$$T \left(\frac{\partial (a/RT)}{\partial T} \right)_\rho = -\tau \left(\frac{\partial \alpha}{\partial \tau} \right)_\delta, \tag{3.46}$$

$$T^2 \left(\frac{\partial^2 (a/RT)}{\partial T^2} \right)_\rho = 2\tau \left(\frac{\partial \alpha}{\partial \tau} \right)_\delta + \tau^2 \left(\frac{\partial^2 \alpha}{\partial \tau^2} \right)_\delta, \tag{3.47}$$

$$T\rho \left(\frac{\partial^2 (a/RT)}{\partial \rho\, \partial T} \right) = -\tau\delta \left(\frac{\partial^2 \alpha}{\partial \delta\, \partial \tau} \right), \tag{3.48}$$

and
$$T^2 \rho \left(\frac{\partial^3 (a/RT)}{\partial \rho\, \partial T^2} \right) = 2\tau\delta \left(\frac{\partial^2 \alpha}{\partial \delta\, \partial \tau} \right) + \tau^2 \delta \left(\frac{\partial^3 \alpha}{\partial \delta\, \partial \tau^2} \right). \tag{3.49}$$

After splitting up α into α^o and α^r, derivatives of α^o with respect to δ can be simplified using the relations given in Table 3.1.

Starting from Eq. 3.11, the reduced internal energy is

$$\frac{u}{RT} = -T \left(\frac{\partial (a/RT)}{\partial T} \right)_\rho \tag{3.50}$$

and with Eq. 3.46 this relation can be rewritten as

3.2 Calculating Thermodynamic Properties from the Helmholtz Energy

$$\frac{u}{RT} = \tau \left(\frac{\partial \alpha}{\partial \tau}\right)_\delta = \tau\left\{\left(\frac{\partial \alpha^\circ}{\partial \tau}\right)_\delta + \left(\frac{\partial \alpha^r}{\partial \tau}\right)_\delta\right\}. \quad (3.51)$$

With $h = u + pv$ the relation for the reduced enthalpy becomes

$$\frac{h}{RT} = \frac{p}{RT\rho} + \frac{u}{RT} = 1 + \delta\left(\frac{\partial \alpha^r}{\partial \delta}\right)_\tau + \tau\left\{\left(\frac{\partial \alpha^\circ}{\partial \tau}\right)_\delta + \left(\frac{\partial \alpha^r}{\partial \tau}\right)_\delta\right\}. \quad (3.52)$$

Starting from a well known formulation for the entropy, $s = -(\partial a/\partial T)_v$, the relation for the reduced entropy becomes

$$\frac{s}{R} = -\left(\frac{\partial (T\,a/RT)}{\partial T}\right)_v = -\frac{a}{RT} - T\left(\frac{\partial (a/RT)}{\partial T}\right)_v$$

$$= -\alpha^\circ - \alpha^r + \tau\left\{\left(\frac{\partial \alpha^\circ}{\partial \tau}\right)_\delta + \left(\frac{\partial \alpha^r}{\partial \tau}\right)_\delta\right\}. \quad (3.53)$$

Establishing the required relations for the reduced Helmholtz energy is more difficult for properties which are defined as derivatives at constant values of variables which are not identical with temperature or density. Two simple mathematical relations which are always helpful to deal with such problems are

$$\left(\frac{\partial X}{\partial Y}\right)_Z = \left(\frac{\partial X}{\partial Y}\right)_W + \left(\frac{\partial X}{\partial W}\right)_Y \left(\frac{\partial W}{\partial Y}\right)_Z \quad (3.54)$$

and

$$\left(\frac{\partial X}{\partial Y}\right)_Z = -\frac{(\partial Z/\partial Y)_X}{(\partial Z/\partial X)_Y}. \quad (3.55)$$

Starting from the definition of the isobaric heat capacity, $c_p = (\partial h/\partial T)_p$, the relation for the reduced isobaric heat capacity can be derived according to

$$\frac{c_p}{R} = \left(\frac{\partial (T\,h/RT)}{\partial T}\right)_p = \frac{h}{RT} + T\left(\frac{\partial (h/RT)}{\partial T}\right)_p$$

$$= \frac{h}{RT} + T\left(\frac{\partial (h/RT)}{\partial T}\right)_\rho + T\left(\frac{\partial (h/RT)}{\partial \rho}\right)_T \left(\frac{\partial \rho}{\partial T}\right)_p$$

$$= \frac{h}{RT} + T\left(\frac{\partial (h/RT)}{\partial T}\right)_\rho - T\left(\frac{\partial (h/RT)}{\partial \rho}\right)_T \frac{(\partial p/\partial T)_\rho}{(\partial p/\partial \rho)_T}. \quad (3.56)$$

In the next step the known relations for h/RT and $p/\rho RT$ are introduced and Eq. 3.56 is converted to reduced variables. Using the mathematically correct notation, the resulting relations become somewhat cumbersome. Thus, at this point it is helpful to introduce the abbreviations

$$\alpha_\delta^r = \left(\frac{\partial \alpha^r}{\partial \delta}\right)_\tau, \qquad \alpha_\delta^o = \left(\frac{\partial \alpha^o}{\partial \delta}\right)_\tau,$$

$$\alpha_{\delta\delta}^r = \left(\frac{\partial^2 \alpha^r}{\partial \delta^2}\right)_\tau, \qquad \alpha_{\delta\delta}^o = \left(\frac{\partial^2 \alpha^o}{\partial \delta^2}\right)_\tau,$$

$$\alpha_\tau^r = \left(\frac{\partial \alpha^r}{\partial \tau}\right)_\delta, \qquad \alpha_\tau^o = \left(\frac{\partial \alpha^o}{\partial \tau}\right)_\delta,$$

$$\alpha_{\tau\tau}^r = \left(\frac{\partial^2 \alpha^r}{\partial \tau^2}\right)_\delta, \qquad \alpha_{\tau\tau}^o = \left(\frac{\partial^2 \alpha^o}{\partial \tau^2}\right)_\delta,$$

$$\alpha_{\delta\tau}^r = \left(\frac{\partial^2 \alpha^r}{\partial \delta \partial \tau}\right), \text{ and } \alpha_{\delta\tau\tau}^r = \left(\frac{\partial^3 \alpha^r}{\partial \delta \partial \tau^2}\right). \tag{3.57}$$

Mixed derivatives of α^o are always equal to 0 and are therefore not considered. With these abbreviations Eq. 3.56 leads to

$$\frac{c_p}{R} = 1 + \delta\alpha_\delta^r + \tau(\alpha_\tau^o + \alpha_\tau^r) + \frac{T_r}{\tau}\left(\frac{\partial(1+\delta\alpha_\delta^r + \tau(\alpha_\tau^o + \alpha_\tau^r))}{\partial \tau}\right)_\delta \left(\frac{d\tau}{dT}\right)$$

$$-\frac{T_r}{\tau}\left(\frac{\partial(1+\delta\alpha_\delta^r + \tau(\alpha_\tau^o + \alpha_\tau^r))}{\partial \delta}\right)_\tau \left(\frac{d\delta}{d\rho}\right) \frac{RT_r\rho_r \left(\frac{\partial(\delta/\tau(1+\delta\alpha_\delta^r))}{\partial \tau}\right)_\delta \left(\frac{d\tau}{dT}\right)}{RT_r\rho_r \left(\frac{\partial(\delta/\tau(1+\delta\alpha_\delta^r))}{\partial \delta}\right)_\tau \left(\frac{d\delta}{d\rho}\right)}. \tag{3.58}$$

and with

$$\left(\frac{d\delta}{d\rho}\right) = \frac{1}{\rho_r} \quad \text{and} \quad \left(\frac{d\tau}{dT}\right) = -\frac{T_r}{T^2} = -\frac{\tau^2}{T_r}, \tag{3.59}$$

this results in

3.2 Calculating Thermodynamic Properties from the Helmholtz Energy

$$\frac{c_p}{R} = 1 + \delta\alpha_\delta^r + \tau(\alpha_\tau^o + \alpha_\tau^r) - (\delta\alpha_{\delta\tau}^r + \alpha_\tau^o + \alpha_\tau^r + \tau(\alpha_{\tau\tau}^o + \alpha_{\tau\tau}^r))\tau$$

$$-(\alpha_\delta^r + \delta\alpha_{\delta\delta}^r + \tau\alpha_{\delta\tau}^r) \frac{\delta\left(-\dfrac{(1+\delta\alpha_\delta^r)}{\tau^2} + \dfrac{\delta\alpha_{\delta\tau}^r}{\tau}\right)(-\tau)}{\dfrac{1}{\tau}(1+2\delta\alpha_\delta^r + \delta^2\alpha_{\delta\delta}^r)}$$

$$= -\tau^2(\alpha_{\tau\tau}^o + \alpha_{\tau\tau}^r) + (1+\delta\alpha_\delta^r - \delta\tau\alpha_{\delta\tau}^r)$$

$$-\delta(\alpha_\delta^r + \delta\alpha_{\delta\delta}^r + \tau\alpha_{\delta\tau}^r)\frac{1+\delta\alpha_\delta^r - \delta\tau\alpha_{\delta\tau}^r}{1+2\delta\alpha_\delta^r + \delta^2\alpha_{\delta\delta}^r}. \quad (3.60)$$

Reducing the expression $(1+\delta\alpha_\delta^r - \delta\tau\alpha_{\delta\tau}^r)$ to higher terms by $(1+2\delta\alpha_\delta^r + \delta^2\alpha_{\delta\delta}^r)$ and summarising the two fractions in the resulting equation finally leads to

$$\frac{c_p}{R} = -\tau^2(\alpha_{\tau\tau}^o + \alpha_{\tau\tau}^r) + \frac{(1+\delta\alpha_\delta^r - \delta\tau\alpha_{\delta\tau}^r)^2}{1+2\delta\alpha_\delta^r + \delta^2\alpha_{\delta\delta}^r}. \quad (3.61)$$

The definition of the speed of sound is $w = v\sqrt{-(\partial p/\partial v)_s} = \sqrt{(\partial p/\partial \rho)_s}$. Starting from this definition, the reduced speed of sound becomes

$$\frac{w^2}{RT} = \left(\frac{\partial\left(\rho\dfrac{p}{\rho RT}\right)}{\partial\rho}\right)_s = \frac{p}{\rho RT} + \rho\left(\frac{\partial\left(p/\rho RT\right)}{\partial\rho}\right)_s$$

$$= \frac{p}{\rho RT} + \rho\left(\frac{\partial\left(p/\rho RT\right)}{\partial\rho}\right)_T + \rho\left(\frac{\partial\left(p/\rho RT\right)}{\partial T}\right)_\rho\left(\frac{\partial T}{\partial\rho}\right)_s$$

$$= \frac{p}{\rho RT} + \rho\left(\frac{\partial\left(p/\rho RT\right)}{\partial\rho}\right)_T - \rho\left(\frac{\partial\left(p/\rho RT\right)}{\partial T}\right)_\rho\frac{(\partial s/\partial\rho)_T}{(\partial s/\partial T)_\rho}. \quad (3.62)$$

Introducing reduced variables and the known relations for $p/\rho RT$ and s/R, Eq. 3.62 becomes

Table 3.9. Definitions of common thermodynamic properties and their relation to the reduced Helmholtz energy α

Property and definition	Relation to α and its derivatives (see Eq. 3.57)
Pressure $p(T,\rho) = -(\partial a/\partial v)_T$	$\dfrac{p}{\rho RT} = 1 + \delta\alpha^r_\delta$
Derivatives of Pressure	
$\left(\dfrac{\partial p}{\partial \rho}\right)_T$	$\left(\dfrac{\partial p}{\partial \rho}\right)_T = RT\left(1 + 2\delta\alpha^r_\delta + \delta^2\alpha^r_{\delta\delta}\right)$
$\left(\dfrac{\partial p}{\partial T}\right)_\rho$	$\left(\dfrac{\partial p}{\partial T}\right)_\rho = R\rho\left(1 + \delta\alpha^r_\delta - \delta\tau\alpha^r_{\delta\tau}\right)$
Fugacity coefficient $\ln(\varphi(T,p)) = \int_0^p \left[\dfrac{v}{RT} - \dfrac{1}{p}\right] dp_T$	$\ln(\varphi) = \alpha^r + \delta\alpha^r_\delta - \ln(1 + \delta\alpha^r_\delta)$
Entropy $s(T,\rho) = -(\partial a/\partial T)_v$	$\dfrac{s}{R} = \tau(\alpha^o_\tau + \alpha^r_\tau) - \alpha^o - \alpha^r$
Internal energy $u(T,\rho) = a + Ts$	$\dfrac{u}{RT} = \tau(\alpha^o_\tau + \alpha^r_\tau)$
Isochoric heat capacity $c_v(T,\rho) = (\partial u/\partial T)_v$	$\dfrac{c_v}{R} = -\tau^2(\alpha^o_{\tau\tau} + \alpha^r_{\tau\tau})$
Enthalpy $h(T,p) = u + pv$	$\dfrac{h}{RT} = 1 + \tau(\alpha^o_\tau + \alpha^r_\tau) + \delta\alpha^r_\delta$
Isobaric heat capacity $c_p(T,p) = (\partial h/\partial T)_p$	$\dfrac{c_p}{R} = -\tau^2(\alpha^o_{\tau\tau} + \alpha^r_{\tau\tau}) + \dfrac{(1 + \delta\alpha^r_\delta - \delta\tau\alpha^r_{\delta\tau})^2}{1 + 2\delta\alpha^r_\delta + \delta^2\alpha^r_{\delta\delta}}$
Gibbs energy $g(T,p) = h - Ts$	$\dfrac{g}{RT} = 1 + \alpha^o + \alpha^r + \delta\alpha^r_\delta$
Speed of sound $w(T,p) = \sqrt{(\partial p/\partial \rho)_s}$	$\dfrac{w^2}{RT} = 1 + 2\delta\alpha^r_\delta + \delta^2\alpha^r_{\delta\delta} - \dfrac{(1 + \delta\alpha^r_\delta - \delta\tau\alpha^r_{\delta\tau})^2}{\tau^2(\alpha^o_{\tau\tau} + \alpha^r_{\tau\tau})}$
Joule–Thomson coefficient $\mu(T,p) = (\partial T/\partial p)_h$	$\mu R\rho = \dfrac{-(\delta\alpha^r_\delta + \delta^2\alpha^r_{\delta\delta} + \delta\tau\alpha^r_{\delta\tau})}{(1 + \delta\alpha^r_\delta - \delta\tau\alpha^r_{\delta\tau})^2 - \tau^2(\alpha^o_{\tau\tau} + \alpha^r_{\tau\tau})(1 + 2\delta\alpha^r_\delta + \delta^2\alpha^r_{\delta\delta})}$
Second thermal virial coefficient $B(T) = \lim\limits_{\rho \to 0}(\partial(p/(\rho RT))/\partial \rho)_T$	$B\rho_r = \lim\limits_{\delta \to 0} \alpha^r_\delta$
Third thermal virial coefficient $C(T) = \dfrac{1}{2}\lim\limits_{\rho \to 0}(\partial^2(p/(\rho RT))/\partial \rho^2)_T$	$C\rho_r^2 = \lim\limits_{\delta \to 0} \alpha^r_{\delta\delta}$
Second acoustic virial coefficient[a] $\beta_a(T) = \lim\limits_{\rho \to 0}\left(\partial(w^2/(k_s^o RT))/\partial \rho\right)_T$	$\beta_a \rho_r = \lim\limits_{\delta \to 0}\left[2\alpha^r_\delta - 2\dfrac{k_s^o - 1}{k_s^o}\tau\alpha^r_{\delta\tau} + \dfrac{(k_s^o - 1)^2}{k_s^o}\tau^2\alpha^r_{\delta\tau\tau}\right]$

[a] $k_s^o = c_p^o/c_v^o$ is the isentropic expansion coefficient of the ideal gas

3.2 Calculating Thermodynamic Properties from the Helmholtz Energy

$$\frac{w^2}{RT} = 1 + \delta\alpha_\delta^r + \frac{\delta}{\rho_r}\left(\frac{\partial\left(1+\delta\alpha_\delta^r\right)}{\partial\delta}\right)_\tau \frac{\mathrm{d}\delta}{\mathrm{d}\rho}$$

$$-\frac{\delta}{\rho_r}\left(\frac{\partial\left(1+\delta\alpha_\delta^r\right)}{\partial\tau}\right)_\delta \frac{\mathrm{d}\tau}{\mathrm{d}T} \frac{R\left(\frac{\partial\left(-\alpha^\circ - \alpha^r + \tau\left(\alpha_\tau^\circ + \alpha_\tau^r\right)\right)}{\partial\delta}\right)_\tau \frac{\mathrm{d}\delta}{\mathrm{d}\rho}}{R\left(\frac{\partial\left(-\alpha^\circ - \alpha^r + \tau\left(\alpha_\tau^\circ + \alpha_\tau^r\right)\right)}{\partial\tau}\right)_\delta \frac{\mathrm{d}\tau}{\mathrm{d}T}}. \quad (3.63)$$

Evaluation and reorganisation of this relation finally yields

$$\frac{w^2}{RT} = 1 + 2\delta\alpha_\delta^r + \delta^2\alpha_{\delta\delta}^r - \frac{\left(1+\delta\alpha_\delta^r - \delta\tau\alpha_{\delta\tau}^r\right)^2}{\tau^2\left(\alpha_{\tau\tau}^\circ + \alpha_{\tau\tau}^r\right)}. \quad (3.64)$$

For other properties, the relations to the Helmholtz energy can basically be derived in the same way – to do so is easy from a mathematical point of view, but it requires time and patience. Table 3.9 summarises the definitions of common thermodynamic properties and their relations to the reduced Helmholtz energy; Table 3.10 gives rules for the calculation of some coefficients which are common especially in engineering applications (see also Jacobsen et al., 1997).

Table 3.10. Definitions of common coefficients and rules for their calculation

Coefficient and definition	Rules for calculations
Isothermal throttling coefficient $\delta_T(T,p) = (\partial h/\partial p)_T$	$\delta_T \rho = 1 - \dfrac{1+\delta\alpha_\delta^r - \delta\tau\alpha_{\delta\tau}^r}{1+2\delta\alpha_\delta^r + \delta^2\alpha_{\delta\delta}^r}$
Isothermal expansion coefficient $k_T(T,p) = -v/p\,(\partial p/\partial v)_T$	$k_T = \dfrac{\rho}{p}\left(\dfrac{\partial p}{\partial \rho}\right)_T$
Isentropic expansion coefficient $k_s(T,p) = -v/p\,(\partial p/\partial v)_s$	$k_s = \dfrac{\rho w^2}{p}$
Volume expansivity $\alpha_v(T,p) = 1/v\,(\partial v/\partial T)_p$	$\alpha_v = \dfrac{1}{\rho}\left(\dfrac{\partial p}{\partial \rho}\right)_T \left(\dfrac{\partial p}{\partial T}\right)_\rho$
Isothermal compressibility $\kappa_T(T,p) = -1/v\,(\partial v/\partial p)_T$	$\kappa_T = \dfrac{1}{\rho}\left(\dfrac{\partial p}{\partial \rho}\right)_T^{-1} = \dfrac{k_T}{p}$
Isentropic compressibility $\kappa_s(T,p) = -1/v\,(\partial v/\partial p)_s$	$\kappa_s = \dfrac{1}{\rho w^2} = \dfrac{k_s}{p}$

3.2.2 Properties of the Vapour-Liquid Equilibrium Phases

The equilibrium of the vapour phase ($''$) and liquid phase ($'$) can be described by the conditions

$$T' = T'', \tag{3.65}$$

$$p' = p'', \tag{3.66}$$

and
$$g'_i = g''_i, \tag{3.67}$$

where Eq. 3.67 is considered once for each substance i which is involved in the phase equilibrium. For a pure substance the Gibbs energy can be calculated as

$$\frac{g}{RT} = 1 + \alpha^\circ + \alpha^r + \delta\alpha^r_\delta, \tag{3.68}$$

where the derivatives are abbreviated according to Eq. 3.57 again. With Eq. 3.65 the condition of equal Gibbs energy yields

$$\alpha^\circ(\tau,\delta') + \alpha^r(\tau,\delta') + \delta\alpha^r_\delta(\tau,\delta') = \alpha^\circ(\tau,\delta'') + \alpha^r(\tau,\delta'') + \delta\alpha^r_\delta(\tau,\delta''). \tag{3.69}$$

This relation is usually rearranged as

$$\left(1 + \delta\alpha^r_\delta(\tau,\delta'')\right) - \left(1 + \delta\alpha^r_\delta(\tau,\delta')\right) + \alpha^\circ(\tau,\delta'') - \alpha^\circ(\tau,\delta') = \alpha^r(\tau,\delta') - \alpha^r(\tau,\delta'') \tag{3.70}$$

and using Eq. 3.30 and the simple density dependence of α°, Eq. 3.70 becomes

$$\frac{p_s}{RT}\left(\frac{1}{\rho''} - \frac{1}{\rho'}\right) - \ln\left(\frac{\rho'}{\rho''}\right) = \alpha^r(\tau,\delta') - \alpha^r(\tau,\delta''). \tag{3.71}$$

The first two equilibrium conditions, Eqs. 3.65–3.66, can be rewritten as

$$p(T,\rho') = p(T,\rho'') \tag{3.72}$$

and thus

$$\delta'\left(1 + \delta'\alpha^r_\delta(\tau,\delta')\right) = \delta''\left(1 + \delta''\alpha^r_\delta(\tau,\delta'')\right). \tag{3.73}$$

Simultaneous solution of Eq. 3.71 and Eq. 3.73 yields the thermal properties of the coexisting vapour and liquid phase; the required iterative procedure is described in Sect. 3.3.5.

Independently of each other, Wagner (1970) and Bender (1970) developed algorithms which allowed data for the phase equilibrium to be included during the development of multiparameter equations of state. This step enabled not only the calculation of accurate phase equilibrium data directly from an equation of state but also made it possible to calculate consistent caloric properties for the liquid, since Eq. 3.71 establishes the physical link between caloric properties in the gas and liquid phase. Likewise, if wide range equations of state do not represent the phase equilibria accurately, they cannot be expected to yield accurate caloric properties in the liquid phase.

3.2 Calculating Thermodynamic Properties from the Helmholtz Energy

Once the thermal properties are determined, the basic caloric properties like enthalpy, entropy or internal energy can be calculated using the relations for the homogeneous region (see Table 3.9) with T and ρ' for the saturated liquid and with T and ρ'' for the saturated vapour. For example, the enthalpy of the saturated liquid is

$$\frac{h'}{RT} = 1 + \tau\left(\alpha_\tau^o(\tau,\delta') + \alpha_\tau^r(\tau,\delta')\right) + \delta'\alpha_\delta^r(\tau,\delta') \tag{3.74}$$

with $\delta' = \rho'/\rho_r$. Since α_τ^o does not depend on density (see Table 3.1), the enthalpy of evaporation is

$$\frac{\Delta h_{ev}}{RT} = \frac{h'' - h'}{RT} = \tau\left(\alpha_\tau^r(\tau,\delta'') - \alpha_\tau^r(\tau,\delta')\right) + \delta''\alpha_\delta^r(\tau,\delta'') - \delta'\alpha_\delta^r(\tau,\delta'). \tag{3.75}$$

The speed of sound in the saturated liquid and in the saturated vapour can be calculated like single phase properties but with T, ρ' and T, ρ'', respectively. In general the same is true for derived properties such as heat capacities or expansion coefficients, but one has to be aware that these properties are limiting values for approaching the phase boundaries from the homogeneous phases. When calculating, for example, $c_p' = c_p(T,\rho')$, this value corresponds to the isobaric heat capacity of the homogeneous liquid for the limit $T \to T_s$, but only for $T < T_s$. If heat is supplied to a saturated liquid under isobaric conditions, evaporation starts at constant temperature – the quantity $c_p(T,\rho')$ is completely meaningless for this process. No isobaric heat capacity can be defined within the two phase region, since any change of temperature results in a change of pressure, too. For the same reason, $c_p(T,\rho')$ does not correspond to c_σ, the heat capacity "along the saturated liquid line". The heat capacity of a system which contains saturated liquid and an infinitesimally small amount of saturated vapour is given by

$$c_\sigma(T) = T\left(\frac{\partial s}{\partial T}\right)_\sigma = c_v + T\left(\frac{\partial p}{\partial T}\right)_v \left(\frac{\partial v}{\partial T}\right)_\sigma. \tag{3.76}$$

In terms of the reduced Helmholtz energy, this becomes

$$\frac{c_\sigma}{R} = -\tau^2\left(\alpha_{\tau\tau}^o + \alpha_{\tau\tau}^r\right) + \frac{1 + \delta\alpha_\delta^r - \delta\tau\alpha_{\delta\tau}^r}{1 + 2\delta\alpha_\delta^r + \delta^2\alpha_{\delta\delta}^r}\left[\left(1 + \delta\alpha_\delta^r - \delta\tau\alpha_{\delta\tau}^r\right) - \frac{\rho_r}{R\delta}\frac{dp_s}{dT}\right]\bigg|_{\delta=\delta'}. \tag{3.77}$$

The first derivative of the vapour pressure with respect to temperature can be calculated according to

$$\frac{dp_s}{dT} = \frac{\rho''\cdot\rho'}{\rho''-\rho'} R\left[\ln\left(\frac{\rho''}{\rho'}\right) + \alpha^r(\tau,\delta'') - \alpha^r(\tau,\delta') - \tau\left(\alpha_\tau^r(\tau,\delta'') - \alpha_\tau^r(\tau,\delta')\right)\right]. \tag{3.78}$$

The difference between $c_p(T,\rho')$ and c_σ is very often negligible at temperatures around or below the normal boiling temperature since dp_s/dT is small there, but it becomes relevant for higher temperatures.

When approaching the saturated liquid line from within the two phase region, the isochoric heat capacity,

$$c_{v,2} = \left(\frac{\partial u_2}{\partial T}\right)_v = T\left(\frac{\partial s_2}{\partial T}\right)_v = T\left[\frac{ds(T,\rho')}{dT} - \frac{dp_s}{dT}\frac{dv'}{dT}\right], \quad (3.79)$$

becomes

$$\frac{c_{v,2}}{R} = -\tau^2\left(\alpha^o_{\tau\tau} + \alpha^r_{\tau\tau}\right) - \frac{T}{R\rho'^2}\left.\frac{\left[\left(\frac{\partial p}{\partial T}\right)_\rho - \frac{dp_s}{dT}\right]^2}{\left(\frac{\partial p}{\partial \rho}\right)_T}\right|_{\delta=\delta'}. \quad (3.80)$$

The required pressure derivatives can be calculated from the relations given in Table 3.9. For the saturated vapour line, the same relation holds with $\rho = \rho''$ and $\delta = \delta''$, respectively.

In general, it is useful to completely avoid derived properties when dealing with states on the phase boundary. To rely on basic properties like enthalpy, entropy, or internal energy avoids erroneous results which can otherwise result from a thermodynamically incorrect use of derived properties, even if the used relations would be correct for homogeneous states.

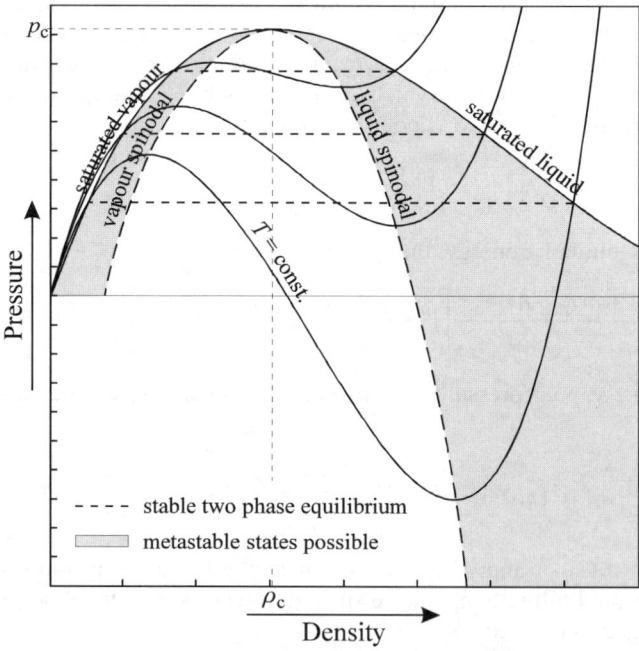

Fig. 3.2. Phase equilibria and metastable states in a p,ρ diagram as calculated from a cubic equation of state.

3.2.3 Properties in the Vapour-Liquid Two Phase Region

When calculating properties for states between the saturated liquid and vapour, one has to distinguish between stable states with two coexisting phases and metastable states which correspond either to a subcooled vapour phase or to a superheated liquid phase.

Metastable states are possible between the homogeneous phase and the corresponding spinodal as shown in Fig. 3.2. For these states, the first order stability criterion is violated – the entropy of the metastable system is smaller than the entropy of the corresponding stable two phase system. Nevertheless, such states can be reached either by fast expansion processes in nozzles or steam turbines (see for instance Moore and Sieverding, 1976) or by superheating or expanding liquids under very special conditions (see Hayward, 1971; for a comprehensive overview on metastable liquids see Debenedetti, 1996). States between the vapour and liquid spinodal violate the second order criterion $(\partial p/\partial \rho)_T > 0$ for mechanical stability and can therefore not be observed on macroscopic time scales. Properties within the metastable region can be calculated according to the relations given for the homogeneous region in Table 3.9, but care has to be taken due to two facts:

- In practice, metastable states are restricted to states rather close to the phase boundary; even very fast expansions do not reach the vapour spinodal. For water the limits of metastability have been experimentally investigated but little is known about other substances, especially when focusing on the vapour side.
- In many cases, multiparameter equations of state may yield misleading results even for states which can be reached in practical processes. Thus, before calculating results for states far within the metastable region, it may be useful to check the location of the spinodals by solving $(\partial p(T,\rho)/\partial \rho)_T = 0$ from the corresponding equation of state (see also Sect. 3.3.5.1).

The stable two phase system is a mixture of two equilibrium phases. The relative amount of mass in each of the phases is characterised by the vapour fraction

$$x = \frac{m''}{m' + m''} \qquad (3.81)$$

which becomes 1 for states on the saturated vapour line and 0 for states on the saturated liquid line. Where appropriate, the liquid fraction

$$x^* = \frac{m'}{m' + m''} = 1 - x \qquad (3.82)$$

can be used alternatively. The vapour fraction can be calculated for given values of T and ρ or y according to

$$x = \frac{1/\rho(T,x) - 1/\rho'(T)}{1/\rho''(T) - 1/\rho'(T)} = \frac{y(T,x) - y'(T)}{y''(T) - y'(T)}, \qquad (3.83)$$

where y can be any specific or molar property such as h, s, u or v. Rewriting Eq. 3.83, values for ρ or y can be calculated using

$$y(T,x) = y'(T) + x(y''(T) - y'(T)) \tag{3.84}$$

and

$$\rho(T,x) = \left[\frac{1}{\rho'(T)} + x\left(\frac{1}{\rho''(T)} - \frac{1}{\rho'(T)} \right) \right]^{-1}, \tag{3.85}$$

respectively. In Eqs. 3.83–3.85 the saturation temperature T_s can be replaced by other suitable properties, such as the vapour pressure p_s. It is important to realise that properties of the two phase system can never be calculated directly from the equation of state – they always have to be calculated using a combination of properties of the coexisting phases.

Working with derived properties is generally not helpful in the two phase region. The isobaric heat capacity is not defined and the speed of sound is not a definite property since its value depends on the distribution of the phases. The only exception is the isochoric heat capacity of the two phase system

$$c_{v,2}(T,x) = c'_{v,2}(T) + x\left(c''_{v,2}(T) - c'_{v,2}(T)\right) \tag{3.86}$$

which can be measured in experimental set ups (see for instance Magee and Ely, 1986) and which is often used to deduce values of c_σ.

In technical applications, Eq. 3.83 is sometimes used to define an equivalent vapour fraction for metastable states, where $y(T,x)$ corresponds, e.g., to a specific enthalpy in the metastable region. For a stable two phase system, Eq. 3.83 yields unequivocal results, since all properties y are composed from the properties of the stable phases according to Eq. 3.84. This is not true for metastable states, where different values of x can be found for the same state depending on the property which is used in Eq. 3.83. Thus, this concept is questionable. If it is unavoidable to use x in order to describe metastable states the property which was used to calculate x has to be noted.

3.3 Iterative Procedures

In Sect. 3.2, the description of procedures focused on calculations at given values of temperature and density – thus it was assumed that the relations given in Tables 3.9 and 3.10 can be applied directly since the values of the independent variables of the equation of state are known. In practice, the given variables for calculations in the single phase region are usually temperature and pressure and in engineering applications very often pressure and enthalpy or entropy. Thus, iterative procedures have to be used to determine the values of temperature and density which correspond to the known variables. Phase equilibria are usually calculated at given values of temperature or saturation pressure and iterative procedures have to be used to determine the unknown properties.

Iterative procedures like the *regula falsi* or *Newton-Raphson* algorithms are included in most of todays textbooks on mathematics. But their application to wide-range equations of state requires both special care to avoid numerical problems and certain strategies for the determination of starting values. Therefore, the formulation of reliable iterative procedures is often the most crucial problem when setting up program packages for the evaluation of equations of state. To overcome typical problems, this section proposes reasonable structures for the required procedures.

Although it is true that all thermodynamic properties can be calculated directly from an equation of state in form of the Helmholtz energy, the need for starting values makes the use of ancillary equations almost unavoidable. For general applications, a reasonable set of ancillary equations includes simple correlations for vapour pressure and saturated vapour and liquid density as functions of temperature as well as a cubic equation of state for the calculation of starting values in the single phase region. Better starting values always speed up iterative procedures and for certain applications, table look up methods, special correlations for the calculation of starting values, or an appropriate use of given or previously calculated data may be very helpful to improve the quality of starting values. However, such procedures are always specialised and cannot be described here; this section is restricted to procedures which can be used for different tasks without special preconditions.

Iterative procedures are usually formulated by finding the set of independent variables where a residuum becomes 0. For given values of pressure p and temperature T, the iteration of the corresponding density is equivalent to the iterative solution of

$$RES(\rho) = p_{\text{calc}}(T,\rho) - p = 0. \qquad (3.87)$$

Such problems are usually solved with regula falsi or Newton-Raphson algorithms. The Newton-Raphson algorithm is fast but it needs the first derivative of the residuum which is sometimes difficult to calculate in an analytical way; if the derivative has to be calculated numerically, the whole algorithm becomes too slow. Thus the regula falsi algorithm is more convenient for general application.

The regula falsi is defined by the arithmetic rule

$$x_{m+1} = x_m - RES(x_m) \frac{x_m - x_{m-1}}{RES(x_m) - RES(x_{m-1})} \qquad (3.88)$$

where two estimated values x_1 and x_2 are required to start the procedure. x_1 and x_2 do not need to encircle the solution but for the reasons discussed below reasonable results can only be guaranteed in this case. The Newton-Raphson algorithm replaces the slope of the secant in Eq. 3.88 by the derivative of RES at x_m and needs therefore only a single starting value x_1.

For a sufficiently well behaved residual function Eq. 3.88 cannot produce misleading results if the solution lies between x_m and x_{m-1}. But in the next step x_{m+1} and x_m may define an interval which does not contain the root and a reasonable result cannot be guaranteed. Basically, two kinds of problems arise: x_{m+1} may become negative which results in numerical problems or x_{m+1} may correspond to a value within the two phase region or the solid phase, where unphysical solutions of the equation of state can be found. Usually such problems are encountered in the critical region, close to the phase boundary or in the liquid region at low temperatures, where the density difference between liquid at saturation and melting pressure becomes small. Figure 3.3 illustrates a typical problem and shows that the

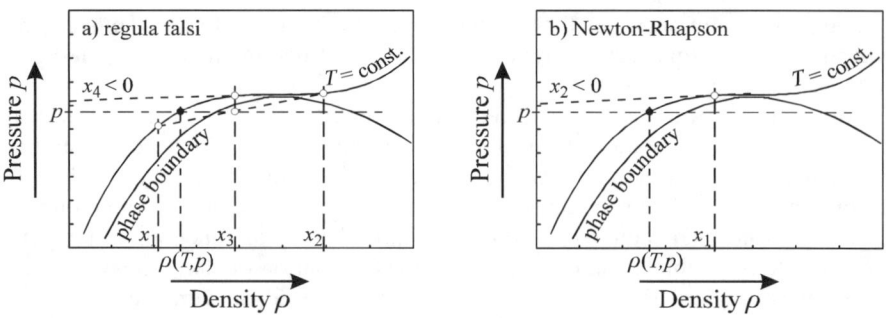

Fig. 3.3. Possible errors resulting from the regula falsi (a) and Newton-Rhapson (b) algorithm when iterating a density for given values of pressure and temperature in the extended critical region.

Newton-Raphson algorithm is even more vulnerable to these kinds of problems than the regula falsi.

To make the regula falsi algorithm more stable Eq. 3.88 can be replaced by

$$x_{m+1} = x_m - RES(x_m) \frac{x_m - x_{m-2}}{RES(x_m) - RES(x_{m-2})} \quad (3.89)$$

for $RES(x_m) \cdot RES(x_{m-1}) > 0$ and $RES(x_m) \cdot RES(x_{m-2}) < 0$; for the next process, x_{m-1} has to be replaced by x_{m-2} in this case. In this way convergence is guaranteed even for difficult problems like the one shown in Fig. 3.3 if the root is bounded on both sides but for standard situations the whole procedure is slowed down. To minimise this disadvantage, the modified regula falsi algorithm can be combined with a nested interval step after applying Eq. 3.89. If the absolute values of $RES(x)$ are quite different at both ends of the new interval, unequally divided intervals yield a further improvement of convergence.

3.3.1 Calculations Based on Temperature and Pressure

The calculation of densities at given values of temperature and pressure is the most common iterative procedure for single phase states. When using the modified regula falsi algorithm described above and the residuum defined in Eq. 3.87, the remaining problem is the determination of starting values for the density.

In general, first estimates for the density can be determined from a cubic equation of state such as the well known equation of Soave (1972). This equation reads

$$p(T,v) = \frac{RT}{v-b} - \frac{a(T)}{v(v+b)} \quad (3.90)$$

with

$$b = 0.08664 \frac{RT_c}{p_c},$$

$$a = 0.42747 \frac{R^2 T_c^2}{p_c} \left(1 + m\left(1 - \left(\frac{T}{T_c}\right)^{0.5}\right)\right)^2$$

and
$$m = 0.480 + 1.574\omega - 0.176\omega^2$$

where ω is the acentric factor

$$\omega = -\log\left(\frac{p_s(T/T_c = 0.7)}{p_c}\right) - 1. \tag{3.91}$$

In terms of the compression factor $Z = p/\rho RT$, this equation can be written as

$$Z^3 - Z^2 + (A - B - B^2)Z - AB = 0. \tag{3.92}$$

with
$$A = \frac{ap}{R^2 T^2}$$

and
$$B = \frac{bp}{RT}.$$

With $Y = Z - 1/3$ this form results in the reduced cubic equation

$$Y^3 + rY + q = 0 \tag{3.93}$$

with
$$r = \frac{3(A - B - B^2) - 1}{3}$$

and
$$q = -\frac{2}{27} + \frac{1}{3}(A - B - B^2) - AB.$$

The discriminant of this reduced cubic equation is

$$D = (r/3)^3 + (q/2)^2. \tag{3.94}$$

For $D \geq 0$, Eq. 3.93 has one real root and for $D < 0$ there are three real roots, where the largest value of Y corresponds to the vapour and the smallest one to the liquid phase solution; this physically expected result is illustrated in Fig. 3.4. Under certain conditions three roots can be found again at very high temperatures ($T > \approx 3\ T_c$), where two roots correspond to unphysical solutions with negative densities. Table 3.11 summarises the solutions of Eq. 3.93. From the definition of Y, the density which corresponds to the i-th solution can be calculated according to

$$\rho_{\text{est},i} = \frac{p}{RT(Y_i + 1/3)}. \tag{3.95}$$

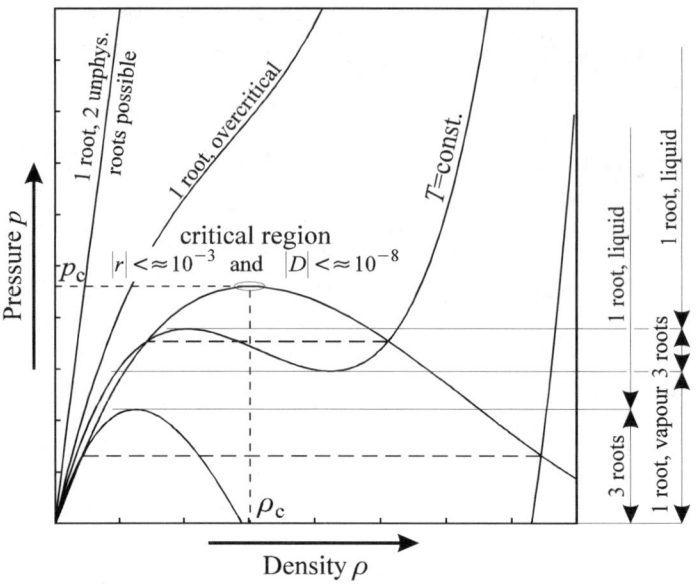

Fig. 3.4. Results for $\rho(T,p)$ as calculated from the Soave equation.

When more than one root is possible, the given pressure must be compared with the vapour pressure in order to decide whether the stable phase is in the vapour ($p < p_s(T)$) or in the liquid ($p > p_s(T)$). In general this can be achieved by comparison with the vapour pressure equation, but where $|p - p_s(T)|$ becomes smaller than the combined uncertainty of vapour pressures calculated from the ancillary equation and from the equation of state, the solution for $p_s(T)$ has to be calculated from the equation of state (see Sect. 3.3.5) to guarantee consistent results.

At the critical point both D and r become zero and the relations given in Table 3.11 are undefined. For $|D| < \approx 10^{-8}$ and $|r| < \approx 10^{-3}$ the critical density can be used

Table 3.11. Solutions of the reduced cubic equation, Eq. 3.93

Discriminant	Solution
$D \geq 0$	$u = \sqrt[3]{-\dfrac{q}{2} + \sqrt{D}}$ and $Y_1 = u - \dfrac{r}{3u}$
$D < 0$	$\theta = \sqrt{-\dfrac{r^3}{27}}$, $\varphi = \arccos\left(-\dfrac{q}{2\theta}\right)$ and $Y_1 = 2\sqrt[3]{\theta}\cos\left(\dfrac{\varphi}{3}\right)$, $Y_2 = 2\sqrt[3]{\theta}\cos\left(\dfrac{\varphi}{3} + \dfrac{2\pi}{3}\right)$, $Y_3 = 2\sqrt[3]{\theta}\cos\left(\dfrac{\varphi}{3} + \dfrac{4\pi}{3}\right)$

as the estimated density for $T > T_c$ and the saturated vapour or liquid density for $T < T_c$. When the equations of state do not match the values of the critical parameters exactly, the critical parameters resulting from the equation of state should be used for reasons of consistency.

For certain substances the Soave equation yields single phase densities which are either significantly larger than the saturated vapour density or significantly smaller than the saturated liquid density. If these densities are used to determine the starting values, the iteration may result in densities in the unstable part of the two phase region where multiparameter equations are not defined and may yield physically incorrect results (see Sect. 3.3.5.1). Thus, it is better to replace the result of the cubic equation by the density at saturation as calculated from the corresponding ancillary equation in this case. For similar reasons, it is useful to replace starting values for the density which are larger than the saturated liquid density at the triple point temperature by the saturated liquid density. Where only calculations in the vapour phase or at far supercritical temperatures are intended, the simple thermal equation of state of the ideal gas, Eq. 3.8, can be used instead of a cubic equation of state.

Based on the estimated density, the required starting values for the density iteration can be determined according to

$$\rho_{1,2} = \rho_{est} \cdot (1 \pm a). \tag{3.96}$$

A reasonable value for a is 0.05, except in the liquid region where smaller values are more appropriate.

3.3.2 Calculations Based on Pressure and Density

Calculations based on pressure and density are very uncommon by themselves since it is unusual to describe states using this combination of variables both in technical and scientific applications. However, there are important exceptions: The calculation of so called Hugoniot curves (see Sect. 4.6.1) requires the iterative determination of the temperature at specified pressure and density. Furthermore, this procedure is needed as part of other, more complex iterations as used mainly in technical applications.

The residuum for iterations based on pressure and density is almost identical with the one given in Eq. 3.87, except for the fact that the residuum in the condition

$$RES(T) = p_{calc}(T, \rho) - p = 0. \tag{3.97}$$

is a function of temperature. In the homogeneous region, Eq. 3.97 has only one root and the derivative $dRES / dT$ is not zero at any point − thus, the whole iteration is less complicated than the one described above.

For $p < p_c$ it should be verified that the given values of pressure and density correspond to a point in the single phase region. To evaluate the criterion $\rho''(p) < \rho < \rho'(p)$ based on the ancillary equations, the saturation temperature $T_s(p)$

must be determined from the vapour pressure equation. From the nearly linear plot of vapour pressure curves on a $\log(p)$ vs. $1/T$ diagram, one finds the relation

$$T_{s,est} = \left[\frac{1}{T_c} - \frac{1/T_t - 1/T_c}{\log(p_c/p_t)} \cdot \log(p/p_c)\right]^{-1} \qquad (3.98)$$

which can be used to determine starting values for this simple iteration. If either $|\rho''(p) - \rho|$ or $|\rho'(p) - \rho|$ is smaller than the combined uncertainty of the ancillary equations and the equation of state, the test for $\rho''(p) < \rho < \rho'(p)$ has to be repeated with equilibrium densities calculated from the equation of state (see Sect. 3.3.5) to guarantee consistent results. For states in the two phase region T_s, ρ' and ρ'' have to be determined from the equation of state. With these values the vapour fraction x and all other properties can be determined according to Eqs. 3.83 and 3.84.

For single phase states, a first estimate for the temperature is required to start the iteration. This first estimate can be calculated from a cubic equation of state, but it is easier to use the classical van der Waals equation at this point since the temperature dependence of the Soave equation is already quite complicated and requirements with regard to accuracy are low. With

$$a = \frac{27}{64}\frac{R^2 T_c^2}{p_c} \quad \text{and} \quad b = \frac{RT_c}{8p_c}$$

the estimated temperature becomes

$$T_{est} = \frac{(p + a\rho^2) \cdot (1/\rho - b)}{R}. \qquad (3.99)$$

When the van der Waals equation yields temperatures which are smaller than the saturation temperature, T_{est} has to be replaced by T_s. For densities above the saturated liquid density at the triple point temperature, the van der Waals equation may yield temperatures below T_t. In this case T_{est} has to be replaced by the melting temperature or, since melting pressure equations are sometimes not available and the melting line is usually very steep, by T_t.

Based on the first estimate for the temperature, the modified regula falsi algorithm can be started with

$$T_{1,2} = T_{est} \cdot (1 \pm a); \qquad (3.100)$$

an appropriate value for a is 0.05. When the smaller of the starting values becomes less than the saturation temperature, it should be replaced by T_s.

3.3.3 Calculations Based on Pressure and Enthalpy

In engineering applications calculations based on pressure and enthalpy or pressure and entropy are very common. Unfortunately it is difficult to estimate starting values for temperature and density from caloric properties, since the relations between these properties and a cubic equation of state cannot be evaluated directly. Thus, a different strategy has to be used if only the ancillary equations described above are available. Any additional information on starting values which may result, for example, from characteristics of the calculated process, speeds up the whole procedure considerably.

For calculations based on pressure and enthalpy the condition for finding the corresponding combination of pressure and temperature can be written as

$$RES(T) = h_{\text{calc}}(T,p) - h = 0. \tag{3.101}$$

Since h can be calculated only as a function of temperature and density in the homogeneous region, every evaluation of the residuum in Eq. 3.101 implies an additional iteration of $\rho(T,p)$, see Sect. 3.3.1.

For $p > p_c$ the enthalpy $h(p,\rho_c)$ can be calculated based on an iteration of $T_0 = T(p,\rho_c)$, see Sect. 3.3.2. For $h > h(p,\rho_c)$ the temperature $T(p,h)$ is larger than T_0 and further temperatures can be calculated as $T_{i+1} = (1+a)\, T_i$ where 0.2 is usually an appropriate value for a. This procedure has to be continued until the temperature $T(p,h)$ is bounded, i.e., until $RES(T_i) \cdot RES(T_{i+1}) < 0$. T_i and T_{i+1} are then used as starting values for the modified regula falsi algorithm. For $h < h(p,\rho_c)$ the same procedure can be used with decreasing temperatures, e.g., with $a = -0.1$. In this case it should be checked after each step, that T_{i+1} is larger than the melting temperature at the given pressure. If no melting pressure equation is available the melting temperature can be replaced by the triple point temperature. For $h < h(p,T_m)$ the given combination of enthalpy and pressure lies outside the range of validity of the equation of state.

For $p < p_c$ it should be checked again whether the given values of pressure and enthalpy correspond to a point in the two phase region. To evaluate the criterion $h'(p) < h < h''(p)$, the saturation temperature $T_s(p)$ has to be calculated from the vapour pressure equation first. With T_s, the densities of saturated vapour and liquid can be determined from the ancillary equations and the corresponding enthalpies can be calculated according to Sect. 3.2.2. If either $|h''(p) - h|$ or $|h'(p) - h|$ is smaller than the combined uncertainty of the ancillary equations and the equation of state, the test has to be repeated with equilibrium densities calculated from the equation of state (see Sect. 3.3.5) to guarantee consistent results. For states in the two phase region T_s, ρ', ρ'', h' and h'' have to be determined from the equation of state. With these values the vapour fraction x and all other properties can be determined according to Eqs. 3.83 and 3.84, respectively.

The same procedure as that described above can be used for states at $p < p_c$ outside of the two phase region. For $h > h''(p)$ the temperature T_0 is replaced by $T_s(p)$; a value of 0.2 can be used for a again. For $h < h'(p)$ the temperature $T(p,h)$ is smaller than $T_s(p)$. Since the temperature difference between melting and saturation becomes small at low pressures, a self adapting step width is useful for the search algorithm in this case. Temperature steps with

$$T_{i+1} = \left(1 - \frac{T_s(p) - T_m(p)}{J\, T_s(p)}\right) \cdot T_i \tag{3.102}$$

and $J \cong 5$ have proved to be reasonable. Where no melting pressure equation is available the melting temperature can be replaced by the triple point temperature.

Iterations based on pressure and enthalpy are generally slow. However, this is usually insignificant when used on state-of-the-art computers. In many cases this procedure is needed for technical calculations where better estimates for the initial temperatures are possible. If calculations based on pressure and enthalpy are used for iterative process calculations with multiple function calls, problem dependent algorithms for the initial estimates should be considered.

3.3.4 Calculations Based on Pressure and Entropy

For calculations based on pressure and entropy, the same problems as those described in Sect. 3.3.3 arise from the fact that no starting values for the temperature can be estimated directly from the entropy. Thus, the same procedure used for pressure and enthalpy can be applied if no problem dependent initial estimates for the temperature are available. In the relations given in Sect. 3.3.3 the enthalpy can simply be replaced by the entropy. The condition Eq. 3.101 becomes

$$RES(T) = s_{calc}(T, p) - s = 0 \qquad (3.103)$$

where every evaluation of the residuum implies an iteration of $\rho(T,p)$. With regard to computing time, the statements made above hold for calculations based on pressure and entropy, too.

3.3.5 Calculation of Phase Equilibria

Several methods have been proposed to calculate phase equilibria from multiparameter equations of state. For equations explicit in the Helmholtz energy, the methods used cover algorithms reaching from the classical equal-area method in terms of pressure which is well known from pressure explicit equations (see Wagner, 1970/1972; Bender, 1970/1973) to recent minimum area methods in terms of the Helmholtz energy which were originally developed for phase equilibrium calculations for mixtures (see Elhassan et al., 1996/1997). The equal area method in terms of pressure can be used to set up reliable iterative procedures but it makes no use of the advantages of an equation in terms of the Helmholtz energy which allows phase equilibrium calculations without integration. Minimum area methods in terms of the Helmholtz energy avoid this disadvantage, but they are difficult to handle in combination with today's multiparameter equations of state. The reasons for this will be explained in detail in Sect. 3.3.5.1. The procedure which is described in this section relies directly on an iterative solution of the phase equilibrium condition in terms of the Helmholtz energy as formulated in Eqs. 3.71 and 3.73. With regard to its iterative set-up, it is similar to the Maxwell equal area rule but it avoids the iteration which is implied by this procedure. Again, the procedure is optimised for reliable use in general applications.

For various reasons, the iteration of phase equilibria is less stable than iterations in the homogeneous region. Good starting values and an ongoing control of the iterative process are essential for reliable results especially if phase equilibria are calculated for temperatures close to T_c – phase equilibrium calculations will usually not use standard procedures for the required iteration, but will incorporate other unique iterative procedures.

In order to evaluate Eqs. 3.71 and 3.73, a complete set of thermal equilibrium properties, T, p_s, ρ' and ρ'', is required. For a given temperature T estimates for ρ' and ρ'' can be calculated from the equation of state according to Sect. 3.3.1 if a sufficiently accurate starting value for p_s is known; the estimates for $\rho'(T)$ and $\rho''(T)$ which are required for the iteration can be calculated from the corresponding ancillary equations instead of the Soave equation. Thus, when starting the procedure it is necessary to estimate a sufficiently accurate value for p_s which allows an iteration of densities both for the vapour and the liquid phase. At low temperatures the Maxwell loop is steep and any vapour pressure equation will yield estimates for p_s which allow solutions in both phases. But when approaching the critical temperature, the Maxwell loop becomes extremely flat and small inconsistencies between the vapour pressures calculated from the equation of state and the ancillary equation may result in situations where only a single density can be found irrespective of the starting values used for $\rho'(T)$ and $\rho''(T)$. This problem is illustrated in Fig. 3.5.

If only a single solution can be found for $\rho'(T,p_{s,est})$ and $\rho''(T,p_{s,est})$, the current estimate for p_s has to be revised. Close to the critical temperature, where this problem occurs frequently, the pressure on the critical isotherm, $p(T,\rho_c)$, is usually a good estimate for the vapour pressure; another possible choice is to use the average between the maximum and the minimum pressure found between $\rho'(T)$ and $\rho''(T)$, where the densities of the coexisting phases can be estimated from the ancillary equations. Whichever strategy is used to revise the value for p_s, it is important to check whether reasonable values were found for $\rho'(T,p_{s,est})$ and $\rho''(T,p_{s,est})$ not only in the first step but after each step of the iteration. For such tests the criterion $\rho'(T,p_{s,est}) > \rho_c$ and $\rho''(T,p_{s,est}) < \rho_c$ is stricter than the simple criterion

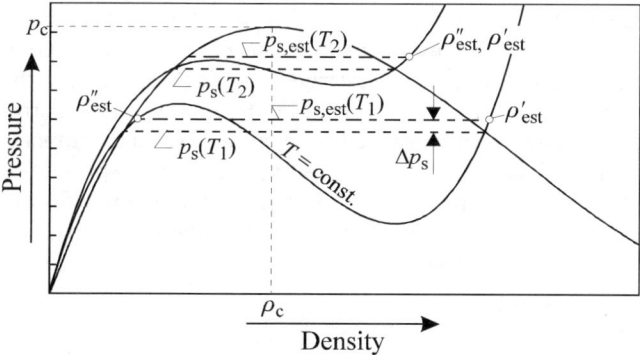

Fig. 3.5. Inaccurate estimates for the vapour pressure can result in situations where only a single value can be found for $\rho'(T,p_{s,est})$ and $\rho''(T,p_{s,est})$.

$\rho'(T,p_{s,\text{est}}) \neq \rho''(T,p_{s,\text{est}})$ for reasons which will become obvious in Sect. 3.3.5.1.

If reasonable values have been found for $\rho'(T,p_{s,\text{est}})$ and $\rho''(T,p_{s,\text{est}})$, these values can be used as estimates for ρ'_{i-1} and ρ''_{i-1} together with the corresponding pressure $p_{s,i-1}$ and a second starting value for the vapour pressure can be calculated from

$$p_{s,i} = RT \frac{\alpha^r(\tau,\delta'_{i-1}) - \alpha^r(\tau,\delta''_{i-1}) + \ln(\rho'_{i-1}/\rho''_{i-1})}{1/\rho''_{i-1} - 1/\rho'_{i-1}}. \qquad (3.104)$$

Again, the corresponding values for ρ'_i and ρ''_i are determined from the equation of state as $\rho'(T,p_{s,i})$ and $\rho''(T,p_{s,i})$. With these data, a regula falsi algorithm can be started to find the pressure for which

$$RES(p_s) = \frac{p_s}{RT}\left(\frac{1}{\rho''} - \frac{1}{\rho'}\right) - \ln\left(\frac{\rho'}{\rho''}\right) - \left(\alpha^r(\tau,\delta') - \alpha^r(\tau,\delta'')\right) = 0. \qquad (3.105)$$

If Eq. 3.105 holds, both Eqs. 3.71 and 3.73 are fulfilled since the densities are calculated from Eq. 3.73 in each step. With the residuum defined by Eq. 3.105 the regula falsi step becomes

$$p_{s,i+1} = p_{s,i} + (p_{s,i-1} - p_{s,i})\frac{RES(p_{s,i})}{RES(p_{s,i}) - RES(p_{s,i-1})}. \qquad (3.106)$$

Close to the critical point, a problem arises with respect to the criterion for termination of the iteration. Under certain conditions the value of $RES(p_s)$ may already be very small, while the corresponding densities of the equilibrium phases still change considerably from step to step. For stable results it is advantageous to combine the usual criterion for termination, $|RES(p_s)| < \varepsilon_p$, with an additional criterion of the type

$$\left(\frac{p_{s,i} - p_{s,i-1}}{p_{s,i}}\right)^2 + \left(\frac{\rho'_i - \rho'_{i-1}}{\rho'_i}\right)^2 + \left(\frac{\rho''_i - \rho''_{i-1}}{\rho''_i}\right)^2 < \varepsilon^2_{\text{comb.}}. \qquad (3.107)$$

For certain applications it becomes necessary to calculate phase equilibria depending on other variables such as p_s and less often ρ' or ρ''. While such function calls are infrequent, the phase equilibrium calculation can be a simple iteration based on the temperature dependent procedure described above. Depending on the given variable, a first estimate for T_s can be calculated by iteration from the corresponding ancillary equation. However, this procedure requires too much computing time since the information which is given for one of the equilibrium properties is not used in the phase equilibrium iteration itself. Where such function calls are frequent, the procedure given above should be adapted to the problem directly.

3.3.5.1 Multiple Maxwell Loops

A reliable calculation of phase equilibria from different multiparameter equations of state requires numerically stable algorithms since such equations behave very differently and quite often unexpected results exist within the two phase region.

While the metastable regions can be described with sufficient accuracy as an extrapolation from the stable single phase regions, no reliable information has ever been available for properties in the unstable region between the spinodals (see Fig. 3.2); equations of state could never be constrained to some kind of "true" behaviour. At a reduced temperature of $T/T_c = 0.5$, this means for methane, that equations of state can be constrained to data at densities up to $\rho'' = 0.420$ kg m^{-3} (usually no gas phase data are available at all at such low densities) and for densities from $\rho' = 445.3$ kg m^{-3} to $\rho(T,p_m) = 457.0$ kg m^{-3}. Thus, at this temperature the density range where no information is available is almost 40 times larger than the range where information is available. Under these circumstances multiparameter equations of state cannot be expected to follow a certain plot within the unstable region – they are simply not valid in this region.

As an example, Figure 3.6 shows typical plots of pressure and Helmholtz energy as calculated from different equations of state for an isothermal pass through the two phase region. The plotted isotherm corresponds to the 160 K isotherm of methane ($T/T_c = 0.84$). The cubic equation proposed by Peng and Robinson (1976), which is shown in Fig. 3.6a, yields the physically expected single Maxwell loop for the pressure and in the Helmholtz plot it can be seen that the Helmholtz energy of the stable equilibrium is always smaller than the Helmholtz energy of the corresponding meta- and unstable states. Thus, the stability criterion is fulfilled for the two phase system.

For the same isotherm, the simultaneously optimised Helmholtz equation with 12 terms (see Sect. 6.2.2) which is shown in Fig. 3.6b yields a second, physically unexpected Maxwell loop. When using iterative procedures like the one described above, multiple solutions for the phase equilibrium criterion become possible but only the outermost one describes the "true" equilibrium. To avoid calculations of incorrect phase equilibria it is important to realise that the relevant outermost solution of $p(T,\rho) = p_s(T)$ fulfils the mechanical stability criterion

$$\left(\frac{\partial p}{\partial \rho}\right)_T > 0 \tag{3.108}$$

while the next solution always fails with respect to this criterion. In this way, misleading solutions can be identified during the phase equilibrium calculation and the iteration of the corresponding density can be repeated with revised starting solutions (ρ_{est} has to be smaller for saturated vapour densities and larger for saturated liquid densities). Compared to the cubic equation of state the plot of the Helmholtz energy shows an unexpected oscillation but nevertheless the phase equilibrium is the stable solution with the smaller Helmholtz energy for all states between the phase boundaries.

In Fig. 3.6c the same isotherm is shown as calculated from the reference quality Helmholtz equation of Setzmann and Wagner (1991) with 40 terms. Compared to Fig. 3.6b the additional Maxwell loop is much more pronounced but still this does not interfere with numerically stable iterative phase equilibrium procedures. A remarkable difference becomes visible in the Helmholtz plot, where this equation mimics pseudo-stable states within the unstable region, for which the Helmholtz energy becomes smaller than for the mixture of the equilibrium phases. Using a

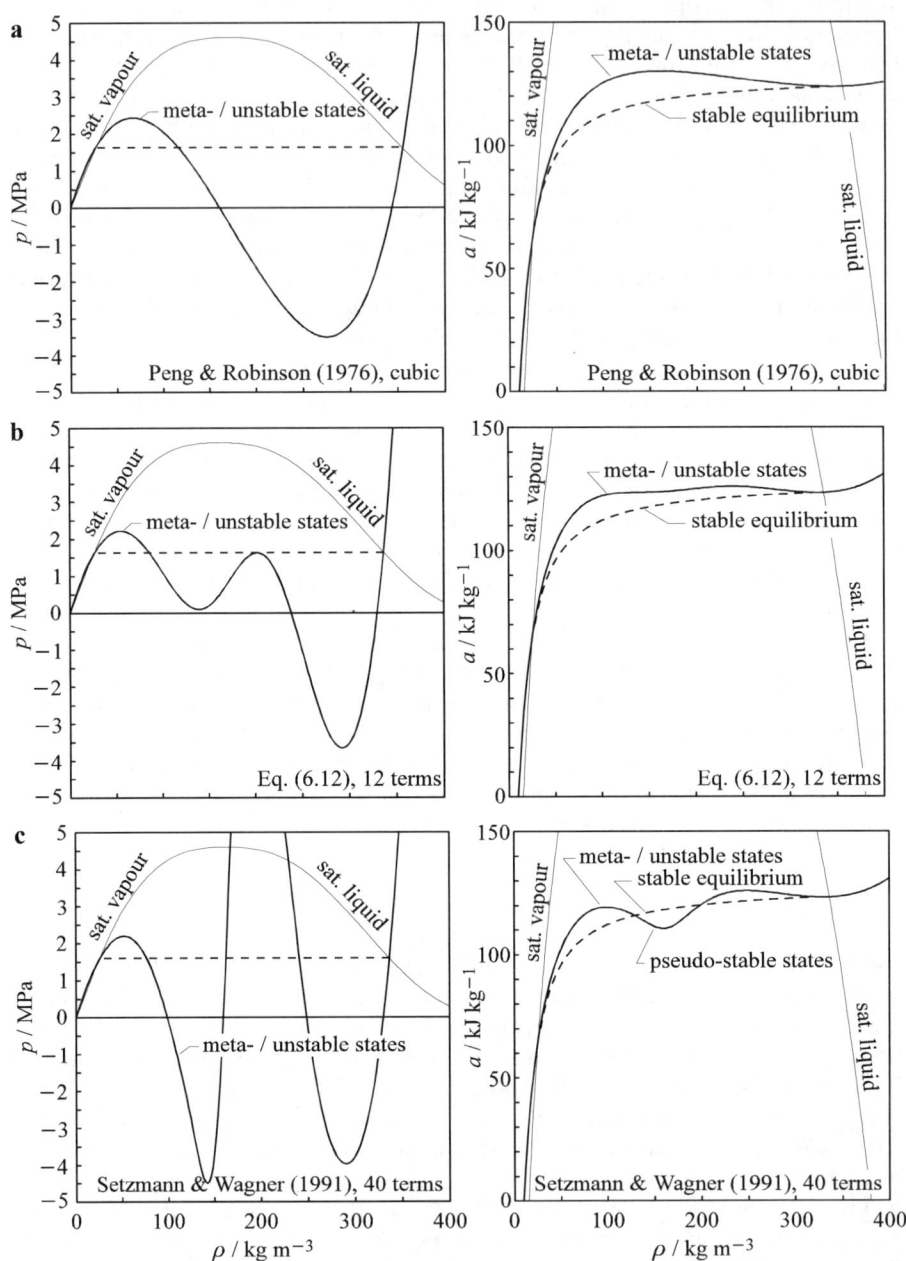

Fig. 3.6. Isothermal plots of pressure and Helmholtz energy in the meta- and unstable regions as calculated from three different equations of state for the 160 K isotherm of methane.

Helmholtz energy based minimum area method (see Eubank et al., 1992; Elhassan et al., 1996) the phase equilibrium calculation would fail in this case.

Basically, the problem of multiple Maxwell loops is well known since multi-parameter equations of state are used to calculate phase equilibria. More recently Elhassan et al. (1997) contributed the Helmholtz energy point of view. So one may ask, why has the problem not yet been solved. In fact, there seem to be two different reasons:

There has been little reason to do so. Algorithms which are able to calculate phase equilibria even under unusual conditions have been developed by different groups working on multiparameter equations of state and can be found in scientific literature (see e.g. Mills et al., 1980) as well. Thermodynamic properties of metastable states can usually be predicted from multiparameter equations very well and properties for states in the unstable region have no physical meaning anyway. Thus, no problems arise for pure substances. With regard to mixtures, multiparameter equations have been applied mainly for the calculation of single phase properties or relatively simple phase equilibria (see e.g. Lemmon et al., 1992; Tillner-Roth, 1993) for which iterative procedures can be formulated easily. In the meantime accurate multiparameter approaches have been extended to more complex systems such as ammonia / water (Tillner-Roth and Friend, 1998) as well, and numerically stable phase equilibrium calculations can still be realised. Keeping in mind that additional constraints with regard to the behaviour in the unstable region would make empirical equations either longer or more complex if the same accuracy has to be realised in the homogeneous regions, there has not been a pressing need to apply such constraints yet. But this situation may change in the future, if new concepts (see e.g. Sects. 6.2 and 7.2.2) enable the use of accurate multiparameter equations of state for a broader variety of pure substances and mixtures.

An improvement is more complicated than one might expect. Very little is known about thermodynamic properties in the metastable areas and almost nothing about unstable states. Empirical multiparameter equations are in error obviously in this region, but the plot of cubic equations, which appears to be physically correct, is in error as well. In the example given in Fig. 3.6, the Peng-Robinson equation yields a saturated liquid density which is too large by 6% and, even worse, the derivative $(\partial p / \partial \rho)_T$ is too small by more than 40% for states on the saturated liquid line. The "true" Maxwell loop has to be steeper and, since the Maxwell equal area criterion still has to be fulfilled, probably deeper than the one predicted by typical cubic equations of state. This general tendency was supported by molecular dynamics simulations which were recently carried out for meta- and unstable states by two different groups (Fischer, 1999). Thus, it is not yet clear to which plot an equation of state should be constrained in the meta- and unstable regions. To constrain it to a physically incorrect plot would certainly affect the representation of properties on the phase boundary and in the homogeneous regions. For this reason, recent work (Elhassan et al., 1997; Span, 1998a) has focused on the stability condition in terms of the Helmholtz energy,

$$a_{\text{calc}}(T,\rho) > a_{\text{two phase}}(T,\rho). \tag{3.109}$$

If this condition is fulfilled, problems with the calculation of mixture phase equilibria should be avoided in combination with recent phase equilibrium procedures even though multiple Maxwell loops can still be encountered. But again, reliable values for $a(T,\rho)$ are not available and to constrain equations of state to the inequality condition given in Eq. 3.109 over the whole unstable region is, if at all, possible only when using non-linear algorithms for the development of the functional form of the equation, see Sect. 4.4.6. Thus, no general breakthrough can be reported yet.

To solve these problems will certainly be one of the topics of future work on multiparameter equations of state. But in the foreseeable future users will have to rely on equations of state which show multiple Maxwell loops and which violate Eq. 3.109 for unstable states. Thus, suitable algorithms for phase equilibrium calculations are one of the key features of any program package for the evaluation of such equations.

4 Setting up Multiparameter Equations of State for Pure Substances

When setting up multiparameter equations of state in the form of Eq. 3.5, two separate tasks have to be fulfilled – formulations have to be developed both for α^o and for α^r. Except for the fact that in both cases least squares fitting techniques are used, which are explained in Sect. 4.1, these tasks require very different approaches.

Since the development of an equation for α^r depends on the equation for α^o this formulation has to be developed or at least finalised first. Theoretically, the contributions to the Helmholtz energy of the ideal gas are well known and empirical simplifications are usually restricted to simple fits of known functional forms to accurate data for a single property in order to establish formulations which are suitable for a broad variety of applications. Section 4.2 summarises the basics of the underlying theory and common simplifications.

The difficulty lies with the development of a formulation for α^r. Until now, no theory has been available which describes the residual contribution to the Helmholtz energy with sufficient accuracy for the whole range of fluid states. Thus, the development of formulations for α^r has to rely almost exclusively on experimental results; Sect. 4.3 explains how formulations for the residual Helmholtz energy can be developed based on data for different thermodynamic properties.

In this context, the assessment of the available experimental data is one of the most crucial points. When setting up equations of state for certain technical applications, the question which has to be answered for each data set is "are these data accurate enough to fulfil the formulated requirements?". For reference equations of state, this question has to be replaced by the more difficult question "how accurate are these data?" since it is the aim to represent all data sets within their experimental uncertainty. Today, estimated uncertainties are usually given in articles on experimental investigations and are generally becoming more reliable on average. Nevertheless, one still has to rely on data without or with overly optimistic information on experimental uncertainties. In these cases, an understanding of the applied experimental techniques is often necessary to find satisfactory answers. Books on experimental techniques (see for instance McCullough and Scott, 1968; Le Neindre and Vodar, 1975; Eder, 1981/1983; Goodwin et al., 2000) are useful as an introduction, but they cannot replace an intense discussion with experienced experimentalists.

One of the characteristic features of the development of multiparameter equations of state is that little is known about their functional form from theory. While the basic fitting techniques which are in use today have been known since the late

70s, there has been continual research on algorithms used for optimising the functional form of such equations. The progress in this area was essential to achieve the accuracies which are typical for modern equations of state. The most important optimisation algorithms are summarised in Sect. 4.4 together with some general rules for setting up reasonable functional forms.

Finally, Sects. 4.5 and 4.6 deal with two aspects which are relevant, especially when setting up reference equations of state: Accurate representation of properties in the critical region and reasonable extrapolation to very high pressures and temperatures.

4.1 Linear and Nonlinear Fitting

The need for fitting empirical formulations to data arises from the fact that the "true" physical relation $\xi(\mathbf{X},Y) = 0$ with its K independent variables $\mathbf{X}$ and the dependent variable Y (e.g. $X_1 = T$, $X_2 = \rho$ and $Y = p$ for $p(T,\rho)$) is either not known or unsuitable for a certain application. Thus, an empirical formulation $\zeta(\mathbf{X},Y,\mathbf{n}) \approx 0$ with the parameter vector $\mathbf{n}$ has to be developed which approximates the true physical relation. The deviation from $\zeta = 0$ is called "residuum". Usually the "true" values of $\mathbf{X}$ and Y are not known as well and have to be replaced by data sets $\mathbf{x}$ and y which are subject to experimental uncertainties.

The task of fitting algorithms is to find the parameter vector $\mathbf{n}$ which yields the best approximation of $\zeta(\mathbf{x},y,\mathbf{n}) = 0$. For certain applications, algorithms have been developed which minimise the largest value of the residuum (see for instance Jing and Fam, 1987) but usually the approximation with the highest statistical probability of $\mathbf{n}$ is regarded as the best one. This is the basic idea of the *maximum-likelihood method* introduced by Fisher (1934) which aims to maximise the likelihood function and can be written as

$$L(\mathbf{n}) = \prod_{m=1}^{M} \frac{1}{\sqrt{2\pi\sigma_m^2}} \exp\left(-\zeta^2(\mathbf{x}_m, y_m, \mathbf{n})/(2\sigma_m^2)\right) \tag{4.1}$$

under four conditions:

- The errors of the experimental results are normally distributed with an expected value of 0 both for the dependent and the independent variables.
- The errors of the variables are independent of each other.
- The experimental uncertainties are known and values for the corresponding variances can be given.
- The error of approximation is negligible compared to the uncertainty of the experimental data.

In Eq. 4.1, σ_m^2 corresponds to the total variance which considers the uncertainties both of the independent and dependent variables; for details on the weighting of experimental data see Sect. 4.3.4. Equation 4.1 can be traced back to the classical method of Gauß (1809) which minimises the weighted sum of squares of the residua of all M experimental data,

$$\chi^2(\mathbf{n}) = \sum_{m=1}^{M} \left(\zeta(\mathbf{x}_m, y_m, \mathbf{n}) / \sigma_m\right)^2. \tag{4.2}$$

Although it is questionable whether the conditions given above are really fulfilled when setting up multiparameter equations of state, the resulting procedure for the determination of the parameter vector **n** is widely used – in fact the maximum-likelihood method has never been used without these assumptions to set up multi-parameter equations of state.

To minimise the sum of squares, Eq. 4.2, for a function with I adjustable parameters, n_i corresponds to the determination of the minimum of a function in an I-dimensional space. The necessary condition for such a minimum is

$$\left(\frac{\partial \chi^2}{\partial n_i}\right)_{\mathbf{x}, y, n_j \neq n_i} = 0 \tag{4.3}$$

for all parameters n_i.

A linear set-up for $\zeta(\mathbf{x}, y, \mathbf{n})$ is given if the parameters n_i are linear coefficients of terms $a_i(\mathbf{x})$ and if the relation between the measured property y or a suitable combination $a_0(\mathbf{x},y)$[1] and the a_i is linear. In this case $\zeta(\mathbf{x}, y, \mathbf{n})$ can be written as

$$\zeta(\mathbf{x}, y, \mathbf{n}) = a_0(\mathbf{x}, y) - \sum_{i=1}^{I} n_i \, a_i(\mathbf{x}) \approx 0, \tag{4.4}$$

where the form of the I functions a_i can be very different, see for instance Eqs. 4.35 – 4.37. With Eq. 4.4, the sum of squares becomes

$$\chi^2(\mathbf{n}) = \sum_{m=1}^{M} \left(a_0(\mathbf{x}_m, y_m) - \sum_{i=1}^{I} n_i \, a_i(\mathbf{x}_m)\right)^2 \cdot \frac{1}{\sigma_m^2}. \tag{4.5}$$

The derivative of χ^2 with respect to the i-th coefficient n_i which is required in Eq. 4.3 can be written as

$$\left(\frac{\partial \chi^2}{\partial n_i}\right)_{\mathbf{x}, y, n_j \neq n_i} = \sum_{m=1}^{M} -2 a_i(\mathbf{x}_m) \cdot \left(a_0(\mathbf{x}_m, y_m) - \sum_{j=1}^{I} n_j \, a_j(\mathbf{x}_m)\right) \cdot \frac{1}{\sigma_m^2} \tag{4.6}$$

or after recombination as

$$\left(\frac{\partial \chi^2}{\partial n_i}\right)_{\mathbf{x}, y, n_j \neq n_i} = -2 \sum_{m=1}^{M} \left(\frac{a_i(\mathbf{x}_m)}{\sigma_m} \cdot \frac{a_0(\mathbf{x}_m, y_m)}{\sigma_m} - \sum_{j=1}^{I} n_j \, \frac{a_i(\mathbf{x}_m)}{\sigma_m} \cdot \frac{a_j(\mathbf{x}_m)}{\sigma_m}\right). \tag{4.7}$$

With Eq. 4.7, the condition for the minimum of the sum of squares, Eq. 4.3, becomes

---

[1] For $p\rho T$ data the residuum may be based, e.g., on $a_0(\mathbf{x},y) = Z = p/(\rho R T)$ instead of $a_0(y) = p/(\rho_n R T_n)$.

$$\sum_{m=1}^{M}\left(\frac{a_i(\mathbf{x}_m)}{\sigma_m}\cdot\sum_{j=1}^{I}n_j\frac{a_j(\mathbf{x}_m)}{\sigma_m}\right)=\sum_{m=1}^{M}\frac{a_i(\mathbf{x}_m)}{\sigma_m}\cdot\frac{a_0(\mathbf{x}_m,y_m)}{\sigma_m}. \qquad (4.8)$$

Equation 4.8 defines a system of I normal equations which can be written in matrix notation as

$$\mathbf{A}\,\mathbf{N}=\mathbf{Q} \qquad (4.9)$$

where the vector $\mathbf{N}$ holds the I coefficients n_j which have to be determined. The $I{\times}I$ matrix $\mathbf{A}$ consists of typical elements of the form

$$a_{ij}=\sum_{m=1}^{M}\left(\frac{a_i(\mathbf{x}_m)\,a_j(\mathbf{x}_m)}{\sigma_m^2}\right) \qquad (4.10)$$

and the elements of the vector $\mathbf{Q}$ can be written as

$$q_i=\sum_{m=1}^{M}\left(\frac{a_i(\mathbf{x}_m)\,a_0(\mathbf{x}_m,y_m)}{\sigma_m^2}\right). \qquad (4.11)$$

Linear systems of normal equations can be solved using the well known *Gauß algorithm* in the form proposed by Cholesky (Benoit, 1924). Details on such algorithms can be found in textbooks on matrix operations such as Zurmühl (1964) and Zurmühl and Falk (1992) but today the required subroutines can be obtained from numerical program packages as well.

Where necessary, constraints can be considered using the method of Lagrangian multipliers as proposed by Hust and McCarty (1967). For each of the constraints, an equation of the form

$$\zeta(\mathbf{x}_c,y_c,\mathbf{n})=a_0(\mathbf{x}_c,y_c)-\sum_{i=1}^{I}n_i\,a_i(\mathbf{x}_c)=0 \qquad (4.12)$$

has to be fulfilled exactly. With these additional conditions the sum of square which has to be minimised becomes

$$\chi^2(\mathbf{n})=\sum_{m=1}^{M}\left(a_0(\mathbf{x}_m,y_m)-\sum_{i=1}^{I}n_i\,a_i(\mathbf{x}_m)\right)^2\cdot\frac{1}{\sigma_m^2}$$
$$+\sum_{c=1}^{C}\lambda_c\left(a_0(\mathbf{x}_c,y_c)-\sum_{i=1}^{I}n_i\,a_i(\mathbf{x}_c)\right). \qquad (4.13)$$

Setting the derivative of χ^2 with respect to a coefficient n_i equal to zero yields

$$\sum_{m=1}^{M}\left(\frac{a_i(\mathbf{x}_m)}{\sigma_m}\cdot\sum_{j=1}^{I}n_j\frac{a_j(\mathbf{x}_m)}{\sigma_m}\right)+\sum_{c=1}^{C}-\frac{\lambda_c}{2}a_i(\mathbf{x}_c)=\sum_{m=1}^{M}\frac{a_i(\mathbf{x}_m)}{\sigma_m}\cdot\frac{a_0(\mathbf{x}_m,y_m)}{\sigma_m}. \qquad (4.14)$$

The derivatives of χ^2 with respect to $\lambda'_c=-\lambda_c/2$ yield C additional normal equations of the form

$$\sum_{i=1}^{I} n_i \, a_i(\mathbf{x}_c) = a_0(\mathbf{x}_c, y_c). \tag{4.15}$$

In matrix notation, the resulting problem can be written as

$$\begin{bmatrix} \mathbf{A} & \mathbf{C}^T \\ \mathbf{C} & \mathbf{0} \end{bmatrix} \begin{bmatrix} \mathbf{N} \\ \lambda' \end{bmatrix} = \begin{bmatrix} \mathbf{Q} \\ \mathbf{Q}_c \end{bmatrix} \tag{4.16}$$

where elements of the $C \times I$ matrix $\mathbf{C}$ have the form

$$c_{c,i} = a_i(\mathbf{x}_c) \tag{4.17}$$

and the C elements of the vector $\mathbf{Q}_c$ correspond to the constrained values

$$q_c = a_0(\mathbf{x}_c, y_c). \tag{4.18}$$

Solving Eq. 4.16 yields the set of coefficients n_i for the least squares correlation which fulfils the C constraints exactly.

The linear procedure described above fails either if the relation between a_0 and a_i becomes non-linear or if the functional form of the a_i implies a non-linear relation between different parameters n_j. Examples of these two types of non-linearity are the residua

$$\zeta(\mathbf{x}, y, \mathbf{n}) = a_0(\mathbf{x}, y) - \sum_{i=1}^{I} n_i \, a_i(\mathbf{x}) + \left(n_i \frac{\partial a_i(\mathbf{x})}{\partial x_1} \right)^2 \approx 0 \tag{4.19}$$

and

$$\zeta(\mathbf{x}, y, \mathbf{n}) = a_0(\mathbf{x}, y) - \sum_{i=1}^{I} n_i \, a_i(\mathbf{x}, n_{I+i}) \approx 0, \tag{4.20}$$

respectively. For such residua, the derivatives $(\partial \zeta / \partial n_i)_{\mathbf{x}, y, n_j \neq n_i}$ depend on the parameter vector $\mathbf{n}$ and so does the matrix $\mathbf{A}$ and the vector $\mathbf{Q}$ in Eq. 4.9 and Eq. 4.16, respectively – the system of normal equations becomes non-linear. Such non-linear systems of normal equations can be solved only by iterative procedures which require a set of starting values $\mathbf{n}_s$ for the parameter vector $\mathbf{n}$. The quality of the starting values is a crucial point since even sophisticated algorithms may not find the absolute minimum of the sum of squares without a suitable vector $\mathbf{n}_s$. Problem dependent strategies have to be applied in order to estimate the required starting values as accurately as possible. But even with good starting values, convergence cannot be guaranteed for problems with complex intercorrelations between different parameters as they are typical, e.g., for a simultaneous adjustment of reducing parameters and coefficients, see Sect. 7.2.2, or for a simultaneous fit of coefficients and exponents of a single term, see Sects. 4.5.3 and 4.5.4.

Different strategies have been developed for the solution of non-linear systems of normal equations; the corresponding algorithms can be found in advanced textbooks and numerical program packages again. An algorithm which has proved to be very suitable for the development of multiparameter equations of state is the Marquardt (1963) algorithm in the modified form proposed by Fletcher (1971).

The application of this algorithm was described in detail by Ahrendts and Baehr (1979b / 1981b).

4.2 Describing the Helmholtz Energy of the Ideal Gas

As already discussed in Sect. 3.1.1.1, the description of the Helmholtz energy of the ideal gas essentially depends on equations for the heat capacity of the ideal gas. Based on equations with the general form

$$\frac{c_p^o}{R} = n_0^* + \sum_{i=1}^{I_{\text{Pol}}} n_i^* T^{t_i^*} + \sum_{k=1}^{K_{\text{PE}}} m_k \left(\frac{\theta_k}{T}\right)^2 \frac{e^{(\theta_k/T)}}{\left(e^{(\theta_k/T)} - 1\right)^2} \qquad (4.21)$$

(see also Eq. 3.9), the steps which are necessary to obtain an equation for the Helmholtz energy of the ideal gas were also discussed in Sect. 3.1.1.1. Thus, this section focuses on the development of correlations for the heat capacity of the ideal gas, which can either be based on results from statistical thermodynamics and molecular constants measured by spectroscopic means or on experimental results for caloric properties in the low density gas phase. Although both approaches are very different, they are discussed in a single section here, since a basic understanding of the theoretical background helps to avoid major mistakes when using the empirical approach, and since the more theoretical approach often results in fitting Eq. 4.21 to special data as well. Any explanations with regard to statistical thermodynamics will be highly simplified and restricted to results which are useful for the topic of this section; for a more detailed and theoretically based description see Lucas (1991).

From classical mechanics, a body which consists of n mass points can be described by $3n$ pairs of position and momentum co-ordinates. Thus, in an ideal gas, a molecule consisting of n atoms has *$3n$ degrees of freedom* (the momentum co-ordinates) for movements which can contribute to the dynamic energy state of the molecule. It is this dynamic energy state which can be observed as internal energy of an ideal gas on a macroscopic scale and which is hence responsible for the ideal gas heat capacity.

Assuming a rigid molecule without internal rotation, the $3n$ degrees of freedom correspond to translation, external rotation and vibrational modes of the atoms in the molecule, where neither the rotational nor the vibrational modes must move the centre of mass of the molecule. When assuming that these effects do not interact with each other, the heat capacity can be written as

$$c_{p,\text{simp}}^o = c_{v,\text{simp}}^o + R = c_{\text{trans}} + c_{\text{rot}} + c_{\text{vib}} + R \qquad (4.22)$$

for this simplified model.

In an ideal gas, each molecule has 3 degrees of freedom for translational movements which are, by definition, fully excited since the molecules must not interact with each other or with any force fields. Each translational degree of freedom contributes ½ R to the heat capacity and thus we find the well known relation

$$c_{\text{trans}} = \frac{3}{2}R. \qquad (4.23)$$

The assumption of fully excited translational modes becomes questionable in the limit of very low temperatures, but it can be shown that it is in general valid for temperatures of $T > 10^{-10}$ K; the only restrictions apply for the quantum gases helium and hydrogen. Monoatomic molecules of noble gases have only the three translational degrees of freedom, and thus their heat capacity becomes completely independent of temperature.

Molecules which consist of 2 or more atoms can rotate around 2 (linear molecules) or 3 (non-linear molecules) principle axis. When assuming that molecules behave like rigid rotators, i.e., the molecule stretches neither with increasing rotational speed nor with increasing vibrations and that therefore the moment of inertia does not depend on temperature, each fully excited mode of external rotation contributes ½ R to the heat capacity. Therefore the contributions for c_{rot} are

for linear molecules
$$c_{\text{rot}} = \frac{2}{2}R \qquad (4.24)$$

and for non-linear molecules
$$c_{\text{rot}} = \frac{3}{2}R. \qquad (4.25)$$

In most cases, it can be assumed that the rotational modes are fully excited in the temperature range where multiparameter equations of state are usually applied, i.e., for temperatures above the triple point of the corresponding substance. For substances with small molecules or for applications at very low temperatures, this assumption should be verified. The i-th rotational mode can be regarded as fully excited if the temperature is much larger than the characteristic temperature of the rotational mode, $\theta_{\text{rot},i}$ ($T > \approx 2\,\theta_{\text{rot},i}$). The characteristic temperature is given by

$$\theta_{\text{rot},i} = \frac{h^2}{8\pi^2 k I_i} = \frac{hcA_i}{k} \qquad (4.26)$$

with
$h = 6.6260755 \cdot 10^{-34}$ J s   (Planck constant),
$k = 1.380658 \cdot 10^{-23}$ J K^{-1} (Boltzmann constant),
and
$c = 2.99792458 \cdot 10^{10}$ cm s^{-1} (speed of light in vacuum).

For linear molecules, the relevant moments of inertia, I_i, and the relevant rotational constants, A_i, are the same for both rotational modes due to reasons of symmetry. The rotation around the bonding axis does not contribute to the dynamic energy state since the corresponding principal moment of inertia is almost 0. For non-linear molecules, the smallest value for I_i or the largest value for A_i has to be considered. Evaluation of Eq. 4.26 yields $\theta_{\text{rot,max}} = 40.11$ K for water, $\theta_{\text{rot,max}} = 7.54$ K for methane, $\theta_{\text{rot,max}} = 2.36$ K for difluoromethane (refrigerant 32), and $\theta_{\text{rot,max}} = 0.49$ K for chlorodifluoromethane (refrigerant 22); the influence of the size and the shape of the molecules is obvious. Listings of molecular constants can be found in scientific literature (see Herzberg and Herzberg, 1972).

Besides translation and external rotation, $3n - 5$ modes of vibration have to be considered for rigid linear molecules. For rigid non-linear molecules $3n - 6$ modes have to be considered. These vibrational modes cannot be regarded as fully excited at ambient temperatures and are therefore responsible for the temperature dependence of the ideal gas heat capacity. To consider the influence of internal vibrations it is usually assumed that the molecules behave like harmonic oscillators. Although this is strictly not correct since the intramolecular forces depend linearly on the elongation of the atomic bonds, the harmonic oscillator model yields good results for moderate temperatures.

The Bose-Einstein statistic yields the so called "Planck-Einstein" function

$$\frac{c_i}{R} = \left(\frac{\theta_i}{T}\right)^2 \frac{e^{(\theta_i/T)}}{\left(e^{(\theta_i/T)} - 1\right)^2} \tag{4.27}$$

for the contribution of harmonic oscillations to the heat capacity, where θ_i is the characteristic temperature of the i-th vibrational mode. The plot of Eq. 4.27 is shown in Fig. 4.1. The characteristic temperatures can be determined from spectroscopic results for the corresponding wavenumbers, v_i, according to

$$\theta_i = \frac{hcv_i}{k}. \tag{4.28}$$

Extensive listings of fundamental vibrational frequencies for various molecules are published in scientific literature (see for instance Shimanouchi, 1972; Shimanouchi, 1977; Shimanouchi et al., 1978; Shimanouchi et al., 1980). As experimental results, wavenumbers are of course subject to experimental uncertainties. However, typical uncertainties in v_i have only a small influence on calculated heat capacities.

Thus, assuming a rigid molecule with harmonic vibrations and excited rotational modes, the heat capacity of the ideal gas becomes

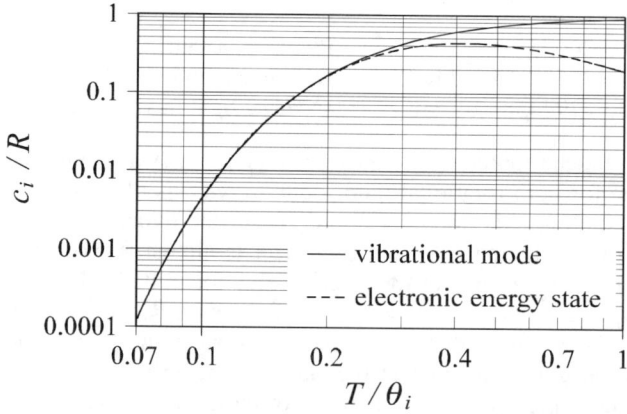

Fig. 4.1. Contribution of a single vibrational mode and a single electronic energy state to the heat capacity of the ideal gas. For the contribution from electronic energy, a degeneracy of 1 has been assumed both for the ground state and for the first excited state (see Lucas, 1991).

$$\frac{c^o_{p,\text{simp}}}{R} = \underbrace{(1+c_{\text{trans}}+c_{\text{rot}})}_{=c_0}+\sum_{i=1}^{I_{\text{vib}}} m_i \left(\frac{\theta_i}{T}\right)^2 \frac{e^{(\theta_i/T)}}{\left(e^{(\theta_i/T)}-1\right)^2}, \quad (4.29)$$

where m_i is equal to 1 except for multiple vibrational modes with identical wave numbers which may occur in symmetrical molecules. For multiple modes, m_i corresponds to the order of the mode and the total number of vibrational modes I_{vib} has to be reduced correspondingly.

For simple molecules at ambient temperatures, Eq. 4.29 yields an accurate description of the isobaric heat capacity of the ideal gas. Unfortunately, the simple rigid rotator, harmonic oscillator model fails at high temperatures and for complex molecules. For these cases, additional contributions have to be considered for excited states of the electron shell, for internal rotation and finally for corrections to the rigid rotator, harmonic oscillator model. With these contributions the relation for the isobaric heat capacity of the ideal gas is

$$c^o_p = c^o_{p,\text{simp}} + c_{\text{el}} + c_{\text{irot}} + c_{\text{cor}}. \quad (4.30)$$

The contribution from excitation of the electron shell, c_{el}, becomes relevant for temperatures where

$$\frac{\theta_{\text{el},1}}{T} >> 1 \quad (4.31)$$

is no longer fulfilled. For practical applications, this equation becomes

$$\frac{\theta_{\text{el},1}}{T} > \approx 10 \quad \text{for} \quad \frac{c_{\text{el},1}}{c^o_p} < 0.1\%; \quad (4.32)$$

see also Fig. 4.1. The characteristic temperature of the lowest electronic state can be calculated from the energy level of the lowest non-ground state, $\varepsilon_{\text{el},1}$, or the corresponding wavenumber, $\nu_{\text{el},1}$, according to

$$\theta_{\text{el},1} = \frac{\varepsilon_{\text{el},1}}{k} = \frac{hc\nu_{\text{el},1}}{k}. \quad (4.33)$$

For monoatomic species such as noble gases, typical values for the wavenumber of the lowest electronic state are in the order of 50,000 cm^{-1} to 160,000 cm^{-1}, resulting in characteristic temperatures of 72,000 K to 230,000 K. Thus, the contribution from excited electronic states is usually negligible for these substances. Extensive tabulations of atomic energy levels are given by Moore (1971) and Kelly (1987). For diatomic or polyatomic molecules, the levels of single electronic states are sometimes much lower. The oxygen molecule (O_2) for example has a single (experimentally uncertain) electronic energy level with a characteristic temperature of about 10,000 K and one energy level of nitric oxide (NO) has a characteristic temperature which is even as low as 174 K. For diatomic or polyatomic molecules, the concept that excitation of electronic energy levels has no significant influence in the temperature range which is relevant for multiparameter equations of state has to be verified for each case. Tables of molecular energy levels are given by Herzberg (1950, 1966). Where necessary, the contribution of excited electronic states can be considered according to the relations given by Lucas (1991); a simple correlation which takes excited electronic states into account was published by Wagner et al. (1982).

For polyatomic molecules, groups of atoms may rotate against each other within the molecule, where the "internal rotation" replaces one vibrational mode of the molecule. The way in which the internal rotation is considered as a contribution to the heat capacity depends on the potential energy barrier, $U_{ir,max}$, which is used to characterise the type of rotation.

If $U_{ir,max}/kT$ is large (an order of 10 or more), the potential energy barrier is so high that only a few molecules can rotate internally around the bonding axis. The result is an oscillation around the minimum of potential energy, which can be treated as a harmonic oscillation like the other vibrational modes, if the wavenumber of the oscillation is known. This behaviour can be observed for example for ethene where the double bond between the carbon atoms avoids a real internal rotation.

If $U_{ir,max}/kT$ is sufficiently small, the internal rotation can be treated as an additional free rotational mode, which results in a contribution of ½ R to the heat capacity of the ideal gas.

Unfortunately, $U_{ir,max}/kT$ is neither "large" nor "small" in most cases, and rotations around single bonds between groups of atoms have to be treated as hindered internal rotations. An algorithm for treating such rotational modes is given by Lucas (1991), but this algorithm is far too complex to be introduced in a simple correlation for the isobaric heat capacity. Furthermore, the required spectroscopic data are often either difficult to assess or not available at all. Thus, in many cases an accurate theoretical calculation of contributions from internal rotation is possible only for scientists who are specialised in this kind of calculations, if it is possible at all.

For corrections to the rigid rotator, harmonic oscillator model the situation is very similar. At very high temperatures or for molecules which consist of heavy atoms with weak bonds, the results from the rigid rotator, harmonic oscillator model do not match experimental results for caloric properties in the ideal gas state. The necessary corrections to the model have to account mainly for three effects:

- *Centrifugal stretching* leads to an increase of the moments of inertia of the molecule and changes therefore the energetic contribution from the rotational modes.
- With increasing amplitudes the *anharmonicity of intramolecular vibrations* becomes important.
- With increasing amplitudes the average distance between atoms in a molecule increases due to the vibrational anharmonicities. The increased distances result in an increase of the moments of inertia and therefore in a *rotational-vibrational coupling*.

Algorithms which account for these corrections on the basis of spectroscopic data are well developed (see, Pennington and Kobe, 1954; Schäfer, 1960; Gurvich et al., 1989). But again, these algorithms are far too complex to be included directly into simple equations for the isobaric heat capacity and the required spectroscopic data are very often either difficult to assess or not available at all.

Due to the difficulties which are related to internal rotations and to necessary corrections to the rigid rotator, harmonic oscillator model, it is usually more convenient to fit the parameters of the equation for the isobaric heat capacity of the ideal gas, Eq. 4.9, to tabulated values if these data were calculated using the necessary corrections to $c_{p,\text{simp}}^o$. Extensive tabulations which are frequently used as the basis for correlation equations were published by Gurvich et al. (1989), by Garvin et al. (1987), by Chase et al. (1985), by the Thermodynamic Research Center at the Texas A&M University (TRC, 1972–1993), or for individual substances in the Journal of Physical and Chemical Reference Data (see Chao et al., 1973; Chen et al., 1975).

To fit the parameters of Eq. 4.9 to data for the isobaric heat capacity of the ideal gas, a simple residuum of the form

$$\zeta^2 = \left(\frac{c_{p,\text{calc}}^o - c_{p,\text{meas}}^o}{\sigma_{\text{meas}}}\right)^2 \tag{4.34}$$

can be used, where the subscript "meas" corresponds to the tabulated values. An accurate assignment of σ_{meas} is difficult in this case, since calculated values do not have an experimental uncertainty in the sense discussed in Sect. 4.3.4. However, since Eq. 4.9 is fitted only to data of a single property, the value of σ_{meas} is arbitrary, as long as it is chosen in a consistent way for all values of $c_{p,\text{meas}}^o$. Quite often it is convenient to set σ_{meas} equal to the aspired accuracy of the correlation Eq. 4.9. The required derivatives of the residuum with respect to the adjustable parameters (see Sect. 4.1) become

$$\left(\frac{\partial \zeta}{\partial c_0}\right) = \frac{R}{\sigma_{\text{meas}}}, \tag{4.35}$$

$$\left(\frac{\partial \zeta}{\partial n_i^*}\right) = T^{t_i} \frac{R}{\sigma_{\text{meas}}}, \tag{4.36}$$

$$\left(\frac{\partial \zeta}{\partial m_k}\right) = \left(\frac{\theta_k}{T}\right)^2 \frac{e^{(\theta_k/T)}}{\left(e^{(\theta_k/T)} - 1\right)^2} \frac{R}{\sigma_{\text{meas}}}, \tag{4.37}$$

and

$$\left(\frac{\partial \zeta}{\partial \theta_k}\right) = m_k \frac{e^{(\theta_k/T)}}{T} \left[\frac{\left[2\frac{\theta_k}{T} + \left(\frac{\theta_k}{T}\right)^2\right]}{\left(e^{(\theta_k/T)} - 1\right)^2} - \frac{2\left(\frac{\theta_k}{T}\right)^2 e^{(\theta_k/T)}}{\left(e^{(\theta_k/T)} - 1\right)^3}\right] \frac{R}{\sigma_{\text{meas}}}. \tag{4.38}$$

While c_0, the n_i^*, and the m_k alone can be fitted with linear algorithms, the θ_k can be adjusted only with non-linear algorithms. When using Planck-Einstein functions, the characteristic temperatures of the vibrational modes of the molecule are useful starting solutions for the θ_k. In this case the m_k should be set equal to 1. For polyatomic molecules with a large number of vibrational modes, groups of modes with similar wavenumbers can be combined in a single term with an average value for θ_k and a value for m_k which corresponds to the number of combined modes. The temperature exponents t_i^* of the polynomials can either be fitted nonlinearly or optimised by a suitable algorithm (see Sect. 4.4), but usually they are simply chosen by trial and error.

Where only a limited range of validity is required, pure polynomial forms of Eq. 4.9 without any Planck-Einstein functions are widely used. The constant c_0 and the n_i^* of the equations are simply fitted to the data and the extrapolation behaviour of the correlations is not examined at lower or higher temperatures. Such equations are typical for Helmholtz equations for the halogenated hydrocarbon refrigerants.

When fitting Eq. 4.9 to data with a broad temperature range, the constant c_0 is usually set equal to the contributions from translation, external rotation, and where applicable, from free internal rotation, c_{trans}, c_{rot}, and c_{irot}. In this way, Eq. 4.9 yields reasonable values for the isobaric heat capacity in the limit of low temperatures, at least if the temperature exponents t_i^* of the polynomials are chosen to be positive.

For a better extrapolation behaviour at high temperatures, it is useful to avoid polynomial terms completely. In this case, the contribution from vibration and all corrections to $c_{p,\text{simp}}^\text{o}$ are expressed only by a summation of Planck-Einstein functions. To achieve reasonable extrapolation at high temperatures, all the coefficients m_k should be positive and their sum should roughly be equal to the number of vibrational modes of the considered molecule. Often the lowest values of θ_k which result from the fit can be assigned to the corresponding values from spectroscopic data, see Eq. 4.28, with only small deviations, while different vibrational modes with high characteristic temperatures can be combined in a single empirical term with a correspondingly large value of m_k – this experience justifies the starting solution proposed above.

Wherever possible, theoretical results for the heat capacity of the ideal gas should be compared to experimental results for caloric properties in the low density gas phase in order to exclude possible errors in the interpretation of underlying spectroscopic data. Traditionally, experimental values for isobaric heat capacities in the ideal gas state were determined by extrapolation of data for the gas phase heat capacity according to

$$c_p^\text{o} = \lim_{p \to 0} c_p(T, p). \tag{4.39}$$

Unfortunately, experimental results for the isobaric heat capacity in the low density gas phase are rather uncertain due to the small masses (mass flows) in the calorimeters. Under these conditions, small absolute errors in the determined heat flow result in considerable uncertainties in the measured heat capacities. While ideal gas heat capacities which were calculated from statistical thermodynamics and spectroscopic data have typical uncertainties of $\Delta c_p^o/c_p^o < \pm 0.1\%$ for moderate temperatures and simple molecules, good experimental results have typical uncertainties of $\Delta c_p^o/c_p^o < \pm 0.3\% - \pm 1.0\%$. Thus, experimental results according to Eq. 4.39 are very often unsuitable to test Eq. 4.9 on a high level of accuracy.

The situation has improved with the introduction of sound speed measurements from spherical resonators. These devices allow extrapolations of the ideal gas speed of sound,

$$w^o = \lim_{p \to 0} w(T,p), \qquad (4.40)$$

with typical uncertainties of $\Delta w^o/w^o < \pm 0.01\%$. For ideal gases, the isobaric heat capacity can be calculated from the speed of sound according to

$$c_p^o = \left(\frac{1}{R} - \frac{T}{w^{o2}}\right)^{-1}, \qquad (4.41)$$

and an error analysis yields

$$\left|\frac{\Delta c_p^o}{c_p^o}\right| = \left|2\left(\frac{c_p^o}{R} - 1\right)\frac{\Delta w^o}{w^o}\right|. \qquad (4.42)$$

Thus data from speed of sound measurements with spherical resonators are very suitable to test results from theoretical calculations, especially for small molecules or low temperatures where c_p^o/R is small.

For more complex molecules like halogenated hydrocarbon refrigerants, the available spectroscopic information is very often insufficient to calculate accurate heat capacities from statistical thermodynamics. Ab initio calculations have been proposed to solve this problem (see for instance Lucas et al., 1993). However, these results have not yet been considered when setting up multiparameter equations of state. For these fluids, it is common to fit Eq. 4.9 directly to experimental results in exactly the same way as discussed above. Where data from spherical resonators are available, this method is not necessarily disadvantageous with regard to accuracy, but with regard to the covered temperature range. Based on apparatuses used today, temperatures from the triple point of typical cryogenic substances up to about 450 K can be covered by spherical resonator measurements (see for instance Estrada-Alexanders and Trusler, 1995). Usually however, reliable data are available only in smaller temperature ranges and a direct fit to these data restricts the range of validity of the equation for the Helmholtz energy of the ideal gas. This method may be acceptable for equations which are intended to become technical standards without application at higher temperatures, but it is unsatisfactory for setting up scientific property standards. For chlorodifluoromethane (refrigerant 22) Wagner et al. (1993; see also Marx et al., 1992) fitted only the necessary (but not sufficiently investigated) anharmonicity corrections to experimental data while maintaining the results from the rigid rotator, harmonic oscillator model as a theoretical background. This mixed approach yielded a better extrapolation to high temperatures than simple fits to experimental data.

4.3 Describing the Residual Helmholtz Energy

As discussed before, recent multiparameter equations of state are usually set-up in the form of the reduced Helmholtz energy (see for instance Eq. 3.5), where the residual part of the Helmholtz energy, α^r, is determined from an empirical representation of experimental data. However, Helmholtz energies cannot be measured directly and adopting the procedures described in Sect. 4.1 to such an equation requires a *multiproperty approach*. Fits which involve data for derived properties have already been used to set-up pressure explicit equations (see for instance Hust and McCarty, 1967; Wagner, 1970/1972; Bender, 1970/1973, Jacobsen and Stewart, 1973) for a long time, but with the introduction of Helmholtz equations they became mandatory. For equations in form of the Helmholtz energy, Ahrendts and Baehr (1979a/1981a) summarised the theoretical background of the required procedures, but their application was still restricted to only a few properties (see Ahrendts and Baehr, 1979b/1981b). During the following years the use of multiple properties was extended more and more, reaching its current maximum for the new reference equation for water developed by Pruß and Wagner (1995; see also Wagner and Pruß, 1997/2000) which was fitted to data of 16 different properties.

The basic idea of the multiproperty approach and its application to Helmholtz equations is described in Sect. 4.3.1. In Sects. 4.3.2–4.3.3 the required residua are discussed and Sect. 4.3.4 describes the determination of the variances of experimental data which are needed to apply the relations given in Sect. 4.1.

4.3.1 The Multiproperty Approach

Following the nomenclature used in Sect. 4.1, the usual formulation for the residual part of the Helmholtz energy, Eq. 3.25, can be written as

$$\alpha^r(\mathbf{x}) = \sum_i^I n_i\, A_i(\mathbf{x}) \qquad (4.43)$$

with $x_1 = \tau = T_r/T$ and $x_2 = \delta = \rho/\rho_r$. In Eq. 4.43 the $A_i(\mathbf{x})$ correspond to $\tau^{t_i}\delta^{d_i}$ for polynomial terms and to $\tau^{t_i}\delta^{d_i}\exp(-\gamma_i\delta^{p_i})$ for exponential terms; more complex functional forms are used in reference equations of state for an improved representation of properties in the critical region, see Sects. 4.5.3 and 4.5.4. However, at no point in Sect. 4.1 was it assumed that the $a_i(\mathbf{x})$ in Eq. 4.4 have the same functional form. Thus, the $A_i(\mathbf{x})$ in Eq. 4.43 can be identified with the $a_i(\mathbf{x})$ in Eq. 4.4.

The fact that the $a_0(\mathbf{x},\mathbf{y})$ and $a_i(\mathbf{x})$ may correspond to very different functional forms makes it possible to fit equations of state to data of multiple derived properties. Usually, the most important information on the thermodynamic surface of pure substances results from data for the $p\rho T$ relation. When introducing Eq. 4.43 into the relation between compression factor and reduced Helmholtz energy, Eq. 3.30, the relation for $p(T,\rho)$ reads

4.3 Describing the Residual Helmholtz Energy

$$\frac{p}{\rho RT} = 1 + \delta \left(\frac{\partial \alpha^{\mathrm{r}}}{\partial \delta}\right)_\tau = 1 + \delta \sum_{i=1}^{I} n_i \left(\frac{\partial A_i(\tau,\delta)}{\partial \delta}\right)_\tau \tag{4.44}$$

and the corresponding residuum $\zeta_{p\rho T}$ can be written in an *explicit linear* form as

$$\zeta_{p\rho T} = \left(\frac{p - \rho RT}{\rho^2 RT}\right) - \sum_{i=1}^{I} n_i \left(\frac{1}{\rho_{\mathrm{r}}}\left(\frac{\partial A_i}{\partial \delta}\right)_\tau\right). \tag{4.45}$$

Thus, $p\rho T$ data can be used in the linear algorithm described in Sect. 4.1 with

$$a_{p\rho T,0} = \frac{p - \rho RT}{\rho^2 RT} \tag{4.46}$$

and

$$a_{p\rho T,i} = \frac{1}{\rho_{\mathrm{r}}}\left(\frac{\partial A_i}{\partial \delta}\right)_\tau. \tag{4.47}$$

In the same way, the residuum

$$\zeta_{c_v} = \left(\frac{c_v}{R} + \tau^2 \alpha_{\tau\tau}^{\mathrm{o}}\right) - \sum_{i=1}^{I} n_i \left(-\tau^2 \left(\frac{\partial^2 A_i}{\partial \tau^2}\right)_\delta\right) \tag{4.48}$$

can be derived from the relation between isochoric heat capacity and Helmholtz energy. With

$$a_{c_v,0} = \frac{c_v}{R} + \tau^2 \alpha_{\tau\tau}^{\mathrm{o}} \tag{4.49}$$

and

$$a_{c_v,i} = -\tau^2 \left(\frac{\partial^2 A_i}{\partial \tau^2}\right)_\delta, \tag{4.50}$$

data for the isochoric heat capacity can be used in linear fits too and since no restrictions have been made with regard to the functional forms of a_i and a_0 in Sect. 4.1 they can be used together with $p\rho T$ data in a single fit. To illustrate this fact, the sum of squares according to Eq. 4.5 can be rewritten as

$$\chi^2(\mathbf{n}) = \sum_{p=1}^{P} \sum_{m=1}^{M_p} \left(a_0(\mathbf{x}_{p,m}, y_{p,m}) - \sum_{i=1}^{I} n_i\, a_{p,i}(\mathbf{x}_{p,m})\right)^2 \cdot \frac{1}{\sigma_{p,m}^2} \tag{4.51}$$

for a multiproperty fit which considers P different properties. Equations 4.9–4.11 change correspondingly and the consideration of constraints is not affected by the reformulation of Eq. 4.5.

From Eq. 4.51 the importance of the variance $\sigma_{p,m}^2$ becomes obvious. In a multiproperty fit, residua of different properties are only comparable if they are reduced by a suitable quantity. This quantity is usually the experimental uncertainty of the data, see Sect. 4.3.4; for the development of equations of state for technical applications, the experimental uncertainty may be replaced by the accuracy which is aspired for the corresponding property, see Sect. 6.2.

Problems arise, e.g., for the consideration of enthalpy differences $\Delta h = h_2(T_2,p_2) - h_1(T_1,p_1)$. Starting from Eq. 3.52 and introducing the abbreviations from Eq. 3.57, the expression for enthalpy differences becomes

$$\frac{h(T_2,p_2)}{RT_2} - \frac{h(T_1,p_1)}{RT_1} = \left[\tau(\alpha_\tau^\circ + \alpha_\tau^r) + \delta\alpha_\delta^r\right]_2 - \left[\tau(\alpha_\tau^\circ + \alpha_\tau^r) + \delta\alpha_\delta^r\right]_1 \quad (4.52)$$

and the corresponding residuum can be written as

$$\zeta_{\Delta h} = \left(\frac{h(T_2,p_2)}{RT_2} - \frac{h(T_1,p_1)}{RT_1} - \left[\tau\alpha_\tau^\circ\right]_2 + \left[\tau\alpha_\tau^\circ\right]_1\right)$$

$$- \sum_{i=1}^{I} n_i \left(\left[\tau\left(\frac{\partial A_i}{\partial \tau}\right)_\delta + \delta\left(\frac{\partial A_i}{\partial \delta}\right)_\tau\right]_2 - \left[\tau\left(\frac{\partial A_i}{\partial \tau}\right)_\delta + \delta\left(\frac{\partial A_i}{\partial \delta}\right)_\tau\right]_1\right). \quad (4.53)$$

This residuum yields an expression for $a_{\Delta h,i}$ which seems to be suitable for linear fits, namely

$$a_{\Delta h,i} = \left[\tau\left(\frac{\partial A_i}{\partial \tau}\right)_\delta + \delta\left(\frac{\partial A_i}{\partial \delta}\right)_\tau\right]_2 - \left[\tau\left(\frac{\partial A_i}{\partial \tau}\right)_\delta + \delta\left(\frac{\partial A_i}{\partial \delta}\right)_\tau\right]_1 \quad (4.54)$$

but since enthalpy differences are usually measured as a function of temperature and pressure, Eq. 4.53 is an *implicit linear* relation. For a consistent evaluation of Eq. 4.53, the densities which correspond to T_1, p_1 and T_2, p_2 have to be calculated from the equation of state using the current parameter vector **n**. Since new values for ρ_1 and ρ_2 exercise an influence on **n** and vice versa, this is only possible for iterative procedures as they are typical for nonlinear algorithms like the one proposed by Fletcher (1971).

Other properties lead to *implicit nonlinear* residua. For the speed of sound one finds, e.g.,

$$\zeta_{w^2} = \left(\frac{w^2}{RT} - 1\right) - \sum_{i=1}^{I} n_i \left(2\delta\left(\frac{\partial A_i}{\partial \delta}\right)_\tau + \delta^2\left(\frac{\partial^2 A_i}{\partial \delta^2}\right)_\tau\right)$$

$$- \frac{\left(1 + \sum_{i=1}^{I} n_i \left(\delta\left(\frac{\partial A_i}{\partial \delta}\right)_\tau - \delta\tau\left(\frac{\partial^2 A_i}{\partial \delta \partial \tau}\right)_\tau\right)\right)^2}{\tau^2\left(\frac{\partial^2 \alpha^\circ}{\partial \tau^2}\right)_\delta + \sum_{i=1}^{I} n_i \left(\tau^2\left(\frac{\partial^2 A_i}{\partial \tau^2}\right)_\delta\right)}. \quad (4.55)$$

From the fraction in Eq. 4.55, the coefficients cannot be isolated and thus any derivative of ζ with respect to a coefficient n_i will contain the complete vector of coefficients, **n**. Nonlinear algorithms are required to minimise sums of squares which contain contributions from such properties unless linearisation procedures are applied which are described in the following section.

4.3.2 Defining Residua for Linear Algorithms

Linear algorithms can be used directly to determine the coefficient vector **n** if only data are used in the fit which lead to *explicit linear* residua. Usually these are data for the $p\rho T$ relation, for differences of the internal energy u, for the isochoric heat capacity c_v, for the second and third virial coefficient B and C, and for the second acoustic virial coefficient β_a. The residua for these properties are given in Table 4.1 except for those of the $p\rho T$ relation and the isochoric heat capacity which were given in Eqs. 4.45 and 4.48, respectively. In Table 4.1, the subdivision into a_0 and a_i which was shown in detail in Eqs. 4.45–4.47, 4.48–4.50, and 4.53–4.54 is indicated by brackets.

Data for properties which result in *implicit linear* residua such as enthalpy differences cannot be considered exactly when using linear algorithms, but the implicit dependency can be transformed into an explicit one by precorrelation of the density which corresponds to the given values of temperature and pressure. Since the uncertainty of calculated densities is usually much smaller than the uncertainty of enthalpy differences, the corresponding densities can be calculated from preliminary equations of state without significantly affecting the representation of the enthalpy differences. Nevertheless, this method does not yield exactly consistent results and the resulting equation has to be compared both to the original $h(T,p)$ data with ρ calculated from the equation and to the precorrelated $h(T,\rho)$ data to prove that the influence of the precorrelation is negligible. The residuum for enthalpy differences was given in Eq. 4.55. Enthalpies of evaporation can be considered as $h(T,\rho'') - h(T,\rho')$ where the densities of the equilibrium phases can be precorrelated either from a preliminary equation of state or from the corresponding ancillary equations. Again, for the resulting equation, it should be proved that the influence of this precorrelation is negligible.

Implicit linear relations are also found for the thermal properties of the vapour-liquid phase equilibrium, p_s, ρ' and ρ''. As written in Eqs. 3.71 and 3.72, the phase equilibrium condition yields implicit residua since all three properties are contained in both conditions. Data for a single property can only be used in combination with iterative procedures where the other properties are calculated from the equation of state depending on the current vector of coefficients, **n**. However, Wagner (1970/1972) and Bender (1970/1973) found that the phase equilibrium condition can be resolved into three explicit linear conditions, if values for all three properties are available. The residua of these conditions read

$$\zeta_{\text{VLE},1} = \left(\frac{p_s}{RT}\left(\frac{1}{\rho''} - \frac{1}{\rho'}\right) - \ln\left(\frac{\rho'}{\rho''}\right)\right) - \sum_{i=1}^{I} n_i \left(A_i(\tau,\delta') - A_i(\tau,\delta'')\right), \qquad (4.56)$$

$$\zeta_{\text{VLE},2} = \left(\frac{p_s - \rho' RT}{\rho'^2 RT}\right) - \sum_{i=1}^{I} n_i \frac{1}{\rho_r} \left(\frac{\partial A_i(\tau,\delta')}{\partial \delta}\right)_\tau, \qquad (4.57)$$

and

$$\zeta_{\text{VLE},3} = \left(\frac{p_s - \rho'' RT}{\rho''^2 RT}\right) - \sum_{i=1}^{I} n_i \frac{1}{\rho_r} \left(\frac{\partial A_i(\tau,\delta'')}{\partial \delta}\right)_\tau, \qquad (4.58)$$

Table 4.1 Residua for properties which result in explicit linear relations.[a]

Property	Residuum	a_0	$\sum n_i a_i$
$\Delta u(T_2,\rho_2,T_1,\rho_1)$	$\zeta_{\Delta u} =$	$\left(\dfrac{u_2}{RT_2} - \dfrac{u_1}{RT_1} - [\tau\alpha_\tau^o]_2 + [\tau\alpha_\tau^o]_1\right) -$	$\sum_{i=1}^{I} n_i \left(\left[\tau\left(\dfrac{\partial A_i}{\partial \tau}\right)_\delta\right]_2 - \left[\tau\left(\dfrac{\partial A_i}{\partial \tau}\right)_\delta\right]_1 \right)$
$B(T)$	$\zeta_B =$	$(B\rho_r)$	$\lim\limits_{\delta \to 0} \sum_{i=1}^{I} n_i \left(\dfrac{\partial A_i}{\partial \delta}\right)_\tau$
$C(T)$	$\zeta_C =$	$(C\rho_r^2)$	$\lim\limits_{\delta \to 0} \sum_{i=1}^{I} n_i \left(\dfrac{\partial^2 A_i}{\partial \delta^2}\right)_\tau$
$\beta_a(T)$	$\zeta_{\beta_a} =$	$(\beta_a \rho_r)$	$-\lim\limits_{\delta \to 0} \sum_{i=1}^{I} n_i \left(2\left(\dfrac{\partial A_i}{\partial \delta}\right)_\tau - 2\dfrac{k_s^o - 1}{k_s^o}\tau\left(\dfrac{\partial^2 A_i}{\partial \delta \partial \tau}\right) \right.$ $\left. + \dfrac{(k_s^o-1)^2}{k_s^o}\tau^2\left(\dfrac{\partial^3 A_i}{\partial \delta \partial \tau^2}\right) \right)$

[a] Except for $p\rho T$ data and isochoric heat capacites for which the residua were given in Eqs. 4.45 and 4.48, respectively.

where Eq. 4.56 corresponds to the Maxwell criterion itself while Eqs. 4.57 and 4.58 represent the condition that the vapour pressure has to result from the $p\rho T$ relation both for T,ρ' and T,ρ''. Since values for all three properties are usually not available for exactly the same temperature from experiments, the data set for the thermal properties at phase equilibrium can be calculated from the corresponding ancillary equations – thus accurate ancillary equations for p_s, ρ' and ρ'' are extremely important. This step becomes a problem for substances where the data situation for one of the properties, usually for the saturated vapour density, is scarce. In this case saturated vapour densities can be calculated from a simple equation of state for the gas phase and from the ancillary equation for the vapour pressure as $\rho'' = \rho(T, p_s)$. The result of the precorrelation has to be checked by comparison of the resulting equation of state with original data.

The necessary precorrelations become more complicated for implicit nonlinear data. For speeds of sound, Jacobsen et al. (1986a) introduced the residuum

$$\zeta_{w^2} = \left(\frac{w^2}{RT} \cdot \frac{1}{\gamma_{w^2}} - 1\right) - \sum_{i=1}^{I} n_i \left(2\delta\left(\frac{\partial A_i}{\partial \delta}\right)_\tau + \delta^2\left(\frac{\partial^2 A_i}{\partial \delta^2}\right)_\tau\right) \quad (4.59)$$

with
$$\gamma_{w^2} = \frac{c_p}{c_v} = 1 - \frac{(1 + \delta\alpha_\delta^r - \delta\tau\alpha_{\delta\tau}^r)^2}{\tau^2(\alpha_{\tau\tau}^o + \alpha_{\tau\tau}^r)\cdot(1 + 2\delta\alpha_\delta^r + \delta^2\alpha_{\delta\delta}^r)}. \quad (4.60)$$

This linearised residuum results directly from the relation

4.3 Describing the Residual Helmholtz Energy

$$w^2(T,p) = \frac{c_p}{c_v} \cdot \left(\frac{\partial p}{\partial \rho}\right)_T \qquad (4.61)$$

which was used to calculate speeds of sound from pressure explicit equations of state. To use speed of sound data in a linear fit, the precorrelation factor γ_{w^2} has to be calculated from a preliminary equation of state and precorrelated densities have to be used to make Eq. 4.59 explicit. The calculation of γ_{w^2} makes high demands on the accuracy of the preliminary equation, especially if speed of sound data have to be precorrelated in regions where no accurate data are available for the isochoric or isobaric heat capacity. These high demands have lead to problems with regard to the representation of highly accurate speed of sound data from measurements with spherical resonators, see Sect. 4.4.6. Tegeler et al. (1997/1999) proposed a different residuum, where speed of sound data are linearised as

$$\zeta_{w^2} = \left(\frac{w^2}{RT} - 1 - \gamma_1\right) - \sum_{i=1}^{I} n_i \left((2+\gamma_2)\delta\left(\frac{\partial A_i}{\partial \delta}\right)_\tau + \delta^2\left(\frac{\partial^2 A_i}{\partial \delta^2}\right)_\tau - \gamma_3 \delta\tau\left(\frac{\partial^2 A_i}{\partial \delta \partial \tau}\right)\right) \qquad (4.62)$$

with $\qquad \gamma_1 = \dfrac{1}{-\tau^2\left(\alpha_{\tau\tau}^\text{o} + \alpha_{\tau\tau}^\text{r}\right)}, \qquad \gamma_2 = \dfrac{2 + \delta\alpha_\delta^\text{r}}{-\tau^2\left(\alpha_{\tau\tau}^\text{o} + \alpha_{\tau\tau}^\text{r}\right)},$

and $\qquad \gamma_3 = \dfrac{2 + 2\delta\alpha_\delta^\text{r} - \delta\tau\,\alpha_{\delta\tau}^\text{r}}{-\tau^2\left(\alpha_{\tau\tau}^\text{o} + \alpha_{\tau\tau}^\text{r}\right)}. \qquad (4.63)$

This more complex precorrelation enhances the benefit of linearised speed of sound data since Eq. 4.62 contains information on the mixed second derivative of the Helmholtz energy which was lost in Eq. 4.59. However, Eq. 4.62 proved to be superior only in regions where highly accurate speed of sound data were available and where accurate preliminary equations could be formulated.

For the isobaric heat capacities, Saul and Wagner (1989) introduced the linearised residuum

$$\zeta_{c_p} = \left(\frac{c_p}{R} + \tau^2\alpha_{\tau\tau}^\text{o} - \varepsilon_{c_p}\right) - \sum_{i=1}^{I} n_i \left(-\tau^2\left(\frac{\partial^2 A_i}{\partial \tau^2}\right)_\delta\right) \qquad (4.64)$$

with $\qquad \varepsilon_{c_p} = \dfrac{\left(1 + \delta\alpha_\delta^\text{r} - \delta\tau\alpha_{\delta\tau}^\text{r}\right)^2}{\left(1 + 2\delta\alpha_\delta^\text{r} + \delta^2\alpha_{\delta\delta}^\text{r}\right)}. \qquad (4.65)$

The precorrelation factor ε_{c_p} and the density which corresponds to the given values of temperature and pressure have to be precorrelated from a preliminary equation of state again in order to transform the implicit nonlinear relation for the isobaric heat capacity into an explicit linear relation.

Experimental results for the speed of sound and the isobaric heat capacity are published not only for homogeneous states but also for the saturated liquid and vapour. These data can be considered according to the relations given above with the only exception that the density has to be precorrelated from the corresponding ancillary equation or from an equation of state (see Sect. 3.3.5) as a function of temperature only. When comparing an equation of state to the original data, the corresponding densities have to be calculated from this equation for consistent results.

Data for the heat capacity of the saturated liquid cannot be used directly in linear algorithms. These data can be converted into data for the isobaric heat capacity of the saturated liquid according to the relation

$$c'_p(T) = c_\sigma(T) - \frac{T}{\rho'^2} \cdot \frac{\left(\frac{\partial p}{\partial T}\right)_\rho \left(\frac{\mathrm{d}p_s}{\mathrm{d}T}\right)}{\left(\frac{\partial p}{\partial \rho}\right)_T}. \tag{4.66}$$

Up to reduced temperatures of about $T/T_c = 0.95$ the error in c'_p which is implied by this conversion should be clearly less than $\pm 0.5\%$ and therefore smaller than the uncertainty of typical data for the heat capacity of the saturated liquid if Eq. 4.66 is calculated from a good preliminary equation (see also Span, 1993).

Until now, no direct linearisation has been proposed for data for the Joule-Thomson coefficient. In most cases, such data are only of minor importance. When the data are essential for a certain substance they can be used as data for isenthalpic pressure changes ($\Delta h(T_2, p_2, T_1, p_1) = 0$). The corresponding values of T_1, p_1 and T_2, p_2 can be calculated from

$$p_{1,2} = p_\mu \pm \Delta p/2 \quad \text{and} \quad T_{1,2} = T_\mu \pm \mu(T_\mu, p_\mu) \cdot \Delta p/2 \tag{4.67}$$

where T_μ and p_μ correspond to temperature and pressure of the published Joule-Thomson data point. Published experimental data for the differential Joule-Thomson coefficient were usually measured as integral Joule-Thomson coefficients over a finite pressure difference Δp. When this pressure difference is given it should be used in Eq. 4.67; otherwise a reasonable value must be estimated.

4.3.3 Defining Residua for Nonlinear Algorithms

With respect to the required residua and derivatives, there is, at least on principle, little difference between linear and nonlinear fitting routines. The sum of squares is still defined by Eq. 4.2, and Eq. 4.3 still defines the necessary condition for the minimum of the sum of squares. The first difference is that the fitted parameters do not need to be linear coefficients when using nonlinear fitting algorithms. Formally the required derivative of the sum of squares has to be given in a more general way as

$$\left(\frac{\partial \chi^2}{\partial n_i}\right)_{\mathbf{x},y,n_j \neq n_i} = \sum_{p=1}^{P} \sum_{m=1}^{M_p} \frac{2}{\sigma_{p,m}^2} \zeta(\mathbf{x}_{p,m}, y_{p,m}, \mathbf{n}) \cdot \left(\frac{\partial \zeta(\mathbf{x}_{p,m}, y_{p,m}, \mathbf{n})}{\partial n_i}\right)_{\mathbf{x},y,n_j \neq n_i}. \quad (4.68)$$

The simplification of this relation, Eq. 4.6 extended to a multiproperty fit, is still valid in most cases. Within certain numerical limits, Eq. 4.68 leads to different results only if nonlinear parameters such as exponents or parameters within exponential expressions are fitted.

For *explicit linear* properties, neither the residua nor their derivatives change when using them in combination with nonlinear algorithms; Eqs. 4.45, 4.48 and the residua which are summarised in Table 4.1 can be used without changes. The given (Eqs. 4.47 and 4.50) or indicated derivatives of the residua change only if fitted parameters require more general formulations, see above.

For *implicit* data precorrelations are no longer necessary since $\rho(T,p,\mathbf{n})$ can be calculated directly from the equation of state. Since the calculated density depends on the current parameter vector $\mathbf{n}$ in this way, an additional contribution has to be considered when calculating the derivative of the residuum with respect to a parameter n_i. For an implicit property $z(T,p)$,[2] the required derivative of the residuum becomes

$$\left(\frac{\partial \zeta_z}{\partial n_i}\right)_{\mathbf{x},y,n_j \neq n_i} = -\left(\left(\frac{\partial z}{\partial n_i}\right)_{T,\rho,n_j \neq n_i} + \left(\frac{\partial z}{\partial \rho}\right)_{T,\mathbf{n}} \cdot \left(\frac{\partial \rho}{\partial p}\right)_{T,\mathbf{n}} \cdot \left(\frac{\partial p}{\partial n_i}\right)_{T,\rho,n_j \neq n_i}\right) \quad (4.69)$$

with

$$\left(\frac{\partial p}{\partial n_i}\right)_{T,\rho,n_j \neq n_i} = \rho RT \, \delta \left(\frac{\partial \alpha_\delta^r}{\partial n_i}\right)_{T,\rho,n_j \neq n_i}. \quad (4.70)$$

The term $(\partial \rho/\partial p)_{T,\mathbf{n}}$ can be replaced by $(\partial p/\partial \rho)_{T,\mathbf{n}}^{-1}$, see Table 3.9. The required derivative of z with respect to density can be calculated based on the relations given in Table 3.9 with

$$\left(\frac{\partial z}{\partial \rho}\right)_{T,\mathbf{n}} = \frac{1}{\rho_r} \left(\frac{\partial z}{\partial \delta}\right)_{\tau,\mathbf{n}}. \quad (4.71)$$

For common implicit properties, these derivatives are summarised in Table 4.2. The resulting relations involve third derivatives of α^r which are not given in the corresponding tables here. While these derivatives are still simple for pure polynomial and exponential terms, they become rather complex for hard sphere terms, Gaussian bell shaped terms and nonanalytical terms, see Tables 3.6, 3.7, and 3.8, respectively. Although analytic derivatives are advantageous in principle, the use of numerical derivatives is justified in this case to avoid these complex expressions. When using data for enthalpy differences, the contribution from $\rho(T,p,\mathbf{n})$ has

---

[2] Since reduced properties are used in the corresponding residua, $z(T,p)$ has to correspond to a reduced property in Eqs. 4.69 and 4.71, and in Table 4.2, too.

Table 4.2. Derivatives of implicit properties with respect to density

Property, reduced	Derivative $(\partial z/\partial \rho)_{T,n}$
Enthalpy	$\left(\dfrac{\partial(h/RT)}{\partial \rho}\right)_{T,n} = \dfrac{1}{\rho_r}\cdot\left(\tau\alpha^r_{\delta\tau}+\alpha^r_\delta+\delta\alpha^r_{\delta\delta}\right)$
Isobaric heat capacity	$\left(\dfrac{\partial(c_p/R)}{\partial \rho}\right)_{T,n} = \dfrac{1}{\rho_r}\cdot\left[-\tau^2\alpha^r_{\delta\tau\tau}+\dfrac{2\left(\alpha^r_\delta+\delta\alpha^r_{\delta\delta}-\tau\alpha^r_{\delta\tau}-\delta\tau\alpha^r_{\delta\delta\tau}\right)}{1+2\delta\alpha^r_\delta+\delta^2\alpha^r_{\delta\delta}}\right.$ $\left.\cdot\left(1+\delta\alpha^r_\delta-\delta\tau\alpha^r_{\delta\tau}\right)-\dfrac{\left(2\alpha^r_\delta+4\delta\alpha^r_{\delta\delta}+\delta^2\alpha^r_{\delta\delta\delta}\right)}{\left(1+2\delta\alpha^r_\delta+\delta^2\alpha^r_{\delta\delta}\right)^2}\right.$ $\left.\cdot\left(1+\delta\alpha^r_\delta-\delta\tau\alpha^r_{\delta\tau}\right)^2\right]$
Speed of sound	$\left(\dfrac{\partial(w^2/RT)}{\partial \rho}\right)_{T,n} = \dfrac{1}{\rho_r}\cdot\left(2\alpha^r_\delta+4\delta\alpha^r_{\delta\delta}+\delta^2\alpha^r_{\delta\delta\delta}-\left(1+\delta\alpha^r_\delta-\delta\tau\alpha^r_{\delta\tau}\right)\right.$ $\cdot\dfrac{2\left(\alpha^r_\delta+\delta\alpha^r_{\delta\delta}-\tau\alpha^r_{\delta\tau}-\delta\tau\alpha^r_{\delta\delta\tau}\right)}{\tau^2\left(\alpha^o_{\tau\tau}+\alpha^r_{\tau\tau}\right)}$ $\left.+\dfrac{\alpha^r_{\delta\tau\tau}\left(1+\delta\alpha^r_\delta-\delta\tau\alpha^r_{\delta\tau}\right)^2}{\tau^2\left(\alpha^o_{\tau\tau}+\alpha^r_{\tau\tau}\right)^2}\right)$
Joule-Thomson coeff.	$\left(\dfrac{\partial(\mu R\rho)}{\partial \rho}\right)_{T,n} = \dfrac{1}{\rho_r}\cdot\left(-\dfrac{1}{\psi}\left(\alpha^r_\delta+3\delta\alpha^r_{\delta\delta}+\delta^2\alpha^r_{\delta\delta\delta}+\tau\alpha^r_{\delta\tau}+\delta\tau\alpha^r_{\delta\delta\tau}\right)\right.$ $+2\dfrac{\varphi}{\psi^2}\left(\alpha^r_\delta+\delta\alpha^r_{\delta\delta}-\tau\alpha^r_{\delta\tau}-\delta\tau\alpha^r_{\delta\delta\tau}\right)\cdot\left(1+\delta\alpha^r_\delta-\delta\tau\alpha^r_{\delta\tau}\right)$ $-\tau^2\dfrac{\varphi}{\psi^2}\alpha^r_{\delta\tau\tau}\left(1+2\delta\alpha^r_\delta+\delta^2\alpha^r_{\delta\delta}\right)$ $\left.-\tau^2\dfrac{\varphi}{\psi^2}\left(\alpha^o_{\tau\tau}+\alpha^r_{\tau\tau}\right)\left(2\alpha^r_\delta+4\delta\alpha^r_{\delta\delta}+\delta^2\alpha^r_{\delta\delta\delta}\right)\right)$
with	$\psi = \left(1+\delta\alpha^r_\delta-\delta\tau\alpha^r_{\delta\tau}\right)^2-\tau^2\left(\alpha^o_{\tau\tau}+\alpha^r_{\tau\tau}\right)\left(1+2\delta\alpha^r_\delta+\delta^2\alpha^r_{\delta\delta}\right)$ $\varphi = \delta\alpha^r_\delta+\delta^2\alpha^r_{\delta\delta}+\delta\tau\alpha^r_{\delta\tau}$

to be considered both for ρ_1 and ρ_2. For data on the phase boundary, Eq. 4.70 has to be replaced by

4.3 Describing the Residual Helmholtz Energy

$$\left(\frac{\partial p_s}{\partial n_i}\right)_{T,n_j \neq n_i} = \frac{\rho' + \rho''}{\rho' \cdot \rho''} RT \left(\left(\frac{\partial \alpha_\delta^r(\tau,\delta')}{\partial n_i}\right)_{\tau,n_j \neq n_i} - \left(\frac{\partial \alpha_\delta^r(\tau,\delta'')}{\partial n_i}\right)_{\tau,n_j \neq n_i}\right) \quad (4.72)$$

and the other derivatives involved in Eq. 4.69 have to be evaluated either for $\rho'(T)$ or for $\rho''(T)$.

Nonlinear data can be used directly in combination with nonlinear algorithms, no linearisation is required. For the speed of sound the corresponding residuum was already given in Eq. 4.55. The derivative of Eq. 4.55 with respect to n_i reads

$$\left(\frac{\partial \zeta_{w^2}}{\partial n_i}\right)_{T,\rho,n_j \neq n_i} = -2\delta\left(\frac{\partial \alpha_\delta^r}{\partial n_i}\right) - \delta^2\left(\frac{\partial \alpha_{\delta\delta}^r}{\partial n_i}\right) - \frac{\left(\frac{\partial \alpha_{\tau\tau}^r}{\partial n_i}\right) \cdot \left(1 + \delta \alpha_\delta^r - \delta \tau \alpha_{\delta\tau}^r\right)^2}{\tau^2 \left(\alpha_{\tau\tau}^o + \alpha_{\tau\tau}^r\right)^2}$$

$$+ \frac{2\left(\delta\left(\frac{\partial \alpha_\delta^r}{\partial n_i}\right) - \delta\tau\left(\frac{\partial \alpha_{\delta\tau}^r}{\partial n_i}\right)\right) \cdot \left(1 + \delta \alpha_\delta^r - \delta \tau \alpha_{\delta\tau}^r\right)}{\tau^2 \left(\alpha_{\tau\tau}^o + \alpha_{\tau\tau}^r\right)} \quad (4.73)$$

where the list of fixed parameters is omitted for the derivatives of α with respect to n_i. The residuum of the isobaric heat capacity,

$$\zeta_{c_p} = \left(\frac{c_p}{R} + \tau^2 \alpha_{\tau\tau}^o\right) - \left(-\tau^2 \alpha_{\tau\tau}^r + \frac{\left(1 + \delta \alpha_\delta^r - \delta\tau \alpha_{\delta\tau}^r\right)^2}{1 + 2\delta \alpha_\delta^r + \delta^2 \alpha_{\delta\delta}^r}\right) \quad (4.74)$$

yields the derivative

$$\left(\frac{\partial \zeta_{c_p}}{\partial n_i}\right)_{T,\rho,n_j \neq n_i} = \tau^2 \left(\frac{\partial \alpha_{\tau\tau}^r}{\partial n_i}\right) + \frac{\left(2\delta\left(\frac{\partial \alpha_\delta^r}{\partial n_i}\right) + \delta^2\left(\frac{\partial \alpha_{\delta\delta}^r}{\partial n_i}\right)\right) \cdot \left(1 + \delta \alpha_\delta^r - \delta\tau \alpha_{\delta\tau}^r\right)^2}{\left(1 + 2\delta \alpha_\delta^r + \delta^2 \alpha_{\delta\delta}^r\right)^2}$$

$$- \frac{2\left(\delta\left(\frac{\partial \alpha_\delta^r}{\partial n_i}\right) - \delta\tau\left(\frac{\partial \alpha_{\delta\tau}^r}{\partial n_i}\right)\right) \cdot \left(1 + \delta \alpha_\delta^r - \delta\tau \alpha_{\delta\tau}^r\right)}{1 + 2\delta \alpha_\delta^r + \delta^2 \alpha_{\delta\delta}^r}. \quad (4.75)$$

Data for the Joule-Thomson coefficient can be considered using the residuum

$$\zeta_\mu = \mu R \rho + \frac{\delta \alpha_\delta^r + \delta^2 \alpha_{\delta\delta}^r + \delta\tau \alpha_{\delta\tau}^r}{\left(1 + \delta \alpha_\delta^r - \delta\tau \alpha_{\delta\tau}^r\right)^2 - \tau^2 \left(\alpha_{\tau\tau}^o + \alpha_{\tau\tau}^r\right)\left(1 + 2\delta \alpha_\delta^r + \delta^2 \alpha_{\delta\delta}^r\right)} \quad (4.76)$$

and its derivative

$$\left(\frac{\partial \zeta_\mu}{\partial n_i}\right)_{T,\rho,n_j \neq n_i} = \frac{\delta\left(\frac{\partial \alpha_\delta^r}{\partial n_i}\right) + \delta^2\left(\frac{\partial \alpha_{\delta\delta}^r}{\partial n_i}\right) + \delta\tau\left(\frac{\partial \alpha_{\delta\tau}^r}{\partial n_i}\right)}{\left(1+\delta\alpha_\delta^r - \delta\tau\alpha_{\delta\tau}^r\right)^2 - \tau^2\left(\alpha_{\tau\tau}^o + \alpha_{\tau\tau}^r\right)\left(1+2\delta\alpha_\delta^r + \delta^2\alpha_{\delta\delta}^r\right)}$$

$$-\frac{\left(\delta\alpha_\delta^r + \delta^2\alpha_{\delta\delta}^r + \delta\tau\alpha_{\delta\tau}^r\right)}{\left(\left(1+\delta\alpha_\delta^r - \delta\tau\alpha_{\delta\tau}^r\right)^2 - \tau^2\left(\alpha_{\tau\tau}^o + \alpha_{\tau\tau}^r\right)\left(1+2\delta\alpha_\delta^r + \delta^2\alpha_{\delta\delta}^r\right)\right)^2}$$

$$\cdot \left\{2\left(\delta\left(\frac{\partial \alpha_\delta^r}{\partial n_i}\right) - \delta\tau\left(\frac{\partial \alpha_{\delta\tau}^r}{\partial n_i}\right)\right)\cdot\left(1+\delta\alpha_\delta^r - \delta\tau\alpha_{\delta\tau}^r\right) - \tau^2\left(\frac{\partial \alpha_{\tau\tau}^r}{\partial n_i}\right)\right.$$

$$\cdot\left(1+2\delta\alpha_\delta^r + \delta^2\alpha_{\delta\delta}^r\right)^2 - 2\tau^2\left(\alpha_{\tau\tau}^o + \alpha_{\tau\tau}^r\right)$$

$$\left.\cdot\left(2\delta\left(\frac{\partial \alpha_\delta^r}{\partial n_i}\right) + \delta^2\left(\frac{\partial \alpha_{\delta\delta}^r}{\partial n_i}\right)\right)\cdot\left(1+2\delta\alpha_\delta^r + \delta^2\alpha_{\delta\delta}^r\right)\right\}. \quad (4.77)$$

The vapour-liquid phase equilibrium condition can be considered in exactly the same way as described for linear algorithms, see Eqs. 4.56–4.58. However, Ahrendts and Baehr (1979b/1981b) proposed a direct nonlinear approach for the thermal properties at phase equilibrium, which does not require a complete set of equilibrium properties at the same temperature. For this approach the required residua and their derivatives become

$$\zeta_{p_s} = \frac{p_s - p_{s,\text{calc}}(T,\mathbf{n})}{\rho_r R T_r}, \quad (4.78)$$

$$\left(\frac{\partial \zeta_{p_s}}{\partial n_i}\right)_{T,n_j \neq n_i} = -\frac{T}{T_r \rho_r}\cdot\frac{\rho' + \rho''}{\rho' \cdot \rho''}\left(\left(\frac{\partial \alpha_\delta^r(\tau,\delta')}{\partial n_i}\right) - \left(\frac{\partial \alpha_\delta^r(\tau,\delta'')}{\partial n_i}\right)\right), \quad (4.79)$$

$$\zeta_{\rho'} = \frac{\rho' - \rho'_{\text{calc}}(T,\mathbf{n})}{\rho_r}, \quad (4.80)$$

$$\left(\frac{\partial \zeta_{\rho'}}{\partial n_i}\right)_{T,n_j \neq n_i} = -\frac{1}{\rho_r}\left(\left(\frac{\partial p_s}{\partial n_i}\right)_{T,n_j \neq n_i} - \left(\frac{\partial p}{\partial n_i}\right)_{T,\rho',n_j \neq n_i}\right)\cdot\left(\frac{\partial \rho}{\partial p}\right)_T, \quad (4.81)$$

$$\zeta_{\rho''} = \frac{\rho'' - \rho''_{\text{calc}}(T,\mathbf{n})}{\rho_r}, \quad (4.82)$$

4.3 Describing the Residual Helmholtz Energy

and
$$\left(\frac{\partial \zeta_{\rho''}}{\partial n_i}\right)_{T,n_j \neq n_i} = -\frac{1}{\rho_r}\left(\left(\frac{\partial p_s}{\partial n_i}\right)_{T,n_j \neq n_i} - \left(\frac{\partial p}{\partial n_i}\right)_{T,\rho'',n_j \neq n_i}\right) \cdot \left(\frac{\partial \rho}{\partial p}\right)_T. \quad (4.83)$$

This procedure requires considerably more computation time than the linearised solution, since the phase equilibrium has to be calculated from the equation of state for every single residuum. However, it is particularly advantageous for substances where the data situation for one of the properties is scarce. This is because it does not constrain the equation of state to a questionable ancillary equation (see Marx et al., 1992; Wagner et al., 1993). An advantage for substances with very accurate data sets is the improved consideration of phase equilibria very close to the critical point, where the decoupled residua of the nonlinear approach allow a more precise weighting of the data. Thus, nonlinear direct fits to phase equilibrium data are state-of-the-art for group 1 reference equations of state (see Chap. 5) and for equations of state which are designed to describe substances with a poor data situation (see Sect. 6.2).

4.3.4 Assigning Weights to Experimental Data

The definition of the likelihood function, Eq. 4.1, by Fischer introduces the variance σ_m^2 of the experimental data. This variance can be found as well in the definition of the sum of squares of the multiproperty approach, Eq. 4.51, where $\sigma_{p,m}^2$ takes into account the difference between the variances calculated for different properties p. The quotient $1/\sigma_{p,m}^2$ is referred to as the "weight" of a data point.

A problem arises from the fact that usually only a single result is reported for each measured point – thus, its variance cannot be determined according to the definitions given by the statistical theory. And even if repeated measurements are reported, the resulting variance should not be used for $\sigma_{p,m}^2$ since the systematic errors are often larger than the normally distributed random errors in experimental set-ups. When working on reference equations of state, it is useful to identify $\sigma_{p,m}^2$ with an estimated experimental uncertainty of the corresponding point or, more precisely, with the experimentally caused uncertainty of the corresponding residuum. In this case, the weighted contribution of a data point to the sum of squares becomes smaller than 1 if the point is represented within its experimental uncertainty and the assessment of results becomes comparably easy.

The problems which are related to the estimation of an experimental uncertainty were already mentioned in the introduction to this chapter. Estimates published by the experimentalists themselves are very helpful but unfortunately are often overly optimistic, especially for older data sets. When setting up a reference equation of state, the framework of weighting has to be based on a detailed analysis of the available data sets. This analysis has to consider both the experimental techniques applied and comparisons between different data sets, where possible even between data sets for different properties. Since Helmholtz equations yield consistent results for all properties by definition, preliminary equations can be powerful tools to detect inconsistencies. However, this kind of data analysis requires a profound knowledge regarding restrictions of the functional forms used (especially when assessing data in the critical region) and a lot of experience both with the data set and with setting up equations of state. This is the reason why highly accurate reference equations are usually the result of long-term projects which comprise dozens or even hundreds of preliminary equations.

When calculating weights, it is important to realise that $\sigma_{p,m}^2$ does not correspond just to the uncertainty of the dependent variable y; the uncertainties of the independent variables $\mathbf{x}$ must be considered as well. A rigorous way to do so is to rewrite Eq. 4.51 in a more general way as

$$\chi^2(\mathbf{n}) = \sum_{p=1}^{P} \sum_{m=1}^{M_p} \left(\left(\frac{\zeta_p(\mathbf{X},Y,\mathbf{n})}{\left(\partial \zeta_p/\partial y\right)_{\mathbf{x}} \sigma_y} \right)^2 + \sum_{k=1}^{K} \left(\frac{X_k - x_k}{\sigma_{x_k}} \right)^2 \right) \qquad (4.84)$$

where $\mathbf{X}$ and Y are the "true" values of the variables again, see Sect. 4.1. Since $\mathbf{X}$ and Y are not known, Eq. 4.84 cannot be evaluated directly. Methods to estimate $\zeta_p(\mathbf{X},Y,\mathbf{n})$ and the (X_k-x_k) from the observed $\zeta_p(\mathbf{x},y,\mathbf{n})$ were discussed in the literature and software is available, e.g., for weighted orthogonal distance regressions, but an easier solution is used in general practice today. With the assumption that errors in the different variables are independent of each other, that $(Y-y)$ and all (X_k-x_k) are small enough to neglect the curvature in ζ_p, and that ζ_p is a continuously differentiable function, Eq. 4.51 becomes equivalent to Eq. 4.84 with

$$\sigma_{p,m}^2 = \left(\left(\frac{\partial \zeta_p}{\partial y} \right)_{\mathbf{x}} \sigma_y \right)^2 + \sum_{k=1}^{K} \left(\left(\frac{\partial \zeta_p}{\partial x_k} \right)_{y,x_{i \neq k}} \sigma_k \right)^2. \qquad (4.85)$$

This relation is known as "*law of error propagation*" and results from the combined uncertainty of y and $\mathbf{x}$ when breaking a Taylor expansion after the linear terms (see Ahrendts and Baehr, 1979b/1981b). The assumptions which lead to Eq. 4.85 are usually fulfilled for experimental data of thermodynamic properties; there are only two kinds of problems which make different approaches necessary.

Especially *close to the critical point*, the curvature of ζ_p becomes too large to be neglected for some properties. Figure 4.2 shows a plot of the isochoric heat capacity of carbon dioxide in the immediate vicinity of the critical point as calculated from the Helmholtz equation of Span and Wagner (1996). When calculating $\sigma_{c_v,m}$ for the plotted point with the given value of σ_T, which is overstated here for better illustration, the resulting contribution from the uncertainty in temperature, $\sigma_{c_v,T}$, is overestimated for temperatures above the measured one $(T > T_m)$ and underestimated for temperatures below the measured one $(T < T_m)$. This effect could be avoided if the curvature of c_v were considered. But when taking into account the curvature, different variances would be obtained for $T > T_m$ and $T < T_m$ and the linear procedures discussed here are not able to consider different variances for a single data point. Thus, the value of σ_T has to be reduced to use the data point in a reasonable way. This is possible in certain situations where the distance from the critical temperature is known more accurately than the absolute temperature. In this case, the given temperatures have to be adjusted by the difference between the critical temperature which was selected for the equation of state and the value

4.3 Describing the Residual Helmholtz Energy

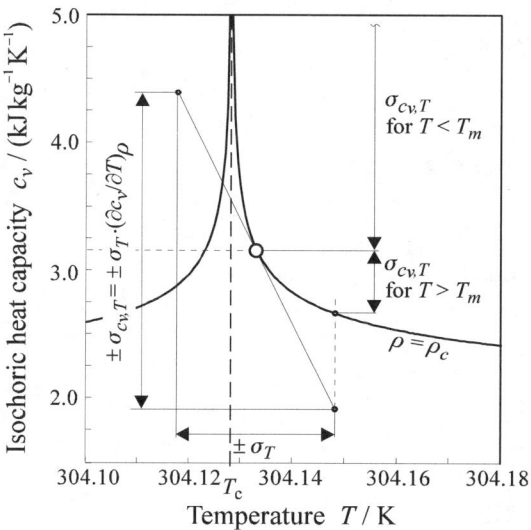

Fig. 4.2 Plot of the isochoric heat capacity of carbon dioxide in the immediate vicinity of the critical point. The strong curvature leads to incorrect results when using the law of error propagation, Eq. 4.85, to determine $\sigma_{c_v,T}$.

which results from the respective data set, $\Delta T = T_{c,\text{eq.}} - T_{c,m}$, where ΔT has to be smaller than the original value of σ_T. The adjusted data can be used with a smaller value of $\sigma_{T,\text{adj.}}$ since only the uncertainty of the difference from the critical temperature has to be considered here.

Close to the phase boundary implicit data may cause problems, since the phase boundary results in discontinuous plots of $\zeta_p(T,p)$. Using the speed of sound in carbon dioxide as an example, Fig. 4.3 shows a data point in the liquid phase. The linearisation implied by Eq. 4.85 is justified, the curvature of the isotherms can be neglected when calculating the contribution of the pressure uncertainty to $\sigma_{w,m}$. However, the pressure difference $p_m - p_s(T_m)$ is smaller than the uncertainty of the vapour pressure calculated from the equation of state, which is overstated in the figure again. Thus, the data point may be regarded as a vapour point with a much smaller calculated speed of sound in a nonlinear fit, see solution (B). In this case, the contribution to the sum of squares becomes very large and the whole fit may become unstable, oscillating between two solutions with the corresponding data point in the liquid and in the vapour phase, respectively. To avoid such problems, it is useful to start the iteration of the density belonging to T_m and p_m with the precorrelated value for $\rho(T_m, p_m)$ from the linearised data set when calculating the nonlinear residuum ζ_{w^2}. If the precorrelated value corresponds to a liquid density, the result of the iteration will also correspond to a liquid state even though this state may be metastable, see solution (A). In this way problems with implicit data close to the phase boundary can be avoided, except for data close to the critical point. For temperatures close to T_c, the Maxwell loop becomes very flat (see Sect. 3.3.5) and a metastable solution on the liquid side may become impossible; the

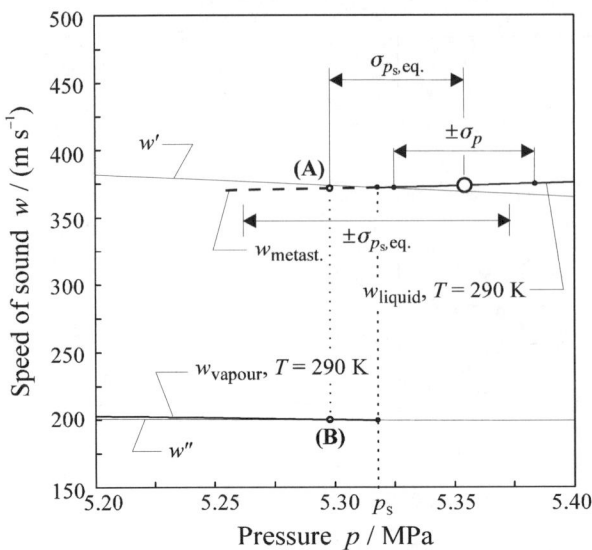

Fig. 4.3 Implicit data close to the phase boundary may cause problems in nonlinear fits. The discontinuity of $\zeta(T,p)$ violates the conditions for use of Eq. 4.85.

iteration results into a vapour density independent of the starting value. In this case, the corresponding data point has to be removed from the data set in order to guarantee stable nonlinear fits.

In Eq. 4.85 the derivative $(\partial \zeta_p/\partial y)_x$ is easy to evaluate, but a preliminary equation of state is required to calculate the derivatives of ζ_p with respect to the x_k. For an initial weighting, the contributions from the uncertainty of the independent variables could be neglected if no suitable equation is available from the literature, but unfortunately this approach fails for $p\rho T$ data. The main contribution to $\sigma_{p\rho T}$ does not result from the pressure, which corresponds to y in Eq. 4.45, but from the independent variable ρ. An initial weighting which is based only on σ_p would lead to an overfitting of liquid data and may not result in an equation which can be used for an improved weighting in the next step. Thus, it is better to use a simple cubic equation or an equation of state adopted from another substance by a simple corresponding states approach for an initial weighting. Data in regions where problems are expected and data for implicit properties should not be used before a reasonable preliminary equation has been established. After this is accomplished, the weighting procedure becomes an iterative process which soon converges – small changes in the preliminary equations result in small changes of the weights, but they have little influence on the whole fit.

To gain additional flexibility, it is advantageous to introduce an additional weighting factor which allows some subjective influence on the weights of the data used. To do so, the expression $(1/\sigma_{p,m})^2$ has to be replaced by $(f_{wt}/\sigma_{p,m})^2$ in Eq. 4.51. The weighting factor f_{wt} has to be equal to 1 in general, but it can be increased for accurate data in regions where experimental results are scarce and it can be reduced in regions with high data density. In this way, overfitting is avoided on the one hand and on the other hand regions which are experimentally investigated less good are considered sufficiently by the fitting algorithms.

Even though the different accuracy of data is considered by the weighting procedure described above, it is not useful to include all data in the used data set when setting up reference equations for substances with extensive data sets. In regions where accurate data from different apparatuses are consistent with each other, there is no need to use less accurate data to set up the equation. To exclude less accurate data avoids negative influences which may result from systematic errors of those data; the procedures described above consider different accuracies formally correct for data with random errors, but negative influences from systematic errors of inaccurate data cannot be excluded. Furthermore, computing times are reduced and comparisons become easier and unequivocal. However, it is important to keep in mind that this step does not mean a disparagement of the excluded data. The whole ranking is influenced more by the data situation in a certain region than by the absolute uncertainty of the data and data sets which are excluded in one region could be very helpful in other regions or for other substances.

Corrections of published data sets are justified in two different situations. The first reason for corrections is that the data analysis reveals an obvious error in the evaluation of the original experimental results. This is usually very difficult, if not impossible, since it requires very detailed information on the experimental procedure. A common exception are data which are based on calibrations with some kind of published primary data. If these data are known to be in error from comparisons with more accurate data, the secondary data can be corrected for this error. Such a procedure is a true *correction* of the published data and its justification is unquestionable. The second reason for corrections is that systematic deviations of a data set from some kind of reference become obvious. Since theoretical models yield highly accurate results only for the low density gas phase (see Sects. 3.1.1.1 and Eq. 4.2), corrections which are based on a theoretically founded reference can usually be applied only to gas phase data. Where such data do not meet well known properties of the ideal gas in the limit of vanishing pressure, the observed differences can be used for corrections; see for instance the correction of isobaric heat capacities of Bender (1982) applied by Smukala et al. (1999). When a data set shows systematic deviations from significantly more accurate data in a sufficiently broad region of overlap, these deviations may be used for corrections as well. This kind of correction is only useful of course if the less accurate data set covers regions which are not covered by the more accurate data; otherwise the less accurate data could simply be omitted from the data set. As an example see the correction which was applied by Tegeler et al. (1997/1999) to the data published by Barreiros et al. (1982). Corrections which are just based on comparisons with other data sets without any experimental background are theoretically questionable and in general the term "*adjustments*" describes this procedure better than the term "corrections". However, even pure adjustments are often justified due to the unsatisfactory way in which systematic errors are taken into account by the applied weighting and fitting algorithms. Uncorrected data sets with significant systematic errors may affect the whole fit even if the data are met within the assigned uncertainty.

When working on equations of state for technical applications (see Sect. 6.2) instead of reference equations of state, the applied weighting strategy has to be changed. A characteristic feature of such equations is that they do not need to represent certain data within their experimental uncertainty. Instead, they should represent thermodynamic properties within uncertainties which are required to fulfil some level of technical demands. Of course, the definition of the required uncertainties is difficult, and to a certain degree subjective, but it is clear that they do not depend on the accuracy which can be achieved in state-of-the-art experimental set-ups. If an uncertainty of $\Delta\rho/\rho \leq 0.2\%$ is regarded as sufficient for example, a state-of-the-art $p\rho T$ point which is weighted with its experimental uncertainty of $\sigma_{p\rho T}/\rho = \pm 0.02\%$ could contribute $(\zeta_{p\rho T}/\sigma_{p\rho T})^2 = 100$ to the sum of squares while it is still met within the required uncertainty. In a different region where the best available data are accurate to $\sigma_{p\rho T}/\rho = \pm 0.1\%$ a point which is met just within the required uncertainty would contribute only $(\zeta_{p\rho T}/\sigma_{p\rho T})^2 = 4$. Since equations of state with simple functional forms, so desired for many technical applications, are usually not able to represent all data within their experimental accuracy the fitting algorithm would concentrate on an improved representation of regions where highly accurate data are available while neglecting regions with less accurate data. To avoid this effect it is better to replace σ_p by the required Δy_p when working on such equations. Only where data with $\sigma_p > \Delta y_p$ have to be considered due to the data situation, should these data be weighted with their experimental uncertainty. In this way data with different uncertainties have the same influence on the fit as long as they are accurate enough to fulfil the demands on the accuracy of the corresponding property.

4.4 Optimising the Functional Form

Until now it has been assumed that the functional form of the equation (the mathematical form of every single term A_i in Eq. 4.43) and the length of the equation (the number of terms I in Eq. 4.43) is known; only the determination of the coefficients n_i was discussed. In fact, the A_i contain additional "*internal*" parameters. For simple polynomial terms these are the exponents d_i and t_i in Eq. 3.25 and for exponential terms, the density exponents in the exponential function, p_i, have to be considered , too. When using more complex functional forms for the description of properties in the critical region, additional parameters have to be determined, see Sects. 4.5.3 and 4.5.4. Furthermore, not only the total number of terms, but also the number of terms in each group of terms, thus the I_{Pol} and I_{Exp} in Eq. 3.25, have to be determined before the n_i can be fitted. This whole procedure is referred to as "*optimisation*" of the functional form.

In the past, the functional form of equations of state was determined by trial and error strategies which were based only on the experience of the correlator. In 1974 the stepwise regression analysis proposed by Wagner (1974) was the first procedure which allowed an objective optimisation of functional forms. Wagner used his procedure to establish a new class of vapour pressure equations (see Wagner, 1974; Wagner and Pollak, 1974) but soon it was applied to equations of state as well. This was the beginning of a continuing development of optimisation algorithms. Although these algorithms have become rather complex, their basic idea is still the same: A regression matrix is set up based on the weighted data set and on an extensive set of terms with different values for the respective internal parameters which is called a "*bank of terms*". Then a mathematical algorithm which is based on objective statistical criteria is applied to select the best combination of terms out of the bank of terms — the terms in the final equation of state are always a subset of the terms in the bank of terms. To guarantee a successful optimisation, the bank of terms has to contain all terms which are regarded as promising candidates for use in the equation of state, or at least a sufficiently extensive set of such terms. The number of terms which are considered in the bank of terms and in the resulting equation is a measure for the complexity of the optimisation process. While Wagner made his first step with the selection of 4 terms from 27 terms, recent reference equations require a selection of about 40 terms out of up to 1000 terms.

Section 4.4.1 gives some criteria for a reasonable set-up of a bank of terms, assuming that only simple polynomial and exponential terms are used in the equation of state; for special critical region terms, the corresponding information is given in Sects. 4.5.3 and 4.5.4. The construction of the required regression matrix is discussed in Sect. 4.4.2 and Sects. 4.4.3–4.4.8 summarise the ongoing efforts to improve the selection algorithms. A very recent algorithm which allows an optimisation based on data for different substances is discussed in Sect. 6.1.

4.4.1 Defining a Bank of Terms

Although the set-up of the bank of terms is crucial when developing an equation of state, little information on reasonable set-ups has been published. Different authors have published banks of terms which they have used to set up a certain equation of state, but these banks of terms can hardly be regarded as universal since different restrictions apply to the bank of terms which are based mainly on the optimisation algorithm used, on the available computing power and on the requirements of the considered substance.

For use with simple optimisation algorithms (see Wagner, 1974; de Reuck and Armstrong, 1979) banks of terms with no more than 100 terms are recommended. Some information on such banks of terms was given by Jacobsen et al. (1986b/c). With the functional forms introduced in Eq. 3.25 such banks of terms can be written as

$$\alpha^r(\tau,\delta) = \sum_{i=1}^{I_{\text{Pol}}} n_i \tau^{t_i} \delta^{d_i} + \sum_{i=I_{\text{Pol}}+1}^{I_{\text{Pol}}+I_{\text{Exp}}} n_i \tau^{t_i} \delta^{d_i} \exp(-\gamma_i \delta^{p_i}) \qquad (4.86)$$

for set-ups which contain only simple polynomial and exponential terms. The only difference between the general formulation of a Helmholtz equation, Eq. 3.25, and Eq. 4.86 is that Eq. 3.25 does not contain all the terms from Eq. 4.86. Thus, I_{Pol}

and I_{Exp} are smaller in Eq. 3.25. However, when establishing a Helmholtz equation of state, the restriction of the bank of terms to 100 terms implies a significant pre-selection of terms which is very likely to affect the optimisation result. Although the optimisation algorithm is guided by objective statistical criteria it cannot compensate for subjective criteria which were used when setting up the bank of terms.

Recent optimisation techniques and enhanced computing powers allow much larger set-ups which gain an increased flexibility and a reduction of subjective influences. For use with such algorithms, Eq. 4.86 can be rewritten as

$$\alpha^r(\tau,\delta) = \sum_{i=I^p_{\min}}^{I^p_{\max}} \sum_{j=J^p_{i,\min}}^{J^p_{i,\max}} n^p_{i,j} \, \tau^{j/T^p_i} \, \delta^{i/D^p}$$

$$+ \sum_{k=1}^{K^e_{\max}} \sum_{i=I^e_{k,\min}}^{I^e_{k,\max}} \sum_{j=J^e_{k,i,\min}}^{J^e_{k,i,\max}} n^e_{i,j,k} \, \tau^{j/T^e_{i,k}} \, \delta^{i/D^e_k} \exp\left(-\gamma_k \delta^{k/P^e}\right). \quad (4.87)$$

For polynomial terms, i/D^p needs to be an integer value larger than 0; otherwise the resulting equation would not be conform to the virial expansion at low densities and higher density derivatives α^r would not become 0 in the limit of vanishing densities. Thus, $I^p_{\min} = 1$ and $D^p = 1$ are mandatory for a reasonable description of the gas region. Common values for the upper limit of i depend on the application. Values up to $I^p_{\max} = 10$ are common and improve the representation of properties in the liquid region, especially at low temperatures. However, recent investigations of the extrapolation behaviour of multiparameter equations of state (see Sect. 4.6.2 or Span and Wagner, 1997) show that values of $I^p_{\max} > 4$ affect the representation of properties at very high temperatures and pressures. Thus $I^p_{\max} = 4$ should be used for reference equations of state where the extrapolation behaviour is considered as crucial. Values of $I^p_{\max} > 4$ are useful for equations of state which are designed for limited technical applications where the extrapolation behaviour to extremely high temperatures and pressures is considered as arbitrary.

The temperature exponent j/T^p_i does not need to be an integer value. Sufficiently large values of T^p_i are important for accurate results since the representation of the thermodynamic surface is very sensitive for changes in the temperature exponents. Values used today range from $T^p_i = 2$ to $T^p_i = 16$ with $T^p_i = 8$ as a good compromise. Negative values for j/T^p_i are useful especially for equations with limited range of validity again; $J^p_{i,\min} = -T^p_i$ is often used for such applications. However, for the corresponding virial coefficient, negative temperature exponents result in increasing absolute values in the limit of very high temperatures where the virial coefficients are expected to approach 0 or a small finite value. Thus, $J^p_{i,\min} = 0$ should be used for reference equations of state. The upper value $J^p_{i,\max}$ is arbitrary in most cases. While large temperature exponents were used in older equations of state recent banks of terms are restricted to $J^p_{i,\max} = 4 \cdot T^p_i$ in most cases. Experience shows that usually only polynomial terms with small values of j/T^p_i are selected by the optimisation algorithms. Especially in combination with restrictions on the number of polynomial terms as discussed in Sect. 4.6.2, set-ups with $J^p_{i,\max} = T^p_i$ and thus with $j/T^p_i \leq 1$ can be useful , too. In this case the required stronger temperature dependencies are shifted to the exponential terms completely.

For *exponential terms*, the same restrictions as discussed above are valid for the parameters D^e, $I^e_{k,\min}$, and $J^e_{k,i,\min}$ since exponential terms contribute to the viral expansion of an equation of state too, where the exponential function has to be considered as

$$\exp\left(-\gamma\,\delta^{k/P^e}\right) = 1 - \frac{\gamma_k\,\delta^{k/P^e}}{1!} + \frac{\left(\gamma_k\,\delta^{k/P^e}\right)^2}{2!} - \frac{\left(\gamma_k\,\delta^{k/P^e}\right)^3}{3!} \pm \ldots; \qquad (4.88)$$

for the same reasons P^e must be an integer value. The upper limits of i/D^e_k and $j/T^e_{i,k}$ are usually much higher for exponential terms. Values for i/D^e_k may reach up to 15 for equations where $I^p_{\max}$ is small; otherwise smaller limitations are reasonable. Special limitations apply only to exponential terms with $k/P^e = 1$ where terms with $i/D^e_1 > \approx 8$ may affect the extrapolation to very high temperatures and pressures. Larger values are useful especially in combination with $k/P^e \geq 4$. In this case 0.5 can be used for D^e_k in order to limit the size of the bank of terms.

Common values for k/P^e reach from 1 to 6, where P^e is equal to 1 in all state-of-the-art Helmholtz equations. Values above $k/P^e = 4$ are useful especially if a good representation of properties in the critical region is attempted without special critical region terms. If such terms are used or if the representation of data in the critical region is considered as less important, $K^e_{\max}$ can be restricted correspondingly. A possible variation of γ_k has never been considered in banks of terms. When the density is reduced with the critical density $\gamma_k = 1$ is generally accepted as the best choice. When different values are used for ρ_r, reasonable values for γ_k can be determined from $\gamma_k \approx (\rho_r/\rho_c)^k$ for $P^e = 1$.

The stepwidth which is used for the temperature exponents is still an important parameter especially for exponential terms with $k/P^e = 1$; values of $2-8$ are useful for $T^e_{i,1}$ while $J^e_{i,1,\max} = 4 \cdot T^e_{i,1}$ is a reasonable upper limit for the temperature exponents. For higher values of k/P^e both the upper limit and the stepwidth has to be increased. Values of $J^e_{i,k,\max} \approx 30 \cdot T^e_{i,k}$ are usual for the highest values of k/P^e and even higher values have been used for single substances. The inverse stepwidth varies usually from $T^e_{i,2} = 2$ to $T^e_{i,k} = 0.5$ for the exponential terms with high upper limits for $j/T^e_{i,k}$. Again, the high temperature exponents improve the representation of properties in the critical region but they have to be used with extreme care. When combined with small values for i/D^e_k these high temperature exponents result in plots of relevant virial coefficients which are clearly too steep in the low temperature limit and which may lead to unreasonable results for properties in the gas phase at low temperatures. In this region experimental data are either scarce or inaccurate in most cases and an erroneous behaviour of an equation of state is difficult to detect.

4.4.1.1 Hard Sphere Terms

Based on the formulation proposed by Saager et al. (1992), the use of equations of state which contain hard sphere terms has been described in detail in Sect. 3.1.3.1. These and similar terms have been used in empirical multiparameter equations of state, but they have never been included in an optimisation algorithm. Numerically stable nonlinear algorithms and good starting solutions are required to determine the adjustable parameters of Eq. 3.37, ρ_r, T_r, and φ – these parameters cannot be determined by optimisation procedures which are based on linear algorithms. Thus, equations of state which use hard sphere terms have to be optimised using the set-up which was formulated in Eq. 3.35. Like the ideal gas contribution, α^o, the contribution from the hard sphere term, α^h_r, has to be determined independently and only the formulation for the remaining residual contribution, α^{r^*}, can be optimised; the residua given in Sect. 4.3.2 have to be adapted to the changed definition of the residual Helmholtz energy.

Theoretically, this approach is promising since it involves additional knowledge from statistical thermodynamics in the establishment of empirical equations of state – corresponding requests are formulated frequently, especially from scientists who are engaged in statistical methods. This point of view is supported by graphs like those given in Fig. 4.4. Figure 4.4a shows the Helmholtz energy surface of methane as calculated from the reference equation of state published by Setzmann and Wagner (1991). States in the two phase region were calculated according to Eq. 3.84 and correspond to the stable phase equilibrium. Figure 4.4b shows the corresponding plot for the residual Helmholtz energy

$$a^r(T,\rho) = a(T,\rho) - a^o(T,\rho) \qquad (4.89)$$

and Fig. 4.4c shows the residual Helmholtz energy

$$a^{r^*}(T,\rho) = a(T,\rho) - a^o(T,\rho) - a^h(T,\rho), \qquad (4.90)$$

of a combined approach which introduces a hard sphere contribution calculated from the corresponding equation of Müller et al. (1996). When comparing Figs. 4.4b and 4.4c, it seems as if the subtraction of the repulsive forces results in a Helmholtz surface which is much easier to fit; correspondingly simpler functional forms could be expected for empirical formulations which describe the residual Helmholtz energy a^{r^*}.

However, the main problem is not to model the Helmholtz surface itself but to model the derivatives which are responsible for the representation of the relevant thermodynamic properties. These derivatives can hardly be seen from absolute plots like the one given in Fig. 4.4. Therefore, Fig. 4.5 shows the corresponding plot for the isobaric heat capacity, c_p, of methane, for the residual isobaric heat capacity, $c_p - c_p^o$, and for the residual isobaric heat capacity after introduction of the hard sphere term, $c_p - c_p^o - c_p^h$. From this figure it becomes obvious that those features of the thermodynamic surface which are difficult to model depend on attractive forces and critical effects, but not on the repulsive forces which are described by the hard sphere term. It is not easier to represent the residual heat capacity given in Fig. 4.4c than it is to represent the one given in Fig. 4.4b. The fact that the residual contribution vanishes in the limit of very high densities when using hard sphere terms implies additional constraints which make the search for a suitable functional form even more difficult.

4.4 Optimising the Functional Form

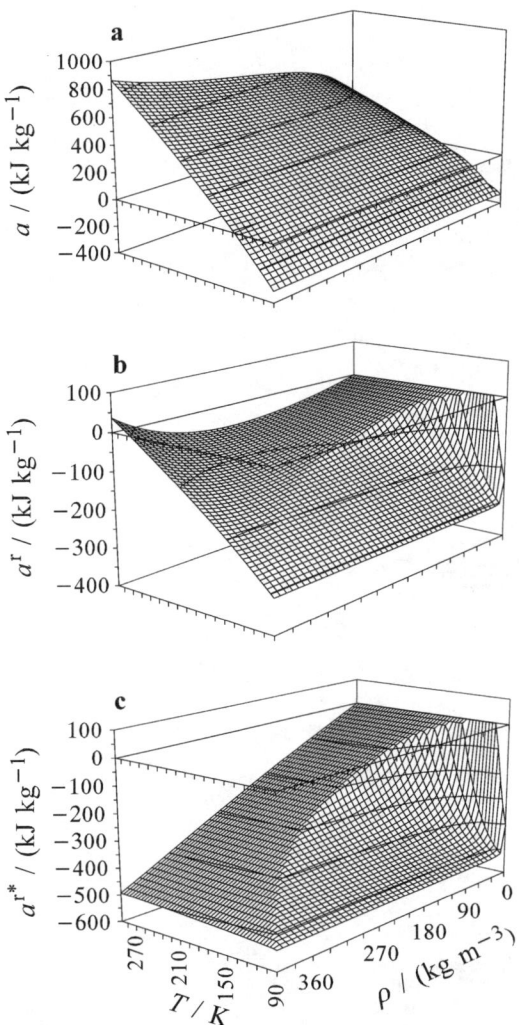

Fig. 4.4. The Helmholtz free energy surface of methane as calculated from the reference equation of state by Setzmann and Wagner (1991). Figure **a** shows the Helmholtz energy a, **b** the residual Helmholtz energy, $a-a^\circ$, and **c** the residual Helmholtz energy, $a-a^\circ-a^h$, for an approach which introduces an additional hard sphere term.

Without any doubt, hard sphere terms are useful in theoretical models which try to describe thermodynamic properties based on molecular parameters of the considered substance; such approaches need to distinguish between the effects of attractive and repulsive forces. But they are of little use for empirical equations of state which are macroscopical approaches, since they make the resulting equation

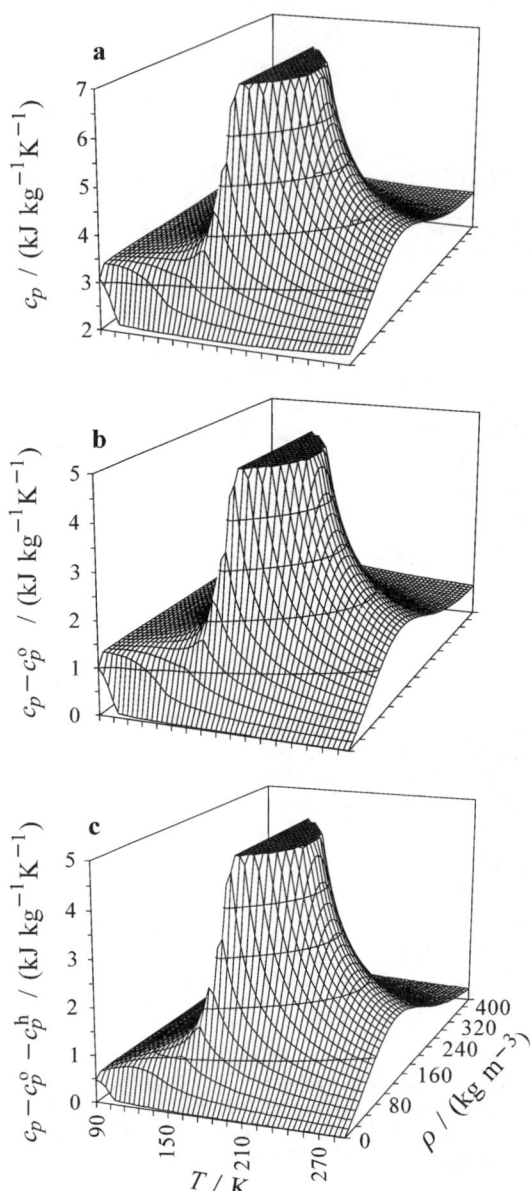

Fig. 4.5. The isobaric heat capacity of methane as calculated from the reference equation of state by Setzmann and Wagner (1991). Figure **a** shows the isobaric heat capacity, c_p, **b** the residual isobaric heat capacity, $c_p - c_p^o$, and **c** the residual isobaric heat capacity, $c_p - c_p^o - c_p^h$, for an approach which introduces an additional hard sphere term.

more complicated and the empirical description of the remaining residual contribution is not simplified with respect to its development or with respect to its application. This example shows a common misunderstanding with regard to the use of theoretical approaches in empirical formulations. The question which has to be answered is not whether an approach is useful or elegant on a theoretical, molecular level – the question is whether it is able to simplify the empirical contribution or whether it is able to improve the results on a macroscopical level. Until now, no theoretical results have been found for describing the residual Helmholtz energy which could be incorporated into empirical equations of state directly. In general, the merit of theoretical approaches is still restricted to an improved knowledge on the qualitative behaviour of certain properties in regions which are difficult to access by experimental means, see for instance Sect. 4.5.1.

4.4.2 Setting Up a Regression Matrix

In general, optimisation procedures rely on linear algorithms for the determination of the required quality criterion, the sum of squares of an equation of state which uses the combination of terms which is investigated in the current optimisation step. In Sect. 4.1, it was shown that the information which is necessary to determine the coefficients of an equation can be written in matrix notation as

$$\mathbf{A}\,\mathbf{N} = \mathbf{Q} \tag{4.91}$$

for linear problems without constraints and as

$$\begin{bmatrix} \mathbf{A} & \mathbf{C}^T \\ \mathbf{C} & \mathbf{0} \end{bmatrix} \begin{bmatrix} \mathbf{N} \\ \boldsymbol{\lambda}' \end{bmatrix} = \begin{bmatrix} \mathbf{Q} \\ \mathbf{Q}_c \end{bmatrix} \tag{4.92}$$

for linear problems with constraints. The same formulations with exactly the same set-up of the different elements (see Eqs. 4.10–4.11 and 4.17–4.18) can be used for optimisation procedures as well, with the only difference being that the parameter I which determines the size of the matrices and vectors corresponds to the number of terms in the bank of terms. The residua which are required to evaluate Eqs. 4.90 or 4.91 correspond to those defined in Sect. 4.3.2.

However, Eqs. 4.90 and 4.91 hold only the information which is necessary to determine the coefficients n_i of the current equation, and not the information on the sum of squares. As can be shown by combination of Eqs. 4.5 and 4.8 (see Wagner, 1974) this information is given by

$$\chi^2 = \sum_{m=1}^{M} \left(\frac{a_0(\mathbf{x}_m, y_m)}{\sigma_m} \right)^2 - \sum_{i=1}^{I} n_i \sum_{m=1}^{M} \left(\frac{a_0(\mathbf{x}_m, y_m) \cdot a_i(\mathbf{x}_m)}{\sigma_m^2} \right) \tag{4.93}$$

or in matrix notation by

$$\chi^2 = S - \mathbf{N}\mathbf{Q}^T \quad \text{with} \quad S = \sum_{m=1}^{M} \left(\frac{a_0(\mathbf{x}_m, y_m)}{\sigma_m} \right)^2 \tag{4.94}$$

and with $\mathbf{Q}^T$ according to Eq. 4.11 if the parameter vector $\mathbf{N}$ of an equation is determined by a linear fit.

Thus, for problems without constraints, the regression matrix

$$\mathbf{B} = \begin{bmatrix} \mathbf{A} & \mathbf{Q} \\ \mathbf{Q}^T & S \end{bmatrix} \quad (4.95)$$

holds all the information which is required to determine both the current parameter vector $\mathbf{N}$ and the current sum of squares χ^2. For problems with constraints, Eq. 4.95 has to be extended correspondingly to

$$\mathbf{B} = \begin{bmatrix} \mathbf{A} & \mathbf{C} & \mathbf{Q} \\ \mathbf{C}^T & \mathbf{0} & \mathbf{Q}_c \\ \mathbf{Q}^T & \mathbf{Q}_c^T & S \end{bmatrix}. \quad (4.96)$$

Since the regression matrix $\mathbf{B}$ is a symmetric $(I+C+1)$ matrix it is sufficient, to supply only half of it.

However, it is useless to solve Eqs. 4.90 or 4.91; for large banks of terms this step would fail due to numerical problems and an equation of state which contains all terms out of the bank of terms makes no sense. To determine both the coefficients and the sum of squares of the current equation of state, which contains only a subset of the I terms in the bank of terms, the regression matrix $\mathbf{B}$ has to be transformed by the optimisation algorithm. The steps which are necessary to do so were explained in detail by Wagner (1974) and are summarised in the following section.

4.4.3 The Stepwise Regression Analysis (SEEQ)

Regression analysises have been used in numerical mathematics for a long time. However, the common procedures were unsuitable even for the solution of simple thermodynamic problems such as the development of vapour pressure equations. Backward regressions fail due to numerical restrictions since they require a solution of Eqs. 4.90 or 4.91 for all terms out of the bank of terms as a starting point and simple forward regressions fail since they are not flexible enough to find the best solution if intercorrelations between different terms in the equation occur. To overcome these problems Wagner (1974) developed the stepwise regression analysis, a forward regression which is able to perform backward steps if the current combination of terms fails with respect to certain advanced statistical criteria. Wagner applied his method to vapour pressure equations first, but soon it was adapted to equations of state (Pollak, 1974/1975). Based on a report to the IUPAC Thermodynamic Tables Centre (Wagner, 1977) de Reuck reprogrammed the algorithm (de Reuck, 1979), applied it to an equation of state (de Reuck and Armstrong, 1979), and introduced the name SEEQ under which the algorithm became known internationally. This algorithm is still in use today in different research groups.

Figure 4.6 shows a flow diagram of the original procedure proposed by Wagner. The algorithm starts with the selection of the most efficient term (1), that is the term which yields the largest reduction to the sum of squares. To do so, the sum of

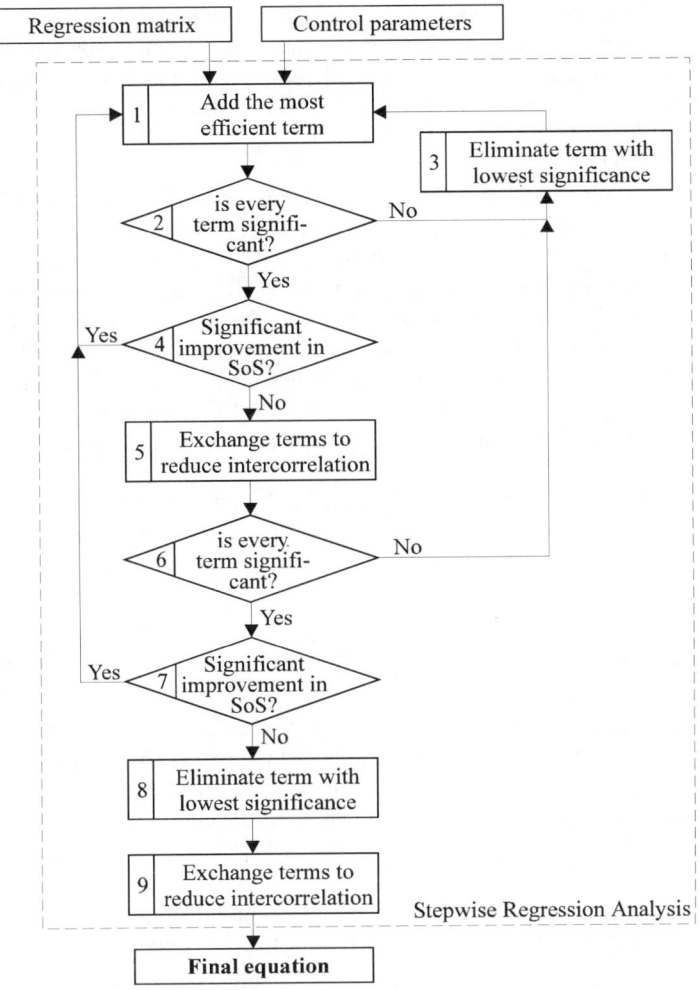

Fig. 4.6. Flow diagram of the stepwise regression analysis developed by Wagner (1974).

squares which results from the addition of every single term in the bank of terms which is not in the current formulation[3] (in the first loop no terms are in the formulation) has to be predicted. This can be done using the relation

$$\chi_n^2 = b_{n,LL} = b_{o,LL} - \frac{b_{L,k}^2}{b_{kk}} \tag{4.97}$$

---

[3] In Sects. 4.4.3–4.4.6 the expression "formulation" refers to the combination of terms which is considered in the current stadium of the optimisation process. The best formulation becomes the functional form of an equation of state when the results are written to a file after the optimisation procedure is finished.

where o refers to the old formulation and n to the formulation which is extended by the k-th term out of the bank of terms. The b_{ij} are elements of the regression matrix **B**, where i refers to the i-th row and j to the j-th column of the matrix. With $L = I + C + 1$, the element b_{LL} refers to the element S which always holds the sum of squares of the current formulation. For a deduction of Eq. 4.97 and the following relations, see Wagner (1974).

After testing all terms which are not in the current formulation, the term k which yields the smallest sum of squares is added to the current formulation and the regression matrix has to be transformed correspondingly. Table 4.3 summarises the relations which are necessary for this transformation as given by Setzmann and Wagner (1989a), where **IN** denotes a vector which is used to identify the terms which are in the current formulation. The element $IN(i)$ is equal to 1 if the i-th term from the bank of terms is in the current formulation and it is 0 if not. The given prescriptions assume that $L \geq i \geq j$, thus that the lower part of the symmetric matrix **B** is used. For the addition of a term to the current formulation, the parameter c becomes equal to 1.

In step 2, the significance of every single term in the current formulation is tested proving whether a coefficient is significantly different from 0 or not by a Student t test. The standard deviation of a coefficient n_i which is in the current formulation is given by

$$\sigma_{n_i} = \sqrt{\frac{b_{LL} b_{ii}}{\nu}} \quad \text{with} \quad \nu = M - N. \tag{4.98}$$

In Eq. 4.98, M is the number of data points which was used when setting up **B** and N is the number of terms in the current formulation. When testing the assumption that a certain coefficient n_i does not deviate from 0 significantly, the student t statistic compares this standard deviation with the value of the coefficient,

$$t_i = n_i / \sigma_i. \tag{4.99}$$

The value of the coefficient n_i corresponds to the element b_{Li} of the regression matrix. The probability that the coefficient n_i is different from 0 is given by

$$S_i = \frac{1}{\sqrt{\pi \nu}} \left(\Gamma\left(\frac{\nu+1}{2}\right) \Big/ \Gamma\left(\frac{\nu}{2}\right) \right) \int_{-t_i}^{t_i} \left(1 + \frac{t^2}{\nu}\right)^{-\frac{\nu+1}{2}} dt \tag{4.100}$$

where Γ is the gamma function. This relation can be evaluated using standard software packages. Wagner proposed rejecting terms if their coefficients do not deviate from 0 with a probability of at least 99.99 %. However, this value is typical only for short correlation equations and smaller values may yield better results for equations of state. To gain more flexibility, it is advantageous to use the corresponding limit as an adjustable parameter of the optimisation procedure (see also Sect. 4.4.5).

If one or more of the terms in the current formulation fail the t test, the term k with the lowest t value is eliminated from the current formulation. In step 3 $IN(k)$ is set equal to zero and the regression matrix is transformed according to the pre-

Table 4.3. Prescriptions for the transformation of the regression matrix **B** when adding ($c=1$) or eliminating ($c=2$) the k-th term from the bank of terms to or from the current formulation

Transformation	Position of the element
$b_{ij} = \dfrac{1}{b_{kk}}$	$i = j = k$
$b_{ij} = \dfrac{b_{ij}}{b_{kk}}$	$i = k, j < k$ or $j = k, i > k$
$b_{ij} = b_{ij} + \dfrac{b_{ik}^2}{b_{kk}}(-1)^{c+IN(i)}$	$i = j > k$ [a]
$b_{ij} = b_{ij} + \dfrac{b_{ki}^2}{b_{kk}}(-1)^{c+IN(i)}$	$i = j < k$
$b_{ij} = b_{ij} + \dfrac{b_{ik} b_{jk}}{b_{kk}}(-1)^c$	$i > k, j > k, i \neq j$
$b_{ij} = b_{ij} + \dfrac{b_{ik} b_{kj}}{b_{kk}}(-1)^c$	$i > k, j < k$
$b_{ij} = b_{ij} + \dfrac{b_{ki} b_{kj}}{b_{kk}}(-1)^c$	$i < k, j < k, i \neq j$

[a] the element b_{LL} is treated as a normal element with $IN(L) = 0$, see Eq. 4.97

scriptions given in Tab. 4.3 with $c=2$; the procedure goes back to the selection of the next term. However, terms which have been rejected immediately before due to a failed t test must not be selected again since this could lead into an endless cyclic exchange of terms.

If all of the terms are significant according to the t test, the significance of the whole formulation is tested (step 4) using the Fisher F statistic. The variance of the current formulation is given by

$$\sigma_{eq}^2 = \frac{b_{LL}}{\nu} \quad \text{with} \quad \nu = M - N. \tag{4.101}$$

From a combination of Eqs. 4.97 and 4.101, the variance of the formulation which is truncated by the term k with the smallest t value can be calculated according to

$$\sigma_{eq-1}^2 = \frac{b_{LL} + b_{Lk}^2/b_{kk}}{\nu + 1}. \tag{4.102}$$

According to the Fisher F statistic, the probability that the difference between σ_{eq}^2 and σ_{eq-1}^2 is not accidental can be calculated as

$$S = \frac{\Gamma(\nu + 0.5)}{\Gamma\left(\dfrac{\nu+1}{2}\right)\Gamma\left(\dfrac{\nu}{2}\right)}(\nu+1)^{(\nu+1)/2}\,\nu^{\nu/2} \int_0^F \frac{f^{(\nu-1)/2}}{(\nu+(\nu+1)f)^{\nu+0.5}}\, df \tag{4.103}$$

where the F value is defined as $F = \sigma_{eq-1}^2/\sigma_{eq}^2$. Again, this relation can be evaluated using common statistical program packages. However, the F statistic introduced by

Wagner when working on vapour pressure equations may yield misleading results for $v > \approx 500$ due to rounding errors. Since more degrees of freedom are common for the establishment of equations of state, where the number of used data, M, may be as large as several thousand, the F statistic has to be replaced by an approximation in this case. For large values of v, the F distribution becomes equal to the single sided normal distribution (see for instance Sachs, 1973). Thus, Eq. 4.103 can be replaced by

$$F_{\min} = \exp\left[z \cdot \sqrt{\frac{2(2v+1)}{v(v+1)}}\right] \qquad (4.104)$$

for $v > \approx 100$ where z is the standard-normal value which corresponds to the chosen value of $S_{\min}$ for the single sided normal distribution. Wagner proposed a value of $S_{\min} = 99.99\%$ for the F test which corresponds to $z = 3.719$. But again different values may be useful depending on the length of the correlation equation. The corresponding standard-normal values can be obtained from tables or statistic packages.

If the current formulation proves to be significant, the next term is added to the formulation (step 1). If the formulation fails the f test, intercorrelation between terms in the formulation is expected. In this case, exchanges of each term in the formulation against each term in the bank of terms (step 5) which is not in the current formulation are tried. The sum of squares which results from an exchange of term n against term m can be predicted according to the prescriptions given in Table 4.4. If one of the tested exchanges reduces the sum of squares, the corresponding exchange is carried out by removing term n (Table 4.3 with $c=2$) and adding term m (Table 4.3 with $c=1$). Subsequently, the significance of every single term and of the formulation as a whole is tested again. If one of the terms in the formulation fails the t test, it is eliminated (step 3) and the procedure continues with the addition of the next term. If the formulation as a whole is significantly better than the formulation which is truncated by the term with the smallest value of S_i (F test), the next term is added as well.

However, if the formulation could not be improved by an exchange of terms or if the improvement was not significant in terms of the F test, the current combination of terms is regarded as too long. In this case, the term with the smallest value of S_i is eliminated from the formulation. Finally, it is again tested whether the truncated formulation can be improved by an exchange of terms; if improvements are possible the corresponding exchanges are carried out.

The equation which results from the stepwise regression analysis is optimised both with respect to its functional form and its length, where the length of the equation is strongly influenced by the value of $S_{\min}$ which is used in the F test. The coefficients n_i of the final equation can be determined from the b_{Li} of the regression matrix.

One characteristic feature of the stepwise regression analysis is that the optimisation algorithm relies only on the regression matrix. Thus, the algorithm can be used for very different problems, as long as a regression matrix of the form which was discussed in Sect. 4.4.2 can be supplied.

Table 4.4. Prescriptions for the prediction of sum of squares which result from an exchange of term n out of the current formulation against term m from the bank of terms as summarised by Setzmann and Wagner (1989a).

Prescriptions for $m > n$	Prescriptions for $m < n$
$c_m = b_{Lm} + \dfrac{b_{Ln} b_{mn}}{b_{nn}}$	$c_m = b_{Lm} + \dfrac{b_{Ln} b_{nm}}{b_{nn}}$
$d_m = b_{mm} + \dfrac{b_{mn}^2}{b_{nn}}$	$d_m = b_{mm} + \dfrac{b_{nm}^2}{b_{nn}}$
$\chi_{-1}^2 = b_{LL} + \dfrac{b_{Ln}^2}{b_{nn}}$	
$\chi_{mn}^2 = \chi_{-1}^2 - \dfrac{c_m^2}{d_m}$	

4.4.3.1 Introduction of a Pairwise Exchange

When using the stepwise regression analysis in the way described above, problems arise from the fact that the exchange of terms cannot handle intercorrelated pairs of terms which are important for the description of the thermodynamic surface. If an exchange of one of these terms is tested according to the prescriptions given in Table 4.4, the sum of squares becomes much larger since the second term becomes inefficient in this way as well and no exchange will be carried out. Nevertheless, another pair of terms which is contained in the bank of terms but not in the current formulation may yield better results – the stepwise regression analysis fails at a point which is important especially for longer correlation equations where such intercorrelated terms are common. Theoretically, it is clear that this problem can be avoided if not just an exchange of single terms is tested, but also an exchange of each pair of terms in the formulation against each pair of terms not in the current formulation. However, such an exchange of pairs of terms has been regarded as impossible due to restrictions of the available computing power.

Bischoff (1988) developed a modified stepwise regression analysis where an exchange of pairs of terms is tested whenever the exchange of single terms was not successful. Lüddecke (1991) compared this modified stepwise regression analysis with a modified stepwise regression analysis without pairwise exchange and with the more complex optimisation algorithm of Setzmann and Wagner (1989a), see Sect. 4.4.5. He found that the stepwise regression analysis with double exchange yields results which are as good as those of the Setzmann and Wagner algorithm but with considerably lower computing times. However, this conclusion is true only for relatively short correlation equations and small banks of terms – Lüddecke tested the selection of up to 12 terms out of banks of terms which contained up to 100 terms.

For an exchange of single terms, the number of combinations which has to be tested can be calculated according to

$$n_{\text{single}} = N \cdot (I - N), \qquad (4.105)$$

where N is the number of terms in the current formulation and I the number of terms in the bank of terms. For an exchange of pairs of terms, the number of possible combinations becomes

$$n_{\text{double}} = \frac{1}{4} N \cdot (N-1) \cdot (I-N) \cdot (I-N-1). \qquad (4.106)$$

To test an exchange of single terms, a maximum of $4 \cdot n_{\text{single}}$ numerical operations is required while the transformations of the regression matrix, which are necessary to exchange two terms, require $(I+C+1) \cdot (I+C+2)$ numerical operations. Thus, with regard to computing time the transformation of the regression matrix is the dominant factor for an exchange of single terms if $I > \approx 4 \cdot N$ holds for the size of the bank of terms. The procedure of Collmann et al. (1996), which is explained in more detail below, requires up to $20 \cdot n_{\text{double}}$ numerical operations to test a pairwise exchange, while the number of operations which are necessary to transform the regression matrix is increased only to $2 \cdot (I+C+1) \cdot (I+C+2)$ since four terms are now involved. Thus, for pairwise exchanges the prediction of sums of squares for possible exchanges is the dominant factor in the computing time.

Table 4.5 summarises results from Eqs. 4.105 and 4.106 for some typical optimisation tasks and gives approximate ratios of the maximum numbers of numerical operations which are required for the prediction of the sums of squares and the following transformation of the regression matrix. For simple optimisation problems, a modified stepwise regression analysis with pairwise exchange of terms may be faster and as successful as the optimisation procedure of Setzmann and Wagner (1989a) with its repeated calls of the regression algorithm (see Sect. 4.4.5). But for longer equations and larger banks of terms, as they are typical for the development of multiparameter equations of state, a single call of a pairwise exchange needs more computing time than the Setzmann and Wagner algorithm. And for such problems, the results of the Setzmann and Wagner algorithm with its heuristic elements are still better than those of a modified regression analysis since the

Table 4.5. Number of possible exchanges and ratio of the maximum number of numerical operations which are required for an exchange of single terms and pairs of terms

| Number of terms in the | | Number of possible exchanges | | Ratio of required |
Formulation	Bank of terms	single exchange	double exchange	numerical operations
4	27	92	1 518	27.1
12	100	1 056	252 648	349.3
20	400	7 600	$1.368 \cdot 10^7$	1 429.8
35	650	21 525	$1.123 \cdot 10^8$	4 402.4
40	1000	38 400	$3.591 \cdot 10^8$	6 210.4

flexibility of a pairwise exchange is not sufficient to resolve the complex intercorrelations in multiparameter equations completely. However, the enormous increase of available computing power makes a modified regression analysis with pairwise exchange interesting again even for the development of highly accurate equations of state with a large number of terms – namely as an integral part of a modified Setzmann and Wagner algorithm (see Sect. 4.4.5.2). This was the focus of the work of Collmann et al. (1996) who developed a procedure similar to the one described below for the required prediction of the sums of squares which result from pairwise exchanges.

A pairwise exchange of terms involves four terms k_i. The corresponding elements of a vector **f** are equal to -1 for those two terms which are in the current formulation and equal to 1 for those two terms which are not yet in the formulation. To predict the sum of squares which results from the pairwise exchange of terms a partial matrix **A** is used which is determined from the regression matrix **B** according to

$$\begin{bmatrix} a_{11} & & & & \\ a_{21} & a_{22} & & & \\ a_{31} & a_{32} & a_{33} & & \\ a_{41} & a_{42} & a_{43} & a_{44} & \\ a_{51} & a_{52} & a_{53} & a_{54} & a_{55} \end{bmatrix} = \begin{bmatrix} b_{k_1 k_1} & & & & \\ b_{k_2 k_1} & b_{k_2 k_2} & & & \\ b_{k_3 k_1} & b_{k_3 k_2} & b_{k_3 k_3} & & \\ b_{k_4 k_1} & b_{k_4 k_2} & b_{k_4 k_3} & b_{k_4 k_4} & \\ b_{L k_1} & b_{L k_2} & b_{L k_3} & b_{L k_4} & b_{LL} \end{bmatrix}. \quad (4.107)$$

The way in which Eq. 4.107 is written assumes $k_1 < k_2 < k_3 < k_4$; where this condition does not hold, the indices of the corresponding b_{ij} have to be changed to b_{ji} if only the lower half of the symmetric regression matrix **B** is supplied to the optimisation program. From the matrix **A**, the sum of squares of a double exchange can be predicted according to the prescriptions given in Table 4.3. However, it is not necessary to transform the complete matrix 4 times. The exchange (addition or elimination) of the term k_1 requires information from all elements of the matrix, but to exchange the second term, k_2, only the matrix elements in the triangle a_{22}–a_{52}–a_{55} are required. Thus, only these 10 elements have to be transformed when exchanging k_1. To exchange the third term requires only the 6 elements of the triangle a_{33}–a_{53}–a_{55} and for the fourth term only the 3 elements of the triangle a_{44}–a_{54}–a_{55} are required. Finally, only the element a_{55} which is equal to the sum of squares has to be transformed when exchanging the fourth term – to predict the sum of squares of a pairwise exchange requires 20 numerical operations with matrix elements. Table 4.6 summarises the prescriptions for the required operations as they result from an adaptation of Table 4.3 to this simpler problem.

To avoid unnecessary pairwise exchanges this routine is called only if the exchange of single terms was not successful. If the predicted sum of squares shows possible improvements of the formulation, the double exchange which results in the smallest sum of squares is carried out. The corresponding two terms are eliminated from the current formulation and the two terms from the bank of terms are added – thus the whole regression matrix has to be transformed four times according to the prescriptions given in Table 4.3. Afterwards, the optimisation procedure continues with testing the potential of further pairwise exchanges. If pairwise ex-

Table 4.6. Prescriptions for the transformation of the working matrix **A**

Transformation	Position of the element
$a_{ij} = a_{ij} - f(k) \cdot f(i) \cdot \dfrac{a_{ik}^2}{a_{kk}}$	$i = j > k$ [a]
$a_{ij} = a_{ij} - f(k) \cdot \dfrac{a_{ik}\, a_{jk}}{a_{kk}}$	$i > k,\ j > k,\ i \neq j$

Conditions

$1 \leq k \leq 4$, $(k+1) \leq i \leq 5$, and $(k+1) \leq j \leq i$

$f(k) = -1$ for terms k which have to be eliminated from the current formulation
$f(k) = 1$ for terms k which have to be added to the current formulation

[a] the element a_{55} is treated as a normal element with $f(5) = 1$

changes cannot improve the formulation any further, the regression procedure continues as shown in Fig. 4.6.

4.4.4 The Evolutionary Optimisation Method (EOM)

When using the stepwise regression analysis for complex optimisation problems, such as the development of an accurate multiparameter equation of state which requires the selection of about 20–40 terms with two independent variables out of banks with 300–1000 terms, it became obvious that a purely deterministic regression analysis is not flexible enough to find global optima. For different sophisticated optimisation problems, similar experiences were reported. It was shown early in the seventies that non-deterministic random search strategies are superior to deterministic procedures since they are able to leave local optima (see Rechenberg, 1973). However, none of the existing random search strategies could be applied to the development of equations of state directly.

Therefore, Ewers and Wagner (1982a/b) developed the *evolutionary optimisation method (EOM)*, a new optimisation algorithm which imitates known principles of biological evolution. The evolutionary principles which formed the theoretical basis of the algorithm were adapted to the development of equations of state wherever such adaptations could improve or simplify the resulting algorithm – it was not the goal of Ewers and Wagner to develop a faithful imitation of biological principles. The characteristic elements of the EOM are

- Optimisation of a *population* of formulations, each represented by a parameter vector which identifies the corresponding terms out of the bank of terms. Usually the population contains 20 formulations.
- Optimisation in *generation cycles*. Usually 150 generations are calculated unless convergence occurs before this number of generations is finished.
- *Mating of diploid individuals* and recombination of descendants from these "parents".
- Change of the "genetic code", the parameter vector, by *mutation*. To speed up the procedure, the probability of mutations is much higher than in biological

systems. The "step width" of mutations is controlled by a teachable procedure which allows different step widths for different terms in the formulation. The terms which are to be changed are selected by a random algorithm.
- Change of the "genetic code" by *crossing over*. Randomly selected parameter vectors of descendants are determined from mixtures of the parameter vectors of the parents. In this way, large changes in the structure of formulations are performed which are essential to leave local minima.
- *Selection* of the descendants with the smallest sums of squares to form the offspring generation.
- Avoidance of "inbreeding" either by limitation of the maximum number of descendants from a single pair of parents or, theoretically more profound, by *division of the population*.

Although all these elements sound familiar, their implementation becomes extremely complicated and the whole algorithm requires much more computing time than deterministic procedures. Moreover, the whole algorithm is very specialised in the development of equations of state. Beside the regression matrix, additional information on predetermined features of the resulting equations of state, such as the total length and the number of terms of certain forms, has to be supplied to the EOM – in this sense the optimisation is incomplete.

Compared to the stepwise regression analysis, the results which were obtained from the evolutionary optimisation method were impressive; the equation of state of Schmidt and Wagner (1985) is still the accepted standard for thermodynamic properties of oxygen. However, this algorithm has never been used by others due to its complicated set-up, its high demands regarding the available computing power, the incomplete optimisation, and its methodical restriction to the development of equations of state.

For the further development of optimisation strategies, an experience gained from the work with the evolutionary optimisation method was decisive: There are always certain combinations of terms which are characteristic for all formulations which yield a good description of a given problem. The stepwise regression analysis is not flexible enough to find these combinations, but when started with the characteristic terms, it yields very good results even for complex problems. This is the basic idea of the optimisation algorithm by Setzmann and Wagner (1989a) which is described in the following section.

4.4.5 The Optimisation Algorithm by Setzmann and Wagner (OPTIM)

In 1989, Setzmann and Wagner (1989a) published a new optimisation algorithm which combines a modified stepwise regression analysis with elements of the evolutionary optimisation method such as mutation and optimisation of a population of correlations. This method was designed to overcome the shortcomings of the EOM – its set-up is simpler, it yields a complete optimisation which includes the length of the resulting correlation and the number of terms of certain forms, and it can be used for very different problems. The name OPTIM was introduced when the corresponding programs were made available for other users (Setzmann and Wagner, 1989b; Setzmann et al., 1990).

Like the stepwise regression analysis, the whole procedure is based only on a regression matrix (see Sect. 4.4.2) and on control parameters. The optimisation procedure starts with reading these data, see the flow diagram which is given in Fig. 4.7. The optimisation starts with the initialisation of the first parental generation (step O1). Each of the NP correlations (typically 6 to 8) in the population is represented by a parameter vector $\mathbf{P}_i$ which holds the position of the corresponding terms in the bank of terms. The NP parameter vectors are initialised by random selection of $N_{\max}$ terms out of the bank of terms, where $N_{\max}$ is an estimate for the maximum length of the correlation. The random selection is repeated NS times for each parameter vector and the combination of terms with the smallest sum of squares is used as a starting solution for the corresponding parameter vector. To determine the sums of squares, the systems of normal equations which correspond to the parameter vectors $\mathbf{P}_i$ have to be constructed from the regression matrix (see for instance Eq. 4.107) and have to be solved as for linear fits, cf. Sect. 4.1. The NP starting solutions form the 0-th parental generation (step O2).

In the next step the formulations in the current generation are modified by "*mutation*" (step O3), a simplified version of the mutation algorithm which was used in the EOM. For each of the NP parameter vectors $\mathbf{P}_i$, a limited number of components $p_{n,\mathrm{old}}$ is selected randomly, where the number of selected terms should be less than $N/2$. New values are selected for these terms randomly, where the procedures "*neighbour*" *mutation* and "*free*" *mutation* are used alternating from parameter vector to parameter vector. When using the procedure neighbour mutation, the selected terms are replaced according to the prescription

$$p_n = p_{n,\mathrm{old}} + z(0,\sigma) \qquad (4.108)$$

where the $z(0,\sigma)$ are normally distributed random numbers which should cover a range from -3 to 3 with an expected value of 0. The mutants which are determined in this way are rather similar to the old formulations. When using free mutation the new values for the selected p_n are determined by random selection from the bank of terms with the only limitation that the selected term must not be in the current formulation; mutants which are determined in this way are very different from the old formulations. For each mutant the quality is determined in the way described above. If the mutant has a smaller sum of squares, the original formulation is replaced by the mutant. The mutation procedure is applied NM times to each parameter vector in the population; Setzmann and Wagner (1989a) give an example to explain the mutation process in more detail.

In the next step (O4), the terms are selected which are considered especially suited for the description of the given problem to use them for a starting solution in a modified stepwise regression analysis. Thus, the terms are identified which are used in different formulations out of the current population. Assuming that the best formulations contain most of the well suited terms, only the $NR+1$ best formulations are considered. From those formulations, the terms are selected which are

4.4 Optimising the Functional Form

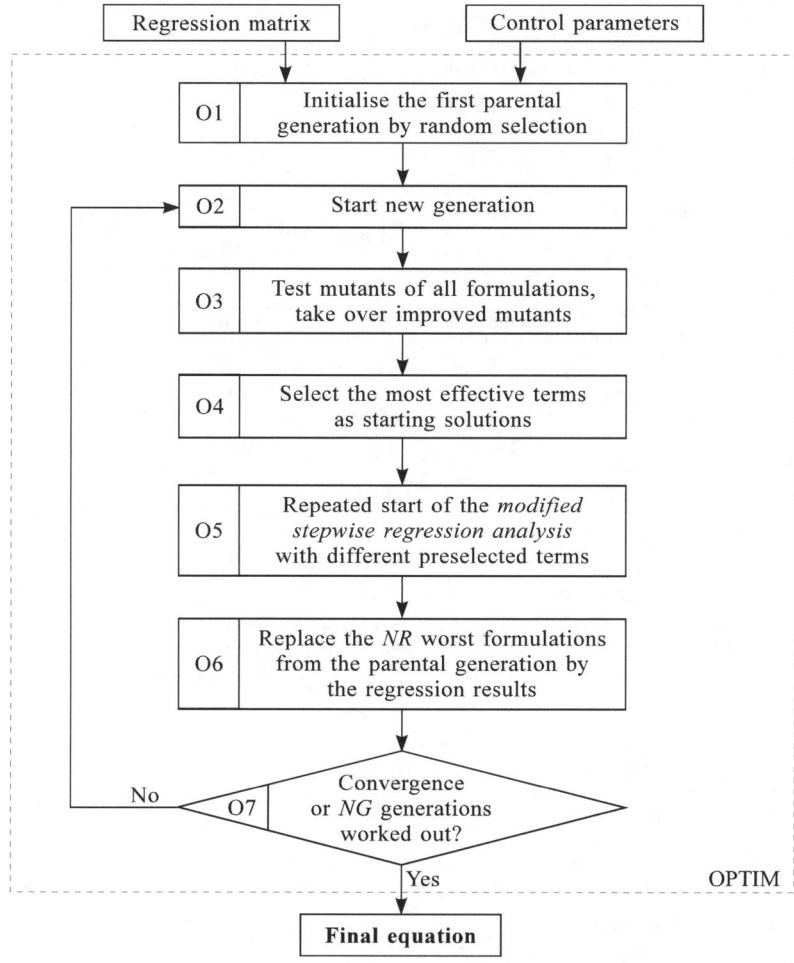

Fig. 4.7. Flow diagram of the optimisation algorithm developed by Setzmann and Wagner (1989a); for details on the modified stepwise regression analysis see Fig. 4.8.

contained in $NR+1$ formulations, NR formulations, and so on down to 2 formulations. With these terms the *modified stepwise regression analysis* is started NR times (step O5).

A detailed flow diagram of the modified regression analysis is given in Fig. 4.8. The regression analysis starts with the addition of the preselected terms (step R1). To do so, step 1 from Fig. 4.6 with the corresponding transformation of the regression matrix, see Table 4.3, is repeated until the starting solution contains all preselected terms. In the next step (R2) the most important term from the bank of terms is added. The sum of squares which results from the addition of a certain term is predicted using Eq. 4.97 again.

110   4 Setting up Multiparameter Equations of State for Pure Substances

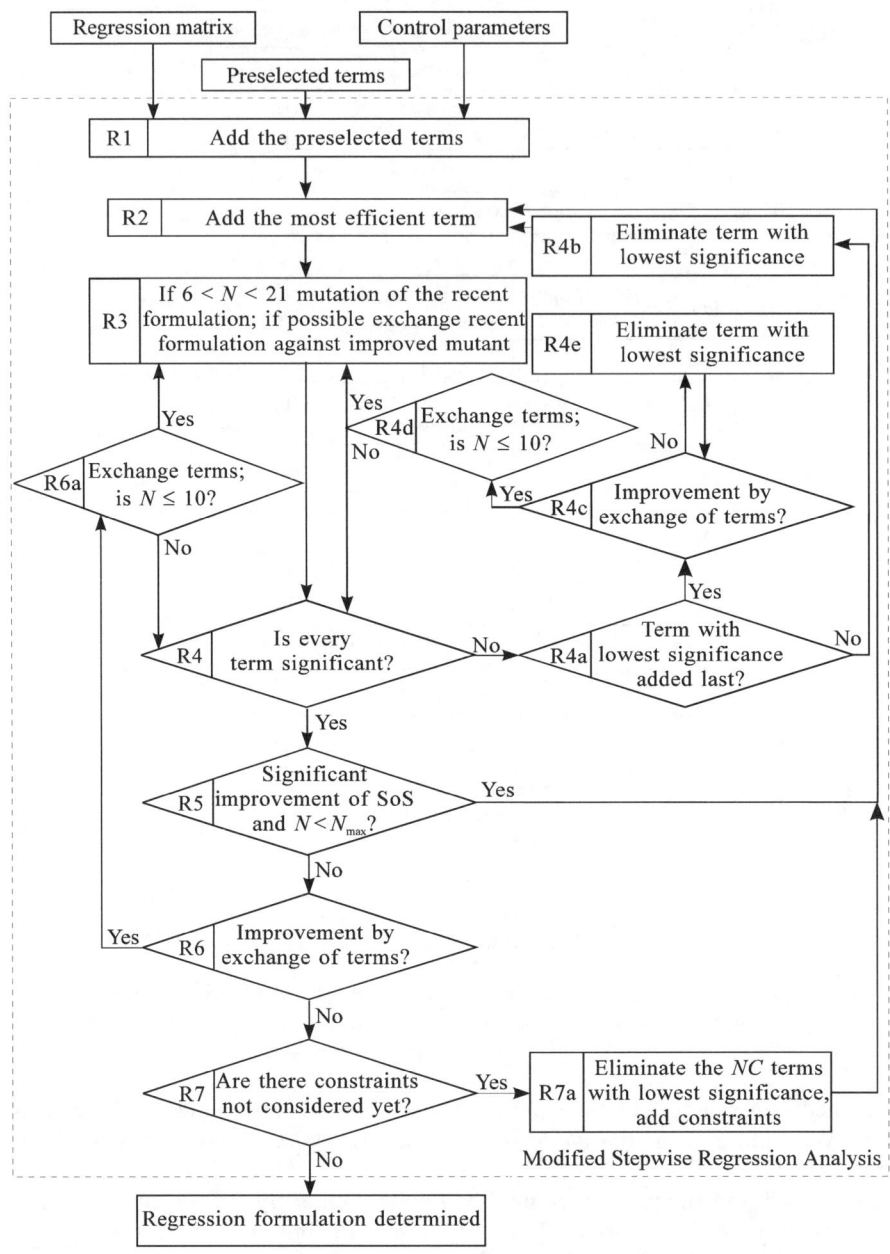

Fig. 4.8. Flow diagram of the modified stepwise regression analysis which is used as an integral part of the optimisation algorithm developed by Setzmann and Wagner (1989a).

After the most important term has been added, the mutation algorithm described above is applied to the current formulation NM times (step R3) in order to gain more flexibility in the deterministic set-up of the regression analysis. However, experience has shown that this step may ruin a good starting solution when applied at an early state and that the probability of successful mutations decreases with increasing length of the formulation. Thus, this step is applied only to formulations with more than 6 and less than 21 terms. The sums of squares of the mutants are determined by linear fits based on the corresponding systems of normal equations. At this point, it is important to use a copy of the regression matrix as the working matrix for the regression analysis since the required matrix for the fit can only be extracted from the unchanged regression matrix. If one of the mutations reduces the sum of squares, the current formulation is exchanged against the mutant and the working matrix is adapted to the new parameter vector by exchanging the necessary terms according to Table 4.3.

Just like the original stepwise regression analysis, the procedure continues with statistical tests for the significance of every single term (step R4, Student t test, see Eqs. 4.98 – 4.100) and for the significance of the formulation as a whole (step R5, Fisher F test, see Eqs. 4.101 – 4.104). If one of the terms in the current formulation fails the t test it is tested (step R4a) whether this is the term which was added in step R2. If this is true, the term must not be again removed without other changes since this would result in an endless exchange of terms.[4] Thus, it is tested whether the formulation can be improved by the step "exchange of terms" (R4c) as described in Sect. 4.4.3; the sums of squares are predicted using the relations which are summarised in Table 4.4. If an improvement by exchange of terms is not possible, the term with the lowest statistical significance (the probability S_i of the t test) is removed from the formulation (step R4e) and an exchange is tried again. If the exchange was successful, the procedure continues either with the step mutation (R3) if less than 11 terms are in the current formulation or otherwise with a new Student t test (step R4) for all of the terms in the formulation.

If the formulation as a whole is significant according to the Fisher F test (step R5) and if a predetermined maximum number of terms is not yet reached, the procedure continues with the addition of the next term (step R2). Thus, like the stepwise regression analysis, OPTIM uses the Fisher F test to optimise the length of correlation equations but an additional upper limit is introduced to guarantee that the formulations which result from different regression runs are comparable with regard to their length. If the formulation fails the F test or if the maximum number of terms is reached, an exchange of terms is tried again (step R6). If the exchange was successful, the procedure continues either with the step mutation (R3) if less than 11 terms are in the current formulation or with the Student t test (step R4). If

---

[4] The procedure which is described here was used in the original Setzmann and Wagner algorithm. However, this procedure cannot avoid cyclic exchanges of 2 or more terms which resulted in endless loops in infrequent cases. Recent modifications of OPTIM additionally use an algorithm which checks how many of the last 10 added terms are among the last 10 eliminated terms too whenever adding a new term. If this ratio exceeds a certain limit (e.g. 7 out of 10) cyclic exchanges are assumed and the regression analysis stops.

the exchange was not successful and if no constraints were formulated, the final formulation is determined.

If the resulting formulation has to fulfil *NC constraints*, the following procedure is applied. The *NC* terms with the smallest significance (t test) are removed from the completely developed regression formulation by using the prescriptions given in Table 4.3. Then the *NC* constraints are added using the same prescriptions (step R7a). The elements of the matrix which contain contributions from the constraints can be treated just like all other elements. However, the constraints are not counted as "terms" when determining the length of the current formulation and they cannot be removed again. After the constraints are added, the procedure continues with step R2 until the final length of the formulation as determined by the Fisher F test or by N_{max} is reached.

After the *n*-th regression run has come to an end, the transformed working matrix is replaced by the original regression matrix and the procedure continues with the (*n*+1)-th regression run using the next combination of preselected terms until all *NR* regression runs are completed. Now the *NR* formulations of the parental generation with the highest sums of squares are replaced by the formulations which resulted from the *NR* regression runs (step O6). To guarantee a sufficient flexibility it is important to replace old formulations even if the sum of squares of the new formulation is larger. However, the number of formulations in the population, *NP*, has to be larger than the number of regression runs, *NR*, in order to preserve the best formulations which have been found before. The new population forms the parental generation of the next generation and the procedure continues with step O2.

The optimisation algorithm ends if either all formulations in the population are identical in step O7 (*convergence*) or if a predetermined number of generations has been worked out. Convergence is common only for simple optimisation problems such as the development of ancillary equations for the vapour pressure or the

Table 4.7. Recommended values for the control parameters which are used in the optimisation procedure by Setzmann and Wagner (OPTIM)

Characterisation of the problem		I	II	III
Maximum number of terms in an equation	N_{max}	≤ 10	≤ 30	≤ 60
Number of terms in the bank of terms	I	≤ 200	≤ 400	≤ 1000
Control parameters				
Number of formulations in the population	*NP*	8	7	6
Number of regression runs in each generation	*NR*	6	5	4
Number of initial random selections[a]	*NS*	250	250	250
Number of mutations	*NM*	60	60	60
Probability value for the Student t test	$S_{i,min}$	0.9999	0.99	0.95
Probability value for the Fisher F test	S_{min}	0.999	0.9	0.8

[a] smaller values are reasonable for *NS* when working with starting solutions

orthobaric densities, but experience has shown that significant improvements are very unlikely after about 10 generations even for complex optimisation problems. Thus, Setzmann and Wagner (1989a) proposed that the algorithm should be stopped generally after $NG = 10$ generations. In the modified OPTIM algorithms which are in use today, NG became an adjustable control parameter. Values of $NG > 10$ are still uncommon but under certain conditions smaller values are very useful.

When using OPTIM in the way which was described above each optimisation run starts from scratch. However, there may be preliminary equations of state which proved to be very suitable for a given problem. To make use of such information the optimisation algorithm can be started with preselected terms. In this case the first regression run starts not with those terms which are present in $NR+1$ formulations of the initial population but with PS preselected terms, where $PS \leq N_{max}$ has to hold. If the preselected terms correspond to a good preliminary equation the first regression run will result in a formulation of at least equal quality and the population will already contain at least one good formulation after the first generation has been worked out. Since the best formulation in the population is never exchanged against regression results, a good starting solution[5] has a strong influence on the development of the whole optimisation process. However, formally the preselected terms are used only in the first regression analysis and can be altered during the optimisation process just like all other terms.

When finalising an equation of state, it may be desirable to restrict the flexibility of the optimisation even further. To do so, recent modifications of OPTIM support a mode where the regression runs of the first generation follow a completely different procedure. In this case, up to 10 regression runs are carried out in the first generation starting with the PS preselected terms, with $PS-2$ preselected terms, with $PS-4$ preselected terms, and so on down to $PS-18$ or 1 preselected terms. The best results of these regression runs are taken over into the current population if they are better than the randomly selected formulations in the parental generation. In this way, the population usually contains only descendants of the starting solution at the end of the first generation and the preceding random selection looses any influence. The equation which results from the optimisation will probably be very similar to the starting solution, but nevertheless this kind of *"fine adjustment"* of the functional form may result in significantly improved sums of squares. For such optimisations, the number of worked out generations can be reduced to $NG \approx 4$ in order to save computing time.

Based on first experiences with the new optimisation algorithm, Setzmann and Wagner (1989a) published proposals for the control parameters of the optimisation process. Based on further experiences and on the increased computing power which has become available in the meantime, some of the values given by Setz-

---

[5] The use of starting solutions is a crucial point since the quality criterion of the optimisation, the sum of squares, is not sufficient to describe the quality of an equation of state completely. Important features like the extrapolation behaviour do not exercise an influence on the sum of squares. Thus, it is sometimes important to preserve results which were found in previous optimisation runs, even if an optimisation without a starting solution could be more successful in finding completely different optima with slightly smaller sums of squares.

mann and Wagner are out of date. Table 4.7 summarises updated recommendations for the most important control parameters. However, just like the values given by Setzmann and Wagner (1989a) these recommendations are only initial guesses and may be altered. The probabilities which are used as limits for the statistical tests have often to be adapted to the specific problem. The value of S_{min} should be harmonised with reasonable values of N_{max} in such a way that formulations which result from the regression analysis have on average 1 or 2 terms less than the maximum length which is defined by N_{max}. The value for $S_{i,min}$ needs to be large enough to initiate exchanges of terms but it must not be too large. Otherwise the procedure may stop early since none of the terms not present in the current formulation passes the t test while the F test indicates that further terms have to be added to the formulation. Thus, good problem specific values for the minimum probabilities can only be found after evaluation of a number of optimisation runs.

4.4.5.1 Adapting OPTIM to Equations of State

One of the basic advantages of the optimisation algorithm of Setzmann and Wagner (1989a) is that it is not restricted to certain problems – any optimisation problem which allows the formulation of a suitable regression matrix (see Sect. 4.4.2) can be dealt with. This flexibility resulted in applications which range from simple equations for the melting pressure (see Wagner et al., 1994) over different kinds of multiparameter equations of state (see Setzmann and Wagner, 1991; Kruse, 1997) up to optimisations of multivariant production processes (see Setzmann and Wagner, 1989b). However, regarding the development of equations of state, the performance of OPTIM can be improved if this versatility is sacrificed; corresponding modifications were introduced by Span (1993).

Adaptations of OPTIM to certain problems generally require additional, problem dependent information on the set-up of the bank of terms. Such information is not needed for the deterministic steps of the modified stepwise regression analysis or for the "free" mutation where all terms in the bank of terms are treated in the same way. But the step *"neighbour" mutation* can be formulated in a more efficient way if additional information on the set-up of the bank of terms is available. Using a section of a very small bank of terms as an example, Fig. 4.9 compares the definition of neighbours as it is used in the algorithm of Setzmann and Wagner (a) with the one used by Span (b). Based only on the position of a term in the bank of terms, the neighbour mutation defined by Eq. 4.108 cannot consider the similarity of terms appropriately when dealing with set-ups where 2 or more parameters are varied. While the similar term No. 40 is not identified as neighbour of term No. 45, the terms 47–48 are treated as neighbours although their contribution is very different from the one of term No. 45. Based on additional information on the different I, J, and K in Eq. 4.87 the definition of neighbours proposed by Span considers similarities of different terms appropriately. The normal distribution which was assumed for $z(0,\sigma)$ in Eq. 4.108 was replaced by an adjustable distribution among neighbours of first, second and third degree and the ratio between neighbour and free mutations also became adjustable. In this way, the non-

4.4 Optimising the Functional Form

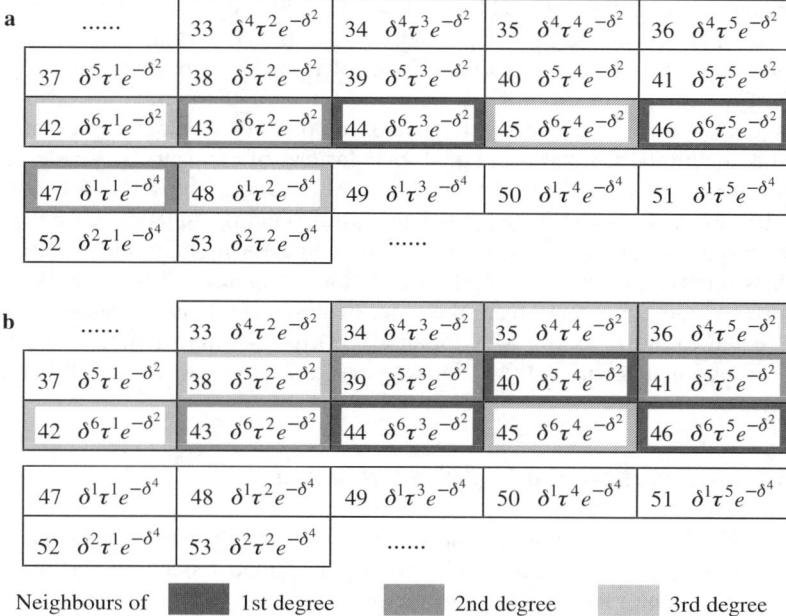

Fig. 4.9. Section out of a small bank of terms as it might be used for the development of an equation of state. The indicated fields are considered as "neighbours" of term No. 45 in the algorithm by Setzmann and Wagner (a) and in the modified procedure proposed by Span (b).

deterministic steps in OPTIM have become more efficient and the flexibility of the procedure can now be adjusted to certain tasks.

Another advancement is related to the fact that the algorithm of Setzmann and Wagner optimises the length of a correlation equation without consideration of limitations regarding the number of certain kinds of terms. However, both the numerical expense which is required to evaluate an equation of state and its numerical stability depend not only on the number of terms in the equation but also on the mathematical structure of these terms; sublimits regarding the maximum number of terms of a certain functional form are useful, e.g., for polynomial terms (see Sect. 4.6.2) and for special critical region terms (see Sects. 4.5.3 and 4.5.4).

During the optimisation process, the information on the set-up of the current formulation which is held in the parameter vector $\mathbf{P}_i$ is transferred to a vector $\mathbf{V}_i$ with I elements in order to simplify necessary interrogations. The i-th element of $\mathbf{V}_i$ is directly related to the i-th term in the bank of terms and is equal to 1 if this term is part of the current formulation and 0 if not. Wherever the addition of a term k from the bank of terms into the current formulation is tested, it is tested first whether v_k is equal to 0. If v_k is equal to 1, the corresponding term is already part of the current formulation and must not be added a second time. To enable the use of upper limits regarding the number of terms of a certain form, the vector $\mathbf{V}_i$ is copied to a vector $\mathbf{V}_i^*$ which contains an "imaginary set-up" of the current formulation. If a term k is added to the current formulation by random selection (step O1 in Fig. 4.7), by mutation (step O3, step R3 in Fig. 4.8), by addition of a term (steps R1, R2), or by exchange of a term (steps R4d, R6a), the values of v_k and v_k^* are changed from 0 to 1. The functional form[6] of term k can be identified using the different I, J, and K in Eq. 4.87. If an upper limit for the number of terms of the corresponding form is reached with the addition of k, all v_j^* which represent terms j with this functional form are set equal to 1. Instead of elements of $\mathbf{V}_i$, elements of $\mathbf{V}_i^*$ are used whenever the addition of a term is tested (steps R1, R2, R3, R4c, R6).[7] In this way no further terms of the corresponding functional form will be added to the formulation. If a term of a functional form which has been blocked before is removed from the current formulation (steps R4b, R4d, R4e, R6a, R7a) the corresponding elements from the bank of terms are released again by setting the v_j^* equal to the v_j. In this way, Span (1993) realised upper limits for terms of certain functional forms without unnecessary restriction of the flexibility of the optimisation procedure.

4.4.5.2 Introduction of a Pairwise Exchange

The use of pairwise exchanges in the stepwise regression analysis has already been explained in Sect. 4.4.3.1. As mentioned, the described procedure was developed by Collmann et al. (1996) for use in the modified stepwise regression analysis as an integral part of OPTIM. Pairwise exchanges are tested if no improvement could be realised by single exchanges in R4c and R6.

However, the use of pairwise exchanges in OPTIM is still a challenge with regard to computing time. For the development of short equations of state for technical applications, where selections of 10 to 12 terms out of banks of terms with less than 600 terms are typical (see Sect. 6.2), pairwise exchanges could be used without limitations and proved to be a very effective supplement to the optimisation algorithm of Setzmann and Wagner. For the development of reference equations of state, where selections of 35 to 40 terms out of banks of terms with up to 1000 terms are typical, the unrestricted use of pairwise exchanges led to insupportable computing times. However, the pairwise exchange proved to be effective especially for those functional forms where strong intercorrelations between the corresponding terms are typical, namely for Gaussian bell shaped terms (see Sect. 4.5.3) and nonanalytical terms (see Sect. 4.5.4). To make use of this advantage, the pairwise exchange can be modified in a way that an exchange is tested only for pairs of terms which belong to certain functional forms. This step requires information on the different I, J, and K in Eq. 4.87 again; the functional forms which have to be considered in a pairwise exchange are determined by corresponding control parameters. Since the number of terms which have to be considered becomes smaller both with respect to the equation and to the bank of terms, the nu-

---

[6] The realised program distinguishes polynomial, Gaussian bell shaped (Sect. 4.5.2), nonanalytic (Sect. 4.5.3), and exponential terms with $e^{-\delta}$, $e^{-\delta^2}$, ..., and $e^{-\delta^6}$.

[7] The only exception are exchanges (R4d, R6a) of 2 terms with the same functional form since the number of terms using the respective functional form is not increased by such exchanges.

merical expense is reduced significantly (c.f. Eq. 4.106) compared to an unrestricted pairwise exchange.

Restricted pairwise exchanges were used as an integral part of OPTIM for the first time when the new reference equation of state for nitrogen (Span et al., 1998b) was developed. The next generation of workstations will show whether an unrestricted pairwise exchange can improve the performance of OPTIM for the development of reference equations of state further.

4.4.6 The Nonlinear Optimisation Algorithms by Tegeler et al.

The modified Setzmann and Wagner algorithm, which was described in the preceding sections, is certainly a very powerful tool for the optimisation of the functional form of an equation of state, but like all other procedures described it has one major disadvantage: it is restricted to linear optimisation problems. This restriction results from four very fundamental features: the way in which the information on the data set is supplied to the optimisation procedure (as a regression matrix) is suitable only for explicit data, the procedures which are used to predict the sum of squares (see Eq. 4.97 and Tables 4.4 and 4.6) and to add and eliminate terms (see Table 4.3) are applicable only for linear fits, and finally, linear fits are used directly in the steps "initialisation" and "mutation". However, this restriction has never been a severe problem, since data for properties which result in implicit linear or implicit nonlinear relations to the Helmholtz energy and its derivatives could be used in linearised form as described in Sect. 4.3.2. The influence of the approximation which is implied by this linearisation could be eliminated by using a cyclic process: the coefficients n_i of the equation which results from the linear optimisation are fitted nonlinearly and directly to all kinds of data; the necessary linearisation of nonlinear data (see Sect. 4.3.2) is repeated with the nonlinearly fitted equation before the linear optimisation algorithm is started again (see Setzmann and Wagner, 1991). This cyclic process is repeated until the difference between the results of linear and nonlinear fit is negligible.

During the development of the recent reference equation for argon (Tegeler et al., 1997/1999) the first highly accurate speed of sound data from spherical resonator measurements at high pressures became available (Ewing and Goodwin, 1992; Estrada-Alexanders and Trusler, 1995). These data reduced the uncertainty of speeds of sounds in the high pressure gas phase by at least one order of magnitude and increased the demands on reference equations of state correspondingly. With the cyclic process described above these data could no longer be represented appropriately.

As an example, Fig. 4.10 shows percentage deviations between values calculated from a preliminary equation of state for argon and highly accurate experimentally determined speeds of sound on the 300 K isotherm. The open symbols show deviations which are calculated between the linearly fitted equation and the linearised data as they are used in linear optimisation procedures. The filled symbols represent the deviations between the original $w(T,p)$ data and values calculated from a nonlinearly fitted equation. The grey lines indicate the experimental uncertainty of the data. OPTIM resulted in a functional form which is able to de-

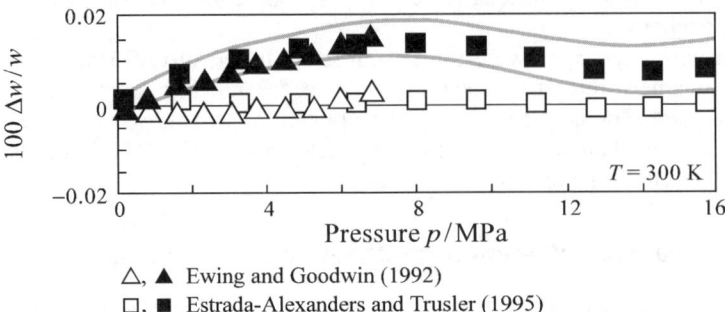

△, ▲ Ewing and Goodwin (1992)
□, ■ Estrada-Alexanders and Trusler (1995)

Fig. 4.10. Percentage deviations $100\,\Delta w/w = 100\,(w_{\text{exp}}-w_{\text{calc}})/w_{\text{exp}}$ between highly accurate experimental results for the speed of sound in argon and values calculated from a preliminary equation by Tegeler et al. (1997/1999). The open symbols correspond to deviations between linearised data and a linearly fitted equation and the full symbols correspond to deviations between the original $w(T,p)$ data and the same equation, fitted nonlinearly directly to the $w(T,p)$ data. The grey lines indicate the experimental uncertainty of the data.

scribe the linearised data very well, but on this level of accuracy the necessary linearisation has distorted the experimental information. Fitted nonlinearly directly to the original data the same equation of state is not able to represent the data within their experimental uncertainty. This status could not be improved significantly by further loops of linear optimisation and nonlinear fit – the need for a nonlinear optimisation algorithm became pressing for the first time.

4.4.6.1 The Nonlinear Quality Criterion

Common optimisation algorithms use a linear quality criterion, the sum of squares which results from a linear fit of an equation which uses the current combination of terms, to come to decisions in different steps of the procedure. OPTIM, for example, uses linear sums of squares during the initialisation (step O1 in Fig. 4.7) to determine the quality of random selected combinations of terms, in the step mutation (O2), and to select the worst formulations in the population (step O6), where only previously determined sums of squares are needed in step O6. The modified stepwise regression analysis which is incorporated in OPTIM uses the linear quality criterion to select the most efficient term from the bank of terms (step R2 in Fig. 4.8), to determine the quality of mutants (step R3), to make decisions regarding single and pairwise exchanges of terms (steps R4c, R6), and to determine the significance of single terms (steps R4, R4b, R4e, R7a) and of the formulation as a whole (step R5). The results of the Student t and Fisher F tests which are used for some of these decisions are advanced statistical criteria which are finally based on the linear quality criterion.

A nonlinear optimisation algorithm needs to use a nonlinear quality criterion, the sum of squares which results from a direct nonlinear fit of an equation which uses the current combination of terms, for all these decisions. However, to replace the linear quality criterion by a nonlinear one has been regarded as almost impossible due to three reasons:

- Iterative nonlinear fits require much more computing time than linear ones, especially when compared to linear fits which are based on a given regression matrix since the required system of linear normal equations (see Sect. 4.1) can be derived from the regression matrix without time-consuming consideration of the original data.
- Nonlinear fits need a good starting solution for the coefficient vector **N** to guarantee safe convergence.
- The methods which are used in the linear optimisation process to predict sums of squares without extensive matrix operations (see Eq. 4.97, Table 4.4, and Table 4.6) cannot be adapted to a nonlinear quality criterion.

Considering the recent development of available computing power, the computing time argument became less compelling at least for a direct comparison between linear and nonlinear fits. Nevertheless, the starting value problem still had to be solved and to replace the fast procedures which are used to predict sums of squares by nonlinear fits completely is still unrealistic today.

For the steps "initialisation" (O1), "mutation" (O2, R3), and "elimination of a term" (R4b, R4e, R7a) the linear quality criterion can be replaced by the nonlinear one directly. The number of nonlinear fits which is required for these procedures is proportional to the number of initial random selections (step O1), to the number of mutations (steps O2, R3), or to the number of terms in the current formulation (steps R4b, R4e, R7a). All of these parameters can be chosen to be less than about 50 and a corresponding number of nonlinear fits is realistic when using recent computers.

To solve the problem of appropriate starting values, a regression matrix which is based on explicit linear data and precorrelated implicit linear and implicit nonlinear data has to be supplied to the optimisation procedure. Based on this regression matrix, starting values for the parameter vector **N** can be determined by linear fits when testing results of initial random selections (step O1). However, the preliminary equation which was used to linearise the data in the initial regression matrix would have a strong influence on the following optimisation steps, if this matrix is used unchanged for the next process. Thus it is important to repeat the linearisation with the current formulation and to rebuild the regression matrix (see Sect. 4.4.6.4) whenever the current formulation is changed during the optimisation process.

The step "mutation" (O2, R3) requires a nonlinear fit for each of the NM tested mutants and the elimination of terms (steps R4b, R4e, R7a) requires $(N-1)$ nonlinear fits for formulations which are truncated by one of the N terms in the formulation, except for the one added last.[8] The required starting values are again determined by linear fits based on the current regression matrix.

The required computing time for direct nonlinear fits becomes insupportable for the step "addition of a term" which requires $I - N$ nonlinear fits to identify the term which yields the best result, where the number of terms in the bank of terms, I, can

---

[8] The elimination procedure is changed significantly in the nonlinear algorithm, see Sect. 4.4.6.2.

be as large as 1000. For the steps "exchange of terms" and "pairwise exchange of terms", the number of possibilities which have to be tested is far larger (see Eq. 4.105 and Eq. 4.106, respectively). To avoid unrealistically large numbers of nonlinear fits, the principle of *linear preselection* is applied:
- Implicit linear and implicit nonlinear data are precorrelated based on the current formulation.
- A linear regression matrix is set up based on the linear and linearised data. All I terms and if necessary C constraints are considered.

These steps are applied in any case after each change of the current formulation in order to be able to calculate an appropriate starting solution for the coefficient vector $\mathbf{N}$, see above.
- The regression matrix is transformed N times according to the prescriptions given in Table 4.3 to add those terms which are in the current formulation.
- The linear quality criterion is determined in the same way as explained for the corresponding step in Sect. 4.4.3 with the only difference that the result is stored for each addition or exchange.
- The possible modifications are arranged in order of increasing linear sum of squares. Since it would require too much memory to store all results for possible exchanges or pairwise exchanges of terms when working with large banks of terms and long equations (see Table 4.5) it is necessary to combine this step with the preceding one.
- Based on this linear ranking list, the results of the NA, NE, or NPE best modifications are examined by nonlinear fits. NA, NE, and NPE are control parameters which can be chosen independently for the steps "addition of a term", "exchange of terms", and "pairwise exchange of terms". This procedure is based on the assumption that the modification which results in the best nonlinear sum of square does not need to be the one with the best linear sum of squares, but that it will be among the ones with the best linear sums of squares with high probability.
- The modification which results in the smallest nonlinear sum of squares is carried out if it improves the current formulation.

If NA, NE, and NPE are chosen large enough, this procedure will find the solution which yields the best nonlinear sum of squares with high probability without the need to carry out an insupportably large number of nonlinear fits.

The effect of the linear preselection is illustrated in Fig. 4.11 which shows results from the development of a gas phase equation for argon (see Tegeler et al., 1997). The tested step is an exchange of terms for a current formulation with 13 terms and a bank of terms with 400 terms. Thus, 5031 possible modifications (4644 if one of the terms in the formulation has been added in the previous step) have to be considered in the linear ranking. The modification which yields the smallest linear sum of squares (position 1 in Fig. 4.11) does not improve the nonlinear sum of squares at all. The second modification improves the nonlinear sum of squares, but the modification with the best nonlinear sum of squares is found on position 23 of the linear ranking list. With increasing rank, the scatter of the resulting nonlinear sums of squares increases, but at the same time the average sum of square increases too – the probability that a better solution can be found some-

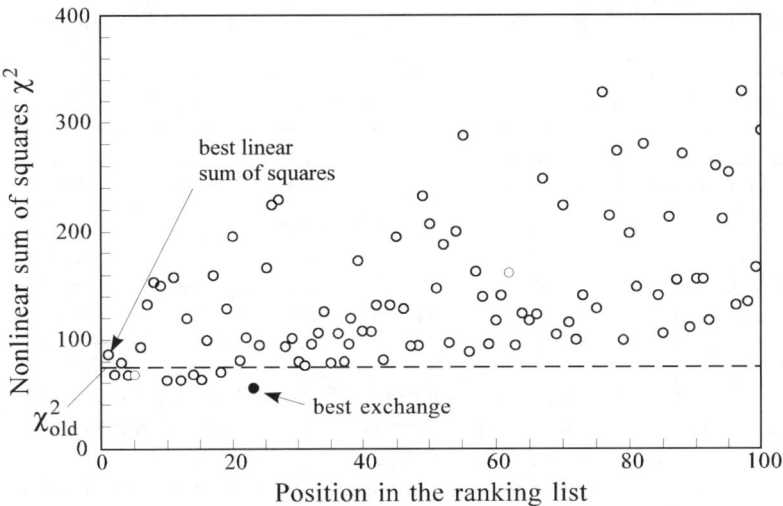

Fig. 4.11. Nonlinear sums of squares which result from exchanges of terms which were placed at the top of a ranking list established by *linear preselection*.

where beyond the 100th position of the ranking list is obviously very small. This assessment has been verified by extensive tests with larger values of *NE* even for problems like this, where the influence of implicit nonlinear data (speeds of sound) on the sum of squares is exceptionally strong. Table 4.8 summarises recommendations for the control parameters *NA*, *NE*, and *NPE* as given by Tegeler et al. (1997).

4.4.6.2 The Nonlinear Stepwise Regression Analysis (NLREG)

Based on the procedures described above, the modified stepwise regression analysis (see Sect. 4.4.5) can be transformed into a quasi nonlinear optimisation algorithm.[9] The only steps which have not been adapted to the nonlinear quality criterion so far are the Student *t* test and the Fisher *F* test which are used to test the significance of single coefficients and of the formulation as a whole in the stepwise regression analysis. In fact, it turned out that these tests are unsuitable for a nonlinear optimisation algorithm. To evaluate the *t* test, the standard deviation of the considered coefficient is required (see Eq. 4.98) which cannot be determined from the regression matrix in the nonlinear case. And without information on the results of the *t* tests, it would be necessary to select the term with the smallest significance by nonlinearly fitting *N* formulations which are truncated by one of the terms in the current formulation each to determine the variance σ^2_{eq-1} (see Eq. 4.102). This

---

[9] In scientific literature the term "nonlinear regression" has been used before repeatedly (see e.g. Kamei et al., 1995; Tillner-Roth, 1996). However, it refers to procedures which simply incorporate nonlinear fits and not to nonlinear optimisation algorithms in these cases.

Table 4.8. Recommended values for the control parameters NA, NE, and NPE of the linear preselection

Characterisation of the problem		I	II	III
Maximum number of terms in an equation	N_{max}	≤ 10	≤ 20	≤ 40
Number of terms in the bank of terms	I	≤ 200	≤ 400	≤ 1000
Recommended number of nonlinear fits				
Addition of a term	NA	50	80	100
Exchange of single terms	NE	80	100	200
Pairwise exchange of terms	NPE	150	200	400

numerical expense was considered as being unjustified and both the t test and the F test were replaced by a simpler procedure. A flow diagram of the nonlinear stepwise regression analysis is given in Fig. 4.12.

Like the modified stepwise regression analysis (see Fig. 4.8) the nonlinear process starts with the addition of the preselected terms (step R1 in Fig. 4.12). Based on the supplied regression matrix, a linear fit is carried out to determine starting values for the coefficient vector **N** of the preselected formulation and the nonlinear quality criterion is determined by a subsequent nonlinear fit. The regression matrix is updated using the resulting formulation for the required precorrelations. If the number of terms in the current formulation exceeds a certain limit (usually $N \geq N_{max} - NC$) the constraints are added to the regression matrix in step R2a for the first time. Wherever the regression matrix is updated in one of the following steps or wherever nonlinear fits are carried out, the constraints are considered, too.

In step R3 the nonlinear stepwise regression analysis continues with the addition of the most efficient term from the bank of terms. To select this term the method of linear preselection (see Sect. 4.4.6.1) is used with nonlinear fits for the NA combinations of terms which result in the best linear sums of squares. After the most efficient term has been added, the significance of the achieved improvement is tested by a simple variance test (step R4). Therefore, the criterion

$$-\frac{v_{new}}{v_{old}} = 1 - \frac{\chi^2_{new}(M-N+1)}{\chi^2_{old}(M-N)} \geq s_{min}, \quad (4.109)$$

with s_{min} as the minimum improvement of the variance which is expected from a formulation with one additional term, M the number of data points used in the nonlinear fit, N the number of terms in the current formulation, χ^2_{new} the nonlinear sum of squares of the current combination of terms, and χ^2_{old} the nonlinear sum of squares of the formulation truncated by the term added last, is tested. If Eq. 4.109 is fulfilled, the procedure continues with the step mutation (R5) as described in Sect. 4.4.5 with the modifications explained in Sect. 4.4.5.1. If Eq. 4.109 is not fulfilled the significance of all terms in the formulation except for the one added last is tested by nonlinear fits of $N-1$ formulations which are truncated by one

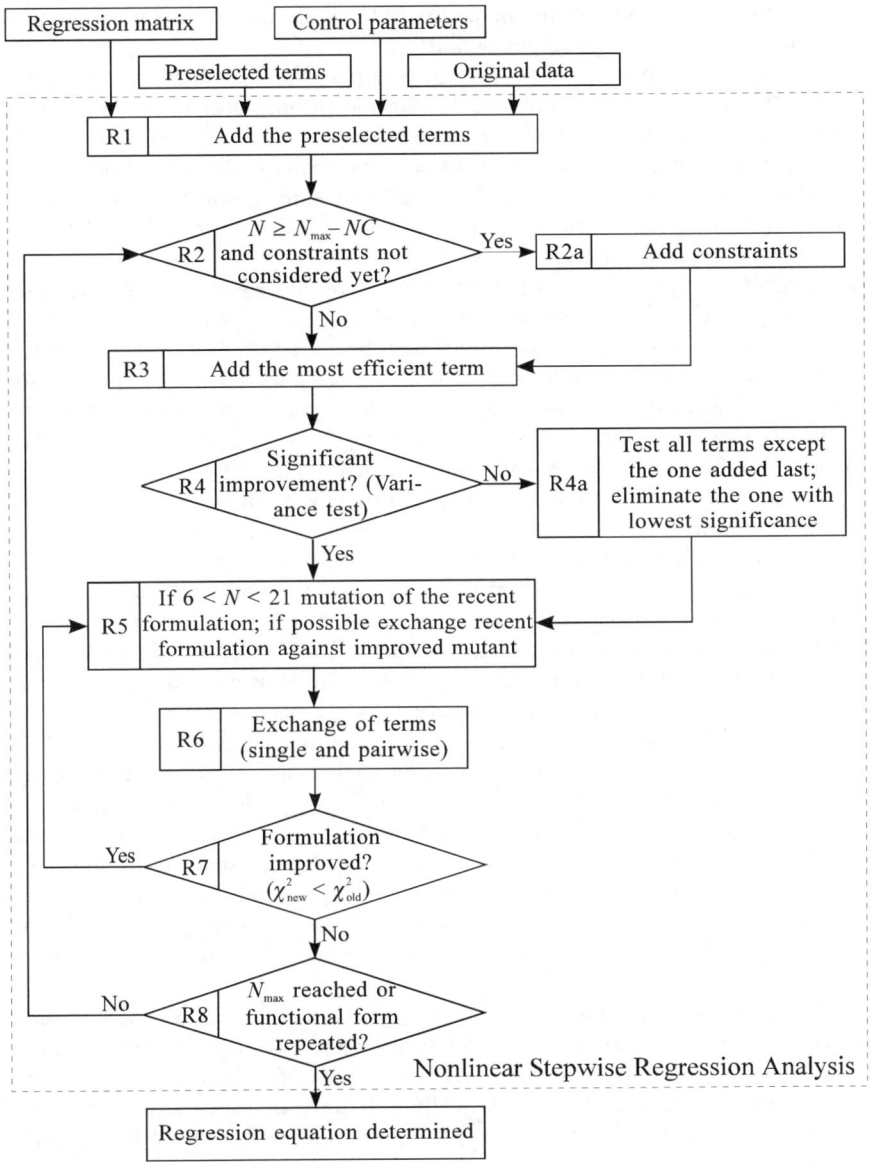

Fig. 4.12. Flow diagram of the nonlinear stepwise regression analysis developed by Tegeler et al. (1997).

term each. The term with the lowest significance is eliminated from the formulation and the procedure continues with the step mutation (R5). The quality of mutants is tested by nonlinear fits with starting solutions determined by preceding linear fits in step R5. If one of the mutants results in an improved sum of squares, the current formulation is exchanged against the mutant.

To test possible improvements by single and pairwise exchanges of terms, the method of linear preselection is used with subsequent nonlinear fits for the NE or NPE combinations of terms which yield the smallest linear sums of squares (step R7). A control parameter is used to decide whether pairwise exchanges are tested or not. If the exchange with the smallest nonlinear sum of squares improves the current formulation, the exchange is carried out and the current formulation is replaced by the modified one.

If the current formulation was improved either by mutation or by an exchange of terms (step R7), the procedure steps back to R5. If not, it continues with step R8 which determines whether a further term should be added or not. The variance test in step R4 can only assess whether the term which was added last has improved the formulation significantly; it cannot test the significance of the formulation as a whole directly since the term which was added last is not necessarily the one with the lowest significance. Thus, a single variance test yields no secure criterion for the determination of the best length of the formulation. However, if the current formulation fails in step R4, the term with the lowest significance (excluding the term which was added last) is eliminated from the current formulation in step R4a. Steps R5 and R6 try to improve the truncated formulation both by mutation and by exchange of terms. If these steps are not successful and if the term which was eliminated last is added again in step R3, the formulation which is tested in step R4 is identical with the previous one and it will fail again. However, this time another term will be removed, since the one added last is always excluded from the variance test. If steps R5 and R6 fail again and if the term eliminated last is added in step R3 again, it is obvious that the regression analysis cannot find a formulation of the current length ($N+1$ terms) which is significant in terms of Eq. 4.109 – the procedure would run into an endless cyclic exchange of two terms. In this case, identical functional forms are found in step R8 repeatedly and the nonlinear regression analysis stops without consideration of the second criterion ($N = N_{max}$). The best truncated formulation with N terms is selected as final regression formulation. In this way, the nonlinear regression analysis is able to optimise the length of correlation equations as well.[10]

Unlike the original stepwise regression analysis (see Sect. 4.4.3), the nonlinear regression analysis depends on preselected terms (step R1) since the corresponding formulation is used to reconstruct the regression matrix for the linear preselection of the first term to be added. As a basically deterministic procedure with only a single non-deterministic element it is not flexible enough to optimise highly accurate equations of state which are based on large banks of terms starting from scratch. However, the nonlinear regression analysis is a very powerful tool to finalise highly accurate equations of state which were developed with OPTIM up to a state where the repeated cycle out of linear optimisation and nonlinear fit cannot improve the equation any further.

---

[10] N_{max} should not be much larger than the optimum length of the formulation; otherwise the way in which constraints are considered (R2) becomes questionable.

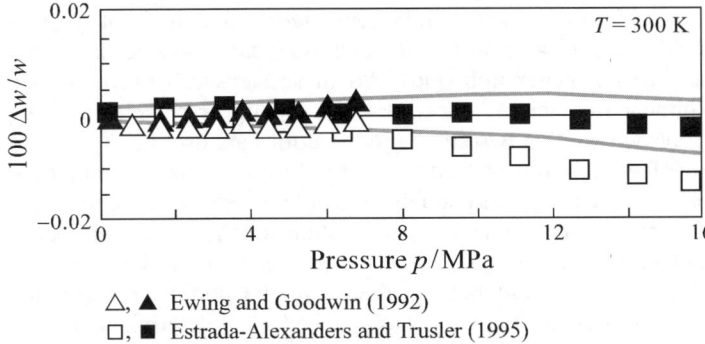

△, ▲ Ewing and Goodwin (1992)
□, ■ Estrada-Alexanders and Trusler (1995)

Fig. 4.13. Percentage deviations $100\,\Delta w/w = 100\,(w_{exp} - w_{calc})/w_{exp}$ between highly accurate experimental results for the speed of sound in argon and values calculated from the best preliminary equation by Tegeler et al. (1997/1999) which resulted from linear optimisation algorithms (open symbols) and from the final equation of Tegeler et al. (filled symbols). The grey lines indicates the experimental uncertainty of the data.

In this sense, the nonlinear regression analysis was used to finalise both the recent reference equations for argon (Tegeler et al., 1997/1999) and nitrogen (Span et al., 1998b). Figure 4.13 illustrates the success of this approach using the speed of sound data for argon which were already shown in Fig. 4.10 as an example. This time the open symbols correspond to deviations between the original $w(T,p)$ data and the best preliminary equation which could be developed by using linear optimisation algorithms in combination with a nonlinear fitting routine. The filled symbols correspond to deviations between the data and the final equation of Tegeler et al. (1997/1999). Only the use of nonlinear optimisation algorithms enabled a representation of these highly accurate speeds of sound within their experimental uncertainty. Further advantages of the nonlinear optimisation algorithm result from the use of experimental phase equilibrium data directly (see Sect. 4.3.3, Eq. 4.72, Eqs. 4.78–4.83) without precorrelation. For well measured substances, the direct consideration of properties at phase equilibrium improved their representation mainly close to the critical point. For other substances, information on the saturated vapour density is very often scarce. In this case the nonlinear stepwise regression analysis avoids the need to use a questionable ancillary equation for the saturated vapour density when precorrelating the linearised residua (see Sect. 4.3.2, Eqs. 4.56–4.58).

4.4.6.3 The Nonlinear Optimisation Algorithm (NLOPT)

The nonlinear regression analysis which was described in the preceding section can be used as an integral part of a nonlinear version (NLOPT) of the modified form of the optimisation algorithm of Setzmann and Wagner (1989a). Basically, there is no need to change the set-up of OPTIM as it was explained in Sect. 4.4.5.

Apart from the steps within the nonlinear regression analysis, the nonlinear quality criterion is needed to select the best randomly generated formulations of the first parental generation (O1 in Fig. 4.7), to test mutants (O3) and to select the worst formulations of the parental generation (O6). All these steps can be performed based on direct nonlinear fits (see also Sect. 4.4.6.1), no linear preselections are needed. However, it turned out that two small modifications are useful.

When using NLOPT without preselected terms, the formulations of the first parental generation are usually not good enough to enable a reasonable reconstruction of the regression matrix. Thus, the starting solution for the regression matrix should not be altered in the step O1 and O2 of the first generation. For the same reason the regression matrix should not be altered in the nonlinear regression analysis (O5) until the current formulation has reached a predetermined length. With regard to the selection of the most effective terms (O4) it transpired that it is advantageous to use all terms which are contained *at least* in *NR*+1 formulations, *NR* formulations, and so on down to 2 formulations instead of those terms which are contained *exactly* in the corresponding number of formulations as starting solutions.

While NLREG can be used for reference equations of state on common workstations without insupportable computing times,[11] NLOPT with its repeated regression runs still places too high demands on the available computing power for such optimisation tasks. Tegeler et al. (1997) used NLOPT successfully to develop a highly accurate equation of state for argon which covers the density range up to $0.5\,\rho_c$ including parts of the extended critical region. This problem required a selection of 19 terms out of a bank of terms with 400 elements (category II in Table 4.8) and was extremely influenced by highly accurate nonlinear data (see Fig. 4.11 which illustrates one step of this optimisation). The routine use of NLOPT for problems of category III remains for future generations of workstations or users who have privileged access to super computers.

4.4.6.4 Speeding Up Nonlinear Optimisation Algorithms

When working with NLREG or NLOPT, the high demands on computing power result mainly from two characteristic steps:
- The repeated set-up of the regression matrix **B** which is based on the original explicit linear data and on linearised data (see Sect. 4.4.2). The regression matrix is needed for the linear steps of the algorithms.
- Nonlinear fits which require vectors and matrices which have to be calculated from the original data.

These steps require a great deal of computing power since they involve summations of all used data for each matrix element.

The contribution of implicit linear and nonlinear data to the required matrices and vectors depends on the precorrelation of the data or on the current parameter

---

[11] For problems of category III (see Table 4.8) computing times of about 5 hours are common for NLREG when using the hardware supported quad precision mode (128 bit representation of matrix elements) of an IBM RS 6000 / 260 workstation.

4.4 Optimising the Functional Form

vectors in an unresolvable way. However, to avoid unnecessary summations, the contribution of explicit linear data can be extracted from an initial regression matrix. To do so, it is necessary to rewrite Eqs. 4.10, 4.11 and 4.93 as

$$a_{ij} = \sum_{m=1}^{M_{impl}} \frac{1}{\sigma_m^2} \cdot \left(\frac{\partial \zeta}{\partial n_i}\right)_{x,y,n_j \neq n_i} \cdot \left(\frac{\partial \zeta}{\partial n_j}\right)_{x,y,n_i \neq n_j} + \sum_{m=M_{impl}+1}^{M_{impl}+M_{expl}} \left(\frac{a_i(\mathbf{x}_m)\, a_j(\mathbf{x}_m)}{\sigma_m^2}\right)$$

$$= \sum_{m=1}^{M_{impl}} \frac{1}{\sigma_m^2} \cdot \left(\frac{\partial \zeta}{\partial n_i}\right)_{x,y,n_j \neq n_i} \cdot \left(\frac{\partial \zeta}{\partial n_j}\right)_{x,y,n_i \neq n_j} + a_{ij,\mathrm{expl}}, \qquad (4.110)$$

$$q_i = \sum_{m=1}^{M_{impl}} \left(\left(\frac{\partial \zeta}{\partial n_i}\right)_{x,y,n_j \neq n_i} \cdot \frac{a_0(\mathbf{x}_m, y_m)}{\sigma_m^2}\right) + \sum_{m=M_{impl}+1}^{M_{impl}+M_{expl}} \left(\frac{a_i(\mathbf{x}_m)\, a_0(\mathbf{x}_m, y_m)}{\sigma_m^2}\right)$$

$$= \sum_{m=1}^{M_{impl}} \left(\left(\frac{\partial \zeta}{\partial n_i}\right)_{x,y,n_j \neq n_i} \cdot \frac{a_0(\mathbf{x}_m, y_m)}{\sigma_m^2}\right) + q_{i,\mathrm{expl}}, \qquad (4.111)$$

and

$$S = \sum_{m=1}^{M_{impl}} \left(\frac{a_0(\mathbf{x}_m, y_m)^2}{\sigma_m^2}\right) + \sum_{m=M_{impl}+1}^{M_{impl}+M_{expl}} \left(\frac{a_0(\mathbf{x}_m, y_m)^2}{\sigma_m^2}\right)$$

$$= \sum_{m=1}^{M_{impl}} \left(\frac{a_0(\mathbf{x}_m, y_m)^2}{\sigma_m^2}\right) + S_{\mathrm{expl}}, \qquad (4.112)$$

respectively. For implicit data, the more general notation is necessary since nonlinear relations between ζ and $\mathbf{N}$ cannot be resolved in the simple way which is used for explicit linear data, see Sect. 4.3.3. The elements of the matrix $\mathbf{C}$ and of the vector $\mathbf{Q}_c$ (see Eqs. 4.17–4.18) depend neither on data nor on preliminary equations of state and do not therefore change during the optimisation process. Thus, the information on explicit linear data and on the constraints can be supplied to nonlinear optimisation procedures as

$$\mathbf{B}_{\mathrm{expl}} = \begin{bmatrix} \mathbf{A}_{\mathrm{expl}} & \mathbf{C} & \mathbf{Q}_{\mathrm{expl}} \\ \mathbf{C}^T & \mathbf{0} & \mathbf{Q}_c \\ \mathbf{Q}_{\mathrm{expl}}^T & \mathbf{Q}_c^T & S_{\mathrm{expl}} \end{bmatrix} \qquad (4.113)$$

or just as one part of the symmetric matrix $\mathbf{B}_{\mathrm{expl}}$. To determine the current regression matrix $\mathbf{B}$ the contribution of the linearised implicit data has to be added to the elements of $\mathbf{B}_{\mathrm{expl}}$ by carrying out the summations according to Eqs. 4.110–4.112.

Algorithms for the solution of non-linear systems of normal equations (see Fletcher, 1971) need a matrix $\mathbf{E}$ which contains the derivatives of all equations with respect to all unknown parameters ($\mathbf{N}$ and λ_c), a vector $\mathbf{F}$ which holds the current values of the functions which have to be solved and the sum of squares χ^2 which results from the current values of $\mathbf{N}$ and λ_c. All these elements depend on $\mathbf{N}$ and λ_c and have to be updated in each step of the iterative procedure.

The nonlinear fits in NLREG and NLOPT involve only those terms from the bank of terms which are part of the current formulation. The contribution of explicit data and constraints can be calculated from the matrix

$$\mathbf{B}^*_{expl} = \begin{bmatrix} \mathbf{A}^*_{expl} & \mathbf{C}^* & \mathbf{Q}^*_{expl} \\ \mathbf{C}^{*T} & 0 & \mathbf{Q}_c \\ \mathbf{Q}^{*T}_{expl} & \mathbf{Q}^T_c & S_{expl} \end{bmatrix} \qquad (4.114)$$

where $\mathbf{B}^*_{expl}$, $\mathbf{A}^*_{expl}$, $\mathbf{C}^*$, and $\mathbf{Q}^*_{expl}$ hold only contributions from terms in the current formulation; $\mathbf{B}^*_{expl}$ can be extracted from $\mathbf{B}_{expl}$ directly (see for instance Eq. 4.107).

Based on $\mathbf{B}^*_{expl}$ the contribution of explicit linear data to $\mathbf{E}$, $\mathbf{F}$, and χ^2 can be calculated according to

$$\mathbf{E} = \begin{bmatrix} \mathbf{A}^*_{expl} & \mathbf{C}_{expl} \\ \mathbf{C}^T_{expl} & 0 \end{bmatrix}, \qquad (4.115)$$

$$\mathbf{F} = \begin{bmatrix} \mathbf{A}^*_{expl} & \mathbf{C}_{expl} \\ \mathbf{C}^T_{expl} & 0 \end{bmatrix} \begin{bmatrix} \mathbf{N}^* \\ \lambda_c \end{bmatrix} - \begin{bmatrix} \mathbf{Q}^*_{expl} \\ \mathbf{Q}_c \end{bmatrix}, \qquad (4.116)$$

and $\qquad \chi^2 = S_{expl} - 2\mathbf{N}^T \mathbf{Q}^*_{expl} + \mathbf{N}^T \mathbf{A}^*_{expl} \mathbf{N} + \mathbf{N}^T \mathbf{C}^* \lambda_c - \mathbf{Q}^T_c \lambda_c, \qquad (4.117)$

respectively. Only the information on implicit linear and nonlinear data has to be added by evaluation of the corresponding summations. In this way the numerical expense for nonlinear fits can be reduced significantly as well.

4.4.7 Automated Optimisation Algorithms

Basically, optimisation algorithms are designed for use by experienced experts only. Simple fits are common features of thermodynamic property packages in process simulators today, but optimisation algorithms are numerically both more flexible and more sensitive. When using optimisation algorithms, inexperienced users are likely to establish unreasonable correlations for one of the following reasons:

- Unreasonable set-ups of the bank of terms (see Sect. 4.4.1) result in unphysical behaviour mostly outside of the region which is covered by accurate data, e.g., for the limit of the ideal gas at vanishing pressures.
- Inadequate weighting of experimental data or undetected inconsistencies in the data set result in an overfitting of certain properties in certain regions while derived properties or properties in other regions are not represented adequately.
- Repeated mathematical operations with large, very often ill conditioned matrices result in numerical problems which have to be detected and avoided by the user.

Thus, it is unlikely that multipurpose optimisation algorithms like those described in the preceding sections will become common tools for broad groups of users in the foreseeable future.

However, for certain tasks, optimisation algorithms can be applied by inexperienced users, too. In view of the problems mentioned above, the corresponding tasks have to be well defined, allowing experts to define an appropriate bank of terms in advance. They have to use data sets which cover a predetermined range of validity with known accuracy and without significant inconsistencies, and which guarantee an appropriate data distribution. And finally the algorithms have to be able to detect and to avoid numerical problems automatically. Until now only a single example for such *"automated" optimisation algorithms* has been published, but the underlying problem is quite typical for a variety of engineering applications and further specialised algorithms may follow.

4.4.7.1 Optimisation of Simplified Equations for Mixtures with Constant Composition

The so called AGA8-DC92 equation of state by Starling and Savidge (1992) is an internationally accepted standard for the calculation of natural gas properties in the gas phase (see also Sect. 8.1.1) which allows calculations for temperatures from 143 K to 673 K and at pressures up to 280 MPa. The set-up of this equation is rather complex due to the fact that broad composition ranges have to be considered for a total of 21 typical natural gas components. For common engineering applications this complex set-up is needed only infrequently since natural gases do not change their composition, e.g., during pipelining processes. Thus, much simpler equations could be used to describe natural gases as pseudo pure fluids, but these equations would be specific for a certain natural gas composition.[12]

To establish such simplified equations, tests were made to determine whether empirical correlations for the compression factor can be fitted to data which were calculated from the AGA8-DC92 equation. For sufficiently broad ranges of validity, simple correlations with given functional form could never represent compression factors calculated from the AGA8-DC92 equation better than within $\Delta Z/Z \leq \pm 0.1\% - \pm 0.2\%$. Since the uncertainty of the AGA8-DC92 equation is estimated to be $\Delta Z/Z \leq \pm 0.1\%$ in the technically most important regions, the additional uncertainty of the simplified correlation affected the overall accuracy of calculated properties significantly.

To overcome this problem, Span et al. (1995b) established an automated optimisation procedure which uses a modified form of the algorithm proposed by Setzmann and Wagner (1989a) to establish accurate simplified equations in form of the compression factor. Characteristic features of this procedure are:
- *Well defined ranges of validity.* Four typical ranges of validity (region 1−4) were defined for the simplified equation which reach from typical pipelining needs to the needs of possible future high pressure storage applications.

---

[12] Usually the composition of a natural gas obtained from a certain source changes only very slowly and simplified equations established once remain valid for months.

- *Problem specific data grids.* For each of the agreed regions, a suitable temperature/pressure grid was established in preceding tests. For the corresponding grid, data are calculated from the AGA8-DC92 equation when establishing a simplified equation for a certain composition. The calculated data are weighted automatically, using the accuracy which is demanded from the simplified equation instead of the uncertainty of the data.
- *Predefined banks of terms.* For each of the regions, an efficient bank of terms was determined in preceding tests with very different natural gas compositions. The used banks of terms could be restricted to pure polynomial terms and to exponential terms with $\exp(-\delta^2)$, see Sect. 4.4.1. In this way, small banks of terms and short computing times could be realised.
- *Predefined starting solutions.* As a result of the preceding tests an equation was found for each of the four regions which yielded good results when fitted to data for typical natural gas compositions ($\Delta Z/Z \leq \pm 0.1\% - \pm 0.2\%$, see above). If these equations are used as starting solutions, the optimisation algorithm finds equations with very accurate functional forms both quickly and reliably.
- *Automated detection of numerical problems.* Numerical problems which may result from rounding errors during the optimisation process are detected automatically by a comparison of the sum of squares calculated from the regression matrix with the one calculated from a direct comparison of the resulting equation with the used data. If these sums of squares do not agree with each other, rounding errors have affected the optimisation process. In this case, the optimisation procedure is repeated with a reduced number of terms in the equation and, if the second attempt fails as well, the data are scattered randomly to simulate a small experimental random error (see the conditions for using the maximum likelihood method in the way it is used, Sect. 4.1 – violations of these conditions are likely to result in ill conditioned matrices). However, it turned out that these precautions are necessary only in very rare situations.

Table 4.9 summarises characteristic parameters for the four regions of validity; allowable gas compositions are defined by the corresponding limits of the AGA8-DC92 equation. Within these ranges of validity, the automated optimisation algorithm establishes simplified equations of state which deviate from the AGA8-

Table 4.9. Definition of the four regions of validity considered by Span et al. (1995b) and characteristic parameters of the corresponding optimisation problems.

Region	Temperature range	Pressure limit	No. of data calc. from AGA8-DC92	Bank of terms polynomial terms	Bank of terms exponential terms	Max. No. of terms in the equation
1	263 K – 338 K	12 MPa	330	65	–	6
2	268 K – 353 K	30 MPa	420	35	63	9
3	273 K – 423 K	60 MPa	576	35	63	13
4	273 K – 473 K	120 MPa	810	35	63	14

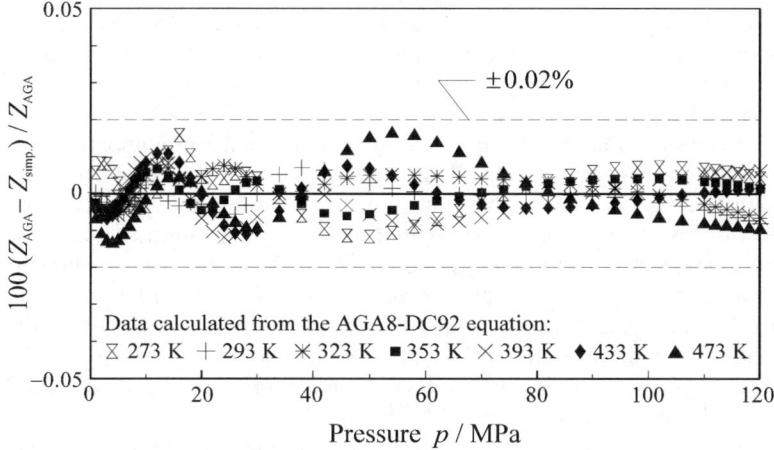

Fig. 4.14. Percentage deviations between compression factors calculated from the AGA8-DC92 equation and from an optimised simplified equation which is valid for region four of a typical natural gas.

DC92 equation by $\Delta Z/Z \leq \pm 0.05\,\%$; for typical natural gas compositions deviations of $\Delta Z/Z \leq \pm 0.02\,\%$ can be realised in general. Figure 4.14 shows deviations between values calculated from the AGA8-DC92 equation and a simplified equation which describes region four of a typical natural gas using the functional form

$$Z(T,\rho) = \frac{p}{\rho RT} = 1 + \sum_{i=1}^{10} n_i \delta^{d_i} \tau^{t_i} + \sum_{i=11}^{14} n_i \delta^{d_i} \tau^{t_i} e^{-\delta^2} \; ; \qquad (4.118)$$

for the values of n_i, d_i, and t_i and the composition of the described natural gas, see Span et al. (1995b).

The optimisation algorithm described above is integrated in a program which contains different direct calculation modes and a composition data base. The user only has to select a natural gas composition and a range of validity to start the optimisation process – all other process parameters are predetermined based on the selected region. Due to some numerical improvements, small banks of terms, and the use of good starting solutions, the necessary computing time could be reduced to less than one minute on recent personal computers even for the development of equations which describe region four. Under such circumstances optimisation algorithms can be applied even by completely inexperienced users.

4.4.8 Independent Developments and Future Perspectives

When reading Sects. 4.4.3 – 4.4.7 one may think of this description of optimisation procedures as unbalanced in some way. However, this impression is wrong – the development of optimisation algorithms for equations of state has been carried out by the group around W. Wagner almost exclusively during the last two decades.

Besides these developments only one completely independent optimisation algorithm for equations of state has been published, namely the "*simulated annealing*" algorithm by Kirkpatrick et al. (1983) and Cerný (1985) which was adapted to equations of state by Shubert and Ely (1995). Shubert and Ely compare equations of state for the refrigerant R134a which were developed with their simulated annealing algorithm and with the simple stepwise regression analysis (see Sect. 4.4.3). Based on this comparison the simulated annealing algorithm seems to be slightly inferior even to the stepwise regression analysis. More precise conclusions cannot be drawn yet, since the algorithm of Shubert and Ely (1995) has not been published in sufficient detail to carry out additional comparisons. Nevertheless, the basic idea of simulated annealing, which is a non-deterministic algorithm like the EOM (see Sect. 4.4.4), is promising and continued work on this topic could stimulate the whole development of optimisation algorithms.

A possibly trend-setting result was reported by Lemmon and Jacobsen (1998/1999b). Lemmon and Jacobsen included the temperature and density exponents of equations of state which were optimised using the stepwise regression analysis into direct nonlinear fits. When using the results of the optimisation algorithm as a starting solution, the nonlinear fit could improve the equations further in this way and the least significant terms in the equations could be eliminated without loss of quality. In terms of optimisation nomenclature this method could be referred to as a "*manual backward regression*". When using recent optimisation algorithms, the advantage of fitted temperature exponents is anticipated partly since these algorithms allow much narrower graduations of the temperature exponents (see Sect. 4.4.1). However, even a very large bank of terms cannot mimic the results of a direct fit completely. Thus, it would be very interesting to test whether this kind of backward regression in combination with a direct nonlinear fit of temperature exponents could be included, e.g., into the nonlinear optimisation algorithm of Tegeler et al. (1997), see Sect. 4.4.6.

Another major challenge with regard to optimisation algorithms is the development of a completely nonlinear algorithm. The algorithm of Tegeler et al. uses the nonlinear quality criterion for all relevant decisions and results in the combination of terms which yields the smallest nonlinear sum of squares. However, both NLREG and NLOPT still depend on linear preselections and cannot be used for problems where important data are in a key position which cannot be linearised – in this sense they are only *quasi nonlinear algorithms*. This restriction becomes a problem, e.g., when using inequalities as constraints (see Eq. 3.109) which can only be considered in nonlinear fits. Work on highly accurate equations of state which avoid unreasonable plots in the instable part of the vapour-liquid two phase region (see Sect. 3.3.5.1) is one of the ambitious topics still pending and thus, the need for a completely nonlinear optimisation algorithm may soon become urgent.

Optimisation algorithms for the development of substance specific functional forms for equations of state will certainly be a matter of ongoing research and the idea of simultaneous optimisation (see Sect. 6.1) has only been recently formulated.

4.5 Describing Properties in the Critical Region

Like all critical points, the gas-liquid critical point of a pure fluid is characterised by the fact that stable equilibria become indifferent. At the critical point, the critical isotherm, the saturated vapour curve and the saturated liquid curve join each other. The isothermal compressibility of the fluid becomes infinite; other properties stay finite, become infinite, or become infinitely small, where the diverging compressibility explains some but not all of these anomalies. Due to gravity and diverging compressibility, systems with finite characteristic length become non-homogeneous when approaching the critical point; density stratifications become a characteristic feature of equilibrium systems. Increasing instability leads to strong responses on small perturbations of an equilibrium system and the critical enhancement of transport properties such as the thermal diffusivity causes long equilibration times. On a microscopic level indifferent equilibria result in properties which are determined not only by interactions between adjoining molecules but by interactions within large clusters of molecules. The characteristic length of these clusters, the correlation length, is still very small on macroscopic scales, but it diverges when asymptotically approaching the critical point.

The region where all these effects are essential for the properties of a fluid is called the *"critical region"*. Its limitations depend on subjective assessment to a certain degree; common values which can be found in literature are of the order $0.99 \leq T/T_c \leq 1.1$ and $0.7 \leq \rho/\rho_c \leq 1.3$. The *extended critical region* which is defined by these limits is useful for an assessment of simple equations of state, but especially with respect to temperature it is far too large to single out the region which is strongly influenced by critical effects. In this sense and for an assessment of advanced equations of state, the limits $0.998 \leq T/T_c \leq 1.01$ and $0.75 \leq \rho/\rho_c \leq 1.25$ are useful to characterise the *critical region*. For certain questions, *asymptotic regions* have to be considered which are even smaller.

The accurate description of thermodynamic properties in the critical region is an interesting challenge for theoreticians, for experimentalists, and last but not least for those who are engaged in the development of empirical equations of state. The theoretical treatment of critical effects resulted in an interesting edifice of thoughts with a high degree of internal consistency. The resulting models do not depend on the substance considered and are able to predict the observed critical effects qualitatively correctly, but an accurate description of properties still requires a considerable number of substance-specific fitted parameters. Experimental work on properties in the critical region has resulted in highly complex adaptations of basically known experimental techniques (see for instance Edwards, 1984; Wagner et al., 1992) as well as in experimental techniques which are applicable only under the conditions found in the critical region (see for instance Straub, 1967). And finally, sophisticated empirical equations of state have been developed both as a reaction to increasing technical interest in accurate data for properties in the extended critical region and to keep pace with experimental and theoretical developments.

In accordance with the orientation of the whole book, this section focuses on the development of empirical equations of state which describe properties in the critical region accurately. The theoretical background is described in Sect. 4.5.1 only as far as necessary to understand both the occurring problems and the semiempiric approaches which are discussed in Sect. 4.5.5. For a more detailed description of the underlying theory see Sengers and Levelt Sengers (1986), Albright et al. (1986), Luettmer-Strathmann et al. (1992), and Sengers (1994).

4.5.1 Predictions from Theory

The key idea of theories which describe critical behaviour is that of critical-point universality. Systems near the critical points are classified in terms of universality classes. Just like magnetic systems close to their critical point, fluids near ordinary critical points are assumed to belong to the universality class of three-dimensional *Ising-like systems*, i.e., systems with short-range forces and a scalar order parameter, e.g., the density of the fluid. Usually such systems are described by a combination of two independent variables, namely a variable h which measures the distance from the coexistence curve and a variable τ which measures the distance from the critical point along this curve. Unfortunately, h and τ are not identical with any known thermodynamic properties – they have to be determined recursively, e.g., from the temperature T and the chemical potential μ.

The renormalisation group theory introduced by Wilson (1974; for an introduction see Wilson, 1979/1986) extended the theoretical understanding of critical phenomena in Ising-like systems. With this theory, it could be shown that an approach which was originally introduced much earlier by Verschaffelt (1896) on a purely empirical basis is valid and that it allows a universal description of systems which belong to the same universality class: along certain paths throughout the critical region thermodynamic properties can be described by so called "*power*

Table 4.10. Examples for power laws describing thermodynamic properties along certain paths throughout the critical region

Power law	Described path	Crit. exp.	Values determined by evaluation of		
			RG theo.[a]	3-pt. eq.[b]	5-pt. eq.[c]
$(\rho' - \rho'') \sim (T_c - T)^\beta$	phase boundary	β	0.326 ± 0.002	0.5	0.25
$\kappa_T \sim (T - T_c)^{-\gamma}$	critical isochore	γ	1.239 ± 0.002	1	1
$\|p - p_c\| \sim \|\rho - \rho_c\|^\delta$	critical isotherm	δ	4.80 ± 0.02	3	5
$c_v \sim \|T - T_c\|^{-\alpha}$	critical isochore	α	0.110 ± 0.003	0	$0 / 0.5^d$

[a] according to Sengers and Levelt Sengers (1986)
[b] at the critical point $(\partial p / \partial \rho)_T = 0$ and $(\partial^2 p / \partial \rho^2)_T = 0$ with $(\partial^3 p / \partial \rho^3)_T > 0$ hold
[c] at the critical point $(\partial^3 p / \partial \rho^3)_T = 0$ and $(\partial^4 p / \partial \rho^4)_T = 0$ with $(\partial^5 p / \partial \rho^5)_T$ 0 hold additiononally to the conditions given in b
[d] $\alpha_2 = 0.5$ when approaching the critical point from the two phase region, $T - T_c < 0$

laws". The non integer exponents of these relations, the so called *"critical exponents"*, are universal within a universality class; only the proportionality factors, the so called *"critical amplitudes"*, depend on the considered system. Table 4.10 summarises the power laws which are most important for the description of thermodynamic properties in the critical region of pure fluids. The values for the corresponding critical exponents which are quoted as results of the renormalisation group theory were taken from Sengers and Levelt Sengers (1986); theoretical values given by other authors agree within the given uncertainties in general.

The evaluation of typical empirical equations of state yields different values for the critical exponents (for a summary, see Table 4.10). When assuming that an equation of state is analytic at the critical point, the pressure which results from this equation can be written as a Taylor expansion according to

$$p(T,\rho) - p_c = \left(\frac{\partial p}{\partial \rho}\right)_T \cdot (\rho - \rho_c) + \left(\frac{\partial p}{\partial T}\right)_\rho \cdot (T - T_c)$$

$$+ \frac{1}{2!} \cdot \left(\left(\frac{\partial^2 p}{\partial \rho^2}\right)_T \cdot (\rho - \rho_c)^2 + 2\left(\frac{\partial^2 p}{\partial \rho \partial T}\right) \cdot (\rho - \rho_c) \cdot (T - T_c) + \left(\frac{\partial^2 p}{\partial T^2}\right)_\rho \cdot (T - T_c)^2 \right)$$

$$+ \frac{1}{3!} \cdot \left(\left(\frac{\partial^3 p}{\partial \rho^3}\right)_T \cdot (\rho - \rho_c)^3 + \ldots \right), \tag{4.119}$$

where the derivatives have to be calculated at the critical point. This Taylor expansion is valid for simple pressure explicit equations such as cubic equations of state as well as for highly accurate empirical equations of state which are explicit in the Helmholtz energy. If the formulation for the Helmholtz energy is analytic, the resulting relation for the pressure (see Eq. 3.30) is analytic too and this is the only assumption behind Eq. 4.119.

For most analytic equations of state, the first two derivatives of pressure with respect to density become 0 at the critical point while the third one is larger than 0 (3-point contact, see also Table 4.10, footnote b); quite often accurate equations are constrained to meet these conditions and $p(T_c, \rho_c) = p_c$ exactly at T_c and ρ_c to reproduce selected critical data exactly. With these conditions, the power series which describes the shape of the critical isotherm $(T - T_c = 0)$ becomes

$$p(T_c, \rho) - p_c = \frac{1}{3!} \cdot \left(\frac{\partial^3 p}{\partial \rho^3}\right)_T \cdot (\rho - \rho_c)^3 + \ldots \tag{4.120}$$

with the leading exponent $\delta = 3$. Where equations of state are designed to realise a 5-point contact (see Table 4.10, footnote c) the corresponding power series becomes

$$p(T_c, \rho) - p_c = \frac{1}{5!} \cdot \left(\frac{\partial^5 p}{\partial \rho^5}\right)_T \cdot (\rho - \rho_c)^5 + \ldots \tag{4.121}$$

with $\delta = 5$. Along the critical isochore the isothermal compressibility becomes

$$\kappa_T(T,\rho_c) = \frac{1}{\rho_c} \left(\frac{\partial p}{\partial \rho}\right)_T^{-1} = \frac{1}{\rho_c} \left(\left(\frac{\partial^2 p}{\partial \rho \partial T}\right) \cdot (T-T_c) + \ldots\right)^{-1} \quad (4.122)$$

and the critical exponent which describes the temperature dependence of κ_T becomes $\gamma = 1$ both for equations with 3- and 5-point contact.

It is slightly more complicated to determine the values for the critical exponent β. With Eq. 4.119, the phase equilibrium condition

$$p(T,\rho') = p(T,\rho'') \quad (4.123)$$

can be rewritten as

$$\left(\frac{\partial^2 p}{\partial \rho \partial T}\right) \cdot (\rho' - \rho_c) \cdot (T-T_c) + \frac{1}{3!}\left(\frac{\partial^3 p}{\partial \rho^3}\right)_T \cdot (\rho' - \rho_c)^3 =$$
$$\left(\frac{\partial^2 p}{\partial \rho \partial T}\right) \cdot (\rho'' - \rho_c) \cdot (T-T_c) + \frac{1}{3!}\left(\frac{\partial^3 p}{\partial \rho^3}\right)_T \cdot (\rho'' - \rho_c)^3, \quad (4.124)$$

or simplified as

$$p_{11}(\rho' - \rho_c)(T-T_c) + \frac{p_{30}}{6}(\rho' - \rho_c)^3 = p_{11}(\rho'' - \rho_c)(T-T_c) + \frac{p_{30}}{6}(\rho'' - \rho_c)^3 \quad (4.125)$$

when considering the limit of vanishing distance to the critical point. The p_{ij} in Eqs. 4.125–4.134 indicate the i-th derivative of p with respect to ρ and the j-th derivative with respect to T. When inserting the power law

$$(\rho' - \rho'') = c_\beta \cdot (T_c - T)^\beta \quad (4.126)$$

as $\quad (\rho' - \rho_c) = c'_\beta \cdot (T_c - T)^\beta \quad \text{and} \quad (\rho'' - \rho_c) = -c''_\beta \cdot (T_c - T)^\beta, \quad (4.127)$

Eq. 4.125 yields

$$p_{11} c'_\beta (T-T_c)^{1+\beta} + \frac{p_{30}}{6} c'^3_\beta (T-T_c)^{3\beta} = -p_{11} c''_\beta (T-T_c)^{1+\beta} - \frac{p_{30}}{6} c''^3_\beta (T-T_c)^{3\beta}, \quad (4.128)$$

and thus $\quad p_{11}(c'_\beta + c''_\beta) + \frac{p_{30}}{6}(c'^3_\beta + c''^3_\beta)(T-T_c)^{2\beta-1} = 0. \quad (4.129)$

This equation can only be fulfilled independently from T if $\beta = 0.5$ holds. For an equation with 5-point contact, Eq. 4.125 becomes

$$p_{11}(\rho' - \rho_c)(T-T_c) + \frac{p_{50}}{120}(\rho' - \rho_c)^5 = p_{11}(\rho'' - \rho_c)(T-T_c) + \frac{p_{50}}{120}(\rho'' - \rho_c)^5, \quad (4.130)$$

and thus Eq. 4.129 becomes

$$p_{11}(c'_\beta + c''_\beta) + \frac{p_{30}}{120}(c'^5_\beta + c''^5_\beta)(T-T_c)^{4\beta-1} = 0, \quad (4.131)$$

resulting in $\beta = 0.25$.

To determine the critical exponent which results from the increase of the isochoric heat capacity in the homogeneous phase, it is not necessary to develop power series. Since the isochoric heat capacity is linked directly to a derivative of the reduced Helmholtz energy (see Eq. 3.14), the isochoric heat capacity has to be analytic for any analytic correlation in terms of the Helmholtz energy. Thus the isochoric heat capacity stays finite at the critical point and $\alpha = 0$ holds for $T > T_c$. For the isochoric heat capacity on the critical isochore of the two phase system at $T < T_c$ the situation becomes more complicated. The isochoric heat capacity has to be calculated from the heat capacities of the saturated vapour and liquid and the contribution

$$\Delta c_v \sim -\frac{T}{R\rho'^2} \cdot \frac{\left[\left(\frac{\partial p}{\partial T}\right)_\rho - \frac{dp_s}{dT}\right]^2}{\left(\frac{\partial p}{\partial \rho}\right)_T} \quad (4.132)$$

has to be considered additionally when investigating the corresponding critical exponent α_2 (see Eq. 3.80). With $p_s(\Delta T) = p(\Delta T, \Delta\rho') = p(\Delta T, \Delta\rho'')$, the power series for the vapour pressure can be written as

$$p_s = p_c + p_{01}(T - T_c) + p_{11} c_{2\beta}(T - T_c)^{1+2\beta} + \ldots \quad (4.133)$$

The exponent 2β results from the fact that the first term in the power series for the coexisting densities which goes with $(T_c - T)^\beta$ is symmetric for ρ' and ρ'' and cancels out; the second term which is antisymmetric goes with $(T_c - T)^{2\beta}$ and determines the corresponding contribution. The value of $c_{2\beta}$ is arbitrary in this context. By examining the required limit, the relation Eq. 4.132 can be rewritten with Eq. 4.133 as

$$\Delta c_v \sim -\frac{1}{(T_c - T)^{2\beta}} \cdot \frac{\left[p_{01} + p_{02}(T_c - T) - p_{01} - c(T_c - T)^{2\beta}\right]^2}{(T_c - T)}$$

$$\sim \frac{(T_c - T)^{4\beta}}{(T_c - T)^{1+2\beta}} \quad (4.134)$$

for $\beta \leq 0.5$ and with c as an arbitrary constant. Thus, the contribution Δc_v becomes independent of infinitesimal temperature changes around the critical temperature for equations with 3-point contact ($\beta = 0.5$) and α_2 remains equal to 0. But for equations with 5-point contact ($\beta = 0.25$) Δc_v diverges with $\alpha_2 = 0.5$.

The relations derived above have been known for decades (see Baehr, 1963) and when both the theoretical understanding of critical effects and the experimental basis became more evident, it was clear that the critical exponents which result from analytic equations of state do not match the theoretical and experimental results.[13] This was the starting point for the development of equations of state which specialise in the description of properties in the critical region.

4.5.1.1 Theoretically Founded Equations of State

To represent thermodynamic properties in the critical region, one needs a functional form for an equation which yields the required nonanalytical behaviour of the considered properties. For this purpose, the Ising model is not very helpful, because an explicit equation of state, in this case an explicit scaling function (see Sengers and Levelt Sengers, 1986), has never been formulated for this model. In practice, empirical closed-form expressions are used therefore as scaled equations of state which conform with the asymptotic behaviour and the symmetry of the Ising model. These formulations require transformations to parametric variables; common transformations use the parametric variables θ and r and are defined implicitly by

$$h = \tilde{\mu} - \tilde{\mu}_0(\tilde{T}) = a\, r^{\beta\delta}\, \theta(1-\theta^2) \tag{4.135}$$

$$\tau = \Delta\tilde{T} + c\,h = r(1-b^2\theta^2) \tag{4.136}$$

with $\quad \tilde{\mu} = \mu\dfrac{\rho_c T_c}{T p_c}, \quad \tilde{T} = -\dfrac{T_c}{T}, \quad \Delta\tilde{T} = 1-\dfrac{T_c}{T},$

$\tilde{\mu}_0$ the analytical background of the chemical potential, a and c as system-dependent parameters, b^2 as a "universal parameter", and the critical exponents β and δ. Based on this implicit transformation, the so called *"simple scaling"* approach leads to a functional form for the equation of state where just two functions $m_0(\theta)$ and $m_1(\theta)$ still need to be determined. The simplest approach for these unknown functions is the *"linear model"*,

$$m_0(\theta) = k_0 \cdot \theta \quad \text{and} \quad m_1(\theta) = k_1 \cdot \theta, \tag{4.137}$$

---

[13] The recent experimental results of Wagner et al. (1992), Kurzeja et al. (1999/2000), and Kurzeja and Wagner (2000) for the "thermal" critical exponents β, γ, and δ, derived from $p\rho T$ measurements in the immediate vicinity of the critical point show clear differences from the values predicted by the renormalisation group theory. These surprising results might have been caused by an extended validity range of the so-called explicit influence of gravity (the implicit influence of gravity, e.g., the averaging errors based on density stratifications, was taken into account when evaluating the experimental $p\rho T$ data); a theoretical solution of the three dimensional Ising model under influence of an outer field like gravity is still pending.
The region where these results were found is by about one order of magnitude smaller than the region where state-of-the-art reference equations fail with respect to the representation of certain caloric properties (see Sect. 4.5.2) and it is by at least two orders of magnitude smaller than the region where simple equations of state fail. Thus, these results do not disprove the thesis that up to now empirical multiparameter equations of state have certain shortcomings with regard to the representation of caloric properties in the critical region. However, they are definitely worth considering when discussing further work on properties in the critical region.

4.5 Describing Properties in the Critical Region

which was proposed by Schofield et al. (1969). This very early scaled equation of state was used by Angus et al. (1976) for their compilation on thermodynamic properties of carbon dioxide. However, simple scaling approaches assume a high degree of symmetry of the critical region and this assumption holds only in a very narrow region around the critical point.

In the following years, *"revised and extended"* scaled equations of state were introduced to overcome these limitations, see for instance Albright et al. (1987a) for carbon dioxide. These equations have been *revised* in the sense that they consider the asymmetry of the phase boundary by using a twisted system of coordinates and *extended* in the sense that they consider first order Wegner corrections (see Wegner, 1972) to the simple power laws and the corresponding scaled equations. Based on comparisons with experimental data, Sengers and Levelt Sengers (1986) estimate that the range where typical revised and extended scaled equations of state are valid is $0.9995 \leq T/T_c \leq 1.03$ and $0.75 \leq \rho/\rho_c \leq 1.25$; more recent comparisons with highly accurate experimental data support these results (see for instance Tegeler et al., 1997/1999).

From an application oriented point of view, the major disadvantages of revised and extended scaled equations of state are their complex structure with recursively defined parametric variables, their small range of validity in combination with a comparatively large number of fitted parameters (usually about 13 system dependent parameters plus a considerable number of "universal" parameters), and the fact that a direct combination with an accurate wide range equation of state seems to be impossible – the analytic background contributions were always calculated from Taylor expansions in terms of critical point parameters and not from common equations of state. The last two disadvantages were addressed by Fox (1983) who introduced a different scaling strategy. Fox redefined the variables of an equation of state by introducing an empirical *"damping function"* which results in a scaling of thermodynamic variables when approaching the critical point; Fig. 4.15 illustrates the basic idea of such an approach. When applied to real systems the resulting relations are as complex and implicit as for the equations described above, but Fox could apply his approach to a Van der Waals like equation of state in the very first step, and outside of the critical region, the model approached the classical limits of this simple equation – a major advantage compared to revised and extended equations of state.

Two different lines of development are based on the idea of Fox: The *"transformation"* approach which has to be considered as semiempirical and which is described in Sect. 4.5.5.2 and the *"crossover"* approach which forms the basis of most of the recent scaled equations of state.

Albright et al. (1986) introduced a theoretically more satisfactory but also more complex *"crossover"* mechanism to rescale the variables of an analytic background function. In their first crossover article, the group around J. V. Sengers used a Van der Waals equation to calculate the analytic background. However, the development later switched back to Taylor expansions at the critical point (see for instance Albright et al., 1987b; Chen et al. 1990a/b). The equations published by Chen et al. (1990b) for carbon dioxide, water and ethane are valid in regions which correspond roughly to $0.96 \leq T/T_c \leq 1.23$ and $0.4 \leq \rho/\rho_c \leq 1.6$; this is a

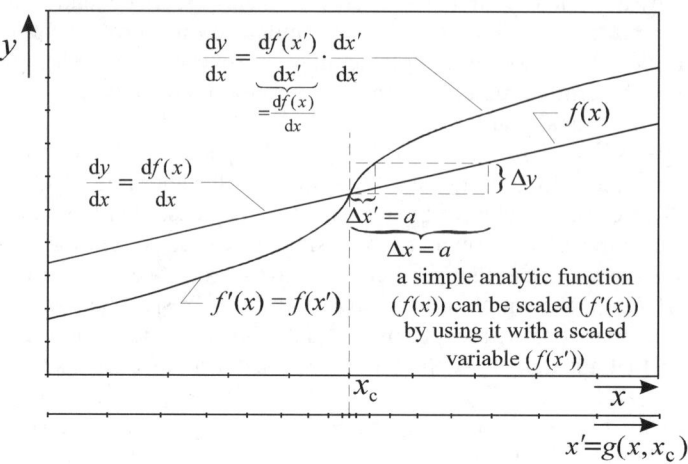

Fig. 4.15. The basic idea of transformation and crossover approaches is to scale an analytic function by using it with scaled variables.

major advantage compared to revised and extended scaled equations. However, with 18 system-dependent fitted parameters, such formulations are even more complex. Outside the region where revised and extended scaled equations are valid, the achieved accuracy is unsatisfactory as compared to recent reference equations of state (see Span and Wagner, 1996).

During the 90's a broad variety of crossover equations was published by different groups both for pure fluids and mixtures – see for instance Kiselev et al. (1991) for water and carbon dioxide, Jin et al. (1992) for methane, Luettmer-Strathmann et al. (1992), who presented the first parametric crossover model for ethane, Kiselev and Sengers (1993) for methane and ethane, Povodyrev et al. (1996) for mixtures of methane and ethane, Kiselev (1998) for a crossover model using a cubic equation of state as background function, and Kiselev et al. (1998) and Edison et al. (1998) for a more general crossover approach which describes mixtures. To discuss all these models would be inappropriate at this point. For the context of this section, three general statements are sufficient:

- The development of scaled correlations for thermophysical properties of pure fluids and mixtures is a lively area of scientific research. Recent work has resulted in a number of interesting models and extended theoretical knowledge on thermodynamic properties of pure fluids to mixtures as well as to transport properties.
- The complexity of the models still hinders practical application. A crucial point is the unavoidable introduction of scaled variables which require implicit definitions and therefore iterative procedures to switch between thermodynamic variables and scaled variables. In combination with the required numerical steps, these implicit dependencies make results unreproduceable and the use of analytic derivatives for the calculation of derived properties becomes

extremely complicated if not impossible. A significant simplification of scaled models seems to be impossible unless physical exactness is sacrificed (see for instance Sect. 4.5.5.3).
- The formal range of validity of scaled models for pure substances may cover the whole fluid region (see for instance Kiselev, 1998), but outside of the critical region, such complex models compare well only to very simple equations of state. The region where their accuracy is comparable to the one of empirical reference equations of state is still very limited. Improvements seem to be possible in combination with more complex background functions but adapting the tools which are commonly used to establish accurate equations of state to the development of background functions is difficult.

Based on this assessment of the practical capabilities of theoretically founded equations of state it becomes interesting to focus on the practical capabilities of empirical equations of state again.

4.5.2 Capabilities of Empirical Multiparameter Equations of State

From the theoretical results presented in Sect. 4.5.1, it is usually concluded that analytic equations of state cannot represent the properties of pure fluids within the critical region since they do not yield correct critical exponents. However, this conclusion is incorrect for most thermodynamic properties if state-of-the-art reference equations of state are considered. This section is intended to illustrate both the capabilities and the limitations of empirical multiparameter equations of state regarding the representation of properties in the extended critical region.

Figure 4.16 shows percentage deviations between experimental data for the vapour pressure and the saturated-liquid and -vapour density of nitrogen and values calculated from the recent reference equation of Span et al. (1998b). The data were measured by Nowak et al. (1997b) and cover the temperature range $0.998 \leq T/T_c \leq 0.9998$. Even close to the critical point, the reference equation is able to represent the most accurate available data within their experimental uncertainty; no reliable experimental data are available closer to the critical point. The nitrogen equation of Span et al. uses modified Gaussian bell shaped terms (see Sects. 3.1.3.2 and 4.5.3) but no nonanalytical terms (see Sects. 3.1.3.2 and 4.5.4) and is thus analytic in the sense discussed in the preceding section. The equation was constrained to a three point contact at the selected critical point which results in a critical exponent of $\beta = 0.5$.

For carbon dioxide, $p\rho T$ data measured by Straub (1972) are available directly on the critical isotherm. Measured with an optical method, the total uncertainty of these data is relatively large but the consistency, which is important to assess the limiting behaviour of equations of state, is impressive. In Fig. 4.17, these data are compared with data which were calculated from the recent reference equation for carbon dioxide, the equation of Span and Wagner (1996); the critical parameters used by Straub were readjusted to coincide with the critical parameters the equation was constrained to. Data which were calculated from other equations of state are represented by dashed and dash-dotted lines.

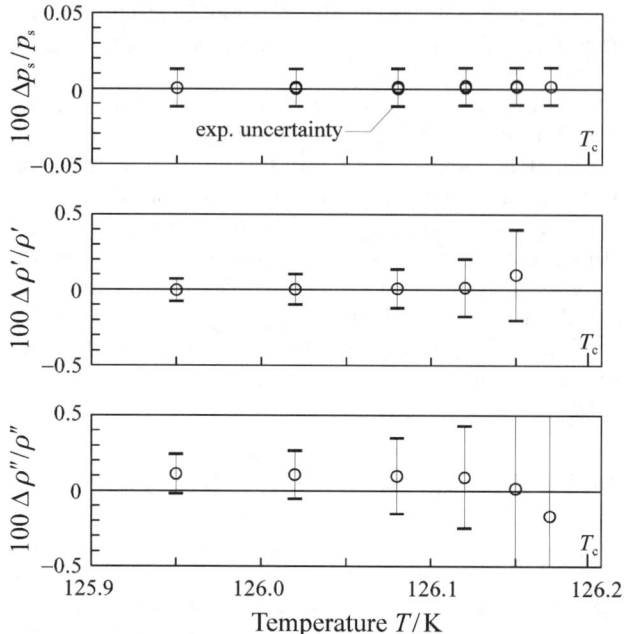

Fig. 4.16. Percentage deviations $100\,\Delta y/y = 100\,(y_{\text{exp}} - y_{\text{calc}})/y_{\text{exp}}$ between data for the vapour pressure and the saturated-liquid and -vapour density of nitrogen measured by Nowak et al. (1997b) and corresponding values calculated from the reference equation of Span et al. (1998b).

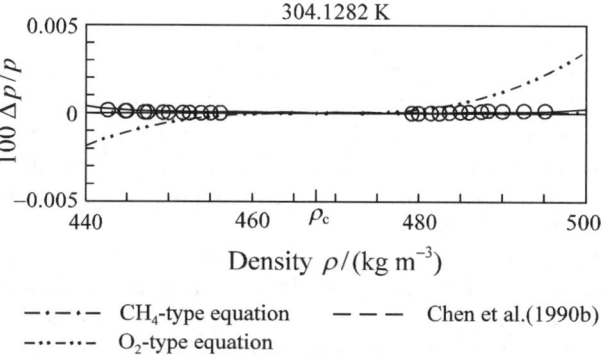

—·—·— CH_4-type equation   — — — Chen et al.(1990b)
······· O_2-type equation

Fig. 4.17. Percentage deviations $100\,\Delta p/p = 100\,(p_{\text{exp}} - p_{\text{calc}})/p_{\text{exp}}$ between $p\rho T$ data on the critical isotherm of carbon dioxide measured by Straub (1972) and values calculated from the reference equation of Span and Wagner (1996). For comparison, values calculated from other equations of state are plotted as dashed and dash-dotted lines.

4.5 Describing Properties in the Critical Region

The equation of Chen et al. (1990b) is a highly complex nonanalytical crossover equation of state which describes the extended critical region of carbon dioxide. The equation which is presented as a CH_4-type equation uses the functional form with four modified Gaussian bell-shaped terms developed by Setzmann and Wagner (1991) for methane and was fitted to the same data set as the equation of Span and Wagner, as was the O_2-type equation which uses the functional form developed by Schmidt and Wagner (1985) for oxygen which contains only simple polynomial and exponential terms. An equation with this functional form was published by Ely (1986) for carbon dioxide. However, since this equation was based on different data sets, it would have been unfair to use it in this comparison. All

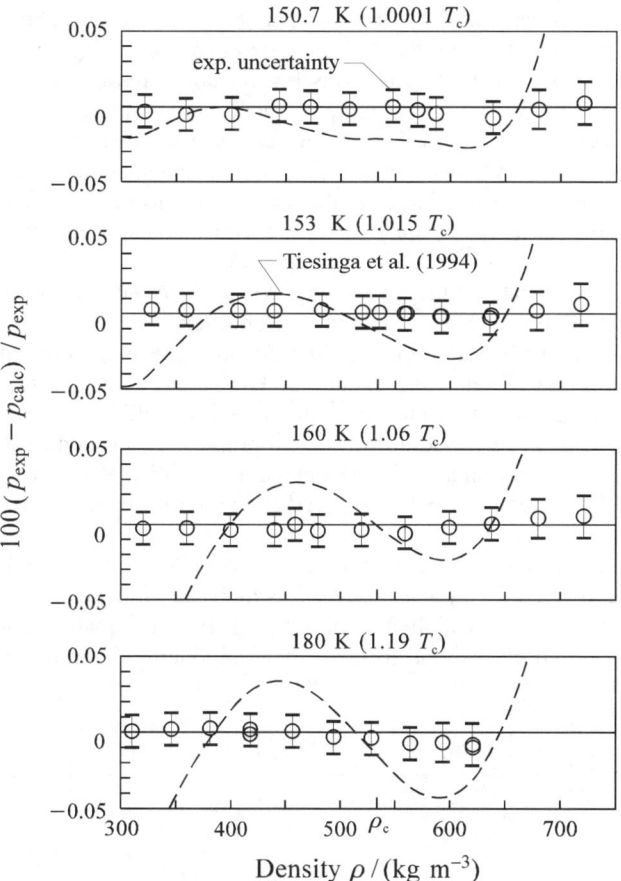

Fig. 4.18. Percentage deviations between $p\rho T$ data in the extended critical region of argon measured by Gilgen et al. (1994) and values calculated from the reference equation of Tegeler et al. (1997/1999). For comparison, values calculated from the revised and extended scaled equation of state by Tiesinga et al. (1994) are plotted as dashed lines.

equations were constrained to the same critical point parameters and none of them were fitted to the data of Straub – these data were used only as a consistency test. Due to the chosen three point contact, the analytic equations of state[14] result in a critical exponent of $\delta = 3$ and should result in a plot of the critical isotherm which is too steep when approaching the critical point. In the extreme resolution chosen for Fig. 4.17, this behaviour becomes obvious for the O_2-type equation, but both the CH_4-type equation and the equation of Span and Wagner predict the plot of the critical isotherm as well as the crossover equation.

As a third example for the representation of thermal properties in the critical region, Fig. 4.18 shows deviations between highly accurate $p\rho T$ data for argon which were measured by Gilgen et al. (1994) and values calculated from the recent reference equation of Tegeler et al. (1997/1999). For comparison, the dashed line represents results of the revised and extended scaled equation of state by Tiesinga et al. (1994). The shown data cover both sub- and overcritical densities throughout the extended critical region. From the fact that the analytic equation of Tegeler et al., which uses modified Gaussian bell-shaped terms but no nonanalytical terms, yields a critical exponent of $\delta = 3$, one would expect that it cannot represent the changing curvature of the isotherms around the critical density. However, the analytic equation represents the best available data within their experimental uncertainty. Even very close to the critical temperature, the scaled equation cannot represent the data appropriately and its uncertainty increases with increasing distance from the critical point.

The surprising capabilities of state-of-the-art analytic equations of state become intelligible when analysing the power laws for thermal properties, see Table 4.10. The power laws which describe orthobaric densities and the pressure on the critical isotherm are formulated in terms of vanishing distances. The values of the critical exponents β and δ describe the way in which the corresponding difference vanishes, but since these exponents are defined at the critical point, their influence on the representation of the data is very small – in the immediate vicinity of the critical point, the remaining differences are small compared to the values of the corresponding properties.

To assess the capabilities of an equation of state in the critical region, it is advantageous to adapt the concept of *"local exponents"* which was introduced by Wagner et al. (1992) for the analysis of their experimental data to equations of state. The local exponents which correspond to the critical exponents β and δ can be calculated according to

$$\beta_1(T) = \lim_{T_1 \to T} \frac{\log\left((\rho' - \rho'')_{T_1} / (\rho' - \rho'')_T\right)}{\log((T_c - T_1)/(T_c - T))} \qquad (4.138)$$

---

[14] Although the equation of Span and Wagner (1996) contains three nonanalytical terms, it behaves basically like an analytic equation with respect to thermal properties, see Sect. 4.5.4.

4.5 Describing Properties in the Critical Region

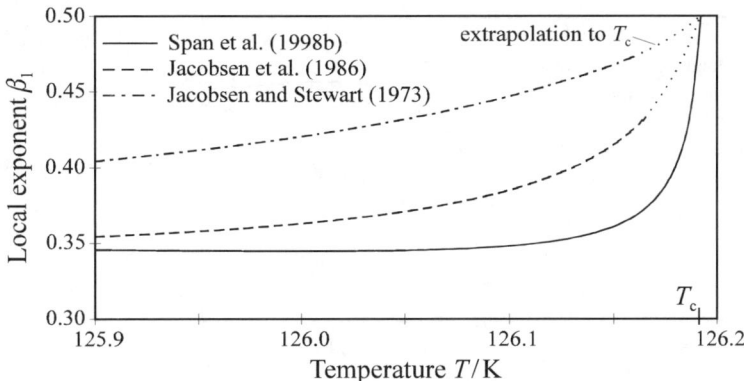

Fig. 4.19. Plots of the local exponent β_1 (see Eq. 4.138) as calculated from three multiparameter equations of state for nitrogen. The plotted temperature range corresponds to $0.998 \leq T/T_c \leq 1$. The necessary Maxwell iteration fails for the older equations close to T_c.

and
$$\delta_1(\rho) = \lim_{\rho_1 \to \rho} \frac{\log\left((p_1 - p_c)_{\rho_1} / (p - p_c)_\rho\right)}{\log((\rho_1 - \rho_c)/(\rho - \rho_c))}, \qquad (4.139)$$

respectively. For $T \to T_c$ or $\rho \to \rho_c$ the local exponents become equivalent to the critical exponents.

Using nitrogen as an example, Fig. 4.19 shows plots of the local exponent β_1 which result from three generations of empirical equations of state. The functional form of the equation of Jacobsen and Stewart (1973) became known as the MBWR equation (see Sect. 3.1.2); this equation is formulated in terms of the compression factor and consists of polynomial and exponential terms, where the exponential terms are restricted to $\exp(-\delta^2)$. The equation of Jacobsen et al. (1986a) is formulated in terms of the reduced Helmholtz energy and uses an optimised functional form which consists of polynomial and exponential terms; with regard to its capabilities in the critical region, this equation is comparable to the O_2-type equation discussed above. And finally, the recent equation of Span et al. (1998b) which is an optimised equation in terms of the reduced Helmholtz energy consisting of polynomial, exponential, and Gaussian bell shaped terms; with regard to its capabilities in the critical region, this equation is comparable to the CH_4-type equation discussed above. It can be seen from Fig. 4.19 that all the equations yield the expected critical exponent $\beta = 0.5$, but their behaviour throughout the critical region is very different. State-of-the-art reference equations of state yield local exponents which are on the order of values which can be derived from experimental data up to a region very close to the critical point. The shift to the classical limit occurs so close to the critical temperature that it does not affect the representation of satu-

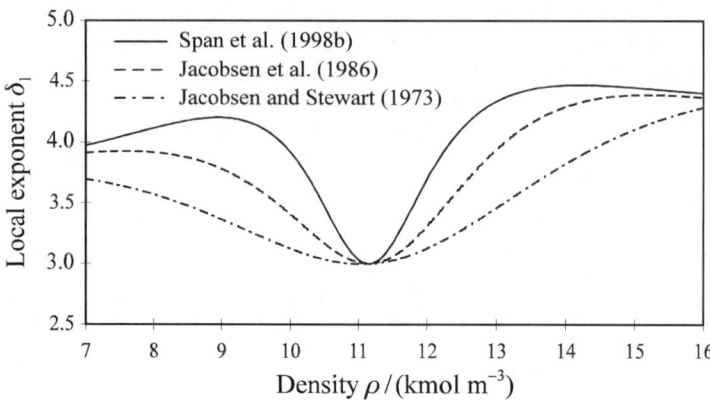

Fig. 4.20. Plots of the local exponent δ_1 (see Eq. 4.139) as calculated from three multiparameter equations of state for nitrogen. The plotted density range corresponds to $0.63 \leq \rho/\rho_c \leq 1.43$.

rated liquid- and vapour-densities significantly.[15] The older equations depart from the experimentally found values for β_1 too quickly and are therefore not able to accurately describe orthobaric densities in the critical region.[16]

Similar results are shown for δ_1 in Fig. 4.20. Again, the region where the local exponent deviates from typical experimentally found values is small for the equation of Span et al. (1998b) and is restricted to a region where $|p-p_c|$ is already very small – the equation is able to represent even the most accurate $p\rho T$ data on the critical isotherm (Nowak et al., 1997a/b) within their experimental uncertainty. For the older equations, the region where δ_1 approaches the classical limit is broader and the description of $p\rho T$ data in the critical region is affected as theoretically predicted.

For the isochoric compressibility along the critical isochore, the situation is the same. Recent equations of state with modified Gaussian bell shaped terms are obviously able to describe experimentally found local exponents γ_1 even better than scaled equations of state which are constrained to the theoretically predicted value of γ by their functional form, see Fig. 4.18.

State-of-the-art analytic multiparameter equations of state describe thermal properties in the critical and extended critical region at least as well as scaled equations of state. This capability is not affected by the classical limiting behaviour of the equations.

For caloric properties the situation becomes more complex. When calculated from accurate equations of state, properties which describe caloric behaviour for

---

[15] In fact the observed shift agrees well with recent results found for fluids under the influence of gravity (see footnote 13). This may be one reason why state-of-the-art multiparameter equations usually describe orthobaric densities in the critical region better than scaled equations.

[16] When calculating β_1 from the older equations the maxwell iteration fails close to the critical point; this is not a characteristic feature of the used functional forms but depends on the data and techniques which were used to fit these equations.

4.5 Describing Properties in the Critical Region

final changes of state such as enthalpy differences Δh or differences of the internal energy Δu are usually hardly influenced by the different description of critical phenomena. The region where the limiting behaviour of recent equations of state becomes relevant is so small that the influence of the effects which will be discussed below simply averages out for typical values of ΔT, Δp and ΔT, $\Delta \rho$, respectively. The properties which have to be examined in more detail are the speed of sound and derived properties such as the heat capacities.

The specific isobaric heat capacity can be expressed as

$$\frac{c_p}{R} = \underbrace{-\tau^2 \left(\alpha^o_{\tau\tau} + \alpha^r_{\tau\tau}\right)}_{\frac{c_v}{R}} + \frac{\overbrace{\left(1+\delta\alpha^r_\delta - \delta\tau\alpha^r_{\delta\tau}\right)^2}^{\sim \left(\frac{\partial p}{\partial T}\right)^2_\rho}}{\underbrace{1+2\delta\alpha^r_\delta + \delta^2\alpha^r_{\delta\delta}}_{\sim \left(\frac{\partial p}{\partial \rho}\right)_T}} \qquad (4.140)$$

where the derivatives of the reduced Helmholtz energy α are abbreviated according to Eq. 3.57. Since $(\partial p/\partial \rho)_T^{-1}$ grows much faster when approaching the critical point than c_v, the specific isobaric heat capacity is dominated by the fraction, which is closely related to the representation of $p\rho T$ data. Thus, an equation which yields an accurate description of the $p\rho T$ surface within the critical region should also yield reliable values of the specific isobaric heat capacity.

The situation is different for the isochoric heat capacity which is given by

$$\frac{c_v}{R} = -\tau^2 \left(\alpha^o_{\tau\tau} + \alpha^r_{\tau\tau}\right) \qquad (4.141)$$

If the second derivative of the residual part of the Helmholtz energy with respect to τ is finite at the critical point, the value of the specific isochoric heat capacity is also finite.[17] An equation which is analytic fails with regard to the representation of the isochoric heat capacity at the critical point. At the same time, analytic formulations will result in finite values of the speed of sound at the critical point since the speed of sound corresponds to

$$\frac{w^2}{RT} = \underbrace{1+2\delta\alpha^r_\delta + \delta^2\alpha^r_{\delta\delta}}_{\sim \left(\frac{\partial p}{\partial \rho}\right)_T} - \frac{\overbrace{\left(1+\delta\alpha^r_\delta - \delta\tau\alpha^r_{\delta\tau}\right)^2}^{\sim \left(\frac{\partial p}{\partial T}\right)^2_\rho}}{\underbrace{\tau^2\left(\alpha^o_{\tau\tau} + \alpha^r_{\tau\tau}\right)}_{-\frac{c_v}{R}}}. \qquad (4.142)$$

---

[17] One may argue, that this behaviour agrees well with the results found for fluids under the influence of gravity (see footnote 13). However, the region where even (analytic) state-of-the-art reference equations of state fail with respect to the representation of c_v is much larger than the region where classical critical exponents were experimentally found.

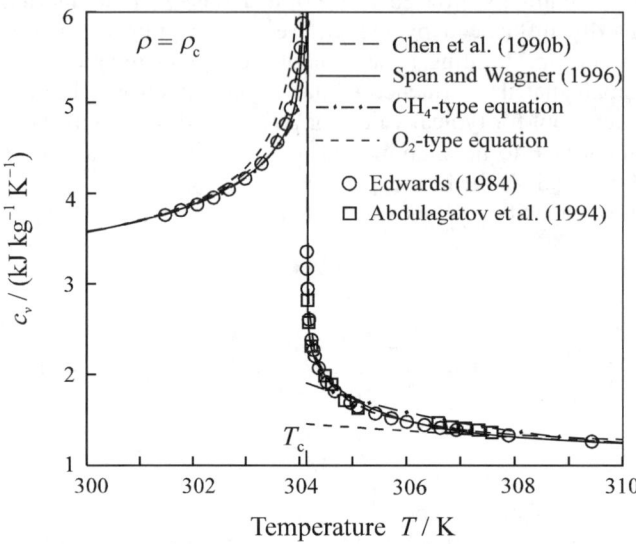

Fig. 4.21. Experimental results for the isochoric heat capacity on the critical isochore of carbon dioxide and values calculated from four equations of state.

At the critical point, the expression $(\partial p / \partial \rho)_T$ becomes zero and $(\partial p / \partial T)_\rho$ is a finite value. Thus, only if c_v becomes infinite, the speed of sound becomes zero.

Figure 4.21 illustrates the limitations of analytic multiparameter equations of state using the representation of experimental results for the isochoric heat capacity on the critical isochore of carbon dioxide as an example. When approaching the critical temperature from the two phase region, all equations yield qualitatively correct behaviour up to temperatures which are very close to T_c. In this region the dominant contribution to the residual isochoric heat capacity results from $\Delta c'_v$ and $\Delta c''_v$ which are described satisfactorily by rather simple analytic equations of state. When approaching the critical temperature from the single phase region the O_2-type equation fails to represent the increasing isochoric heat capacity for temperatures below about $1.01\ T_c$ (≈ 307 K). Equations which use modified Gaussian bell shaped terms like the CH_4-type equation are able to follow the increasing isochoric heat capacity down to about $1.002\ T_c$ (≈ 304.7 K) but as analytic equations of state they cannot result in an infinite isochoric heat capacity at the critical point. Just like the nonanalytical crossover equation by Chen et al. (1990b), equations which use nonanalytical terms (see Sects. 3.1.3.2 and 4.5.4) like the recent reference equation by Span and Wagner (1996) are able to follow the experimentally found strong curvature and result in an isochoric heat capacity which becomes infinite at the critical point.

The corresponding results for the speed of sound are shown in Fig. 4.22. Just 0.5 K above the critical temperature the CH_4-type equation, the equation of Span and Wagner (1996) and the crossover equation by Chen et al. (1990b) still yield very similar results. But when further approaching the critical temperature, the

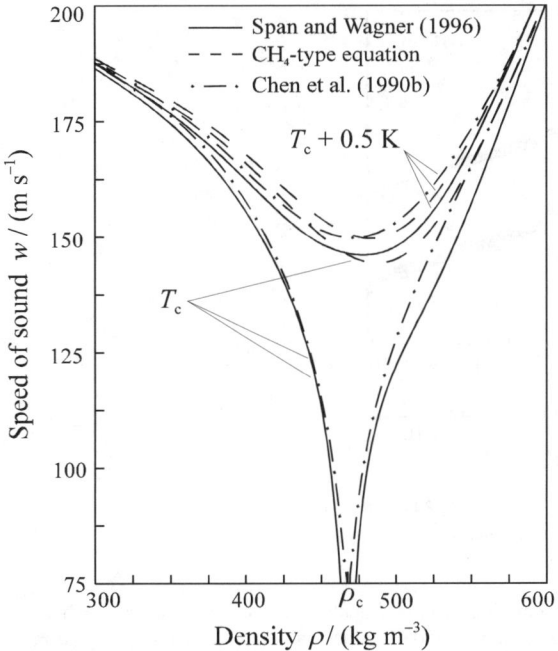

Fig. 4.22. Values for speed of sound in the critical region of carbon dioxide as calculated from three different equations of state.

speeds of sound calculated from the analytic CH_4-type equation remain finite, while the crossover equation and the nonanalytical multiparameter equation result in the theoretically expected vanishing speed of sound at the critical point.

However, for reasons which will be explained in more detail in Sect. 4.5.4, equations which use nonanalytical terms of the form introduced by Span (1993) and Span and Wagner (1996) are still not able to fulfil the theoretical demands on the asymptotic limiting behaviour of the isochoric heat capacity. Figure 4.23 shows the plot of the isochoric heat capacities calculated from the O_2-type equation, from the equation of Span and Wagner (1996) and from the crossover equation by Chen et al. (1990b) in a double logarithmic diagram. Only the crossover equation yields the theoretically expected linear extrapolation of the experimentally found isochoric heat capacities for temperatures asymptotically close to T_c.

Summarising the results regarding the representation of properties in the critical region by multiparameter equations of state, one has to conclude as follows:

- Regarding the representation of thermal properties recent reference equations of state with modified Gaussian bell shaped terms are at least as good as special scaled equations of state. Properly designed equations which use only polynomial and exponential terms are sufficient for most technical applications in the critical region, but with respect to data needs for scientific applications, they tend to yield unsatisfactory results.

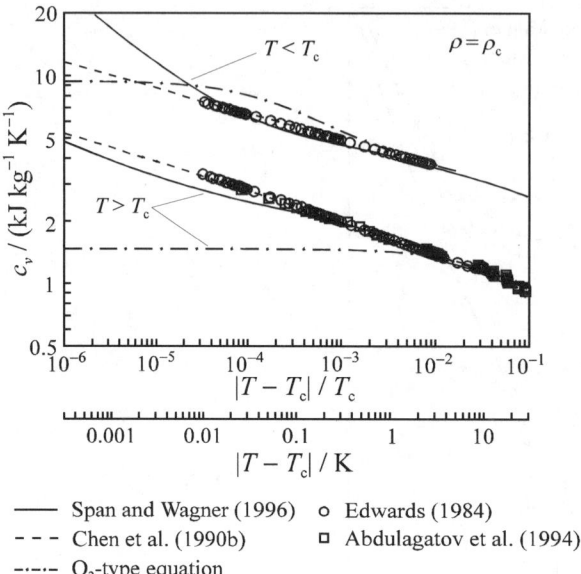

Fig. 4.23. Double logarithmic representation of experimental results for the isochoric heat capacity on the critical isochore of carbon dioxide and of values calculated from three equations of state.

- Recent reference equations of state with nonanalytical terms are able to represent caloric properties in the critical region within the uncertainty of the available data. Equations with modified Gaussian bell shaped terms fail to describe isochoric heat capacities and speeds of sound in the critical region for about $1 \leq T/T_c \leq 1.002$ but they are still sufficient for all technical and scientific data needs. Equations which use only polynomial and exponential terms fail in the critical region for about $0.998 \leq T/T_c \leq 1.01$ and tend to yield unsatisfactory results for advanced data needs.
- Empirical multiparameter equations of state do not fulfil theoretical demands on the limiting behaviour of thermodynamic properties for the asymptotical approximation to the critical point. No theoretical conclusions should be drawn from the limiting behaviour of empirical equations of state.[18]

Based on this assessment, the use of nonanalytical terms has been avoided for the most recent reference equations of state (see Tegeler et al., 1997 / 1999; Span et al., 1998b; Smukala et al., 1999). The numerical expense which is caused by the use of nonanalytical terms for both the user and for those who develop the equation, seems to be justified only for substances where properties in the critical region are of special scientific interest.

---

[18] This is in fact almost self-evident; theoretical conclusions which are drawn from empirical correlations are almost always questionable.

4.5.3 Setting Up Equations with Modified Gaussian Bell Shaped Terms

The modified Gaussian bell shaped terms introduced by Setzmann and Wagner (1991) in the form

$$A_i(\tau,\delta) = n_i \tau^{t_i} \delta^{d_i} \exp\left(-\eta_i(\delta-\varepsilon_i)^2 - \beta_i(\tau-\gamma_i)^2\right) \qquad (4.143)$$

have 6 internal adjustable parameters besides the leading coefficient n_i which can be determined directly in linear or nonlinear fits. The inverse reduced temperature τ and the reduced density δ have to be calculated with the corresponding critical parameters when using modified Gaussian bell shaped terms. The parameter ε_i has always been chosen to be equal to 1 since any other value would result in an undesired asymmetry of the contribution of the term. The parameter γ_i shifts the point where the derivatives of the term become maximal ($\rho/\rho_c = \varepsilon_i$ and $T/T_c = \gamma_i^{-1}$) into the unstable part of the two phase region and has to be larger than 1. The exponents t_i and d_i are typical temperature and density exponents which have to be determined by optimisation algorithms just like those for the polynomial and exponential terms.

Setzmann and Wagner (1991) determined reasonable values for η_i, β_i, and γ_i by nonlinear fits, but the nonlinear relations between these parameters and the residua of the thermodynamic properties resulted in numerically unstable fits. The parameters had to be fitted one by one together with the n_i of all other terms in the equation and the whole procedure became both very time-consuming and unsatisfactory. Thus, the most powerful combination of these parameters was determined by using the optimisation procedure, after a range of reasonable values had been established by nonlinear fits. The extensive derivatives of A_i and its derivatives with respect to η_i, β_i, and γ_i, which are required for nonlinear fits are not needed for the optimisation approach and are therefore not given here. For equations of state without additional nonanalytical terms the following range of parameters turned out to be useful:[19]

$$1 \leq d_i \leq 3$$
$$0 \leq t_i \leq 3$$
$$15 \leq \eta_i \leq 25$$
$$200 \leq \beta_i \leq 375$$
$$1.11 \leq \gamma_i \leq 1.25 \qquad (4.144)$$

In combination with nonanalytical terms Wagner and Pruß (1997/2000) used smaller values for η_i ($10 \leq \eta_i \leq 20$) and for β_i ($150 \leq \beta_i \leq 250$) while Span and Wagner (1996) used parameter ranges which correspond roughly to the limits given in Eq. 4.144.

---

[19] Setzmann and Wagner (1991) additionally used values up to 40 for η_i, 1.07 for γ_i, and 0 for d_i; the recommendations given here are based on later experiences.

While working with modified Gaussian bell shaped terms, it is important to realise that correlated pairs of these terms are especially effective. Since the capabilities of the original Setzmann and Wagner (1989a) algorithm are restricted when optimising such intercorrelated pairs of terms, it is advantageous to use it in its modified form (see Sect. 4.4.5.2) testing double exchanges of the modified Gaussian bell shaped terms in the equation against those in the bank of terms.

4.5.4 Setting Up Equations with Nonanalytical Terms

The nonanalytical terms which were already discussed above and in Sect. 3.1.3.2 are essentially based on a development by Span (1993) and were used in a modified form by Span and Wagner (1996) and Wagner and Pruß (1997/2000). To establish equations of state which use such nonanalytical terms, it is important to understand the basic idea behind these terms. In order to cause the isochoric heat capacity to become infinite and the speed of sound to vanish at the critical point (see footnote 17), in an equation of state it is necessary to introduce nonanalytical terms which yield infinite values for the second derivative of α with respect to τ at the critical point. However, reasonable nonanalytical terms have to fulfil three additional demands:

- The values resulting for $\alpha_{\tau\tau}^{\rm r}$ have to be finite everywhere except at the critical point.
- Singular behaviour of the other second derivatives and all derivatives with respect to δ has to be avoided everywhere.
- Within the δ,τ surface of the critical region, the maximum of $\alpha_{\tau\tau}^{\rm r}$ has to follow the course of the saturated vapour- and saturated liquid line in order to avoid unreasonable maxima of the isochoric heat capacity in the single-phase region.

These demands led to the functional form

$$A_i(\tau,\delta) = n_i\, \delta\, \Delta^{b_i} \exp\left(-e_i(\delta-1)^2 - f_i(\tau-1)^2\right) \qquad (4.145)$$

with
$$\Delta = \left\{(1-\tau) + c_i\left[(\delta-1)^2\right]^{1/(2\beta_i)}\right\}^2 + d_i\left[(\delta-1)^2\right]^{a_i}.$$

The nonanalytical behaviour is introduced by the non-integer exponent b_i of the distance function Δ in combination with the set-up of Δ itself. The reduced density δ in Eq. 4.145 had to be introduced in order to guarantee a completely vanishing contribution when approaching the ideal gas limit and the exponential term damps the influence of the nonanalytical terms outside of the critical region. The derivatives of Eq. 4.145 with respect to τ and δ have been summarised in Table 3.8.

When approaching the critical point, the asymptotically leading terms in $(\partial A_i/\partial\tau)_\delta$ become

$$\lim_{\delta=1,\tau\to 1}\left(\frac{\partial A_i}{\partial\tau}\right)_\delta \sim -C_0\cdot(1-\tau)^{2b_i-1} + \ldots..$$

and
$$\lim_{\tau=1,\delta\to 1}\left(\frac{\partial A_i}{\partial \tau}\right)_\delta \sim -C_1 \cdot (\delta-1)^{2a_ib_i-2a_i+\frac{1}{\beta_i}} - C_2 \cdot (\delta-1)^{\frac{2}{\beta_i}(b_i-0.5)} - \ldots \quad (4.146)$$

where the contribution from the exponential function is considered analogous to Eq. 4.88 and C_0, C_1, and C_2 are arbitrary constants which have no influence on the limiting behaviour. Since the first temperature derivative of A_i has to be finite and continuous at the critical point, the conditions $b_i > 0.5$ and $a_i < (2\beta_i(b_i-1))^{-1}$ can be derived from Eq. 4.146. The second derivative $(\partial^2 A_i/\partial \tau^2)_\delta$ is expected to diverge at the critical point. For a course along the critical isochore, the asymptotically leading term becomes

$$\lim_{\delta=1,\tau\to 1}\left(\frac{\partial^2 A_i}{\partial \tau^2}\right)_\delta \sim C_0 \cdot (1-\tau)^{2b_i-2} + \ldots \quad (4.147)$$

When comparing Eq. 4.147 with the power law for the isochoric heat capacity (see Eq. 3.14 for the calculation of c_v and Table 4.10 for the power law), the relation $\alpha = 2 - 2b_i$ between b_i and the critical exponent α becomes obvious – the asymptotic behaviour of the isochoric heat capacity is determined by the nonanalytical term with the smallest value of b_i. Theoretically, a value of $b_i = 0.945$ would be expected from $\alpha = 0.110$. The coefficient n_i of the leading nonanalytical term has to be negative to result in $c_v(T_c,\rho_c) = +\infty$. With the leading term

$$\lim_{\tau=1,\delta\to 1}\left(\frac{\partial^2 A_i}{\partial \delta^2}\right)_\tau \sim C_0 \cdot (\delta-1)^{2a_i(b_i-1)+\frac{1}{\beta_i}-2} + C_1 \cdot (\delta-1)^{\frac{2}{\beta_i}(b_i-1)+\frac{1}{\beta_i}-2} + \ldots, \quad (4.148)$$

the condition that $(\partial^2 A_i/\partial \delta^2)_\tau$ has to be finite and continuous everywhere yields the relations $\beta_i < (2a_i(1-b_i)+2)^{-1}$ and $b_i > 0.5 + \beta_i$. No further relevant conditions were found when investigating the limiting behaviour of $(\partial^2 A_i/\partial \tau \partial \delta)$, $(\partial^3 A_i/\partial \tau \partial \delta^2)$, and $(\partial^3 A_i/\partial \delta^3)_\tau$; $(\partial^3 A_i/\partial \tau^3)_\delta$ and $(\partial^3 A_i/\partial \tau^2 \partial \delta)$ are expected to become infinite at the critical point. The definition of the distance function Δ and the fact that θ is used only with positive exponents in the derivatives of A_i (see Table 3.8) guarantees that none of the derivatives is discontinuous anywhere but at the critical point. Thus, if the conditions for a_i, b_i, and β_i are fulfilled, Eq. 4.145 fulfils the first two demands formulated above.

The contour of the curve where the contribution of a nonanalytical term to $(\partial^2 \alpha/\partial \tau^2)_\delta$ becomes a maximum is determined by the form of the distance function Δ. The relation

$$(1-\tau) + c_i\left[(\delta-1)^2\right]^{\frac{1}{2\beta_i}} = 0 \quad (4.149)$$

can be interpreted as single sided equivalent of the power law describing the orthobaric densities, see Table 4.10. Using a transformed form of Eq. 4.149,

$$\left|\frac{\rho}{\rho_c} - 1\right| = c_i^{-\beta_i}\left(1 - \frac{T_c}{T}\right)^{\beta_i}, \quad (4.150)$$

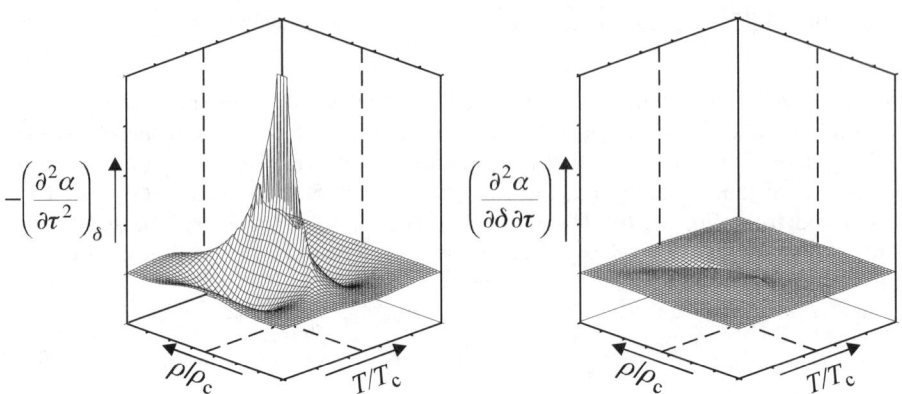

Fig. 4.24. Qualitative illustration of the contribution of a nonanalytical term to the derivatives $(\partial^2 \alpha / \partial \tau^2)_\delta$ and $(\partial^2 \alpha / \partial \tau \, \partial \delta)$ in the critical region.

c_i and β_i can be determined by a nonlinear fit to data for both the saturated-vapour and -liquid density in the immediate vicinity of the critical point. However, since the condition for b_i which was derived above is very sensitive to large values of β_i the value for β_i has to be chosen slightly smaller ($\beta_i \approx 0.30$) than expected from the power law and only c_i can be fitted to orthobaric densities directly.[20] The necessary data can be calculated from the corresponding ancillary equations.

The contribution of a nonanalytical term to $(\partial^2 \alpha / \partial \tau^2)_\delta$ and to $(\partial^2 \alpha / \partial \tau \, \partial \delta)$ is illustrated in Fig. 4.24. For $(\partial^2 \alpha / \partial \tau^2)_\delta$ the sharp maximum which corresponds to the desired increase of the isochoric heat capacity at the critical point becomes visible while the contributions to $(\partial^2 \alpha / \partial \tau \, \partial \delta)$ and to the other first and second derivatives stay very small. Terms like this can be used as an integral part of empirical multiparameter equations of state to obtain a nonanalytical behaviour with respect to $(\partial^2 \alpha / \partial \tau^2)_\delta$.

The parameters a_i, b_i, d_i, e_i, and f_i can be determined from a nonlinear fit theoretically. Reasonable values for b_i, e_i, and f_i were determined by Span and Wagner (1996) by nonlinear fits of preliminary equations. However, as with the Gaussian bell shaped terms (see Sect. 4.5.3), this procedure turned out to be very time-consuming and unsatisfactory since the parameters could only be fitted one at a time due to numerical restrictions. The final parameter combination was determined by means of an optimisation algorithm by Span and Wagner (1996) and

---

[20] One disadvantage of this kind of nonanalytical terms is that the contour defined by Eq. 4.149 can never match the phase boundary calculated from the equation of state exactly. To avoid unreasonable maxima of c_v close to the phase boundary, c_i and β_i have to be chosen in such a way that the described contour lies within the two phase region, as close to the actual phase boundary as possible. Scaled approaches which use the parametric variables defined by Eqs. 4.135–4.136 avoid this problem since the orthobaric densities correspond to $\theta = \pm 1$ per definition. However, the advantage of this set-up of the parametric variables is questionable since their definition restricts the flexibility of the equation; scaled formulations cannot describe orthobaric densities in the critical region as accurately as multiparameter equations of state.

4.5 Describing Properties in the Critical Region

Wagner and Pruß (1997/2000). The nonanalytical terms used in the bank of terms of Span and Wagner (1996) covered the parameter ranges

$$3.0 \leq a_i \leq 4.0$$

$$0.875 \leq b_i \leq 0.925$$

$$0.30 \leq d_i \leq 1.00$$

$$10 \leq e_i \leq 15$$

$$225 \leq f_i \leq 275. \tag{4.151}$$

Slightly different combinations were used by Wagner and Pruß (1997/2000). The extensive derivatives of A_i and its derivatives with respect to a_i, b_i, d_i, e_i, and f_i, which are required for nonlinear fits, are not needed for the optimisation approach and are therefore not given here.

The introduction of nonanalytical terms has significantly improved the capabilities of empirical equations of state with regard to the representation of caloric properties in the critical region, but such equations still have certain limits if an exact fulfilment of the asymptotic power law for the isochoric heat capacity is required as shown in Fig. 4.23. In multiparameter equations of state, the analytic terms contribute decisively towards the description of critical effects. Nonanalytical terms do not replace this contribution close to the critical point but they fill the increasing gap between analytic and nonanalytical behaviour with regard to $(\partial^2 \alpha / \partial \tau^2)_\delta$. Since the relative growth of this gap is much faster than the relative growth of c_v in the region where experimental data are available, efficient values of the exponent b_i are smaller than theoretically expected and result in a critical exponent α which is too large from an asymptotic point of view. Basically, this is an unavoidable problem, but recent semiempiric approaches show that further improvements are possible (see Sect. 4.5.5.3).

A second problem is linked to the density derivatives of Eq. 4.145. When using these kinds of nonanalytical terms, it was assumed that continuous plots of derivatives which do not diverge are sufficient; possible unphysical contributions have to be compensated by other terms in the multiparameter equation of state in this case.

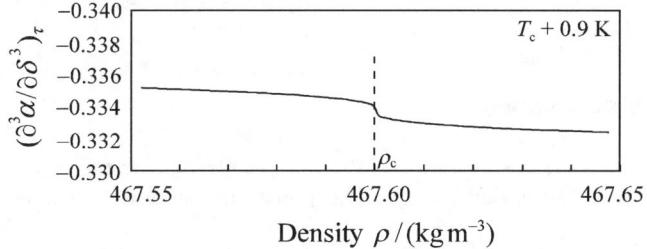

Fig. 4.25. Results for the third density derivative calculated from the equation of Span and Wagner (1996) close to the critical isochore.

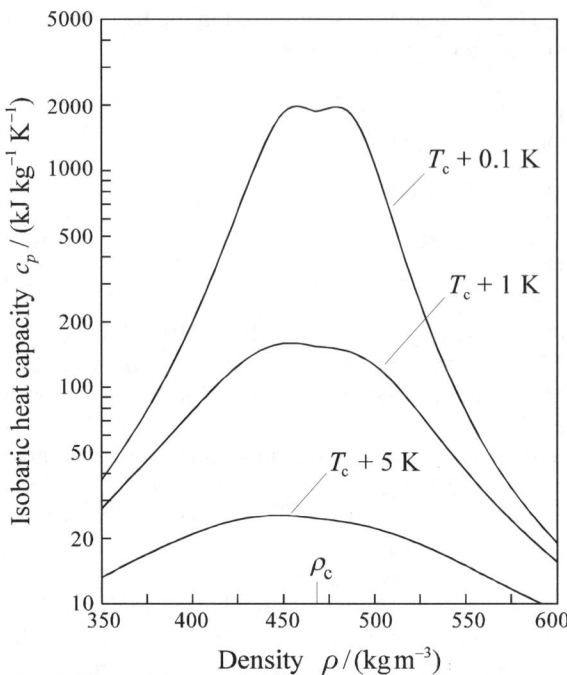

Fig. 4.26. Close to the critical point the sudden changes in higher density derivatives caused by nonanalytical terms result in an unphysical behaviour of the isobaric heat capacity.

However, along the critical isochore, Eq. 4.145 results in changes of higher density derivatives which are small and continuous but so rapid that even modified Gaussian bell shaped terms cannot compensate for this effect. Figure 4.25 shows such a change for the third density derivative of the carbon dioxide equation by Span and Wagner (1996). The observed decrease in $(\partial^3 \alpha / \partial \delta^3)_\tau$ is too small to affect the representation of thermal properties, but it results in an unphysical behaviour of the isobaric heat capacity, see Fig. 4.26.[21] This problem has to be addressed first to further improve empirical nonanalytical terms – for future applications, Eq. 4.145 should be reformulated in a way that its higher density derivatives become well behaved along the critical isochore.

4.5.5 Semiempirical Approaches

Compared to theoretically based descriptions of the critical region, the major advantage of the nonanalytical terms described in the preceding section is that no

---

[21] The isobaric heat capacity is especially sensitive to such effects since the sum in the denominator of Eq. 4.140 is nearly zero close to the critical point and results in the extreme increase of the isobaric heat capacity. Even very small oscillations in α^r_δ or $\alpha^r_{\delta\delta}$ result in visible changes of c_p.

4.5 Describing Properties in the Critical Region

implicit transformations of variables become necessary. With the independent thermodynamic variables (T and ρ) the equation and all its derivatives can be evaluated directly without any iterative steps. For certain purely scientific applications where this feature is regarded as less important, semiempiric approaches can be used in combination with multiparameter equations of state to achieve a nonanalytical limiting behaviour. The following sections summarise these methods just briefly, since the related problems depend mainly on the theoretical approaches used and not on the applied multiparameter equations of state.

4.5.5.1 The Use of Switching Functions

The main idea of switching approaches is that two equations of state, a scaled equation of state for the description of the critical region and an empirical multiparameter equation for the description of the whole fluid region, are linked by a third function, the so called *"switching"* or *"blending function"*. This function combines both equations in such a way that the results of the analytic multiparameter equation are dominant far away from the critical point while the results of the nonanalytical scaled equation dominate close to the critical point.

The first well known application of switching functions was documented in the IUPAC monograph on thermodynamic properties of carbon dioxide which was published by Angus et al. (1976). Angus et al. combined the pressure explicit multiparameter equation of state by Altunin and Gadetski (1971) and the simple scaling equation by Schofield (1969) and Schofield et al. (1969) using the switching function by Chapela and Rowlinson (1974). This approach reads

$$p = f(r) \cdot p_{\text{an.}} + (1 - f(r)) \cdot p_{\text{sc.}} \quad (4.152)$$

with the switching function

$$f(r) = 1 - \left[1 - \exp\left(-\left(\frac{k_1}{r}\right)^{n_1}\right)\right] \cdot \left[1 - \exp\left(-\left(\frac{k_2}{r}\right)^{n_2}\right)\right]. \quad (4.153)$$

The independent parametric variable of the switching function, r, corresponds to a distance to the critical point and is defined by Eqs. 4.135–4.136. n_1, n_2, k_1, and k_2 are adjustable parameters of the switching function. For small values of r the switching function $f(r)$ becomes small (almost 0) and $p_{\text{sc.}}$ dominates in Eq. 4.152. For large values of r the switching function $f(r)$ becomes equal to 1 and $p_{\text{an.}}$ dominates.

This simple switching approach resulted in major problems with regard to derived properties such as heat capacities. To avoid unphysical results, caloric properties tabulated by Angus et al. were partly obtained from graphical interpolations. In the switching region, where $f(r)$ changes from 0 to 1 or vice versa, the derivative of Eq. 4.152 with respect to an arbitrary variable x (in practice ρ or T) becomes

$$\left(\frac{\partial p}{\partial x}\right)_y = f(r) \cdot \left(\frac{\partial p_{\text{an.}}}{\partial x}\right)_y + (1 - f(r)) \cdot \left(\frac{\partial p_{\text{sc.}}}{\partial x}\right)_y + (p_{\text{an.}} - p_{\text{sc.}}) \cdot \left(\frac{\partial f}{\partial x}\right)_y \quad (4.154)$$

where $(\partial f/\partial x)_y$ has to be calculated as $(df/dr)\cdot(\partial r/\partial x)_y$ in this case. The first two terms in Eq. 4.154 correspond to the set-up of Eq. 4.152 and are physically expected, but the last term which depends on differences between the equations and on derivatives of the switching function causes significant unphysical contributions to derived properties especially for higher derivatives.

Wooley (1983) showed that this problem is theoretically unavoidable, but that its impact can be minimised by a suitable construction of the switching function. However, with a total of 15 adjustable parameters in the distance function, the switching function proposed by Wooley was far too complicated for practical application.

The breakthrough in switching was achieved by Hill (1990) who combined an analytic Helmholtz equation of state for water with a revised and extended scaled equation of state. His switching approach reads

$$\alpha = \alpha_{\text{an.}} + f \cdot (\alpha_{\text{sc.}} - \alpha_{\text{an.}}) \qquad (4.155)$$

using the switching function

$$f = 1 - \exp\left(\frac{-1}{\exp\left((\zeta/a_1)^4\right) - 1}\right) \quad \text{with} \quad \zeta = \sqrt{\left(\frac{\rho/\rho_c - 1}{a_2}\right)^2 + \left(\frac{1 - T_c/T}{a_3}\right)^2} \qquad (4.156)$$

with only three adjustable parameters a_1, a_2, and a_3. The success of this comparably simple switching approach is based on two facts: The analytic and the scaled equation of state are harmonised in the sense that they yield very similar results in the switching region – the expression $(\alpha_{\text{an.}} - \alpha_{\text{sc.}})$ becomes small when adapting Eq. 4.154 to a Helmholtz approach. And the range where both equations yield similar results is sufficiently broad due to the use of a revised and extended scaled equation on the one side and a sophisticated multiparameter equation on the other side – the switching range can be extended and the expression $(\partial f/\partial x)_y$ in Eq. 4.154 becomes small. The unphysical contribution in Eq. 4.154 cannot be avoided, but it is for the most part negligible.

The features which make Hills approach successful raises criticism at the same time. Since results calculated from the analytic equation have to be very similar to results calculated from the scaled equation throughout the whole switching region, the analytic equation has to be quite complex and advantages of the combined model are restricted to a very small region around the critical point. In addition, since the scaled equation has to have a rather broad range of validity, complex formulations with a large number of adjustable parameters have to be used. Thus, when using this kind of switching approach, one has to pay a high price for small advantages.

4.5.5.2 The Transformation Approach

The transformation approach introduced by Erickson and Leland (1986) and Erickson et al. (1987) is based on the earlier work of Fox (1983), see Sect. 4.5.1.1.

Erickson and Leland replaced the *"damping function"* introduced by Fox by a more complex formulation with four instead of two adjustable parameters. Using this improved damping function they applied the concept proposed by Fox to MBWR type pressure explicit multiparameter equations of state without refitting the parameters of these equations. The reported results for methane show that the analytic equation could be improved with respect to some typical critical region effects, but major shortcomings become visible, too. Obviously a multiparameter equation which describes critical effects as well as possible with analytic terms is unsuitable as a background equation for a scaling approach.

Erickson et al. (1987) modified the damping function again and fitted its parameters simultaneously with the 32 coefficients of the MBWR type background equation to data for pentane and carbon dioxide. The resulting nonanalytical model avoids in part the shortcomings which became obvious in the preceding report and is said to be superior to an equation which uses the form developed by Schmidt and Wagner (1985) for oxygen (O_2-type equation, see Sect. 4.5.2) in the critical region, while the O_2-type equation is superior outside of the critical region. However, state-of-the-art multiparameter equations of state are far better than O_2-type equations even without nonanalytical terms (see Sect. 4.5.2) and the results of the transformation approach are still not asymptotically correct. It does not seem as if the limited advantages justify the numerical expense implied by the use of transformed variables.

The transformation approach has not been further pursued over the last few years. When taking up this idea again today, one would certainly try to scale analytic multiparameter equations of state using recent crossover mechanisms, see Sect. 4.5.1.1.

4.5.5.3 The Approach of Kiselev and Friend

Recently, Kiselev and Friend (1999) have published a semiempirical approach which is similar to the use of nonanalytical terms described in Sect. 4.5.4. According to the crossover theory developed by Chen et al. (1990a), the renormalised Landau expansion of the Helmholtz energy can be written as

$$\Delta A(\tau, \Delta\eta) = a_{12}\,\hat{\tau}\,Y^{-\frac{\alpha}{2\Delta}}\,\Delta\eta^2\,Y^{-\frac{\gamma-2\beta}{2\Delta}} + a_{04}\,\Delta\eta^4\,Y^{-\frac{\gamma-2\beta}{\Delta}} - K(\rho, \hat{\tau}^2) \qquad (4.157)$$

where K is the *"kernel term"* which provides the expected scaling behaviour of the isochoric heat capacity.

When investigating the recent reference equation for water (Wagner and Pruß, 1997 / 2000) which reads

$$\alpha^r(\tau,\delta) = \sum_{i=1}^{7} n_i\,\tau^{t_i}\,\delta^{d_i} + \sum_{i=8}^{51} n_i\,\tau^{t_i}\,\delta^{d_i}\,\exp(-\gamma_i\delta^{p_i})$$

$$+ \sum_{i=52}^{54} n_i\,\tau^{t_i}\,\delta^{d_i}\,\exp\left(-\eta_i(\delta-\varepsilon_i)^2 - \beta_i(\tau-\gamma_i)^2\right)$$

$$+ \sum_{i=55}^{56} n_i \delta \Delta^{b_i} \exp\left(-e_i(\delta-1)^2 - f_i(\tau-1)^2\right) \quad (4.158)$$

with $\quad \Delta = \left\{(1-\tau) + c_i\left[(\delta-1)^2\right]^{1/(2\beta_i)}\right\}^2 + d_i\left[(\delta-1)^2\right]^{a_i}$

Kiselev and Friend found that this kind of multiparameter equation yields an excellent description of the thermodynamic properties in the critical region, but that the nonanalytical terms ($i = 55, 56$) still do not result in the asymptotically correct limiting behaviour for the heat capacities.[22] Thus, they replaced the nonanalytical terms in Eq. 4.158 by a modification of the kernel term in Eq. 4.157 to improve the asymptotic behaviour of the equation. The modified equation reads

$$\alpha^r(\tau,\delta) = \sum_{i=1}^{7} n_i \tau^{t_i} \delta^{d_i} + \sum_{i=8}^{51} n_i \tau^{t_i} \delta^{d_i} \exp\left(-\gamma_i \delta^{p_i}\right)$$

$$+ \sum_{i=52}^{54} n_i \tau^{t_i} \delta^{d_i} \exp\left(-\eta_i(\delta-\varepsilon_i)^2 - \beta_i(\tau-\gamma_i)^2\right)$$

$$+ K(\rho, \hat{\tau}^2) \quad (4.159)$$

with $\quad \hat{\tau} = \dfrac{T - T_c}{T_c};$

the terms 1–54 remain unchanged. The kernel term in Eq. 4.159 reads

$$K = \frac{1}{2} a_{20} \hat{\tau}^2 \left(Y^{-\frac{\alpha}{\Delta}} - 1\right) + \frac{1}{2} a_{21} \hat{\tau}^2 \left(Y^{-\frac{\alpha}{\Delta}+1} - 1\right)$$

$$- a_{12} \left(\frac{\rho - \rho_c}{\rho_c}\right)^2 \left(Y^{-\frac{\gamma-\alpha-2\beta}{2\Delta}} - 1\right) \quad (4.160)$$

where the crossover function Y is implicitly defined by

$$Y = (q^2/R), \quad (4.161)$$

$$q^2 = \frac{\hat{\tau}}{Gi} + b^2 \left(\frac{\Delta \eta}{Gi^\beta}\right)^2 Y^{\frac{1-2\beta}{\Delta}} \quad (4.162)$$

with $\quad \Delta\eta = \left(\dfrac{\rho_c}{\rho} - 1\right) \cdot \left[1 + v_1\left(\dfrac{\rho_c}{\rho} - 1\right)^2 \exp\left(-v_2\left(\dfrac{\rho_c}{\rho} - 1\right)\right)\right],$

---

[22] This conclusion is undenied, see Sects. 4.5.2 and 4.5.4 or Span and Wagner (1996).

and
$$R = \left(1 + \frac{q^2}{q_0 + q}\right)^2.$$
(4.163)

$\alpha, \beta, \gamma, \Delta, b_r^2$ and q_0 are "universal" constants of this approach and $a_{20}, a_{21}, a_{12}, Gi,$ v_1, and v_2 are parameters which were adjusted to experimental data. To evaluate Eq. 4.160 for a given temperature and density, the equations Eqs. 4.161–4.163 have to be solved simultaneously using an iterative solver. To calculate the derivatives required in thermodynamic property calculations from Eq. 4.160 analytically becomes extremely complicated due to this implicit set-up. Thus, this approach shows typical disadvantages of scaled equations, even though it is still rather simple compared to typical crossover equations.[23] Furthermore, the general approach of Kiselev and Friend, as applied to water, is questionable since both the nonanalytical terms in the original equation and the kernel term in the modified equation contribute to all derivatives of α. Except for $(\partial^2 \alpha / \partial \tau^2)_\delta$ and higher temperature derivatives, these contributions may be small, but they are certainly not completely negligible. Thus, such a kernel term should not be added to an existing equation but it has to be fitted together with all other coefficients of the equation and, if possible, it should be considered when optimising the functional form of the equation. These demands make the whole approach increasingly complex.

However, based on the results reported by Kiselev and Friend, the modified equation seems to be superior to the equation with nonanalytical terms especially with regard to the asymptotic behaviour of the isochoric heat capacity. Since the basic concept of adding a nonanalytical contribution to an analytic multiparameter equation is the same as for the comparably simple nonanalytical terms, this result gives reason to believe that further improvements of empirical nonanalytical terms are possible too, even though the principal restrictions discussed in Sect. 4.5.4 remain valid. Hopefully, such improved nonanalytical terms can preserve their explicit set-up and their relative simplicity.

4.6 Consideration of the Extrapolation Behaviour

Over the years, considerable interest in thermodynamic properties of fluids at very high pressures and temperatures has resulted mainly from applications in geology, petrology and geophysics. Several simple equations of state have been developed especially for these applications. Usually, these equations are valid only in restricted ranges of temperature and pressure and they fail to properly represent accurate experimental data at lower temperatures and pressures. On the other hand, empirical multiparameter equations of state have often failed with respect to extrapolation beyond the temperature and pressure range of the data to which the equations were fitted. Instead of developing special equations of state for very high

---

[23] A major advantage of the kernel term defined by Eqs. 4.160–4.163 is that it depends directly on the thermodynamic variables T and ρ and not on some kind of parametric variables r and θ, as recent crossover equations usually do. In this way a second level of implicit dependencies is avoided.

temperatures and pressures, it would be desirable to improve the extrapolation behaviour of these accurate equations of state in order to describe the entire range of thermodynamic properties of a fluid as accurately as possible with a single equation of state. With this in mind, the extrapolation behaviour of empirical equations of state was one of the main topics of the Fifth International Workshop on Equations of State which took place at the Ruhr-Universität Bochum in 1990. De Reuck (1991) summarised the results of this discussion, which focused mainly on the so called *"ideal curves"* (see Sect. 4.6.3) of pure fluids. Based on these discussions and the report of de Reuck, Span and Wagner investigated the extrapolation behaviour of multiparameter equations of state further and published a comprehensive report on the current status of our knowledge of the extrapolation behaviour of empirical equations of state (Span and Wagner, 1997). Essentially, the set-up of this section follows the set-up of the report by Span and Wagner.

It is well known that recent multiparameter equations of state are able to represent experimental data accurately up to very high temperatures and pressures; for some examples see Chap. 5. However, for the extrapolation discussion it is more important to point out that state-of-the-art equations of state are not flexible enough to follow systematically wrong courses of single data sets in the high-pressure region. During the optimisation of the mathematical form, strongly correlated pairs of terms are automatically replaced by single terms with similar contributions and steps for the optimisation of the length of the equation are used. Although recent reference equations contain usually 35 or more terms, only a few of these terms contribute significantly to the behaviour in the high-pressure region and the flexibility of the equations in this region is therefore very restricted. To illustrate this concept, Fig. 4.27 shows a comparison between experimental data

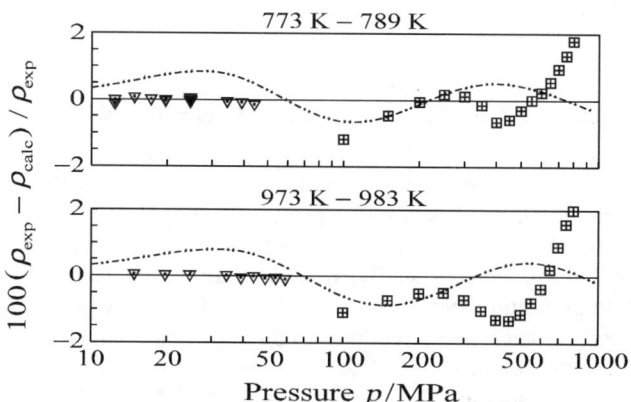

⊞ Shmonov and Shmulovich (1974)   ▽ Vukalovich and Altunin (1962)
-··-··- Sterner and Pitzer (1994)

Fig. 4.27. Percentage deviations of selected $p\rho T$ data at high temperatures and pressures from values calculated from the equation of Span and Wagner (1996). Values calculated from the equation of Sterner and Pitzer (1994) are plotted for comparison.

for carbon dioxide and results from the recent reference equation by Span and Wagner (1996) which is used for the baseline, and from a special high-pressure equation by Sterner and Pitzer (1994, see also Pitzer and Sterner 1995a/b). With its 42 fitted coefficients, the empirical reference equation represents the accurate data at pressures up to 60 MPa much better than the semiempiric equation of Pitzer and Sterner with 28 fitted coefficients, but at pressures above 100 MPa both equations yield similar results and do not follow the faulty course of the data.[24]

In contradiction to common teachings, empirical equations with a carefully optimised mathematical structure are not flexible enough to follow incorrect courses of data sets in the high pressure and temperature region even if these data are the only experimental information which is available in this region. Thus, it can be concluded that they will also be stable enough to yield a reasonable extrapolation behaviour in regions not covered by data. Qualitatively this statement agrees with experiences made during the establishment of other equations of state (see Setzmann and Wagner, 1991; Span and Wagner, 1996; Pruß and Wagner, 1995; Tegeler et al., 1997/1999; Span et al., 1998b/1999; Panasiti et al., 1999; Lemmon et al., 1999; Wagner and Pruß, 2000) but it has not been quantified up to now. It is clear that the limits of the range in which an extrapolation is useful depend on the considered property, on the demanded accuracy, on features of the data set used to establish the equation and finally on features of the equation itself. Thus, a simple answer cannot be expected; systematic studies on this topic are still lacking. However, more systematic information on the extrapolation behaviour of empirical equations of state is available from the following three approaches.

4.6.1 Comparisons with Data Beyond the Range of Primary Data

At pressures and temperatures beyond the range covered by reliable experimental pure fluid data, which are usually used to establish multiparameter equations of state, additional experimental information on fugacities is available particularly for substances of geological interest. The origin of these data are measurements of chemical equilibria and their evaluation depends on sets of thermodynamic data of the other components involved in the chemical equilibrium. The resulting fugacities vary significantly depending on the assumptions made for the other components. Geologists are familiar with the internally consistent sets of thermodynamic data needed for the evaluation of the measured equilibria, but scientists working on reference equations of state are usually not. Thus, it would be valuable to set up a pure-component data base by calculating the corresponding fugacities from the equilibrium data published mainly in geological literature.

For carbon dioxide, which can be used again as an example, only Haselton et al. (1978) have published pure-component fugacities which are derived from the evaluation of their experimental results for the decarbonation of magnesite and calcite. Figure 4.28 compares these data with results calculated from the equations of state of Span and Wagner (1996) and of Sterner and Pitzer (1994). While the

---

[24] For a more detailed discussion on the shortcomings of the data sets of Shmonov and Shmulovich (1974) see Sterner and Pitzer (1994)

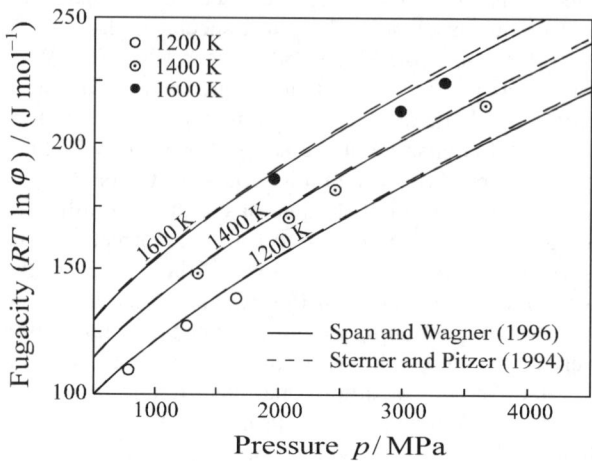

Fig. 4.28. Fugacities of carbon dioxide calculated from the equations of Span and Wagner (1996) and Sterner and Pitzer (1994) at high temperatures. The corresponding experimental results of Haselton et al. (1978) are given as symbols indicating the isotherm to which they belong.

semiempiric equation of Sterner and Pitzer was fitted to an extensive set of fugacities calculated from published equilibria data, the data of Haselton et al. were used only for an assessment of the extrapolation behaviour during the development of the equation of Span and Wagner. Nevertheless, the empirical reference equation, which is fitted to data only up to pressures of 800 MPa and temperatures of 1073 K, yields a slightly better representation of these data up to 1600 K and more than 3600 MPa. This observation clearly supports the claim that the extrapolation behaviour of state-of-the-art multiparameter equations of state is far better than usually expected.

At even higher pressures and temperatures shock-wave measurements of the Hugoniot curve are available for some substances. The evaluation of the Hugoniot relation

$$h_h - h_0 = 0.5(p_h - p_0) \cdot (\rho_0^{-1} + \rho_h^{-1}) \quad (4.164)$$

or

$$u_h - u_0 = 0.5(p_h + p_0) \cdot (\rho_0^{-1} + \rho_h^{-1}) \quad (4.165)$$

yields data for the enthalpy h_h or the internal energy u_h as a function of pressure p_h and density ρ_h at shock wave pressures up to tens of GPa and temperatures of several thousand Kelvins; the index 0 corresponds to the initial state prior to release of the shock wave. Consideration of these data in the development of multiparameter equations of state results in nonlinear relations which require iterative solutions for temperature. For optimised functional forms, nonlinear fits which involve Hugoniot data hardly exercise an influence on the representation of these

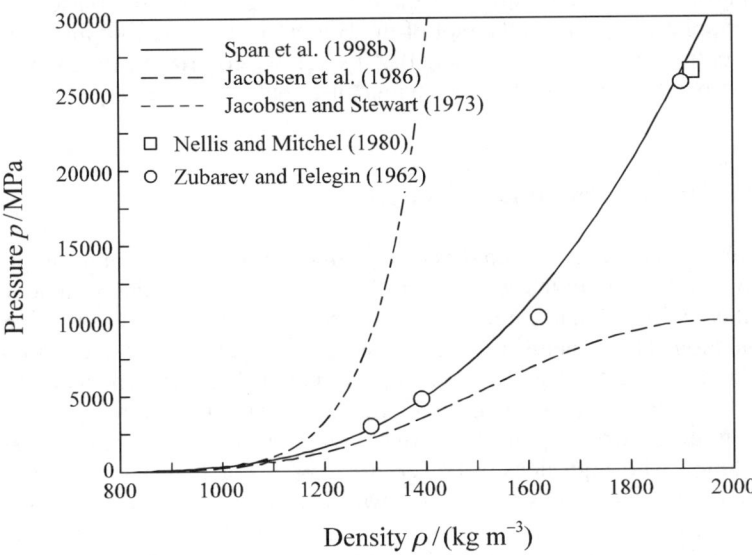

Fig. 4.29. Plot of the Hugoniot curve with $T_0 = 77.5$ K and $h_0 = h'(T_0)$ of nitrogen as calculated from three multiparameter equations of state. The corresponding experimental results are given as symbols.

data – the extrapolation behaviour of equations of state is closely related to their functional form (see also Sect. 4.6.2). Any attempt to use Hugoniot data in linear optimisation algorithms requires a precorrelation of the temperatures belonging to the data points. Since no $p\rho T$ data, which could verify the precorrelated temperatures, are available under these extreme conditions, this approach implies the risk of distorting the experimental information. Thus, data for the Hugoniot curve could only be used for comparisons until recently, but for state-of-the-art equations these comparisons showed very reasonable results up to extreme pressures and temperatures (see for instance Setzmann and Wagner, 1991; Span and Wagner, 1996).

During the development of the new reference equation of state for nitrogen (Span et al., 1998b/1999) Hugoniot data were used directly according to Eq. 4.164 as $\Delta h = h(p_h, \rho_h) - h(p_0, \rho_0)$ in combination with the nonlinear optimisation algorithm of Tegeler et al. (1997/1999; see Sect. 4.4.6). This procedure resulted in an exceptionally good representation of the Hugoniot curve although the Hugoniot data were used with only low weights; their contribution to the sum of squares was less than 1% of the total sum of squares. Figure 4.29 shows the plot of the Hugoniot curve as calculated from three multiparameter equations of state for nitrogen. Obviously, state-of-the-art reference equations of state are not necessarily inferior to special high-pressure equations with regard to the representation of Hugoniot data, but it remains questionable whether the good representation of the data by the recent equation of Span et al. (1998b) is a kind of overfitting or not, since little is known about the accuracy of the data. In addition, when considered as the only extrapolation criterion, the plot of the Hugoniot curve may be misleading (for an example, see Span and Wagner, 1997) – further criteria are required to assess or to improve the extrapolation behaviour of a multiparameter equation of state.

4.6.2 The Influence of the Functional Form

Generally speaking, the terms in empirical equations using the general set-up shown in Eq. 3.25[25] are highly intercorrelated and it is assumed that the behaviour of an empirical equation of state cannot be associated with the behaviour of single terms in the equation. This conclusion is true in most cases, but not for the extrapolation to very high temperatures and very high pressures, which also correspond to high densities. For high densities, the behaviour of the equation is influenced only by polynomial terms and by exponential terms with $\exp(-\delta^1)$, if the exponential terms are combined with high density powers ($d_i \geq \approx 5$). For high temperatures, which correspond to small values of τ, terms with high temperature exponents t_i fade away as well. Under these conditions, which are typical for the region covered by Hugoniot data, one or a small number of leading polynomial terms can be identified which determines the behaviour of the whole equation. Based on this, an attempt was made by Span and Wagner (1997) to formulate demands on the mathematical form of an empirical equation of state in order to ensure a more reasonable extrapolation behaviour. These demands can be summarised in the following way:

- The number of polynomial terms in the equation should be small, if possible less than 10.

Intercorrelations between the polynomial terms affecting the extrapolation to high densities cannot be avoided completely this way, but they are reduced considerably. At the same time, an increased number of exponential terms with $\exp(-\delta^1)$ has to be used to guarantee the necessary flexibility of the equation in the range of intermediate densities. For functional forms like this, usually only one or two terms determine the behaviour of the equation in the range of very high temperatures and densities. The term which is dominant at high densities (high values of δ) and high temperatures (small values of τ) is the polynomial term with the smallest temperature exponent t_i among the terms with the highest density exponents d_i. For this term, the following requirements can be formulated:

- The coefficient n_i has to be positive to yield a positive contribution to the residual pressure.
- The temperature exponent should fulfil the condition $0 < t_i < 1$ since the pressure should increase on an isochore with increasing temperature but the compression factor should decrease.

---

[25] Additional critical region terms included in some reference equations of state do not affect the behaviour in the high pressure, high temperature region.

- The density exponent d_i has to be an integer value and should be equal to 3 or 4. This is a purely empirical finding which results from a study of the compressibilities calculated both from empirical and semiempirical equations of state in the high pressure, high density region (see Span and Wagner, 1997).

These conditions were considered during the development of the new reference equations of state for carbon dioxide (Span and Wagner, 1996), water (Pruß and Wagner, 1995; Wagner and Pruß, 1997/2000), argon (Tegeler et al., 1997/1999), and nitrogen (Span et al., 1998b/1999). For the new equation of state for carbon dioxide, Fig. 4.30 shows the relative contributions of all polynomial terms, of all exponential terms with $\exp(-\delta^{-1})$, and of the leading polynomial term to the residual pressure p^r; the plotted lines correspond to isotherms. In the region where the available Hugoniot data indicate that carbon dioxide is still chemically stable, roughly for reduced densities $4.5 < \rho/\rho_c < 5.5$ and reduced temperatures $5 < T/T_c < 15$, a polynomial term with $d_i = 3$ and $t_i = 0.75$ dominates the behaviour of the equation with a contribution of more than 70% of the total residual pressure. Since this dominant term fulfils the requirements given above, the whole equation behaves reasonably in the high density limit. The negative contribution of the exponential terms in the range of intermediate densities was not desired with respect to extrapolation but unavoidable for the representation of accurate data at lower temperatures; at least with respect to basic properties like pressure, enthalpy, and fugacity it does not affect the extrapolation up to the limits where spontaneous disintegration occurs.

Similar to carbon dioxide, the final mathematical form is a compromise between requirements for representing the data set and the extrapolation behaviour for the

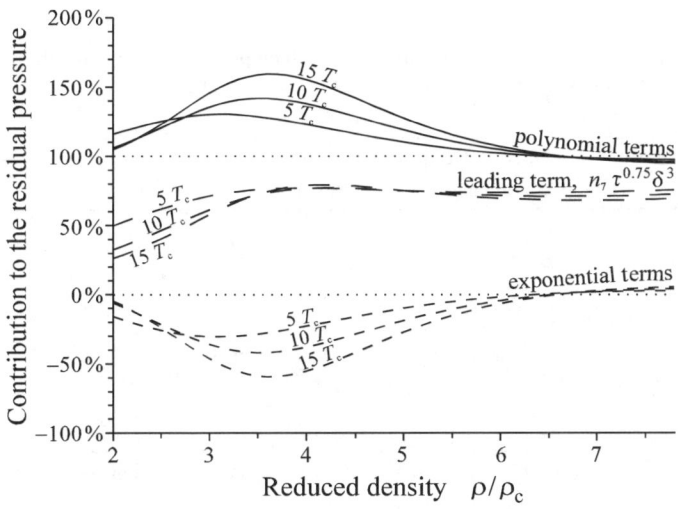

Fig. 4.30. Contribution to the residual pressure of all polynomial terms, of all exponential terms with $\exp(-\delta^1)$, and of the leading polynomial term of the equation of Span and Wagner (1996) for carbon dioxide.

other reference equations as well. But nevertheless, the extrapolation behaviour of an empirical equation of state becomes predictable from its mathematical structure by such investigations and unreasonable behaviour can be avoided. The new equations of state for carbon dioxide, water, argon, and nitrogen yield reasonable results up to extreme temperatures and pressures.

The results discussed above are applicable to formulations in terms of the compression factor as well, since the exponents d_i and t_i of the polynomial terms do not change when the equation for Z is integrated to yield α^r, see Sect. 3.1.2.1.

4.6.3 The Representation of Ideal Curves

Ideal curves are curves along which one property of a real fluid is equal to the corresponding property of the hypothetical ideal gas at the same temperature and density. Based on this very general definition, ideal curves can be defined for almost every property, but usually the discussion is focused on the ideal curves of the compression factor and its first derivatives; these curves are given in Table 4.11 together with their definitions. In the 1960s, there was an intensive investigation of ideal curves in order to specify criteria for a generalised behaviour of pure fluids. Well known results are those of Brown (1960) which were summarised by Rowlinson (1965) and the results of Gunn et al. (1966) and Miller (1970) on the Joule-Thomson inversion curve. Less well known are the very detailed studies of Morsy (1963), Straub (1964), and Schaber (1965) which have been published only in German. More recently, Angus (1983) and de Reuck (1991) gave short summaries of the known characteristics of ideal curves. Although various authors have stated that the representation of ideal curves is a sensitive test for the extrapolation

Table 4.11. The zeroth- and first-order ideal curves of the compression factor and their definition in terms of the compression factor, $Z(T,\rho)$, and of the residual part of the reduced Helmholtz energy, $\alpha(T,\rho)$.

Designation	Definition in terms of the	
	Compression factor	Residual Helmholtz energy
(Classical) Ideal curve	$Z = 1$	$\left(\dfrac{\partial \alpha^r}{\partial \delta}\right)_\tau = 0$
Boyle curve	$\left(\dfrac{\partial Z}{\partial \rho}\right)_T = 0$	$\left(\dfrac{\partial \alpha^r}{\partial \delta}\right)_\tau + \delta \left(\dfrac{\partial^2 \alpha^r}{\partial \delta^2}\right)_\tau = 0$
Joule-Thomson inv. curve	$\left(\dfrac{\partial Z}{\partial T}\right)_p = 0$	$\left(\dfrac{\partial \alpha^r}{\partial \delta}\right)_\tau + \delta \left(\dfrac{\partial^2 \alpha^r}{\partial \delta^2}\right)_\tau + \tau \left(\dfrac{\partial^2 \alpha^r}{\partial \delta \partial \tau}\right) = 0$
Joule inversion curve	$\left(\dfrac{\partial Z}{\partial T}\right)_\rho = 0$	$\left(\dfrac{\partial^2 \alpha^r}{\partial \delta \partial \tau}\right) = 0$

4.6 Consideration of the Extrapolation Behaviour

behaviour of equations of state, systematic investigations have always dealt with results for simple model fluids, with simple equations of state, or with values derived directly from experimental data or from compression factors tabulated for corresponding states approaches. In order to verify whether ideal curves are really useful for assessing the extrapolation behaviour of empirical equations of state, Span and Wagner (1997) compared the ideal curves calculated from equations of state for argon (Stewart and Jacobsen, 1989; Tegeler et al., 1997/1999), nitrogen (Jacobsen and Stewart, 1973; Jacobsen et al., 1986a; Span et al., 1998b/1999), oxygen (Schmidt and Wagner, 1985), methane (Setzmann and Wagner, 1991), ethane (Friend et al., 1991), carbon dioxide (Sterner and Pitzer, 1994; Span and Wagner, 1996), water (Wagner and Pruß, 1997/2000), and helium (Sychev et al., 1984) with each other and with the "theoretical" predictions.

Figure 4.31 shows a typical plot of the ideal curves discussed here in a reduced pressure, temperature diagram with logarithmic axes. The broken lines indicate the limits of the regions where primary data (usually $p\rho T$ data) are available for the corresponding substance. For reference substances with low critical temperatures and pressures like nitrogen and argon, the Boyle, the Ideal, and the Joule-Thomson inversion curve lie completely within the range covered by primary data; for helium, even the Joule inversion curve lies within this range. The situation changes if substances with higher values for the critical temperature and the critical pressure or with a more restricted data set are investigated. For carbon dioxide and methane the Joule-Thomson inversion curve reaches into the extrapolation region; for wa-

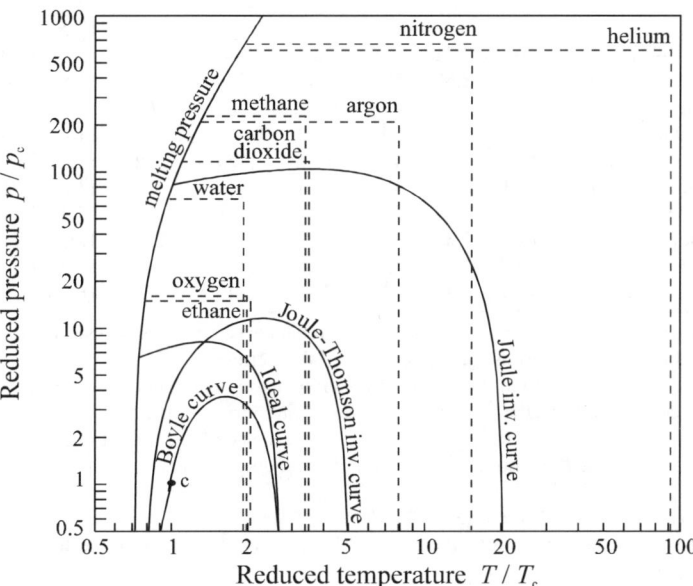

Fig. 4.31. A typical plot of the considered ideal curves in a reduced double-logarithmic pressure, temperature diagram. The dashed lines indicate the regions where primary data are available for the substances considered by Span and Wagner (1997).

ter, oxygen, and ethane the Boyle and the Ideal curve exceed the temperature range and the Joule inversion curve also exceeds the pressure range covered by data.

When considering results of earlier investigations, certain features of the ideal curves should be universal at least for simple substances (see Gunn et al., 1966; Miller, 1970; Schaber, 1965). Table 4.12 summarises the results of a comparison of these "theoretical predictions" with the corresponding values calculated from the equations of state considered by Span and Wagner (1997). For nitrogen and carbon dioxide, the results of different equations are given for comparison.

When analysing the results given in Table 4.12 in more detail, it becomes obvious that the numerical values for the characteristic points of the different ideal curves are useful as criteria for an assessment of the extrapolation behaviour only for simple substances with limited data sets – and for such substances other corresponding states approaches yield reliable results as well. A comparison between the different equations of state for nitrogen and carbon dioxide explains this thesis. Although the accuracy of reference equations of state for nitrogen in the high temperature, high pressure region has been improved substantially since 1973, the

Table 4.12. Characteristic values of the ideal curves as predicted from "theoretical" studies and as calculated from accurate equations of state for helium, argon, methane, oxygen, nitrogen, ethane, carbon dioxide, and water.

Subst.	Equation	ω	$\theta_\mathrm{B}, \theta_\mathrm{I}$ [a]	θ_JT [b]	θ_J [c]	$\delta_\mathrm{B}\|_{T=T_\mathrm{c}}$ [d]	$\delta_\mathrm{I}\|_{T=T_\mathrm{c}}$ [e]	$\pi_{\mathrm{max,JT}} / \theta$ [f]
predicted values		0.00	2.66	5.00	19.5	1.435	2.235	11.8 / 2.25
He	Sychev et al. (1984)	–0.39	4.67	9.03	29.9	1.407	2.201	17.5 / 4.48
Ar	Tegeler et al. (1997)	0.00	2.71	5.07	20.3	1.425	2.201	11.6 / 2.30
CH_4	Setzmann & Wagner (1991)	0.01	2.67	5.14	–[g]	1.428	2.203	11.8 / 2.28
O_2	Schmidt & Wagner (1985)	0.02	2.62	4.90	–[g]	1.425	2.201	11.7 / 2.25
N_2	Span et al. (1998b)	0.04	2.59	4.82	17.4	1.425	2.193	11.6 / 2.20
	Jacobsen et al. (1986a)	0.04	2.60	4.83	16.6	1.423	2.191	11.6 / 2.20
	Jacobsen & Stewart (1973)	0.04	2.58	4.76	16.1	1.429	2.195	11.6 / 2.21
C_2H_6	Friend et al. (1991)	0.10	2.45	4.57	–[g]	1.428	2.196	12.1 / 2.14
CO_2	Span & Wagner (1996)	0.23[h]	2.37	4.45	26.9	1.441	2.232	12.5 / 1.94
	Sterner & Pitzer (1994)	0.23[h]	2.36	4.29	23.8	1.448	2.242	13.4 / 1.90
H_2O	Wagner & Pruß (1997)	0.34	2.35	3.93	7.7	1.540	2.638	18.7 / 1.96

[a] Reduced temperature $\theta = T/T_\mathrm{c}$ at which the Boyle and the Ideal curve end for $\rho = 0$; at this temperature the condition $B(T) = 0$ holds for the second virial coefficient.
[b] Reduced temperature $\theta = T/T_\mathrm{c}$ at which the Joule-Thomson inversion curve ends for $\rho = 0$; at this temperature the condition $dB/dT = B/T$ holds for the second virial coefficient.
[c] Reduced temperature $\theta = T/T_\mathrm{c}$ at which the Joule inversion curve ends for $\rho = 0$; at this temperature the condition $dB/dT = 0$ holds for the second virial coefficient.
[d] Reduced density $\delta = \rho/\rho_\mathrm{c}$ on the Boyle curve for $T = T_\mathrm{c}$.
[e] Reduced density $\delta = \rho/\rho_\mathrm{c}$ on the Ideal curve for $T = T_\mathrm{c}$.
[f] Reduced pressure $\pi = p/p_\mathrm{c}$ and temperature $\theta = T/T_\mathrm{c}$ at the pressure maximum of the Joule-Thomson inversion curve.
[g] Equations without a maximum in $B(T)$ yield no intersection between the Joule inversion curve and the axis $\rho = 0$.
[h] Calculated from an extrapolation of the vapor pressure equation given by Span and Wagner (1996).

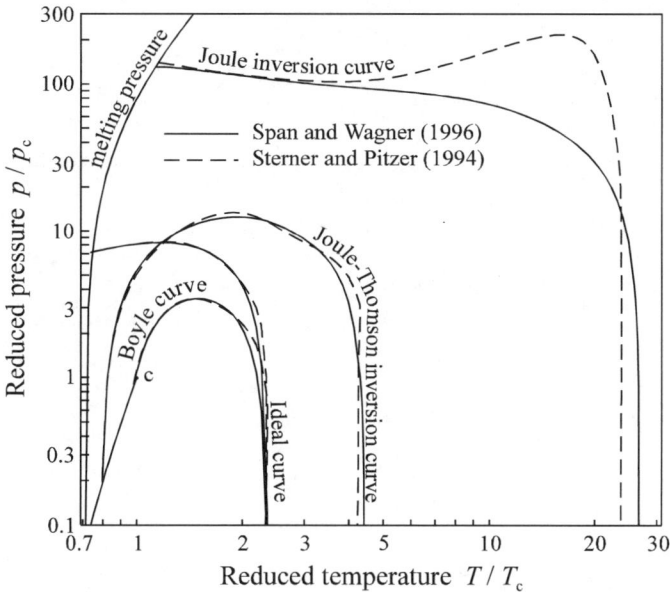

Fig. 4.32. Plots of the zeroth- and first-order ideal curves of the compression factor calculated from the equations of Span and Wagner (1996) and Sterner and Pitzer (1994) for carbon dioxide.

three investigated equations yield very similar results for the characteristic points of the ideal curves; based on these results, no assessment of the equations is possible. For carbon dioxide, the differences between the equations of Span and Wagner (1996) and Sterner and Pitzer (1994) are larger. The difference observed for the maximum pressures on the Joule-Thomson curve would be sufficient for conclusions regarding the reliability of the equations, if such differences would occur for methane or nitrogen. But since carbon dioxide does not belong to the group of simple substances, the observed differences are still not significant enough for an assessment.

Nevertheless, the plot of the ideal curves itself contains important information on the behaviour in the high temperature, high pressure region. To demonstrate the sensitivity of this graphical criterion, Fig. 4.32 shows the plot of the ideal curves of carbon dioxide as calculated from the equation of Span and Wagner (1996) and Sterner and Pitzer (1994). From Fig. 4.27 it can be seen that the equation of Sterner and Pitzer deviates from the data of Vukalovich and Altunin (1962) by up to $\Delta \rho / \rho \approx 1\%$ for $T/T_c \leq 3.5$ and $p/p_c \leq 8$; in Fig. 4.32, these deviations result in visible deformations of the Boyle and the ideal curve. At higher temperatures, stronger deformations of the Joule-Thomson and the Joule-inversion curve occur. Thus, in this order of magnitude, oscillations of an equation of state can be detected easily by an investigation of its ideal curves.

5 The Performance of Multiparameter Equations of State

A detailed assessment of the performance of an equation of state is an ambitious task and requires a sound knowledge both of the features of the equation and of the data set which is available for the corresponding substance. The difficulties which are linked to such assessments become obvious where the performances of different equations of state have to be compared with each other to determine, e.g., accepted standards for the properties of a certain substance. Such detailed studies fill extensive reports (see for example Kilner and Craven, 1996; Klimeck et al., 1996) and cannot be replaced by general assessments in any way. However, on a more general level, equations of state can be assigned to different categories and typical features can be described for equations which belong to a certain category. This approach is useful for inexperienced users, since it helps convey what can be expected from a certain kind of equation of state in general; the user understands the restrictions of the applied equations of state and more detailed investigations on crucial features may follow where necessary.

At this point, it is convenient to introduce three categories of multiparameter equations of state. The first two categories comprise so called *"reference equations of state"*. Reference equations are intended for use both as technical and scientific standards for the thermodynamic properties of a certain substance. Ideally they are able to represent all the experimental data for thermodynamic properties of the corresponding fluid within their respective uncertainty. However, depending both on the available data sets and on the techniques used to establish the equations, the achieved accuracies can be rather different.

- *Group one* reference equations of state are highly accurate multiparameter equations of state which describe the thermodynamic properties of substances with exceptionally high quality data. In general, the accuracy of such equations corresponds to the accuracy which can be reached when using state-of-the-art experimental set-ups – group one equations can only be established for substances where such highly accurate data are available. Once established they can be used for the whole spectrum of advanced technical and scientific applications which cover, e.g., calibration needs, conversions of measured variables,[1] tests of new measurement devices, or tests of physical models. However, group one equations of state can be used for process calculations and other

---

[1] Very often thermophysical properties or parameters of technical processes depend on temperature and density while temperature and pressure are measured – accurate equations of state are required to calculate the corresponding densities without loss of accuracy.

engineering applications without special demands on accuracy as well. Recent software solutions and the computing power available today allow an unre-

Table 5.1. A selection of recent reference equations of state.

Fluid	Reference	Funct. form (no. of coeff.)	Range of validity $T/$K	$p_{max}/$MPa
Group one reference equations of state				
Argon	Tegeler et al. (1997 / 1999)	Helmholtz[a,b,c] (41)	83 – 700	1000
Nitrogen	Span et al. (1998b / 1999)	Helmholtz[a,b,c] (36)	63 – 1000	2200
Carbon dioxide	Span & Wagner (1996)	Helmholtz[a,b,c,d] (42)	216 – 1100	800
Water[e]	Wagner & Pruß (1997 / 2000)	Helmholtz[a,b,c,d] (56)	273 – 1273	1000
Methane[f]	Setzmann & Wagner (1991)	Helmholtz[a,b,c] (40)	90 – 625	1000
Ethylene	Smukala et al. (1999)	Helmholtz[a,b,c] (35)	104 – 450	300
Group two reference equations of state				
Helium	McCarty & Arp (1990)	Pressure[b,g] (32)	2 – 1500	100
Neon	Katti et al. (1986)	Helmholtz[a,b] (29)	24 – 700	700
Xenon[h]	Šifner & Klomfar (1994)	Pressure[i] (63)	161 – 800	350
Oxygen[f]	Schmidt & Wagner (1985)	Helmholtz[a,b] (32)	161 – 800	350
Chlorine[h,j]	Angus et al. (1985)	Pressure[a,b] (18)	180 – 900	25
Ammonia	Tillner-Roth et al. (1993)	Helmholtz[a,b] (21)	195 – 700	1000
Ethane	Friend et al. (1991)	Helmholtz[b,l] (32)	90 – 500	60
Propane	Younglove & Ely (1987)	Pressure[b,g] (32)	85 – 600	100
n-Butane[m]	Younglove & Ely (1987)	Pressure[b,g] (32)	134 – 500	70
Isobutane[h]	Younglove & Ely (1987)	Pressure[b,g] (32)	113 – 600	35
Cyclohexane[h]	Penoncello et al. (1995)	Helmholtz[a,b] (26)	279 – 700	80
Propene[f,h]	Angus et al. (1980)	Pressure[a,b] (21)	87 – 575	1000
Methanol[f]	de Reuck & Craven (1993)	Helmholtz[a,b,k] (44)	175 – 570	800
R11	Marx et al. (1992)	Helmholtz[a,b] (21)	163 – 525	200
R12	Marx et al. (1992)	Helmholtz[a,b] (22)	116 – 525	200
R22	Wagner et al. (1993)	Helmholtz[a,b] (22)	116 – 525	200
R32[n]	Tillner-Roth & Yokozeki (1997)	Helmholtz[a,b] (19)	136 – 435	70
R113	Marx et al. (1992)	Helmholtz[a,b] (18)	237 – 525	200
R123[n]	Younglove & McLinden (1994)	Pressure[b,g] (32)	166 – 500	40
R124	de Vries et al. (1995)	Helmholtz[a,b] (20)	100 – 400	40
R125[h,n,o]	Piao & Noguchi (1998)	Pressure[b] (18)	173 – 475	68
R134a[n]	Tillner-Roth & Baehr (1994)	Helmholtz[a,b] (21)	170 – 455	70
R143a[n]	Lemmon & Jacobsen (1999b)	Helmholtz[a,b] (20)	161 – 500	60
R152a	Outcalt & McLinden (1996)	Pressure[b,g] (32)	162 – 453	35
Air[p]	Lemmon et al. (1999)	Helmholtz[a,b] (19)	60 – 2000	2000

[a] optimised for the corresponding substance (see Sect. 4.4)
[b] uses polynomial and exponential terms
[c] additionally uses Gaussian bell shaped terms (see Sects. 3.1.3.2 and 4.5.3)
[d] additionally uses nonanalytic terms (see Sects. 3.1.3.2 and 4.5.4)
[e] IAPWS-95 formulation; recomended as standard by the International Association for the Properties of Water and Steam (IAPWS)
[f] recomended as standard by the International Union of Pure and Applied Chemistry (IUPAC)
[g] MBWR-type equation using the functional form introduced by Jacobsen and Stewart (1973)

stricted use of highly accurate equations of state in everyday work. As a result of the available experimental techniques, high demands are placed on the accuracy of calculated $p\rho T$, phase equilibrium, and speed of sound data. Typical features of group one reference equations of state are high accuracies in the technically and scientifically most important regions,[2] an accurate description of properties in the critical and extended critical region, and reasonable extrapolation behaviour. Both the work on the experimental foundation and the development of such equations are very time consuming. Thus, group one reference equations of state are only available for a few substances to date.

- *Group two* reference equations of state are also designed as high quality equations for the properties of certain substances but, due to restrictions regarding the available data set and / or the set-up of the equation, they do not satisfy the high demands on group one equations of state. Typical group two equations are less accurate than group one equations,[3] do not use special approaches for the description of the critical and extended critical region, and do not guarantee a reasonable extrapolation behaviour. Such equations of state are still accurate enough for most technical applications and for a variety of scientific applications but they should be used with care where very high accuracies are required (e.g. for calibrations or for scientific and advanced technical applications in the extended critical region). Group two equations are available for many substances. During the last decade, especially the Annex 18 project of the International Energy Agency (IEA) has extended the list of available equations for the technically most important new HCFC and HFC working fluids. However, the number of substances for which group two reference equations are available is

---

[h] equation lags behind recent expectations regarding the accuracy of reference equations of state significantly with respect to several features
[i] uses only polynomial terms and simple, polynomial based critical region terms
[j] recommended as tentative standard by the International Union of Pure and Applied Chemistry (IUPAC)
[k] uses special terms to account for polymerisation in the gas phase
[l] O_2-type equation using the functional form introduced by Schmidt and Wagner (1985)
[m] equation lags behind recent expectations regarding the accuracy of reference equations of state with respect to some features
[n] recommended as standard by the International Energy Agency (IEA), Annex 18
[o] the equation by Piao et al. (1998) is referred to since it is the recommended IEA Annex 18 standard for R125; a state-of-the-art equation was published for R125 by Sunaga et al. (1998)
[p] air can be treated as pseudo pure fluid for calculations in the homogeneous phases; the article of Lemmon et al. (1999) additionally contains a mixture model which allows phase equilibrium calculations

---

[2] The vapour-liquid phase boundary and the homogeneous fluid at temperatures up to 500 K and pressures to 30 MPa are regarded as "technically and scientifically most important regions".
[3] Some of the equations which are regarded as group two reference equations here reproduce all of the data which are available for the corresponding fluid within their experimental uncertainty, just like group one reference equations are expected to do. However, the data sets for the corresponding substance are not as accurate as is expected for reference data – in this sense, a classification as group two equation is not always a devaluation of the equation of state.

still restricted.
- Group three or *technical equations of state* are simple multiparameter equations of state which can be found in many engineering applications. Typical representatives of this group of equations of state range from the very simple original BWR-type equation with 8 terms (Benedict et al., 1940) to the modified BWR equation by Starling (1973) with 12 terms up to the equation developed by Bender (1970) with 19 terms; other types of comparable equations are located between BWR- and Bender-type equations. Such equations are available for a broad variety of substances (see for instance Polt, 1987 and Platzer, 1990 for Bender-type equations), but their accuracy is questionable. The functional forms used do not result from certain technical demands but simply from historical developments – when these equations were established, computing power was a severe limitation both for those who developed the equations and for those who applied them. In addition, the theoretical "know how" on empirical equations of state was still limited. The problem with such equations is that only some users are aware of the restricted accuracy, especially of the BWR-type equation and its simple modifications; the resulting uncertainties may affect typical technical calculations very well.

Table 5.1 summarises information on available reference equations of state. This listing is certainly not complete with respect to group two equations, but it gives a comprehensive overview and covers the most important developments of the last two decades. For substances where different equations are available which conform to the standards given for groups one and two above, only the equation which is considered the best choice is given in general.

While group one is rather homogeneous with regard to the performance of the assigned equations, the second and third group cover equations with very different accuracies. Further subdivisions would be necessary to make these groups more homogeneous as well, but in this case it would become more difficult for the user to assign an equation of state to one of the groups. Thus, it seems more appropriate to introduce only these three groups and to point out the range of performances which are typical for the different groups.

For different substances the following sections show exemplary comparisons between selected experimental data and values calculated from equations of state which are regarded as typical for one of the three groups introduced above. To characterise the performance of group two equations, it is advantageous to rely on substances where highly accurate data are available and group one reference equations have become available recently. In this way, a clear reference for the thermodynamic "truth" is given and uncertainties of the older equations can be assessed easily. It will be shown that these results are typical for other substances as well.

Reference equations are used exactly as published by the authors, except for the fact that changes in the valid temperature scale are taken into account (see Sect. 3.1.1). BWR-type, Starling-type, and Bender-type equations are used as examples for typical technical equations of state. Since these equations were originally fitted to data sets which are regarded as obsolete today and since they were mostly fitted to only data of certain properties covering only parts of the thermodynamic surface, it would be misleading to use the original equations in comparisons. Therefore, these equations were refitted to the data sets described in Sects. 6.2.2–6.2.3. A disadvantage of this approach is that it makes the shown comparisons irreproducible for others,[4] but in this way more general conclusions regarding the performance of certain types of equations of state can be drawn; otherwise the results of comparisons would depend mainly on the data sets used originally and would not be representative for other substances.

5.1 Comparisons with Thermal Properties

With regard to the calculation of $p\rho T$ data, the smallest relative uncertainties are achieved in the liquid phase of water at moderate temperatures and pressures. In this region, the uncertainty of the recent IAPWS standard for general and scientific use, the IAPWS-95 formulation by Wagner and Pruß (1997/2000), reaches from $\Delta\rho/\rho \leq \pm 0.0001\%$ at ambient pressures to $\Delta\rho/\rho \leq \pm 0.003\%$ up to pressures of 100 MPa. The representation of selected, highly accurate data in this region is shown in Fig. 5.1. Although extensive data sets are available for water (see Sato et al., 1991), the uncertainty of thermal properties is rather large outside the liquid region, at least when compared to other reference substances. On average, the available data are old and high values for the critical temperature and pressure ($T_c = 647.096$ K, $p_c = 22.064$ MPa) make the conditions for experimental investigations very inconvenient. Only the liquid phase is easy to access.

When considering the representation of thermal properties over the whole range of fluid states, the best data situation can be found for reference substances with low critical temperatures and pressures, such as nitrogen ($T_c = 126.192$ K, $p_c = 3.396$ MPa), argon ($T_c = 150.687$ K, $p_c = 4.863$ MPa), and methane ($T_c = 190.564$ K, $p_c = 4.599$ MPa). For these substances, highly accurate data are

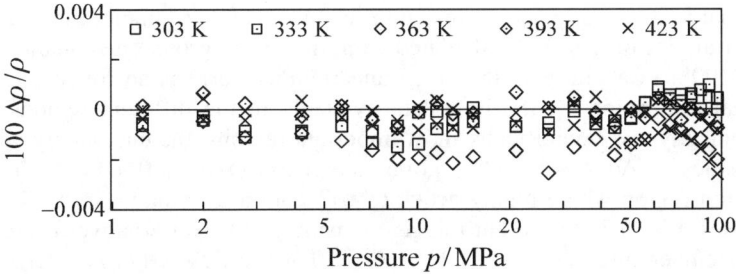

Fig. 5.1. Percentage deviations $100\,\Delta\rho/\rho = 100\,(\rho_{\text{exp}} - \rho_{\text{calc}})/\rho_{\text{exp}}$ between highly accurate experimental results for the density of liquid water by Kell and Whalley (1975) and values calculated from the equation of Wagner and Pruß (1997/2000).

---

[4] To publish the results of the fits in extensive tables is regarded as needless here since a superior type of technical equations of state is introduced in Sect. 6.2.

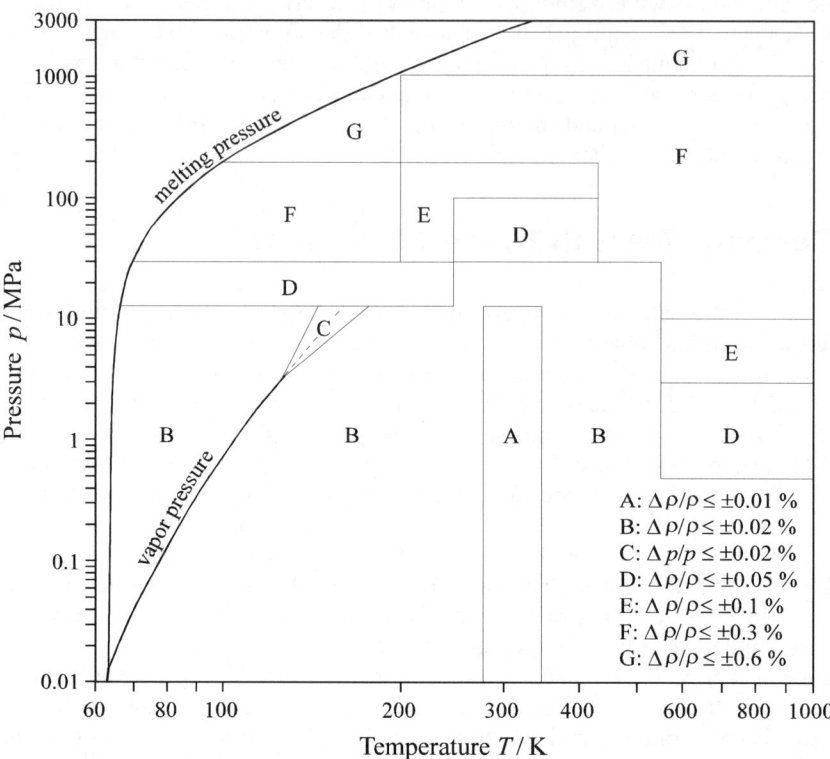

Fig. 5.2. Uncertainties of densities calculated from the recent reference equation of state for nitrogen (Span et al., 1998b).

available up to large reduced temperatures and pressures. As an example, Fig. 5.2 shows the uncertainties of densities calculated from the recent group one equation of Span et al. (1998b) for nitrogen; the given uncertainties correspond roughly to the experimental uncertainty of the best data available in the different regions. Within the technically and scientifically most important regions, the uncertainty of calculated densities is $\Delta\rho/\rho \leq \pm 0.02\%$, going down to $\Delta\rho/\rho \leq \pm 0.01\%$ in the typical calibration region where highly accurate $p\rho T$ data are available from different apparatuses. Fig. 5.3 shows that this assessment is still conservative. From the triple-point temperature ($T_t = 63.151$ K, $T_t/T_c = 0.50$) to 520 K ($T/T_c = 4.12$) results from four different single- and two-sinker densimeters are generally represented within $\Delta\rho/\rho \leq \pm 0.01\%$. Larger deviations occur only in the extended critical region (see below).

Other group one reference equations of state show very similar results with regard to the representation of data from single- and two-sinker densimeters. For argon, Fig. 5.4 shows representative deviations between highly accurate $p\rho T$ data and values calculated from the reference equation of state by Tegeler et al.

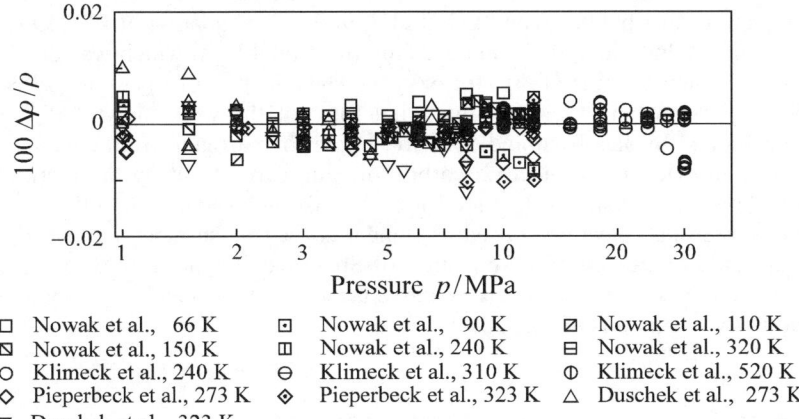

□ Nowak et al., 66 K	⊡ Nowak et al., 90 K	⊠ Nowak et al., 110 K
⊠ Nowak et al., 150 K	⊞ Nowak et al., 240 K	⊟ Nowak et al., 320 K
○ Klimeck et al., 240 K	⊖ Klimeck et al., 310 K	⊕ Klimeck et al., 520 K
◇ Pieperbeck et al., 273 K	◆ Pieperbeck et al., 323 K	△ Duschek et al., 273 K
▽ Duschek et al., 323 K		

Fig. 5.3. Percentage deviations $100\, \Delta\rho/\rho = 100\,(\rho_{\mathrm{exp}} - \rho_{\mathrm{calc}})/\rho_{\mathrm{exp}}$ between experimental data for the density of nitrogen by Nowak et al. (1997a), Klimeck et al. (1998), Pieperbeck et al. (1991), and Duschek et al. (1988) and values calculated from the equation of Span et al. (1998b).

(1997/1999). Again, the data cover the range from the triple-point temperature ($T_{\mathrm{t}} = 83.8058$ K, $T_{\mathrm{t}}/T_{\mathrm{c}} = 0.56$) to 520 K ($T/T_{\mathrm{c}} = 3.45$) and are generally represented within $\Delta\rho/\rho \leq \pm 0.01\%$. The claimed uncertainty of the equation of state ranges from $\Delta\rho/\rho \leq \pm 0.02\%$ to $\Delta\rho/\rho \leq \pm 0.03\%$ within this region, except for the extended critical region.

As a first example of the performance of group two reference equations, Fig. 5.5 shows deviations between highly accurate $p\rho T$ data for carbon dioxide and

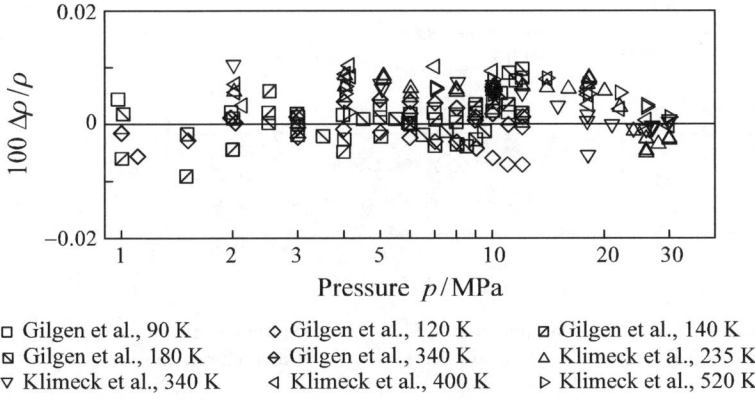

□ Gilgen et al., 90 K	◇ Gilgen et al., 120 K	⊠ Gilgen et al., 140 K
⊠ Gilgen et al., 180 K	◆ Gilgen et al., 340 K	△ Klimeck et al., 235 K
▽ Klimeck et al., 340 K	◁ Klimeck et al., 400 K	▷ Klimeck et al., 520 K

Fig. 5.4. Percentage deviations $100\, \Delta\rho/\rho = 100\,(\rho_{\mathrm{exp}} - \rho_{\mathrm{calc}})/\rho_{\mathrm{exp}}$ between experimental results for the density of argon by Gilgen et al. (1994) and Klimeck et al. (1998) and values calculated from the equation of Tegeler et al. (1997/1999).

values calculated from the group one equation of Span and Wagner (1996) and from the equations of Ely (1986) and Ely et al. (1989). The equation of Ely (1986) is explicit in the Helmholtz energy and uses the functional form which was developed by Schmidt and Wagner (1985) for oxygen. The equation of Ely et al. (1989) is explicit in pressure and uses the functional form of the well known MBWR equation developed by Jacobsen and Stewart (1973) for nitrogen. Both equations resulted from an extensive project on carbon dioxide carried out by the thermophysics division of the National Institute of Standards and Technology (NIST) in Boulder, Colorado and have been used as standards for the thermodynamic properties of carbon dioxide before the equation of Span and Wagner (1996) became available. Based on the definitions given above these equations can be regarded as typical group two equations of state.

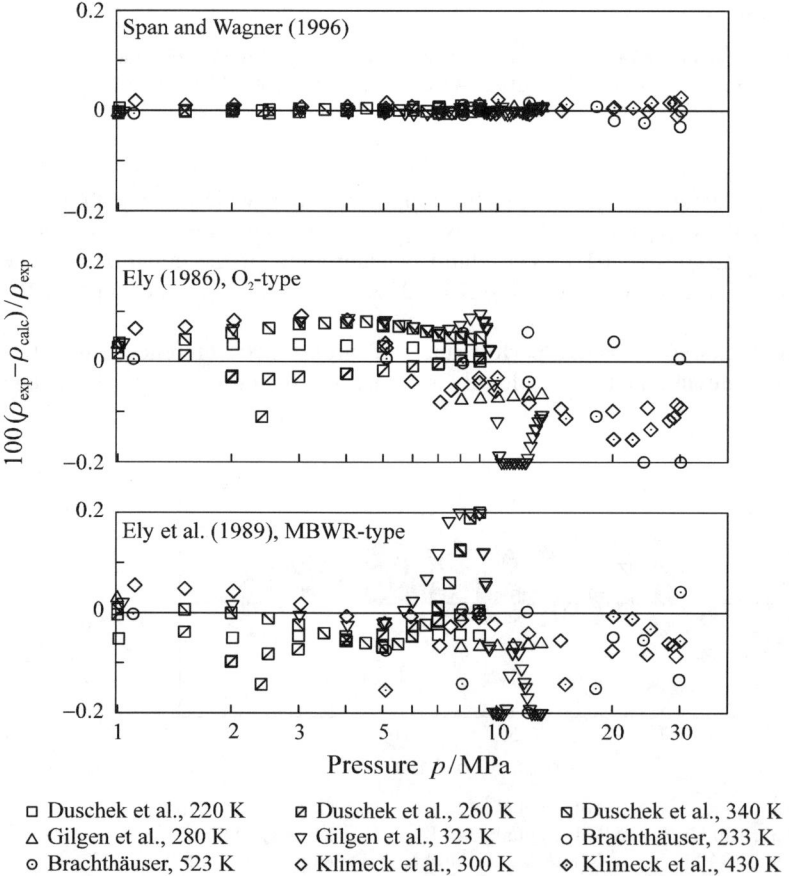

Fig. 5.5. Percentage deviations between experimental results for the density of carbon dioxide by Duschek et al. (1990), Gilgen et al. (1992), Brachthäuser (1993), and Klimeck et al. (2000) and values calculated from different equations of state.

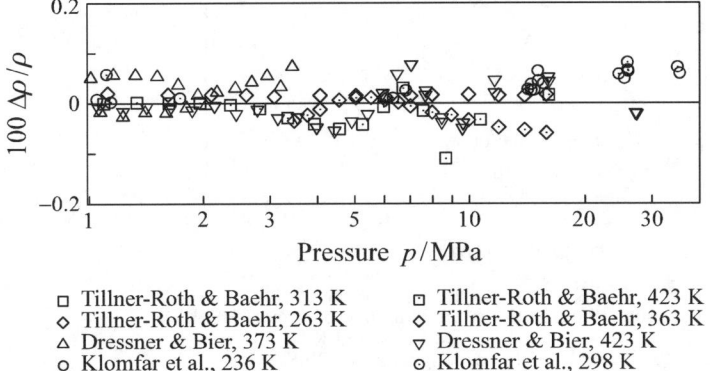

Fig. 5.6. Percentage deviations $100 \, \Delta\rho/\rho = 100 \, (\rho_{\mathrm{exp}} - \rho_{\mathrm{calc}})/\rho_{\mathrm{exp}}$ between experimental results for the density of R134a by Tillner-Roth and Baehr (1992/1993), Dressner and Bier (1993) and Klomfar et al. (1993) and values calculated from the group two reference equation of Tillner-Roth and Baehr (1994).

The group one equation of Span and Wagner represents most of the data within $\Delta\rho/\rho \leq \pm 0.01\%$. For pressures above about 20 MPa, the observed deviations increase up to $\Delta\rho/\rho \approx \pm 0.03\%$; some of the high pressure data of Klimeck et al. (2000) became available after the equation had been finalised. Compared to the same data set, the equations of Ely (1986) and Ely et al. (1989) show typical deviations in the order of $\Delta\rho/\rho \approx \pm 0.1\%$ with maximum deviations of $\Delta\rho/\rho \leq \pm 0.3\%$. Thus, compared to the recent reference equation, the older equations are less accurate by about one order of magnitude; a more detailed comparison would have to distinguish between different regions. Similar comparisons between highly accurate $p\rho T$ data for argon, methane, and nitrogen, the corresponding group one reference equations of state, and older group two equations for these substances show improvements by a factor of 5 to 10 as well – the results found for carbon dioxide are typical.

However, the accuracy of group two equations of state does not need to be assessed primarily for substances where more accurate equations are available but for substances for which group two reference equations are regarded as standards. Such an assessment becomes more complicated since no highly accurate $p\rho T$ data are available for the corresponding substances which could be used as an undisputed reference. Comparisons with the available data sets show significant differences between the performances of the different equations – as already mentioned, a further subdivision of group two would be justified.

As an example for an above average group two equation of state, Fig. 5.6 shows deviations between selected $p\rho T$ data for refrigerant R134a (1,1,1,2-tetrafluoroethane) and values calculated from the reference equation of Tillner-Roth and Baehr (1994). The equation represents reliable data in the technically and scientifically most important regions within $\Delta\rho/\rho \leq \pm 0.1\%$ without exceptions; data in the extended critical region are represented within $\Delta p/p \leq \pm 0.1\%$.

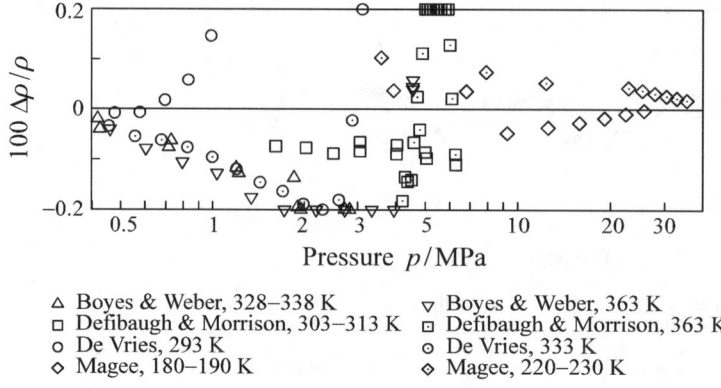

Fig. 5.7. Percentage deviations $100 \, \Delta\rho/\rho = 100 \, (\rho_{\mathrm{exp}} - \rho_{\mathrm{calc}})/\rho_{\mathrm{exp}}$ between experimental results for the density of R125 by Boyes and Weber (1995), Defibaugh and Morrison (1992), de Vries (1997), and Magee (1996) and values calculated from the group two equation of Piao and Noguchi (1998).

These values lag behind typical results of reference equations of state by about a factor of 5 and mark the upper limit for the performance of group two reference equations of state available today. To achieve this level of accuracy, neither advanced optimisation strategies[5] nor special functional forms for an improved description of the critical region are required.

As an example of the performance of a subaverage group two equation of state, Fig. 5.7 shows deviations between selected experimental data for refrigerant R125 (pentafluoroethane) and values calculated from the equation of Piao and Noguchi (1998). An accuracy of $\Delta\rho/\rho \leq \pm 0.1\%$ can be claimed only for the liquid region at low temperatures for this equation; in other regions the maximum deviations between reliable experimental data and calculated densities exceed even $\pm 0.2\%$ in density. The equation of Piao and Noguchi uses a simple pressure explicit functional form which was found by trial and error, without use of optimisation algorithms.

Based on these and further comparisons, group two reference equations of state can generally be expected to result in uncertainties of $\Delta\rho/\rho \leq \pm 0.1\% - \pm 0.2\%$ in the technically and scientifically most important regions.

The single- and two-sinker densimeters which are in use today are restricted to pressures up to 30 MPa and to temperatures up to 520 K. For nitrogen, the resulting increase of uncertainty can be seen in Fig. 5.2, where the uncertainty of the reference equation changes from $\Delta\rho/\rho \leq \pm 0.02\%$ directly to $\Delta\rho/\rho \leq \pm 0.3\%$ in the high temperature and pressure region; accurate measurement devices for such states are urgently needed. Thus, at pressures above 30 MPa or temperatures above

---

[5] The functional form of the equation of Tillner-Roth and Baehr (1994) was optimised using a modified version of the stepwise regression analysis by Wagner (1974), see Sect. 4.4.3.

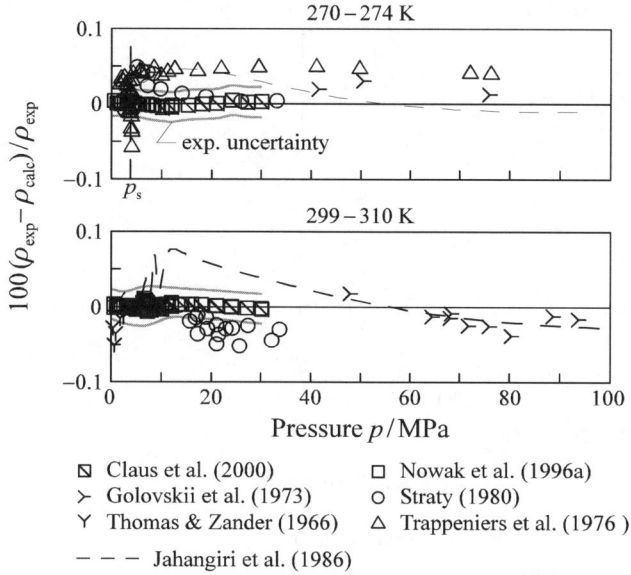

Fig. 5.8. Percentage deviations between experimental results for the density of ethylene and values calculated from the recent group one reference equation of Smukala et al. (1999). Values calculated from the IUPAC recommended group two equation of Jahangiri et al. (1986) are plotted as dashed lines.

520 K there are still no $p\rho T$ data available which can be used to distinguish between group one and good group two reference equations of state.

This is further illustrated in Fig. 5.8, where selected experimental data for ethylene are compared with results from the recent group one equation of Smukala et al. (1999) and from the recommended IUPAC standard, the equation of Jahangiri et al. (1986) which is one of the most accurate group two equations available today. Up to 30 MPa, the two-sinker data of Nowak et al. (1996a) and the single-sinker data of Claus et al. (2000) clearly prove the superiority of the equation of Smukala et al. However, both equations are of very similar standard at higher pressures. When working with group two reference equations at high pressures, one has to be aware that the equations for working fluids are often valid only up to rather low pressures.[6] In general, no suitable measures were taken to guarantee a reasonable extrapolation behaviour when these equations were established and thus they should not be used beyond the indicated range of validity.

At even higher pressures, recent group one equations are superior to common group two equations again due to more suitable functional forms and intense studies of the extrapolation behaviour (see Sect. 4.6). As a typical example, Fig. 5.9 compares selected $p\rho T$ data for argon at high pressures with results calculated

---

[6] For common technical applications the range of validity is absolutely sufficient; for scientific applications both the relatively low temperature and pressure limits can be severe restrictions.

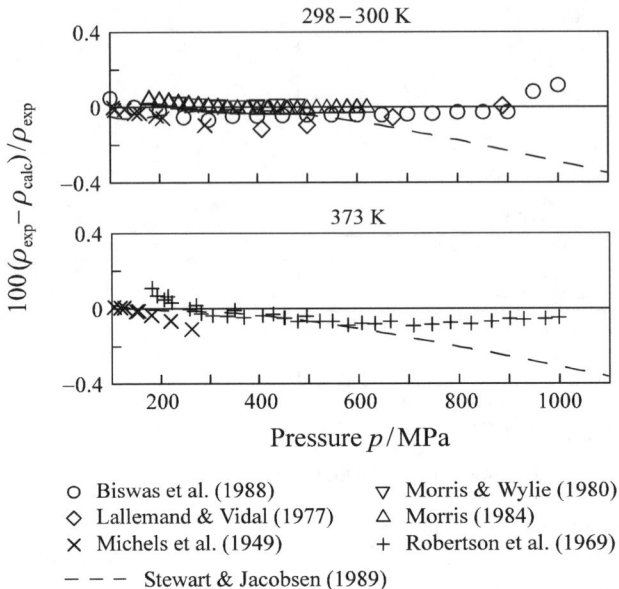

Fig. 5.9. Percentage deviations between experimental results for the density of argon at high pressures and values calculated from the group one reference equation of Tegeler et al. (1997/1999). Values calculated from the group two equation of Stewart and Jacobsen (1989) are plotted as dashed lines.

from the group one equation of Tegeler et al. (1997/1999) and from the group two equation of Stewart and Jacobsen (1989). At pressures above about 600 MPa the older equation results in systematic deviations from the experimental data, which were known at the time in 1989 without exception. However, the representation of data at very high pressures is not directly linked to the accuracy at moderate pressures and thus it is not necessarily linked to a group one assignment. The recent group two equation for air by Lemmon et al. (1999), for example, was designed to yield reasonable results up to very high pressures and compares well with group one equations with respect to this feature, while the data situation for air does not allow comparable accuracies at moderate pressures.

To assess the accuracy of typical technical equations of state, Fig. 5.10 shows deviations between highly accurate $p\rho T$ data for nitrogen, which were already shown in Fig. 5.3, and values calculated from Bender-type, Starling-type, and BWR-type equations of state. Figure 5.11 shows the same comparisons for carbon dioxide using the highly accurate $p\rho T$ data shown already in Fig. 5.5. The presented data cover the technically most interesting regions including gaseous, liquid and supercritical states; data in the critical region which would result in very large density deviations are excluded (see below).

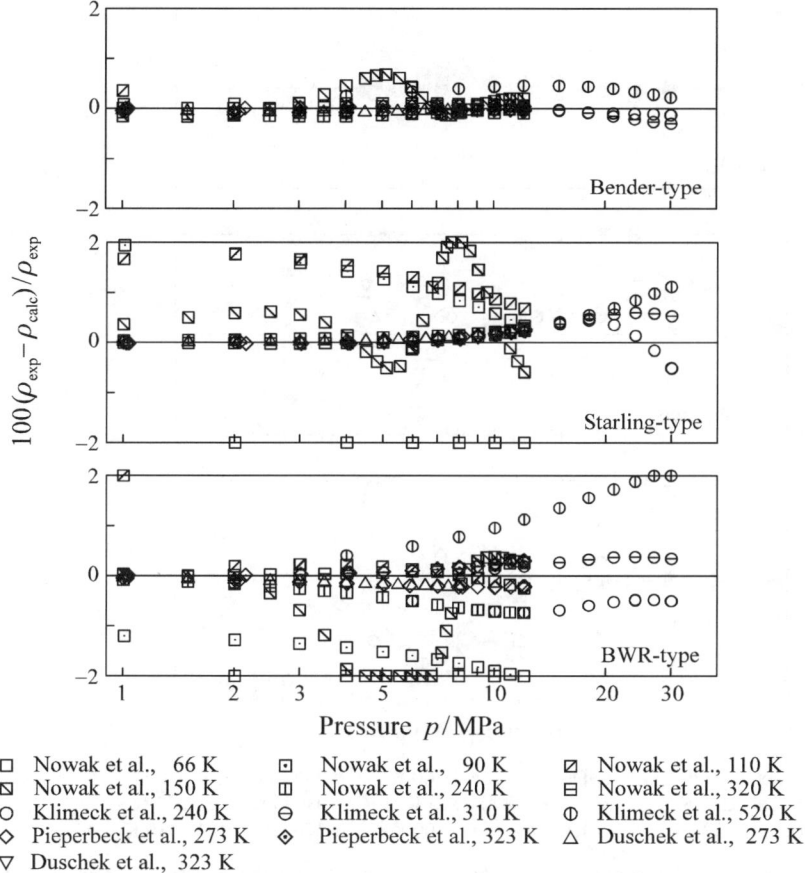

Fig. 5.10. Percentage deviations between experimental data for the density of nitrogen by Nowak et al. (1997a), Klimeck et al. (1998), Pieperbeck et al. (1991), and Duschek et al. (1988) and values calculated from Bender-, Starling-, and BWR-type equations of state.

Apart from the new type of equations presented in Sect. 6.2, the Bender-type equation with its 19 terms can be regarded as the best available "simple" equation of state for technical applications. Both for nitrogen and carbon dioxide the deviations between experimental data and values calculated from the equations of state remain within $\Delta\rho/\rho \approx \pm 0.2\%$ over broad ranges. Larger deviations become obvious for isotherms in the extended critical region (Fig. 5.11, $T = 323$ K, $T/T_c = 1.06$), for typical supercritical isotherms (Fig. 5.10, $T = 150$ K, $T/T_c = 1.19$), for large reduced temperatures (Fig. 5.10, $T = 520$ K, $T/T_c = 4.12$), and generally for the highest plotted pressures. In these regions, the uncertainty of Bender-type equations typically increases to $\Delta\rho/\rho \leq \pm 0.5\%$. Similar results can be found for other substances as well. In general, densities calculated from Bender-

186   5 The Performance of Multiparameter Equations of State

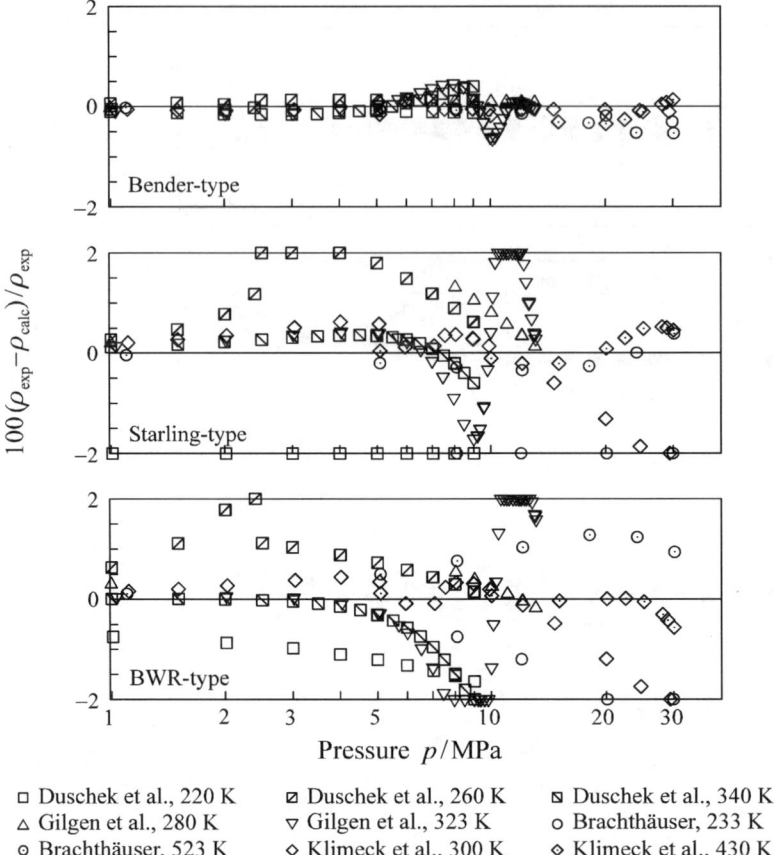

Fig. 5.11. Percentage deviations between experimental results for the density of carbon dioxide by Duschek et al. (1990), Gilgen et al. (1992), Brachthäuser (1993), and Klimeck et al. (2000) and values calculated from Bender-, Starling-, and BWR-type equations of state.

type equations of state are accurate enough for technical applications without special demands on accuracy.

Starling-type equations yield acceptable results for the gas phase and for states at far supercritical temperatures. But even for these states, deviations between accurate $p\rho T$ data and calculated values reach up to about $\Delta\rho/\rho \approx \pm 0.5\%$. In the extended critical region, for typical supercritical states, and for liquid states, Starling-type equations result in deviations in an order of $\Delta\rho/\rho \approx \pm 2\%$; using these equations becomes questionable even for engineering tasks with average requirements with regard to the accuracy of calculated densities. Better results can be achieved when sacrificing the representation of certain properties (e.g., phase equilibrium data) or of data in certain regions (e.g., liquid states) while fitting the equation. Such restricted equations can be useful for certain applications, but they are not a solution for general technical use.

For liquid states, the original BWR-type equation yields slightly better results than the Starling-type equation, but maximum deviations of up to ±2% in density can be observed as well, especially at low temperatures where the liquid isotherms become very steep. While the supercritical fluid is described with about $\Delta\rho/\rho \approx \pm 1\%$ for nitrogen, deviations up to ±2% in density occur for carbon dioxide over the whole range of states. In the extended critical region and for typical supercritical isotherms, the deviations exceed $\Delta\rho/\rho = \pm 2\%$ for both substances.

With respect to the accuracy of calculated densities, the BWR-type equation and its simple modifications, which can still be found in technical applications, lag behind typical group two reference equations by a whole order of magnitude.

In Fig. 5.2, the uncertainty for $\Delta\rho/\rho$ is replaced by an uncertainty of $\Delta p/p \leq \pm 0.02\%$ in the extended critical region. The same effect can be found for other equations of state as well – comprehensive articles will not give uncertainties in the critical region in terms of density. When approaching the critical point, the influence of the uncertainties of measured pressures and temperatures dominates the total uncertainty of measured $p\rho T$ data. With increasing values of $(\partial\rho/\partial p)_T$, uncertainties in pressure result in increasing uncertainties in density and the effect of uncertainties in temperature increases with

$$\left(\frac{\partial \rho}{\partial T}\right)_p = -\left(\frac{\partial p}{\partial T}\right)_\rho \cdot \left(\frac{\partial \rho}{\partial p}\right)_T \sim -c \cdot \left(\frac{\partial \rho}{\partial p}\right)_T. \tag{5.1}$$

Since experimental data build the foundation of each accurate equation of state, the uncertainty $\Delta\rho/\rho$ of the equation increases, where the value of $\Delta\rho/\rho$ depends on the value of $(\partial\rho/\partial p)_T$ for every single point – values for $\Delta\rho/\rho$ cannot be given for certain regions. However, since the influence of uncertainties in temperature is linked to $\Delta p/p$ by the almost constant derivative $(\partial p/\partial T)_\rho \sim c$, the uncertainty in pressure can be given as a more or less constant value throughout the critical region.

The performance of group one reference equations of state with regard to the representation of $p\rho T$ data in the critical region has already been discussed in Sect. 4.5.2. Group one equations of state are able to represent even the most accurate $p\rho T$ data in the critical region within their experimental uncertainty, see for instance Figs. 4.14 and 4.15. As already shown, the estimated uncertainty in the critical region is usually $\Delta p/p \leq \pm 0.02\%$ for such equations. Based on the same set of argon data which was already shown in Fig. 4.15 in higher resolution, Fig. 5.12 illustrates the performance of a good group two equation and of a Bender-type equation in the critical region. The equation of Stewart and Jacobsen (1989) shows deviations up to $\Delta p/p \approx \pm 0.1\%$, except for the highest plotted densities which correspond to the transition range to liquid like states where $(\partial\rho/\partial p)_T$ becomes small again. The observed uncertainty of the Bender-type equation is of the order $\Delta p/p \approx \pm 0.2\%$, with increasing tendency at high densities as well. This value is typical for secondary reference equations with poor performance in the critical

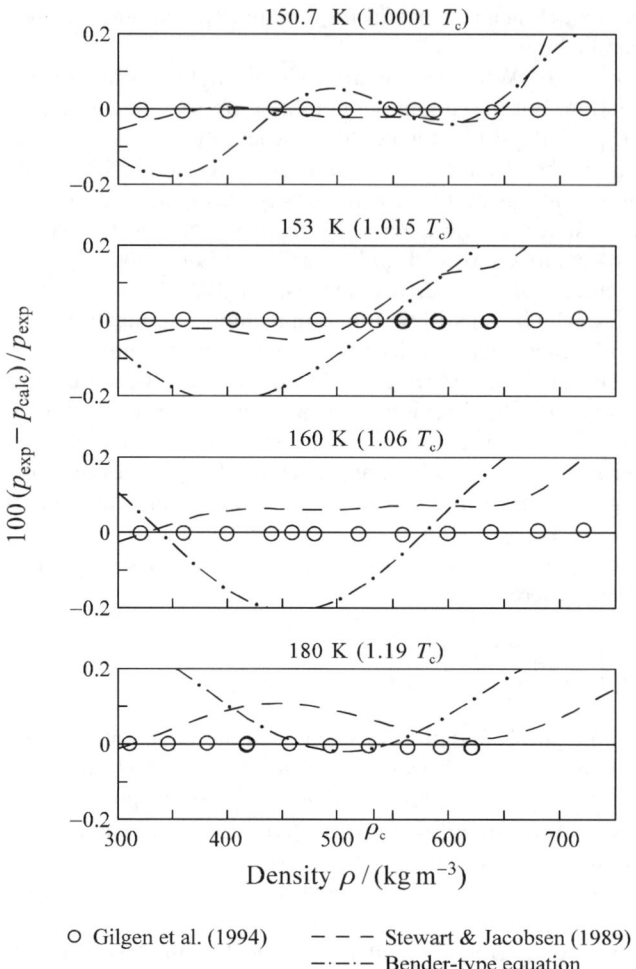

Fig. 5.12. Percentage deviations between experimentally determined pressures in the critical region of argon and values calculated from the group one equation by Tegeler et al. (1997/1999). Values calculated from the group two equation of Stewart and Jacobsen (1989) and from a Bender-type equation are plotted for comparison.

region. Simple technical equations of state like BWR- and Starling-type equations result in typical uncertainties of $\Delta p/p \approx \pm 1\%$.

For two of the selected isotherms (153 K $\approx T_c + 2.3$ K, 160 K $\approx T_c + 9.3$ K),[7] Fig. 5.13 translates the observed pressure deviations into density deviations. At

---

[7] To aid comprehension, the distance from the critical point is expressed in absolute temperature differences here. However, the influence of critical effects depends on the reduced distance $T/T_c - 1$ rather than on absolute temperature distances. In terms of absolute temperatures, the critical region becomes larger for substances with larger critical temperatures.

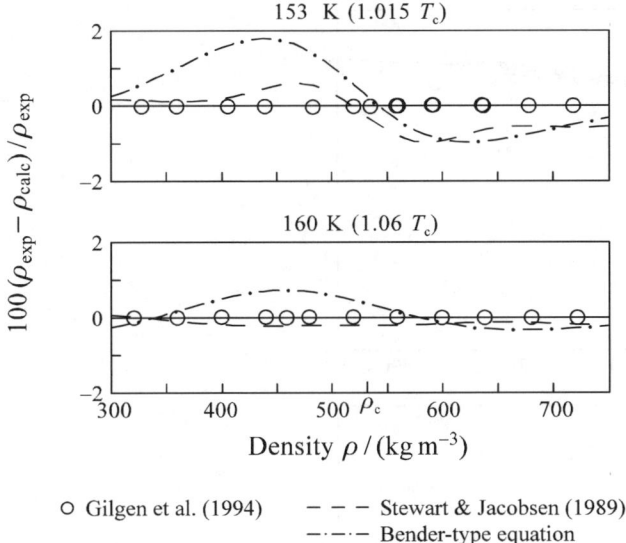

Fig. 5.13. Percentage deviations between experimentally determined densities in the critical region of argon and values calculated from the group one equation by Tegeler et al. (1997/1999). Values calculated from the group two equation of Stewart and Jacobsen (1989) and from a Bender-type equation are plotted for comparison.

2.3 K above the critical temperature, the Bender-type equation results in deviations up to $\Delta\rho/\rho \approx \pm 2\%$ and the equation of Stewart and Jacobsen deviates from the reference equation by up to $\Delta\rho/\rho \approx \pm 1\%$. On this scale, no deviations between the results of the group one reference equation and the experimental data can be observed. At 9.3 K above the critical temperature, the Bender-type equation deviates by $\Delta\rho/\rho \leq \pm 1\%$ while the equation of Stewart and Jacobsen results in $\Delta\rho/\rho \leq \pm 0.3\%$.

From these results it becomes obvious that an accurate calculation of densities in the critical region is only possible with group one reference equations of state, if at all. In the extended critical region, good group two equations of state are sufficient for many scientific and for almost all technical applications[8] but when using subaverage group two equations or good group three technical equations, one has to be aware that large uncertainties may distort the results of typical scientific and technical calculations. Simple technical equations of state such as BWR- and Starling-type equations are completely unsuitable for calculations in the critical and extended critical region.

---

[8] The accurate determination of mass flows in ethylene pipelines is an example of an application where even good secondary reference equations are insufficient for technical calculations in certain temperature and pressure ranges.

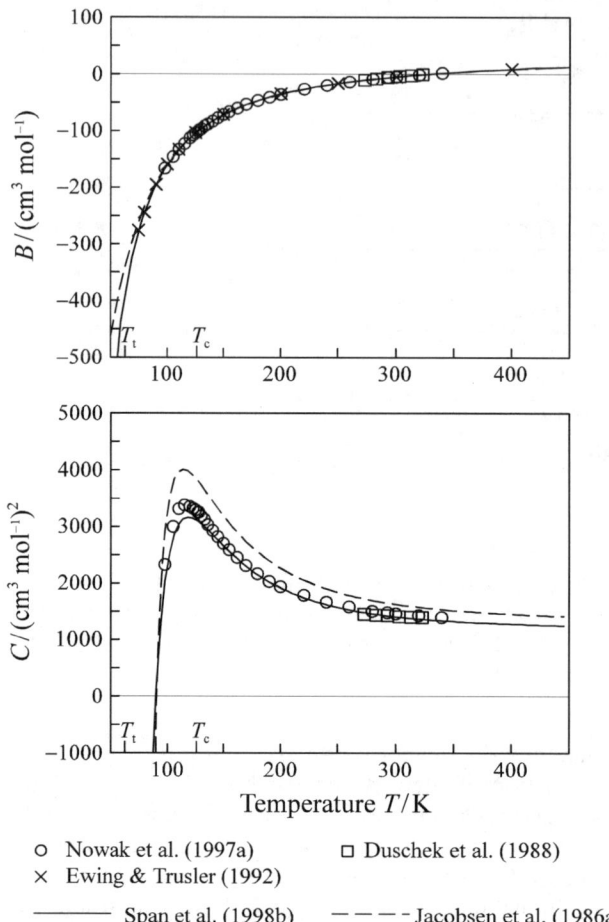

Fig. 5.14. Second and third virial coefficients of nitrogen calculated from experimental data and from the equations of state by Span et al. (1998b) and Jacobsen et al. (1986a).

Besides $p\rho T$ data, *virial coefficients* are often discussed in scientific literature as a second source of information on thermal properties in the gas phase and at gas-like supercritical states. However, virial coefficients are always derived from experimental results on other properties, such as $p\rho T$ data or speeds of sound. Only for the noble gases with very simple monoatomic molecules, theoretical results on virial coefficients are accurate enough to be considered in comparisons regarding the accuracy of reference equations of state (see for example Tegeler et al., 1997/1999). Since equations of state yield consistent solutions for different properties per definition, an equation which represents the original experimental $p\rho T$ or speed of sound data within their experimental uncertainty has to represent the derived virial coefficients within their uncertainty as well. If this statement does not hold, the uncertainty of the virial coefficients has been estimated too high. In fact it is rather difficult to assign uncertainties, especially to third virial coefficients since their uncertainty is linked to the uncertainty of the derived second virial coefficients.

When assessing the performance of reference equations of state, the general behaviour of calculated virial coefficients is usually the only criterion which has to be considered besides the original data. At low temperatures, $p\rho T$ data may be represented within their experimental uncertainty by equations of state which do not result in the theoretically expected steep decrease of the virial coefficients or which even result in increasing virial coefficients at very low temperatures. As an example for qualitatively correct plots, Fig. 5.14 shows the virial coefficients of nitrogen calculated from the equations of state by Span et al. (1998b) and Jacobsen et al. (1986a). At very low temperatures, the deviation between the second virial coefficients calculated from the equation of Jacobsen et al. (1986a) and from the speeds of sound measured by Ewing and Trusler (1992) increases; this deviation coincides with the deviations observed for the corresponding speeds of sound in the following section.

For the third virial coefficients, systematic deviations between values calculated from the $p\rho T$ data of Nowak et al. (1997a) and from the equations of state can be observed especially at low temperatures. However, at these temperatures, the experimental uncertainty of third virial coefficients becomes large even when derived from highly accurate $p\rho T$ data. The deviations between the data of Nowak et al. and values calculated from the equation of Span et al. (1998b) are still within the experimental uncertainty.

For technical equations of state, a qualitatively correct plot of the third virial coefficient at low temperatures has to be regarded as arbitrary. Below the critical temperature the contribution caused by the third virial coefficient rapidly decreases and typical technical accuracies can be achieved both with regard to thermal and caloric properties without consideration of the third virial coefficient.

For the *acoustic virial coefficients,* which can be derived from experimental results for the speed of sound (see for instance Ewing and Trusler, 1992), the same assessment as given for thermal virial coefficients above holds as well. However, when using linear algorithms to optimise the functional form of an equation of state it can be advantageous to make use of second acoustic virial coefficients since they result in linear residua, while speeds of sound need to be linearised (see Sects. 4.3.2 and 4.3.3).

5.2 Comparisons with Caloric Properties

In addition to data for thermal properties, data for the speed of sound play a leading role both for the development and the assessment of equations of state. On the one hand, this property is difficult to describe especially in the liquid or liquid-like high density region since it involves different contributions which are usually not well described by other data. According to Eq. 4.61, the speed of sound becomes

$$w = \sqrt{\frac{c_p}{c_v}\left(\frac{\partial p}{\partial \rho}\right)_T}. \quad (5.2)$$

An accurate and consistent description of speeds of sound requires an accurate description both of thermal and caloric properties; the speed of sound is very sensitive to inconsistencies between data sets for thermal and caloric properties. On the other hand, highly accurate speeds of sound have become available in the gaseous and gas like low density region[9] during the last several years. In this region, it is easier to describe the speed of sound since $(\partial p/\partial \rho)_T$ is usually well known from $p\rho T$ data and the residual contribution to heat capacities is comparably small, but meeting the high accuracies which can be achieved experimentally with spherical resonators is still an ambitious challenge.

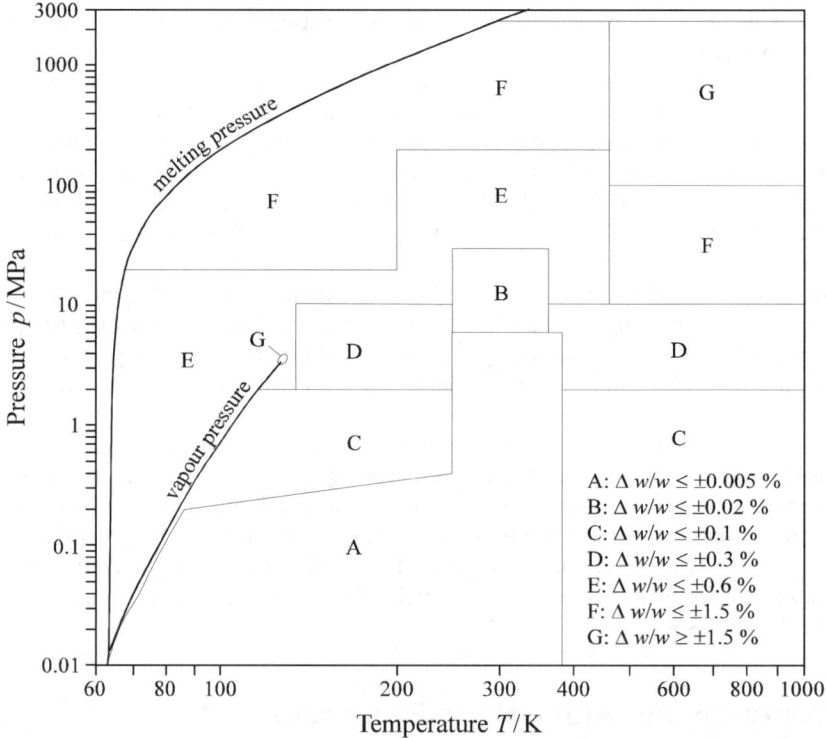

Fig. 5.15. Uncertainties of speeds of sound calculated from the recent group one reference equation of state for nitrogen (Span et al., 1998b).

---

[9] For a detailed description of the theoretical background and of available experimental techniques, see Trusler (1991).

Engineers in particular, frequently ask whether high accuracies in the representation of speeds of sound are actually an important feature of an equation of state. Besides the fact that an accurate description of properties is a scientific goal by itself, one has to be aware of two points to answer this question. In the first place, the speed of sound itself is an important property for some technical applications, such as volume flow determinations with sonic nozzles or with ultrasound flowmeters. In the second place, and this is probably the more important point for most users, the representation of speed of sound data is directly linked to accuracies regarding typical engineering applications. The isentropic expansion coefficient, e.g., can be written as $k_s = \rho w^2 / p$, see Table 3.10. Although the relation between accuracies of calculated speeds of sound and accuracies of calculated isentropic efficiencies has never been investigated in detail, the connection is obvious.

As an example of the performance of recent group one reference equations of state, Fig. 5.15 summarises the uncertainties of speeds of sound calculated from the equation of Span et al. (1998b) for nitrogen. In the gas region and for supercritical states up to about the critical density, highly accurate speed of sound data are available from measurements with spherical resonators. In this region the uncertainty of calculated speeds of sound is as small or even smaller as the uncertainty of calculated densities. To validate the estimated uncertainties, Fig. 5.16 shows the representation of speeds of sound which were measured with three different spherical resonators and which range from temperatures close to the triple point ($T/T_c = 0.63$) up to 373 K ($T/T_c = 2.96$). With the exception of data on a single isotherm measured by Costa Gomes and Trusler (1998a), all of the data are represented by the equation of Span et al. (1998b) within $\Delta w/w \leq \pm 0.005\%$.

Similar results can be found for argon, where data from measurements with different spherical resonators are available as well. With the exception of a few data

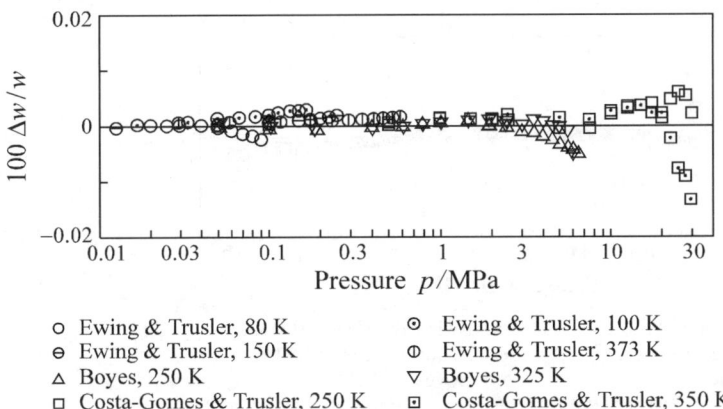

○ Ewing & Trusler, 80 K   ⊙ Ewing & Trusler, 100 K
⊖ Ewing & Trusler, 150 K   ⊕ Ewing & Trusler, 373 K
△ Boyes, 250 K   ▽ Boyes, 325 K
□ Costa-Gomes & Trusler, 250 K   ⊡ Costa-Gomes & Trusler, 350 K

Fig. 5.16. Percentage deviations $100 \, \Delta w/w = 100 \, (w_{\text{exp}} - w_{\text{calc}})/w_{\text{exp}}$ between experimental data for the speed of sound in nitrogen by Ewing and Trusler (1992), Boyes (1992), and Costa Gomes and Trusler (1998a) and values calculated from the group one equation of Span et al. (1998b).

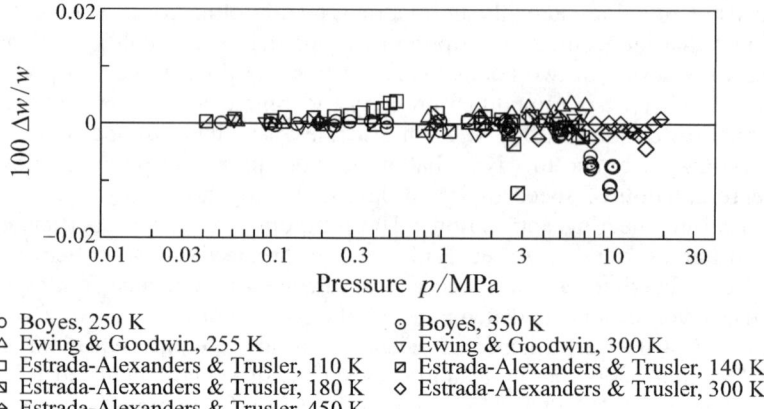

○ Boyes, 250 K ○ Boyes, 350 K
△ Ewing & Goodwin, 255 K ▽ Ewing & Goodwin, 300 K
□ Estrada-Alexanders & Trusler, 110 K ⊠ Estrada-Alexanders & Trusler, 140 K
◨ Estrada-Alexanders & Trusler, 180 K ◇ Estrada-Alexanders & Trusler, 300 K
◆ Estrada-Alexanders & Trusler, 450 K

Fig. 5.17. Percentage deviations $100\,\Delta w/w = 100\,(w_{\text{exp}} - w_{\text{calc}})/w_{\text{exp}}$ between experimental data for the speed of sound in argon by Ewing and Goodwin (1992), Boyes (1992), and Estrada-Alexanders and Trusler (1995) and values calculated from the group one equation of Tegeler et al. (1997 / 1999).

points by Boyes (1992) and some data close to the phase boundary, the data for argon are represented by the equation of Tegeler et al. (1997/1999) within $\Delta w/w \leq \pm 0.005\%$, see Fig. 5.17. This extremely accurate representation of speed of sound data at gaseous and supercritical states became possible only after nonlinear optimisation algorithms had been developed, see Sect. 4.4.6.

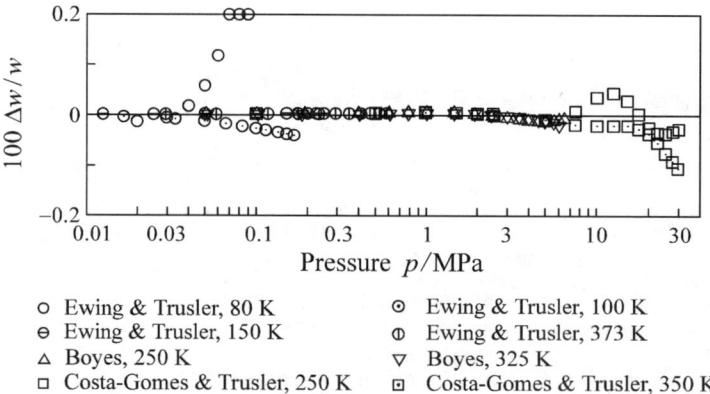

○ Ewing & Trusler, 80 K ⊙ Ewing & Trusler, 100 K
⊖ Ewing & Trusler, 150 K ⊕ Ewing & Trusler, 373 K
△ Boyes, 250 K ▽ Boyes, 325 K
□ Costa-Gomes & Trusler, 250 K ▣ Costa-Gomes & Trusler, 350 K

Fig. 5.18. Percentage deviations $100\,\Delta w/w = 100\,(w_{\text{exp}} - w_{\text{calc}})/w_{\text{exp}}$ between experimental data for the speed of sound in nitrogen by Ewing and Trusler (1992), Boyes (1992), and Costa Gomes and Trusler (1998a) and values calculated from the group two equation of Jacobsen et al. (1986a).

Figure 5.18 shows the representation of the speed of sound data for nitrogen shown already in Fig. 5.16 by the equation of Jacobsen et al. (1986a) which can be regarded as an example of an above average group two equation of state. The comparably large deviations of the data on the isotherm at $T = 80$ K indicate a severe problem of the equation in the low temperature gas region, see also Fig. 5.14 and the discussion on virial coefficients. At higher temperatures the maximum deviations exceed those observed in Fig. 5.16 by a factor of 5–10; this corresponds to the results found for thermal properties in Sect. 5.1.

However, keeping in mind that Jacobsen et al. did not know these data when setting up their equation, the equation represents the supercritical data surprisingly well up to pressures of about 10 MPa. Figure 5.19, which shows the residual contribution to the speed of sound at gaseous and supercritical states, helps to clarify this observation. In the gas phase, the residual contribution to the speed of sound is small, except for states close to the phase boundary where the residual contribution increases strongly for reduced temperatures above about $T/T_c = 0.8$. On isotherms just outside the extended critical region, the speed of sound is difficult to model (see $T/T_c = 1.2$ in Fig. 5.19; for the representation of speeds of sound in the critical region, see Sect. 4.5.2) but at temperatures far above the critical temperature the residual contribution remains small up to pressures of about 10 MPa. Below this limit, speeds of sound are predicted accurately by equations of state which represent $p\rho T$ data accurately. A direct consideration of speeds of sound in nonlinear optimisation algorithms is only necessary to represent highly accurate speed of sound data at higher pressures and close to the phase boundary within their experimental uncertainty. At these states the residual contribution to the speed of sound increases rapidly.

With these results in mind, Fig. 5.20 confirms the assessment of the equation of Piao and Noguchi (1998) for R125 which was characterised as a subaverage group

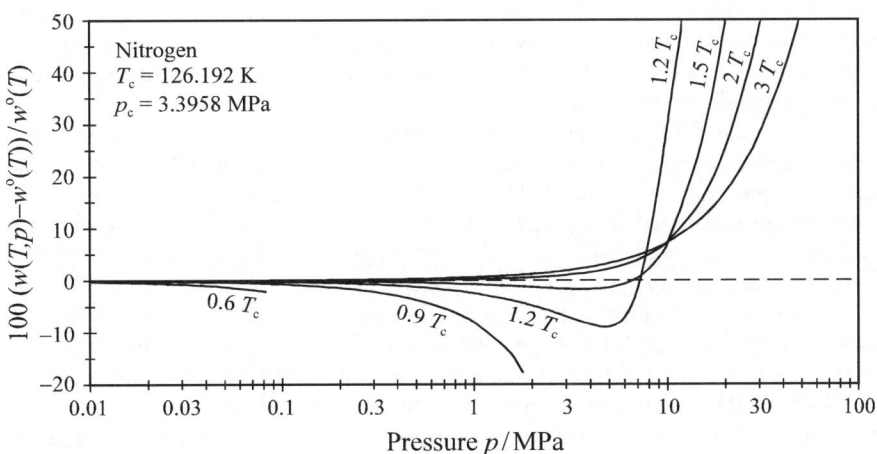

Fig. 5.19. The residual contribution to the speed of sound at gaseous and supercritical states shown for nitrogen as a typical example.

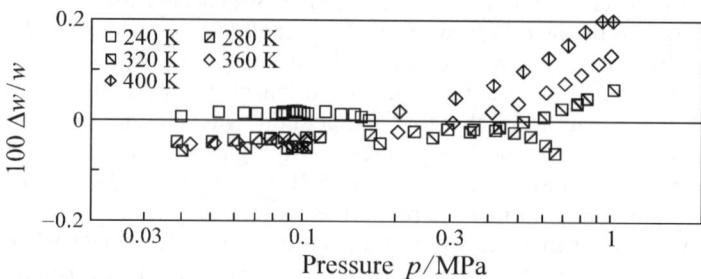

Fig. 5.20. Percentage deviations $100\,\Delta w/w = 100\,(w_{\mathrm{exp}} - w_{\mathrm{calc}})/w_{\mathrm{exp}}$ between experimental data for the speed of sound in R125 by Gillis (1997) and values calculated from the equation of Piao and Noguchi (1998).

two equation of state in Sect. 5.1 based on comparisons with experimental $p\rho T$ data. The maximum deviations in the speed of sound are not larger than those shown in Fig. 5.18, but they occur at pressures up to 1 MPa far away from the phase boundary and at supercritical states where the speed of sound should be described accurately even with $p\rho T$ data alone. The fact that the deviations increase above the critical temperature ($T_{c,\mathrm{R125}} \approx 339$ K) indicates severe shortcomings of the equation. Data at higher pressures would certainly show much larger deviations if they were available for R125. The obvious offset at very low pressures indicates that the problems are not only caused by the pressure explicit equation for the residual contribution but also by the simple quadratic polynomial which is used to describe the caloric properties of the ideal gas. This correlation does not describe the temperature dependence of the ideal gas isobaric heat capacity properly; the cubic polynomial which was used for R125 by Outcalt and McLinden (1995) yields clearly better results.

When assessing equations of state on the basis of speeds of sound, it is always important to be aware of the states at which certain deviations are observed. Deviations of the same order of magnitude lead to very different conclusions if they are observed at different states.

Due to the small residual contributions, technical equations of state are able to describe speeds of sound rather accurately in the gas phase and for supercritical states up to pressures of about 10 MPa in general. As a typical example, Fig. 5.20 shows deviations between the selected set of speed of sound data for nitrogen which was already used in Figs. 5.16 and 5.18 and values calculated from a Bender-type, a Starling-type, and a BWR-type equation of state. Up to 30 MPa, the Bender type equation describes all of the data with an uncertainty of $\Delta w/w \leq \pm 0.5\%$; except for flow measurement applications and calibrations this accuracy can be regarded as sufficient for common technical calculations. Up to about 10 MPa the Starling-type equation is superior to the Bender-type equation in this case, but at higher pressures the observed uncertainties increase to $\Delta w/w \leq \pm 2\%$. As already discussed in Sect. 5.1, Starling-type equations of state have been developed mainly to describe states in the gaseous and supercritical

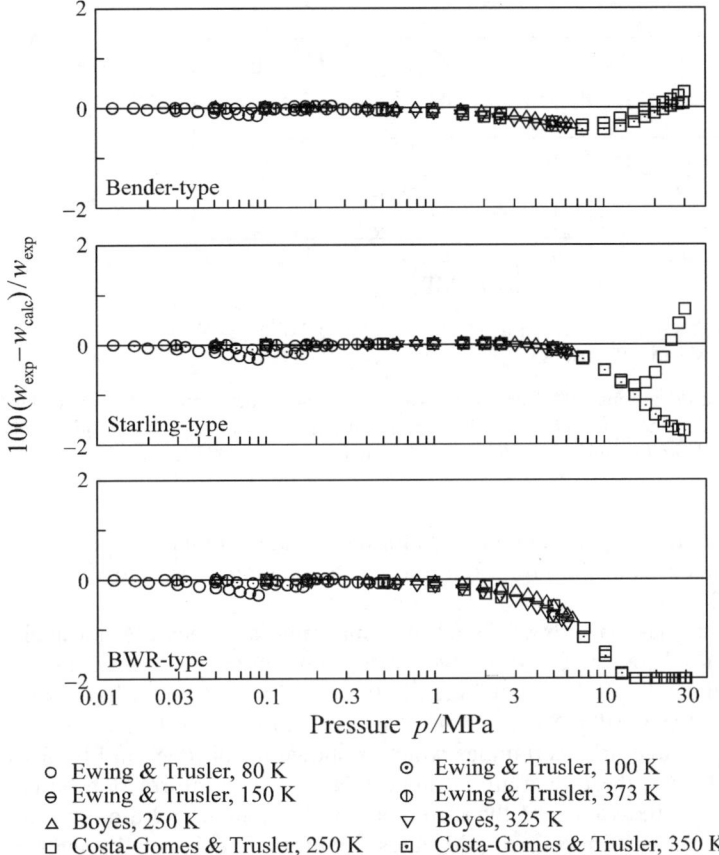

Fig. 5.21. Percentage deviations between experimental data for the speed of sound in nitrogen by Ewing and Trusler (1992), Boyes (1992), and Costa Gomes and Trusler (1998a) and values calculated from Bender-, Starling-, and BWR-type equations of state.

regions at low densities. In these regions the BWR-type equation is clearly inferior to the other technical equations. Systematic deviations can be observed starting at pressures of about 1 MPa and the limit of $\Delta w/w = \pm 0.5\%$ is exceeded at about 4 MPa. For less simple substances, slightly higher deviations can be observed in the corresponding states, but nevertheless simple equations of state are usually sufficient for technical calculations with regard to speeds of sound in the gaseous and gas-like supercritical region.

The whole situation changes drastically when considering speeds of sound in the liquid phase or at supercritical states with liquid-like densities. Theoretically, empirical reference equations of state are able to reproduce speeds of sound under these conditions as accurately as in the gas phase and highly accurate experimental techniques are available. Figure 5.22 verifies this statement showing deviations

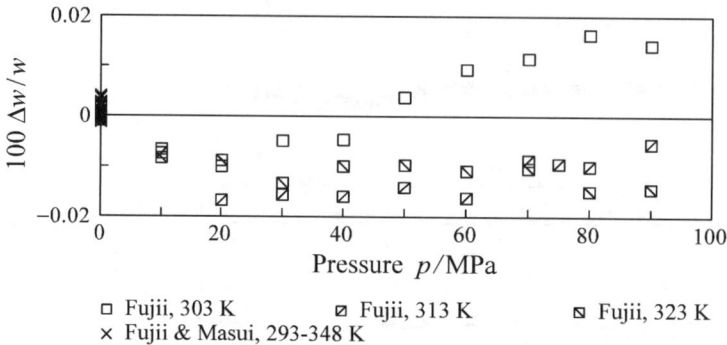

Fig. 5.22. Percentage deviations $100\,\Delta w/w = 100\,(w_{\text{exp}} - w_{\text{calc}})/w_{\text{exp}}$ between experimental data for the speed of sound in the liquid phase of water from Fujii and Masui (1993) and Fujii (1994) and values calculated from the reference equation of Wagner and Pruß (1997/2000).

between highly accurate experimental results for the speed of sound in liquid water and values calculated from the group one reference equation of state by Wagner and Pruß (1997/2000).

However, liquid phase speeds of sound of comparable accuracy are not available for any other substance. For other well measured substances, typical uncertainties of experimental speed of sound data in the liquid and at liquid-like supercritical states are of the order $\Delta w/w \leq \pm 0.2\% - \pm 1\%$ and increase further at very high pressures; the range of uncertainties which is shown for nitrogen in Fig. 5.15 is typical. Nevertheless, these data are an important source of information on caloric properties since the values of the different contributions in 5.2 are not constrained by their ideal gas limits for liquid states and errors in calculated speeds of sound may become much larger than for gaseous or gas-like supercritical states.

Figure 5.23 shows the representation of speeds of sound in nitrogen at liquid and supercritical states up to high pressures. Both the equation of Span et al. (1998b) and the equation of Jacobsen et al. (1986a) represent the plotted high density data within their experimental uncertainty of $\Delta w/w \leq \pm 0.5\% - \pm 1\%$. At very high pressures, the group one equation of Span et al. is superior due to its improved functional form and intense studies of the extrapolation behaviour, see also Sects. 4.6 and 5.1.

However, the good representation of liquid speeds of sound shown for the equation of Jacobsen et al. (1986a) is not self-evident for group two reference equations of state. Besides the representation of properties in the critical region and the extrapolation behaviour, the representation of speeds of sound at liquid states is one of the most sensitive criteria for a distinction between more or less sophisticated equations of state.

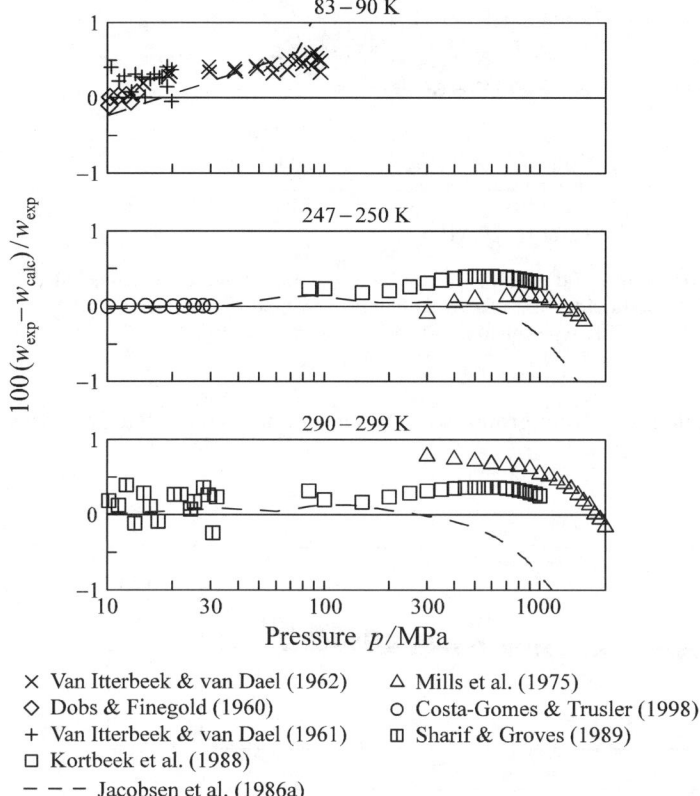

Fig. 5.23. Percentage deviations between experimental data for liquid and supercritical speeds of sound in nitrogen at high pressures and values calculated from the equation of Span et al. (1998b). Values calculated from the equation of Jacobsen et al. (1986a) are plotted as dashed lines for comparison.

As an example, Fig. 5.24 shows deviations between accurate experimental data for the speed of sound in liquid normal butane and values which were calculated from the MBWR-type equation of state by Younglove and Ely (1987), the equation which is widely used as standard for thermodynamic properties of normal butane today. At a temperature of 300 K ($T/T_c \approx 0.71$), the equation represents the data clearly within $\Delta w/w < \pm 1\%$ but at lower temperatures the deviations increase systematically reaching up to $\Delta w/w \approx \pm 3\%$ at 200 K ($T/T_c \approx 0.47$); no data are available at lower temperatures but a further increase of the uncertainty of the equation is expected down to the triple point temperature of about 135 K ($T_t/T_c \approx 0.32$). This behaviour is typical for the widespread MBWR-type equations which tend to yield inaccurate results for caloric properties in the liquid region at low reduced temperatures. Unless high accuracies can be proved by comparisons with accurate caloric data, one has to proceed from the fact that the uncertainty of

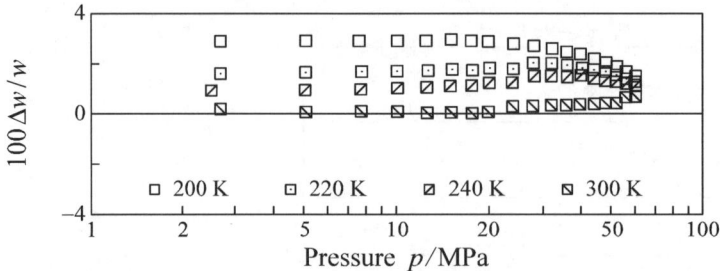

Fig. 5.24. Percentage deviations $100 \, \Delta w/w = 100 \, (w_{\text{exp}} - w_{\text{calc}})/w_{\text{exp}}$ between experimental data for the speed of sound in the liquid phase of normal butane measured by Niepmann (1984) and values calculated from the MBWR-type equation of Younglove and Ely (1987).

caloric properties calculated from group two reference equations of state increases significantly in this region.

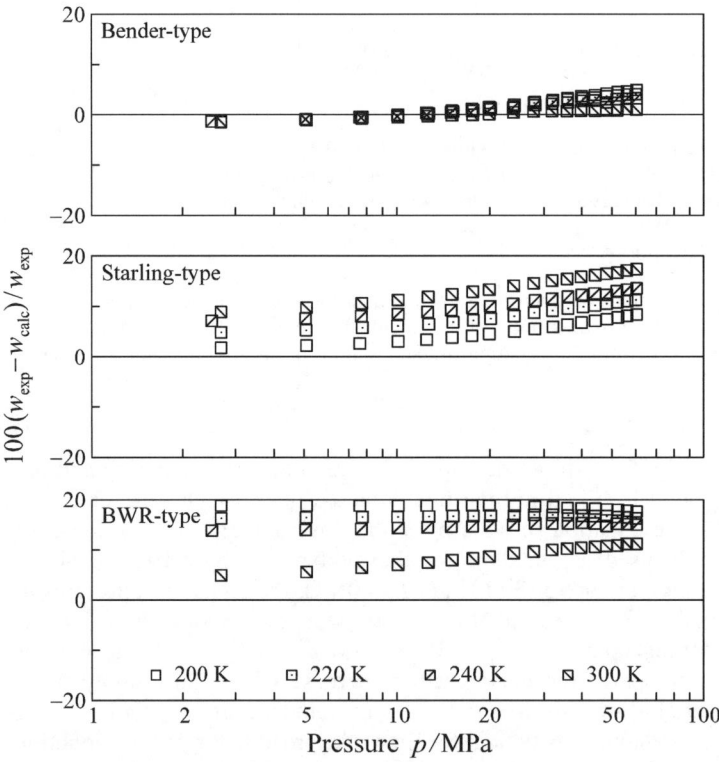

Fig. 5.25. Percentage deviations between experimental data for the speed of sound in the liquid phase of normal butane measured by Niepmann (1984) and values calculated from Bender-, Starling-, and BWR-type equations of state.

Figure 5.25 shows deviations between the speed of sound data which were already shown in Fig. 5.23 and values calculated from Bender-, Starling-, and BWR-type equations of state for normal butane. The deviations observed for values calculated from the Bender-type equation amount to $\Delta w/w \approx \pm 5\%$ at a temperature of 200 K and a pressure of 60 MPa, with increasing deviations to lower temperatures and higher pressures. This result may still be considered as sufficient at least for estimative technical calculations, especially since the deviations are smaller at typical technical pressures. Simpler technical equations of state like the Starling- and BWR-type equations result in typical deviations of $\Delta w/w \approx \pm 10\% - \pm 20\%$ even at low pressures.

In general, simple technical equations of state are completely unsuitable for the calculation of speeds of sound in the liquid phase or at liquid-like supercritical states. Without further examination, accurate results can only be expected from group one equations of state; good group two equations yield accurate results as

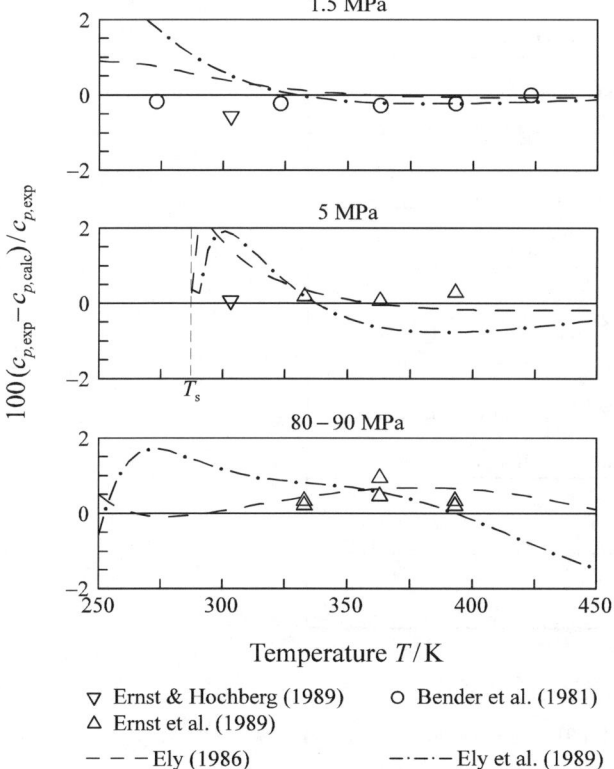

▽ Ernst & Hochberg (1989)   ○ Bender et al. (1981)
△ Ernst et al. (1989)
— — — Ely (1986)   — · — · — Ely et al. (1989)

Fig. 5.26. Percentage deviations between experimental data for the isobaric heat capacity of carbon dioxide and values calculated from the group one equation by Span and Wagner (1996). As examples for the performance of typical group two equations of state values calculated from the O_2-type equation of Ely (1986) and the MBWR-type equation of Ely et al. (1989) are plotted as dashed lines.

well, but this has to be verified for each equation. Bender-type and subaverage group two equations of state may be used for technical applications with low requirements with regard to accuracy.

Experimental data for the isobaric heat capacity are helpful in assessing equations of state at gaseous and supercritical states. Similar to the speed of sound, the residual contribution to the isobaric heat capacity is small at gaseous or gas-like supercritical states. Since the uncertainty of experimental results for heat capacities is much larger than the uncertainty of state-of-the-art speed of sound data in this region, measured isobaric heat capacities do not allow a clear distinction between group one and group two reference equations of state; both are able to describe the available data within their experimental uncertainty. But only sophisticated equations of state are able to follow the more rapid change of the isobaric heat capacity accurately when approaching the phase boundary. Figure 5.26 shows the resulting effect using carbon dioxide as an example. Above about 325 K, the gas-like supercritical data at pressures of 1.5 MPa and 5 MPa are represented within their ex-

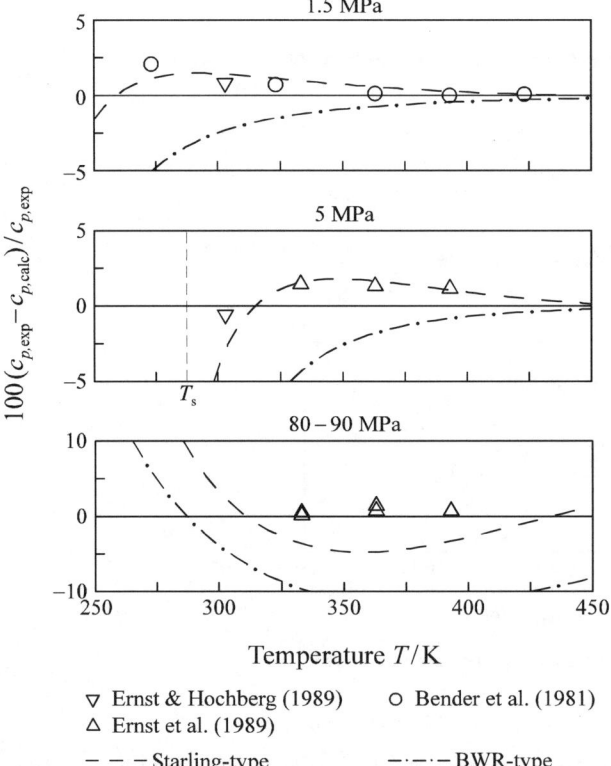

Fig. 5.27. Percentage deviations between experimental data for the isobaric heat capacity and values calculated from a Bender-type equation for carbon dioxide. Values calculated from Starling- and BWR-type equations are plotted as dashed lines.

perimental uncertainty by the recent group one equation of Span and Wagner (1996) as well as by the O_2-type equation of Ely (1986) and, with certain reservations, by the MBWR-type equation of Ely et al. (1989). However, at lower temperatures, the group two equations yield isobaric heat capacities which are clearly too large in this case; the deviations exceed the experimental uncertainties. At pressures of 80–90 MPa which correspond to liquid-like supercritical states, all of the equations are able to represent the available data within their experimental uncertainty.

For many technical applications, isobaric heat capacities at gas-like supercritical states are described accurately enough by technical equations of state as well. Based on comparisons with the same experimental data used above, Fig. 5.27 shows deviations of $\Delta c_p/c_p \leq \pm 2\%$ for values calculated from Bender- and Starling-type equations at pressures of 1.5 MPa and 5 MPa. Larger deviations occur for the Starling-type equation when approaching the phase boundary and for the BWR-type equation at supercritical states. At pressures of 80–90 MPa, it again becomes obvious that simple technical equations do not describe caloric properties at liquid-like supercritical states with acceptable uncertainties. Only the values calculated from the Bender-type equation deviate from the experimental data by $\Delta c_p/c_p < \pm 2\%$; uncertainties of $\Delta c_p/c_p \approx \pm 5\% - \pm 10\%$ are typical for simple technical equations in this region.

However, users are usually interested in enthalpy differences rather than in isobaric heat capacities. For isobaric enthalpy changes, relative uncertainties are a useful measure of uncertainty; the relative uncertainty of isobaric heat capacities is an upper limit for the relative uncertainty of isobaric enthalpy changes in the same

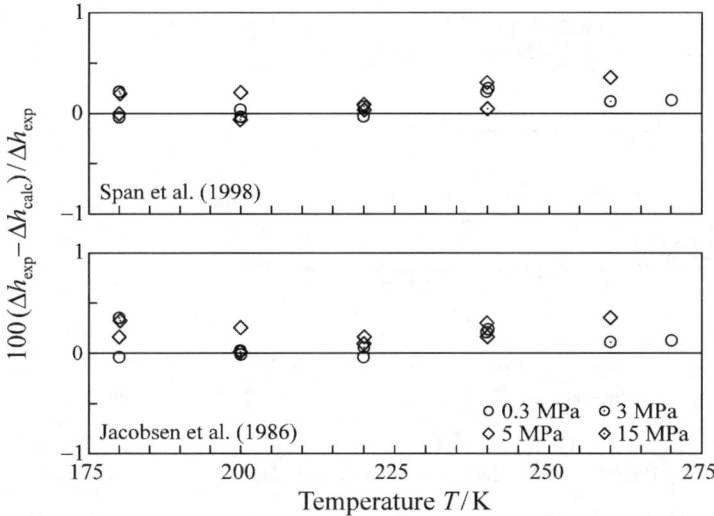

Fig. 5.28. Percentage deviations between experimental data for enthalpy differences in nitrogen measured by Grini and Owren (1997) at supercritical states and values calculated from the equations of Span et al. (1998b) and Jacobsen et al. (1986a).

region. For other processes, uncertainties of enthalpy changes can hardly be given in a general way, since both absolute and relative uncertainties depend on the considered process.

Comparisons with accurate data for isobaric enthalpy changes lead to the same conclusions as discussed above for isobaric heat capacities. Figure 5.28 shows relative deviations between experimental results for isobaric enthalpy differences in supercritical nitrogen and corresponding values calculated from the equations of Span et al. (1998b) and Jacobsen et al. (1986a). In this region, both the reference and the secondary reference equation are able to represent the most accurate data for enthalpy differences within their experimental uncertainty. The corresponding enthalpy differences calculated from Bender- and Starling-type equations deviate from the experimental results by up to $\Delta(\Delta h)/\Delta h \approx \pm 2\%$ while the deviations exceed $\pm 5\%$ for enthalpy differences calculated from a BWR-type equation on the 15 MPa isobar.

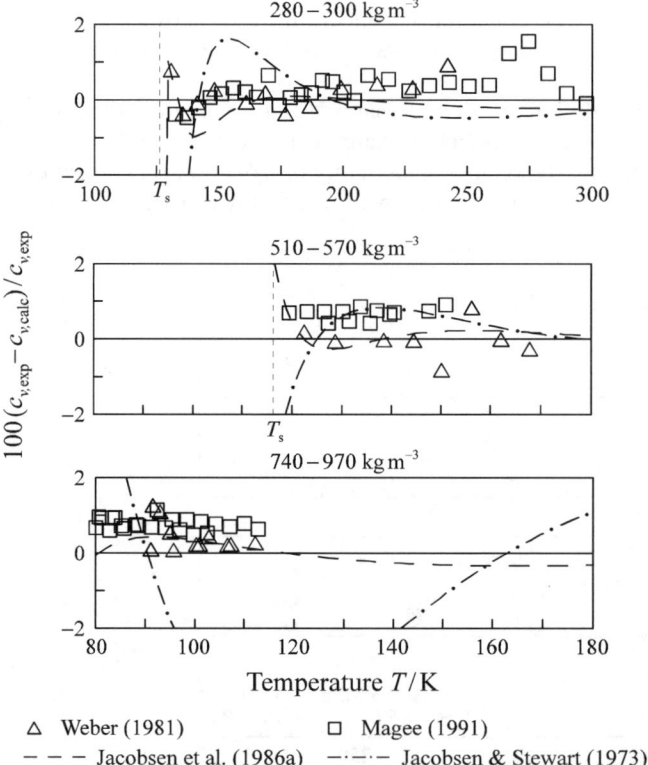

△ Weber (1981)   □ Magee (1991)
– – – Jacobsen et al. (1986a)   –·–·– Jacobsen & Stewart (1973)

Fig. 5.29. Percentage deviations between experimental data for isochoric heat capacities in nitrogen and values calculated from the equation of Span et al. (1998b). Values calculated from the equations of Jacobsen et al. (1986a) and Jacobsen and Stewart (1973) are plotted for comparison.

Due to characteristics of the used quasi isochoric apparatuses, experimental results for the isochoric heat capacity at gaseous or gas-like supercritical states are usually less accurate than experimental results for isobaric heat capacities or enthalpy changes measured in flow calorimeters. But at densities above about the critical density, experimental results on the isochoric heat capacity are an important source of information on caloric properties of fluids.

A typical comparison between experimental results for the isochoric heat capacity and values calculated from group one and two reference equations of state is given in Fig. 5.29; results for nitrogen are again used as example. Both the group one equation by Span et al. (1998b) and the above average group two equation by Jacobsen et al. (1986a) are able to represent the data within their experimental uncertainty of $\Delta c_v / c_v \leq \approx \pm 1\%$. Significant differences between the equations can be observed only for states close to the phase boundary, but the data situation does not allow clear conclusions to be drawn with regard to the performance of the equations in this region.

The density range 280 kg m^{-3} – 300 kg m^{-3} corresponds to reduced densities of $\rho / \rho_c = 0.89 - 0.96$ and its low temperature end is thus typical for the representation of isochoric heat capacities in the extended critical region. Although the theoretical restrictions of analytic equations of state (see Sect. 4.5.1) are undeniable, this comparison shows that shortcomings in the limiting behaviour do not affect practical applications. In the range where accurate data from measurements in equilibrium apparatuses are available for the isochoric heat capacity and where isochoric heat capacities can be relevant in practical calculations, both the group one and the group two reference equation of state represent the most accurate data within their experimental uncertainty.

The results shown for the older MBWR-type equation by Jacobsen and Stewart (1973) are again typical for subaverage group two equations of state. While data at far supercritical states are represented well, this equation predicts too small values for the isochoric heat capacity in the extended critical region, as theoretically expected. However, the most serious shortcoming becomes obvious at typical liquid densities (740 kg m^{-3} – 970 kg m^{-3}), where the MBWR-type equation deviates systematically from the experimental results. Similar problems with MBWR-type equations with respect to the representation of caloric properties at liquid states have already been discussed for the representation of speeds of sound.

Figure 5.30 shows deviations between experimental results for the isochoric heat capacity of nitrogen and values calculated from Bender-, Starling-, and BWR-type equations of state. The Bender-type equation describes the experimental data within $\Delta c_v / c_v \leq \pm 4\%$. The largest deviations occur in the extended critical region and in the liquid phase. Outside of the extended critical region, the typical uncertainty of calculated isochoric heat capacities is $\Delta c_v / c_v \approx \pm 2\%$ at low or medium densities.

For simple technical equations like the Starling- or BWR-type equations, the typical uncertainty of calculated isochoric heat capacities becomes $\Delta c_v / c_v \leq \pm 10\% - \pm 20\%$ both in the extended critical region and at liquid states. At far supercritical states, the uncertainty of the Starling-type equation becomes similar to the uncertainty of the Bender-type equation of state, while BWR-type

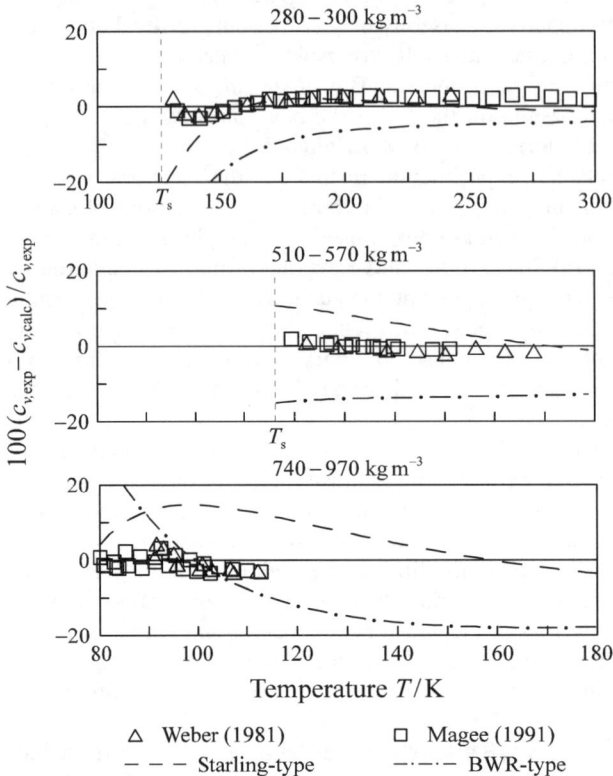

Fig. 5.30. Percentage deviations between experimental data for isochoric heat capacities in nitrogen and values calculated from a Bender-type equation of state. Values calculated from the Starling- and BWR-type equations are plotted for comparison.

equations still result in deviations of $\Delta c_v / c_v \approx \pm 10\%$. Simple technical equations of state prove to be unsuitable for the calculation of caloric properties, especially at liquid and liquid-like supercritical states.

5.3 Properties at Vapour-Liquid Phase Equilibrium

The calculation of thermal and caloric properties of vapour-liquid phase equilibria is one of the most common applications of equations of state both in technical and in scientific calculations. All of the group one and two reference equations of state which are in use today were designed to fulfil the phase equilibrium criteria, see Sect. 3.2.2. Thus, internally consistent sets of thermal and caloric data on the phase boundary can be calculated from the equation of state alone. Ancillary equations for vapour pressure and saturated liquid and vapour density are used only to gen-

5.3 Properties at Vapour-Liquid Phase Equilibrium

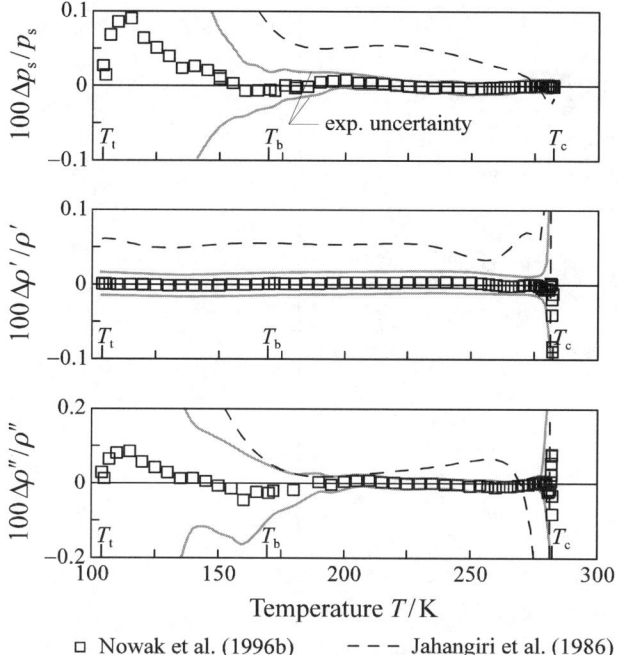

□ Nowak et al. (1996b)   – – – Jahangiri et al. (1986)

Fig. 5.31. Percentage deviations $100\,\Delta y/y = 100\,(y_{\mathrm{exp}} - y_{\mathrm{calc}})/y_{\mathrm{exp}}$ with $y = p_{\mathrm{s}}, \rho', \rho''$ between accurate experimental results for ethylene and values calculated from the equation of state by Smukala et al. (1999). For comparison, values calculated from the equation of Jahangiri et al. (1986) are plotted as dashed lines. The grey lines indicate the uncertainty of the data.

erate starting solutions for the iterative solution of the phase equilibrium criteria, see Sect. 3.3.5.

The most accurate data for thermal equilibrium properties currently available are measurements with two-sinker densimeters. Typical experimental uncertainties of these data are $\Delta p_{\mathrm{s}}/p_{\mathrm{s}} \leq \pm 0.01\% - \pm 0.02\%$ for pressures above $p_{\mathrm{s}} \approx 0.3$ MPa, $\Delta\rho'/\rho' \leq \pm 0.015\%$ for temperatures up to $T_{\mathrm{s}}/T_{\mathrm{c}} \approx 0.99$, and $\Delta\rho''/\rho'' \leq \pm 0.03\%$ for saturated vapour densities above $\rho'' \approx 10$ kg m^{-3} and up to $T_{\mathrm{s}}/T_{\mathrm{c}} \approx 0.98 - 0.99$. The typical performance of group one and group two reference equations of state can be illustrated by comparisons with such data, although these data were usually not available when the group 2 equations were established. As explained before, the shown comparisons do not intend to devaluate the group 2 equations of state, but to give representative examples for the performance of such equations.

Figure 5.31 shows deviations between highly accurate experimental data for ethylene and values calculated from the group one equation of state by Smukala et al. (1999). Just as expected from a reference equation of state, the equation of Smukala et al. is able to represent the data within their experimental uncertainty. At temperatures below the normal boiling point, the deviations between experimental data and calculated vapour pressures and saturated vapour densities in-

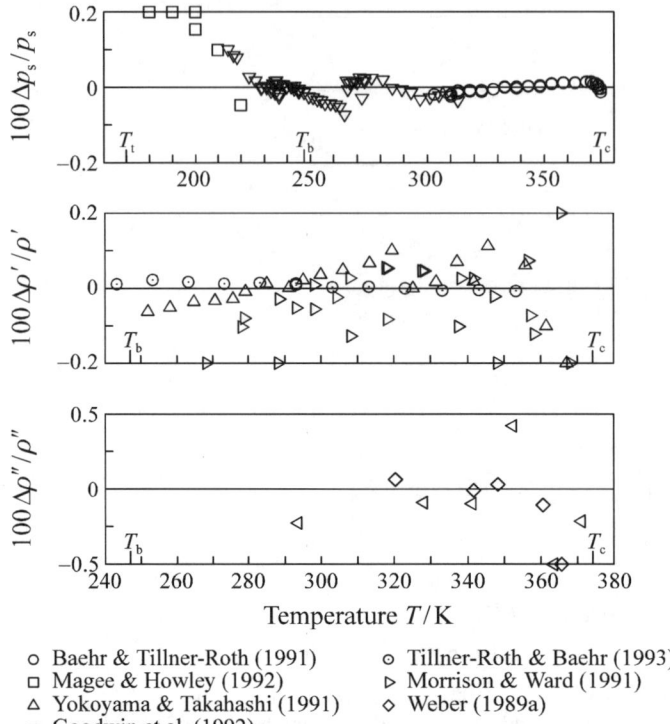

Fig. 5.32. Percentage deviations $100 \, \Delta y/y = 100 \, (y_{\text{exp}} - y_{\text{calc}})/y_{\text{exp}}$ with $y = p_s$, ρ', ρ'' between selected experimental results for refrigerant R134a and values calculated from the equation of state by Tillner-Roth and Baehr (1994).

crease up to ±0.1%, but due to the small pressures and densities these deviations are still far within the experimental uncertainties.[10]

The equation of Jahangiri et al. (1986) can be regarded as an example of a group two reference equation of state with above average representation of thermal properties at phase equilibrium. Above the normal boiling point, vapour pressures are represented within $\Delta p_s/p_s \leq \pm 0.05\%$ and saturated vapour densities are represented within $\Delta \rho''/\rho'' \leq \pm 0.06\%$, except for data above about 270 K where the description of orthobaric densities becomes increasingly difficult due to the onset of critical effects. Saturated liquid densities are represented within $\Delta \rho'/\rho' \leq \pm 0.06\%$ from the triple point temperature up to about 270 K.

However, based on the given data situation, it becomes difficult for other above average group two equations to match the accuracies achieved by Jahangiri et al.

---

[10] The absolute contributions to the experimental uncertainties dominate under these conditions and result in an unavoidable increase of the total relative uncertainty which is indicated by grey lines in Fig. 5.31.

(1986) for ethylene. Figure 5.32 shows the representation of the most accurate data for the vapour pressure and the saturated liquid and vapour densities of refrigerant R134a by the equation of Tillner-Roth and Baehr (1994). For temperatures above the normal boiling point, accurate data for the vapour pressure are represented within $\Delta p_s/p_s \leq \pm 0.05\%$. Based on a comparison with the data of Tillner-Roth and Baehr (1993), the estimated uncertainty of calculated saturated liquid densities is $\Delta\rho'/\rho' \leq \pm 0.05\%$ for temperatures above the normal boiling point, but below the normal boiling point, no measurements with similar uncertainties are available. From the selected saturated vapour densities, an uncertainty of $\Delta\rho''/\rho'' \leq \pm 0.25\%$ can be estimated for temperatures above 290 K. No reasonable assessment is possible below this temperature. Due to the lack of reliable data, large uncertainties in calculated saturated vapour densities are a characteristic feature of typical group two equations. Fortunately, equations of state predict saturated vapour densities reasonably well for reduced temperatures below $T/T_c \approx 0.9$ if they describe vapour pressures and $p\rho T$ data in the gas phase accurately, but the uncertainties are still higher when the description of saturated vapour densities is based on this approach.

Typical subaverage reference equations are still able to represent saturated liquid densities within $\Delta\rho'/\rho' \leq \pm 0.1\%$, at least in limited temperature ranges. But the typical uncertainty of calculated vapour pressures increases to $\Delta p_s/p_s \leq \approx \pm 0.2\%$. Due to larger uncertainties both with regard to $p\rho T$ data in the gaseous phase and to vapour pressures, the uncertainty of calculated saturated

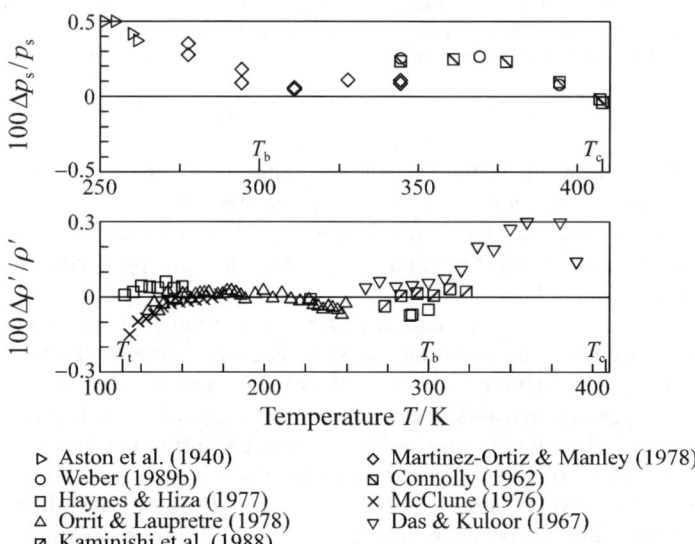

▷ Aston et al. (1940) ◇ Martinez-Ortiz & Manley (1978)
○ Weber (1989b) ◨ Connolly (1962)
□ Haynes & Hiza (1977) × McClune (1976)
△ Orrit & Laupretre (1978) ▽ Das & Kuloor (1967)
◪ Kaminishi et al. (1988)

Fig. 5.33. Percentage deviations $100\,\Delta y/y = 100\,(y_{\text{exp}} - y_{\text{calc}})/y_{\text{exp}}$ with $y = p_s, \rho'$ between selected experimental results for isobutane and values calculated from the equation of state by Younglove and Ely (1987).

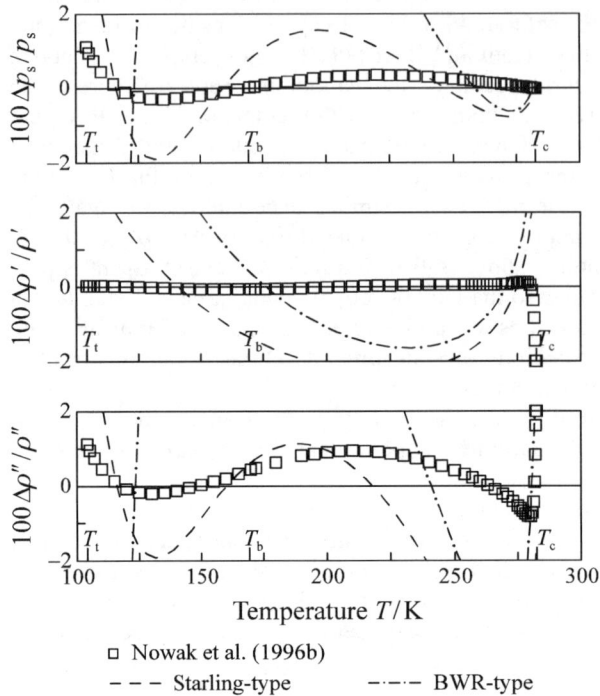

Fig. 5.34. Percentage deviations $100\,\Delta y/y = 100\,(y_{\text{exp}} - y_{\text{calc}})/y_{\text{exp}}$ with $y = p_s$, ρ', ρ'' between accurate experimental results for ethylene and values calculated from a Bender-type equation. For comparison, values calculated from Starling- and BWR-type equations are plotted as dash-dotted lines.

vapour densities increases as well. Reliable data for comparisons are usually not available. Figure 5.33 shows the representation of vapour pressures and saturated liquid densities by such an equation. The MBWR-type equation of Younglove and Ely (1987), which is internationally regarded as standard for the properties of isobutane, is used as an example here.

With regard to the representation of thermal phase equilibrium properties, the typical performance of Bender-type equations of state is comparable to the performance of subaverage group two equations of state which was discussed above; for common technical applications, this accuracy can be regarded as sufficient. However, the phase equilibrium condition was not considered during the development of the functional forms of most of the simple technical equations of state which are in use today. Even when fitted to data sets which contain phase equilibrium data, such equations cannot properly represent phase equilibrium properties.

As an example, Fig. 5.34 shows the representation of thermal phase equilibrium properties by Bender-, Starling- and BWR-type equations of state for ethylene. For vapour pressures and saturated liquid densities, the deviations between experimental data and values calculated from the Bender-type equation are of the expected order. Larger deviations for saturated vapour densities indicate problems with the representation of $p\rho T$ data for gaseous states close to the phase boundary. Both the Starling- and the BWR-type equation result in uncertainties above ±1% for all properties, where the Starling-type equation is superior for vapour pressures and the BWR-type equation is superior for liquid densities. Both equations fail completely with regard to the representation of saturated vapour densities.

For technical applications, the enthalpy of evaporation is the caloric phase equilibrium property needed most frequently. This property is related to the thermal phase equilibrium properties by the well known Clausius-Clapeyron equation,

$$h'' - h' = T(v'' - v') \frac{dp_s}{dT}. \tag{5.3}$$

Due to the multiplication of thermal properties implied by Eq. 5.3 the relative uncertainty of calculated enthalpies of evaporation is larger than the relative uncertainty of thermal equilibrium properties, but it is of the same order of magnitude. Enthalpies of evaporation calculated from reference equations of state are usually more accurate than the available experimental results, except for states with very low vapour pressures where both v'' and dp_s/dT may become increasingly uncertain. Enthalpies of evaporation calculated from simple technical equations of state are only accurate enough for calculations with low demands on accuracy, see Fig. 5.34 for the uncertainty of calculated thermal properties.

For reference equations of state, the representation of derived caloric properties on the phase boundaries is more crucial. Common examples for such properties are

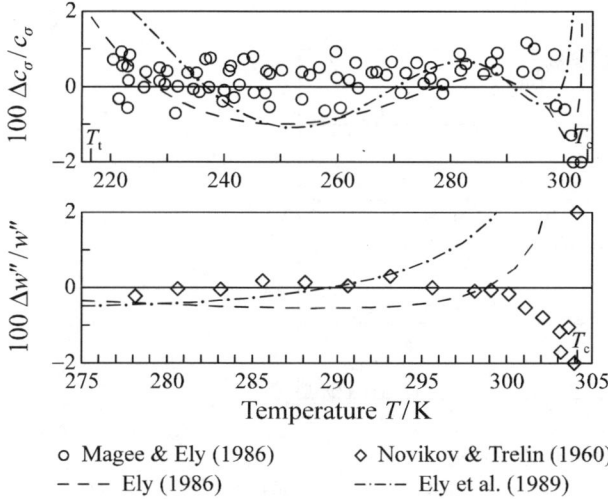

o Magee & Ely (1986)   ◇ Novikov & Trelin (1960)
- - - Ely (1986)   -·-·- Ely et al. (1989)

Fig. 5.35. Percentage deviations $100\,\Delta y/y = 100\,(y_{exp} - y_{calc})/y_{exp}$ with $y = c_\sigma$, w'' between experimental results for carbon dioxide and values calculated from the recent group one equation by Span and Wagner (1996). Results of the equations by Ely (1986) and Ely et al. (1989) are plotted as dash-dotted lines for comparison.

the heat capacity of the saturated liquid, c_σ (see Eq. 3.80), or the speeds of sound in the saturated vapour ($w''(T) = w(T, \rho'')$) and liquid ($w'(T) = w(T, \rho')$). Data for these properties are available for a variety of substances.

Figure 5.35 compares experimental results for c_σ and w'' with values calculated from three different equations of state for carbon dioxide. The recent reference equation by Span and Wagner (1996) reproduces the heat capacities of the saturated liquid within $\Delta c_\sigma / c_\sigma \leq \pm 1\%$ without systematic deviations up to temperatures of $T \approx 301$ K ($T/T_c \approx 0.99$ with $T_c = 304.1282$ K); systematic deviations at higher temperatures remain within the experimental uncertainty of the data, which increases close to T_c. The O_2-type equation by Ely (1986) and the MBWR-type equation by Ely et al. (1989) which can be regarded as typical group two reference equations again result in systematical oscillations, but nevertheless they represent the data within $\Delta c_\sigma / c_\sigma \leq \approx \pm 2\%$ up to $T/T_c \approx 0.99$. Only for temperatures close to the triple point, the MBWR-type equation shows larger systematic deviations.

Speeds of sound in the saturated vapour are represented within $\Delta w'' / w'' \leq \pm 0.6\%$ by the equations of Ely et al. (1989) and Ely (1986) up to temperatures of $T \approx 290$ K and $T \approx 300$ K, respectively. At higher temperatures, these equations result in speeds of sound which are significantly too large. The group one equation describes the data within $\Delta w'' / w'' \leq \pm 0.3\%$ up to $T \approx 300$ K and is able to follow the measured decrease of the speed of sound within $\Delta w'' / w'' \leq \pm 2\%$ up to temperatures of $T/T_c \approx 0.998$; the experimental data at higher temperatures have to be regarded as very uncertain, see Span and Wagner (1996).

As an example of the performance of a subaverage group two equation, Fig. 5.36 shows deviations between experimental results for c_σ and w' and values calculated from the equation of Piao and Noguchi (1998) for R125. The heat ca-

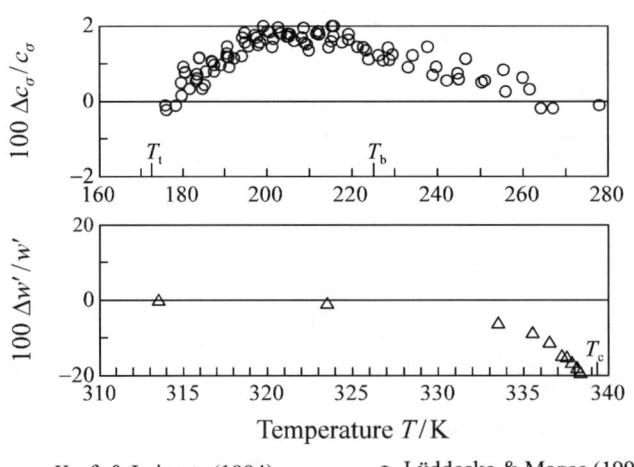

△ Kraft & Leipertz (1994)   ○ Lüddecke & Magee (1996)

Fig. 5.36. Percentage deviations $100 \, \Delta y / y = 100 \, (y_{\text{exp}} - y_{\text{calc}}) / y_{\text{exp}}$ with $y = c_\sigma$, w' between experimental results for R125 and values calculated from the recent group two equation by Piao and Noguchi (1998).

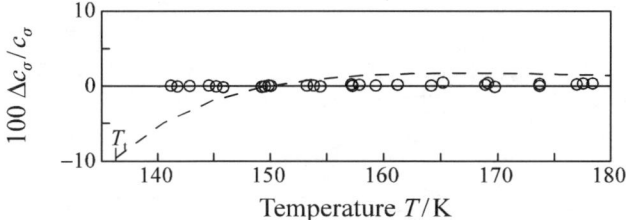

○ Lüddecke & Magee (1996)   − − − Outcalt & McLinden (1995)

Fig. 5.37. Percentage deviations $100\,\Delta c_\sigma/c_\sigma = 100\,(c_{\sigma,\mathrm{exp}} - c_{\sigma,\mathrm{calc}})/c_{\sigma,\mathrm{exp}}$ between experimental results for R32 and values calculated from the recent group 2 equation by Tillner-Roth and Yokozeki (1997). Values calculated from the MBWR-type equation of Outcalt and McLinden (1995) are plotted as a dashed line for comparison.

pacity of the saturated liquid is again represented within $\Delta c_\sigma/c_\sigma \leq \pm 2\%$, but clearly not within the experimental uncertainty of the data which would allow a more accurate description of c_σ than the older data shown for carbon dioxide do. Speeds of sound in the saturated liquid are represented satisfactorily up to $T \approx 325$ K ($T/T_\mathrm{c} \approx 0.96$ with $T_\mathrm{c} = 339.33$ K); at higher temperatures the observed deviations increase systematically up to about -20% at $T/T_\mathrm{c} \approx 0.997$.

Another typical problem is shown in Fig. 5.37 using the heat capacity of saturated liquid refrigerant R32 as an example. The recent reference equation of state by Tillner-Roth and Yokozeki (1997) represents the available data within their experimental uncertainty, but for temperatures below $T \approx 150$ K, the MBWR-type equation by Outcalt and McLinden (1995) deviates systematically from the experimental data. At T_t, the deviation between the equations amounts to $\Delta c_\sigma/c_\sigma \approx 10\%$. Problems regarding the representation of caloric properties in the liquid phase at low reduced temperatures have already been observed for MBWR-type equations of state in Sect. 5.2.

Obviously, both group one and group two reference equations of state are able to represent caloric data on the phase boundary accurately; group one and above average group two reference equations represent available data within their experimental uncertainty. Group one equations are usually superior with regard to the representation of data close to the critical temperature. When high accuracies are demanded for such states, group two equations should only be used if sufficient accuracies can be verified by comparisons with suitable experimental data. When using MBWR-type equations the same is true for liquid states at low (reduced) temperatures.

To give an example of the performance of technical equations of state, Fig. 5.38 shows deviations between the experimental data for carbon dioxide shown in Fig. 5.35 and values calculated from Bender-, Starling- and BWR-type equations. Except for temperatures close to T_c, the Bender-type equation represents heat capacities of the saturated liquid within $\Delta c_\sigma/c_\sigma \leq \pm 2\% - \pm 3\%$ and saturated vapour speeds of sound within $\Delta w''/w'' \leq \approx \pm 1\%$. Again, the performance of the Bender-type equation comes close to the performance of subaverage group two equations

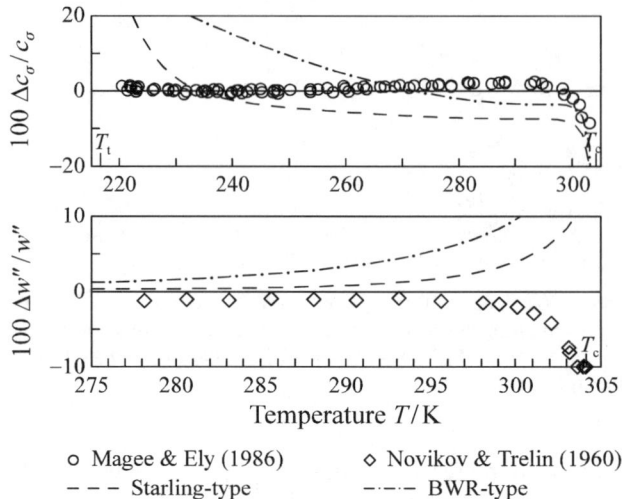

Fig. 5.38. Percentage deviations $100\,\Delta y/y = 100\,(y_{\text{exp}} - y_{\text{calc}})/y_{\text{exp}}$ with $y = c_\sigma$, w'' between experimental results for carbon dioxide and values calculated from a Bender-type equation. Results from Starling- and MBWR-type equations of state are plotted as dash-dotted lines for comparison.

and can be regarded as sufficient for technical applications with average demands on accuracy.

The description of speeds of sound in saturated carbon dioxide vapour by the Starling- and BWR-type equations can still be regarded as acceptable at least at temperatures below $T/T_c \approx 0.9$. At higher temperatures, increasing deviations are observed which exceed $+10\%$ at $T/T_c \approx 0.98$ and $T/T_c \approx 0.99$ for the BWR- and Starling-type equation, respectively. However, with regard to caloric properties of the saturated liquid, the performance of simple technical equations of state again becomes completely unsatisfactory. Typical deviations between experimental results for heat capacities of the saturated liquid and values calculated from these equations are of the order $\Delta c_\sigma/c_\sigma \approx \pm 10\%$. The deviations may even exceed $\pm 20\%$ especially at low temperatures.

5.4 Some General Assessments

Based on the comparisons shown in the preceding sections and on additional comparisons for other substances, it is possible to formulate some kind of general assessment of the performance of typical multiparameter equations of state. As already discussed in the introduction of this chapter, such an assessment is necessarily incomplete and it will always be possible to find examples demonstrating the opposite, but nevertheless it has to be summarised to help inexperienced users become aware of the applicability of certain types of equations of state.

With regard to the accuracy of calculated thermodynamic properties, *group one reference equations of state* (see Table 5.1) represent the highest standard available today. With a typical uncertainty of $\Delta\rho/\rho \leq \pm 0.02\%$ at temperatures up to $T \approx 500$ K and pressures up to $p \approx 30$ MPa and $\Delta p/p \leq \pm 0.02\%$ in the extended critical region, $p\rho T$ data calculated from such equations are more accurate than data calculated from group two equations by a factor of 5–10. The representation of thermal properties at phase equilibrium could be improved by a factor of 3–10 by the development of group one equations and at least for argon (Tegeler et al., 1997/1999) and nitrogen (Span et al., 1998b) similar improvements can be verified for speeds of sound at gaseous and gas-like supercritical states as well. Since fundamental equations of state yield consistent results for different properties by definition, one may also expect significant improvements for other properties, but based on the data available today this expectation is difficult to prove. Typical uncertainties for caloric properties which can be proved by comparisons with experimental data are $\pm 0.1\% - \pm 0.5\%$ at gaseous and gas-like supercritical states and $\pm 0.5\% - \pm 1\%$ at liquid and liquid-like supercritical states. Compared to average group two equations, the advantages of group one equations become obvious especially in the extended critical region, for liquid states at low reduced temperatures, or at gaseous states close to the phase boundary. For higher pressures and temperatures, group one equations and above average group two equations are of the same standard both for thermal and caloric properties but in most cases the range of validity is larger for group one equations and their extrapolation behaviour is more reliable. Thus, at very high pressures and temperatures, group one equations are usually superior again.

With regard to their application, group one reference equations of state are an ideal basis for calibrations of secondary measurement devices, for tests of new measuring methods and for conversions of measured parameters. In scientific use, the verification of physically founded models for thermodynamic properties is another common application. Data calculated from group one equations of state can be used in the critical region without restrictions, but no conclusions with regard to the limiting behaviour at the critical point should be drawn from empirical equations of state (see Sects. 4.5.1 and 4.5.2). A specific technical application of group one equations of state is the highly accurate determination of mass flows. However, based on recent software developments and the computing power available today, group one equations can be used in everyday applications such as process calculations as well. For example, the recent group one equation by Span and Wagner (1996) is used as data base by a variety of scientific laboratories and industrial research and development departments which cooperate on the development of mobile carbon dioxide based air conditioning systems.

The most serious disadvantage of group one equations of state is the limited number of substances for which such equations are available. For example, the existing set of equations is far too small to guarantee a sufficiently narrow grid of reference substances for calibration purposes. The worst situation is encountered for caloric properties in the liquid phase, where highly accurate results are available only for water (Wagner and Pruß, 1997/2000). Additional work on group one equations for other substances is proceeding, but establishing both the required experimental data bases and the equations is a very time-consuming task; quick progress is not to be expected.

Group two, the second group of reference equations of state, is less homogeneous than group one and quite often it will be necessary to check the accuracy of available equations of state if demands on accuracy are high. In the scientifically and technically most important regions, typical uncertainties are of the order $\Delta\rho/\rho \leq \pm 0.1\% - \pm 0.2\%$, except in the extended critical region where the typical uncertainty is $\Delta p/p \leq \pm 0.1\% - \pm 0.2\%$. Calculated values for saturated vapour densities are usually less accurate. Especially at reduced temperatures of $T/T_c > \approx 0.9$ the corresponding uncertainty may exceed $\pm 1\%$ for subaverage group two equations.

In general, the representation of thermal properties by group two reference equations of state can be regarded as sufficient for many scientific and most technical applications; with regard to technical applications, typical exceptions are calibration purposes, etc. Problems are encountered mainly in the extended critical region, where the typical uncertainty in pressure given above results in considerable uncertainties in density.

The situation becomes less uniform for technical calculations based on caloric properties. No problems are to be expected at gaseous, gas-like, and far supercritical states. In these regions, uncertainties of above average group two equations are comparable to those observed for group one equations of state; subaverage group two equations result in typical uncertainties of $\pm 1\% - \pm 2\%$. Increased uncertainties are usually encountered for gaseous states close to the phase boundary and in the extended critical region. For liquid states, above average group two equations are still comparable to group one equations, while subaverage group two equations result in uncertainties up to $\pm 5\%$, especially at low temperatures. Uncertainties of this order may very well affect results of advanced technical calculations. In particular, the widely used MBWR-type equations are sensitive to this kind of problems and need to be tested before they are used for calculations of caloric properties in the liquid phase.

Group three, the group which covers simple technical multiparameter equations of state, is rather inhomogeneous as well. At the upper end of the observed range of accuracies, Bender-type equations come close to subaverage group two equations. For equations which were fitted to extensive data sets, typical uncertainties are $\pm 0.2\% - \pm 0.5\%$ for thermal properties and $\pm 1\% - \pm 2\%$ for caloric properties at gaseous, gas-like, and far supercritical states. At liquid and liquid-like supercritical states the uncertainty of caloric properties increases to $\pm 2\% - \pm 3\%$ and deviations up to $\pm 5\%$ are frequently encountered for liquid states at low temperatures. Typical uncertainties in the extended critical region are of the same order of magnitude. Thus, even the best group three equations become questionable for advanced technical applications in the extended critical region and at liquid states. However, the most serious restrictions result from the numerical instability of Bender-type equations. When fitted to substances where only restricted data sets are available these equations tend to yield unreasonable results especially for derived caloric properties such as heat capacities – this problem will be discussed in more detail in the following section.

Simple technical equations of state such as Starling- or MBWR-type equations of state yield acceptable results only at gaseous- and gas-like supercritical states. In this region, the uncertainties observed for calculated $p\rho T$ data are $\Delta\rho/\rho \leq \pm 0.5\% - \pm 1\%$ and typical uncertainties of $\pm 2\% - \pm 5\%$ are encountered for caloric properties. At supercritical states, the corresponding uncertainties increase to $\pm 2\%$ at high densities and $\pm 5\% - \pm 10\%$ at liquid like densities. For typical liquid states, even larger uncertainties are to be observed; deviations for caloric properties may exceed $\pm 20\%$ especially at low temperatures. When calculated from simple technical equations of state, thermal properties of the vapour-liquid equilibrium phases and caloric properties of the saturated liquid are highly unreliable as well. Thus, results from simple technical equations should only be used for estimative calculations unless processes are restricted to gaseous and gas-like supercritical states with low densities.

6 Generalised Functional Forms

The idea of using functional forms of equations of state in a generalised way, in other words to describe properties of different substances with the same equation of state using just different sets of coefficients, is not new at all. Quite often, functional forms which had been developed for certain substances have been used for other substances as well. On a rather high level of accuracy e.g. Bender- and MBWR-type equations have been used for broad groups of fluids.

The idea of substance-specific functional forms resulted from the use of optimisation algorithms (see Sect. 4.4). During the optimisation process intercorrelated pairs of terms are replaced by single terms out of the bank of terms which result in a similar contribution. In this way, the same accuracy can be achieved with significantly fewer terms and the resulting equation of state becomes numerically more stable – intercorrelated pairs of terms may mimic the contribution of a different term within the range where they were fitted to data, but outside of this range they are likely to result in completely unreasonable contributions. The introduction of optimisation algorithms has not only improved the accuracy of equations of state[1] but also their extrapolation behaviour and their reliability in regions with questionable data situation. The improved numerical stability of optimised functional forms turns into a disadvantage when being applied to different substances. Optimised equations of state lose their superior accuracy when they are fitted to data sets for other substances, unless the considered substances are very similar in terms of reduced properties.

At the same time the substance-specific optimisation of functional forms is both very time consuming and very sensitive to shortcomings of the used data set; the group of substances for which optimised functional forms are available is still small and a fast growth of this group is not to be expected. Nevertheless, accurate thermodynamic data are needed for a multitude of mostly technically important substances. In most cases the data sets which are available for these substances are small and often they cover only certain parts of the fluid surface. Substance-specific optimisation algorithms cannot be used under such circumstances and thus until now, one always had to rely on functional forms which had been developed for other substances with better data sets.

Especially for typical "technical equations of state" such as BWR-, Starling- or Bender-type equations, sets of coefficients are available for a multitude of substances. Polt (1987) and Platzer (1990) e.g. report sets of coefficients for Bender-type equations for a combined total of 51 polar and nonpolar pure substances. The

---

[1] The performance of recent group one and above average group two reference equations could only be achieved with optimised functional forms; see Sect. 5 for examples.

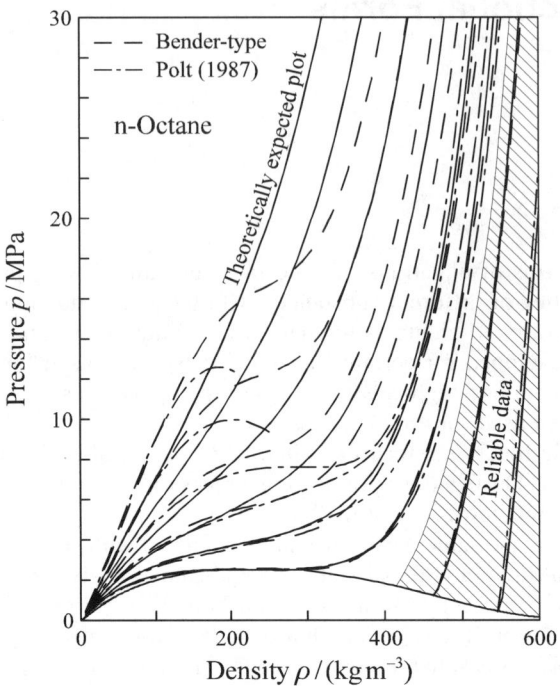

Fig. 6.1. Isotherms in a p,ρ-diagram for n-octane which were calculated from the Bender-type equation published by Polt (1987) and from a Bender-type equation which was refitted to the data set described in Sect. 6.2.2. The Maxwell-like loops which result from the equation by Polt at supercritical temperatures were cut to preserve the clarity of the plot.

comparisons in Chap. 5 showed, that Bender-type equations of state are in general sufficient for most technical applications when fitted to extensive data sets; restrictions became obvious only for certain regions. However, for substances with small data sets this situation changes radically.

Figure 6.1 shows isotherms in a p,ρ-diagram for n-octane which were calculated from the original Bender-type equation published by Polt (1987) and from a Bender-type equation which was refitted to the data set described in Sect. 6.2.2. In the liquid region, where reliable $p\rho T$ data are available up to temperatures of $T \approx 550$ K (see the hatched region in Fig. 6.1), both equations yield at least similar results. At higher temperatures, both equations quickly deviate from each other and from the theoretically expected plot. At supercritical states, both equations yield unphysical results; the equation by Polt (1987) even predicts Maxwell-like loops again at far supercritical temperatures.

At this point one may argue that the misbehaviour of the Bender-type equations is arbitrary since it starts at temperatures where the thermal stability of n-octane is already questionable. However, an unphysical behaviour like the one observed in Fig. 6.1 becomes a serious problem when using such equations in mixture ap-

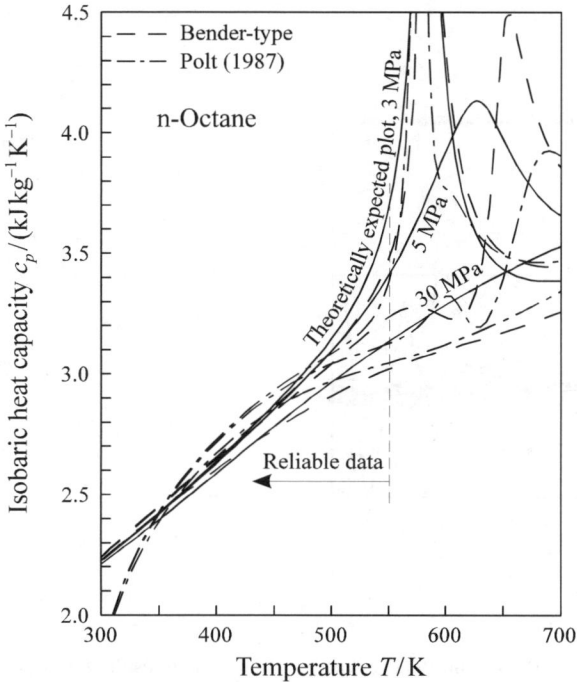

Fig. 6.2. Isobars in a c_p,T-diagram for n-octane which were calculated from the Bender-type equation published by Polt (1987) and from a Bender-type equation which was refitted to the data set described in Sect. 6.2.2.

proaches, where pure component data at reduced temperatures $T/T_c > 1$ are needed very well (see Chap. 8), even if the pure fluid is not stable under such conditions. Moreover, when calculating data for the pure fluid problems arise from the fact that derived caloric properties such as heat capacities or speeds of sound are much more sensitive to numerical problems than $p\rho T$ data. Figure 6.2 shows isobars in a c_p,T-diagram for n-octane which were calculated from the two Bender-type equations again. The equation by Polt (1987) yields unreasonable results over the whole range of plotted states. The refitted Bender-type equation follows the theoretically expected courses up to temperatures of $T \approx 375$ K where it was fitted to isobaric heat capacities. At higher temperatures, still within the range where reliable $p\rho T$ data were available when fitting the equation, it results in unreasonable plots of the isobaric heat capacity as well.

Obviously, sufficiently accurate multiparameter equations of state without optimised functional form are numerically not stable enough to be used for substances with small data sets. This statement becomes intelligible from Fig. 6.3, which shows the values of the coefficients of BWR- and Bender-type equations of

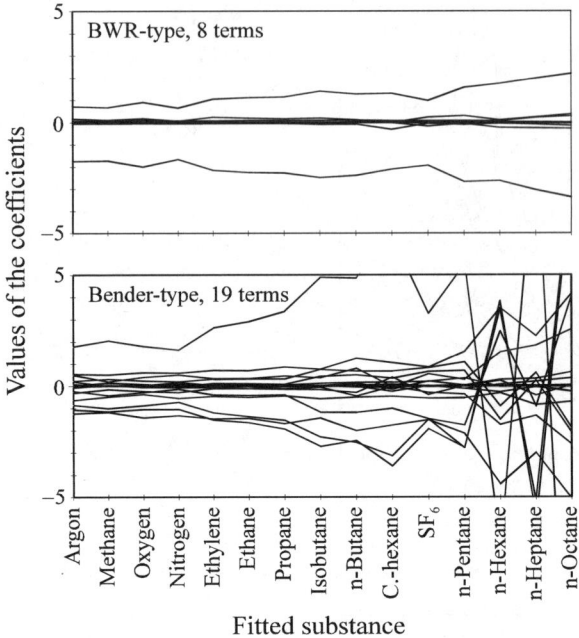

Fig. 6.3. Values of the coefficients of BWR- and Bender-type equations which resulted from fits to data sets of 15 non- and weakly polar fluids. The fluids are ordered by increasing acentric factor, ω.

state which were fitted to data sets for 15 non- and weakly polar fluids,[2] whereby the fluids are ordered by increasing acentric factor, ω.[3] The coefficients of the Bender-type equations are on average rather large – negative and positive contributions compensate for each other, the terms are highly intercorrelated. The plots of the single terms do not have recognisable patterns; especially for the higher alkanes, where the available data sets are small, the values of the coefficients are almost randomly distributed.

For the simple BWR-type equation, the corresponding plot indicates better numerical stability. On average, the values of the coefficients are smaller and the plots of the single terms develop on recognisable patterns which are superimposed only by moderate oscillations. However, in most cases thermodynamic properties

---

[2] For details on the data sets see Sect. 6.2.2. Both equations of state were used in the reduced form according to Eq. 3.24 with $\tau = T_c/T$ and $\delta = \rho/\rho_c$. When using the more common unreduced forms according to Eq. 3.25 or Eq. 3.26 the coefficients differ by several orders of magnitude and comparisons like the one given in Fig. 6.3 become meaningless.

[3] The acentric factor was introduced by Pitzer et al. (1955) as part of an extended corresponding states approach. It describes the deviation between reduced vapour pressures of a fluid and of a nonpolar model fluid with spherical molecules and is defined as $\omega = -\log(p_s(0.7 \cdot T_c)/p_c) - 1$.

calculated from simple BWR-type equations turned out to be unsuitable even for technical applications with moderate demands on accuracy, see Chap. 5.

An accurate calculation of thermodynamic properties based on multiparameter equations of state has always been restricted to relatively few substances for which large data sets are available – this has been the most serious disadvantage regarding the technical application of multiparameter equations of state for pure substances. The following sections will focus on a recent approach, which promises to overcome the described limitation.

6.1 Simultaneous Optimisation of Functional Forms

As explained above, equations of state with substance specific optimised functional form are known to be numerically more stable than e.g. Bender-type equations, but they cannot be transferred to other substances without significant loss of accuracy. To overcome this problem, Span et al. (1998a) developed an optimisation algorithm which considers data sets for different substances simultaneously. The goal of the simultaneous optimisation process is not to determine the functional form which describes the data set for a certain substance as well as possible, but to determine the functional form which yields on average the best results for all considered substances. If the considered substances are a representative subset of a larger group, such as the group of "nonpolar fluids", the resulting functional form should also yield good results for other substances out of this group, too. Moreover, when fitted to substances with small data sets the resulting equations should yield at least reasonable results for states which are not covered by data since their functional form was constrained to data for other substances in the corresponding region. However, this demand will only be fulfilled if the resulting functional form is numerically stable enough; the numerical stability of simultaneously optimised equations of state will be discussed in detail in Sect. 6.2.1.

The simultaneous optimisation algorithm by Span et al. (1998a) is essentially based on the algorithm by Setzmann and Wagner (1989a) which was described in Sect. 4.4.5, including the modifications described in Sects. 4.4.5.1 and 4.4.5.2. A flow diagram with the main steps of the development of an equation of state using either the substance specific or the simultaneous process is given as Fig. 6.4.

Both with respect to data and to constraints, the construction of the regression matrices is completely identical with the procedure described in Sect. 4.4.2 and the general rules given in Sect. 4.4.1 for the set-up of the bank of terms are valid for the simultaneous optimisation algorithm, too. The only difference is that the used values for T_r and ρ_r (see the definition of τ and δ, Eq. 3.1) are no longer arbitrary. To make use of a simple corresponding states similarity for the functional form of the equation the reducing parameters need to correspond to the critical parameters T_c and ρ_c, respectively.[4] All regression matrices which are set up for simultaneous

---

[4] In fact only ρ_r needs to correspond to ρ_c, see Sect. 6.2.1.2. However, differing values for T_r would change the coefficients and would make comparisons like the one shown in Fig. 6.3 meaningless.

224   6 Generalised Functional Forms

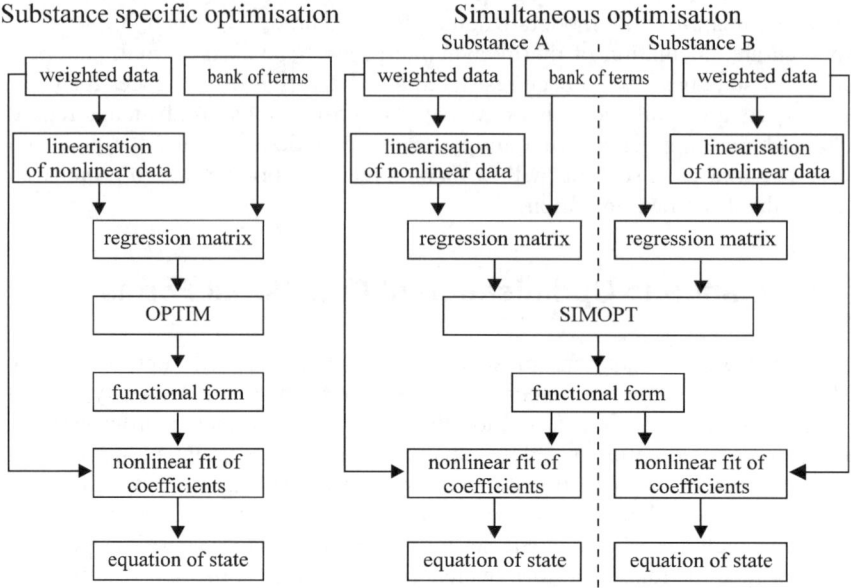

Fig. 6.4. Main steps in the development of an equation of state using either substance specific (for example OPTIM) or simultaneous (SIMOPT) optimisation algorithms.

use have to be based on the same bank of terms since the optimisation algorithm identifies terms only by their position in the bank of terms.

Based on one regression matrix for each of the considered substances, the simultaneous optimisation algorithm, which will be described in detail below, results in the functional form which yields on average the best results for all considered data sets. Finally, the coefficients of the equations of state which use the simultaneously optimised functional form are fitted nonlinearly directly to the data sets for the different substances; the sets of coefficients which result from the linear optimisation procedure can be used as starting solutions. In this way, the coefficients of the resulting equations are completely independent of each other and can also be fitted to data sets of other substances out of the same group of substances. But the functional form of the equations considers all of the information which was available for the substances used during the optimisation process.

The general structure of the simultaneous optimisation algorithm, *SIMOPT*, corresponds exactly to the structure of the algorithm by Setzmann and Wagner (1989a), OPTIM, as was illustrated in Figs. 4.7 and 4.8, including optional tests of pairwise exchanges, see Sect. 4.4.5.2. The codes which will be used in the following paragraphs to identify certain steps of the procedure relate to the corresponding labels used in Fig. 4.7 (O1–O7) and Fig. 4.8 (R1–R7).

The procedure starts with reading one set of control parameters and one regression matrix for each of the J considered substances. In step O1, the initial set of equations in the "population" is determined by repeated random selection of terms. In this step, a quality criterion is needed in order to determine the best formulations[5] which form the starting population. In OPTIM, this quality criterion is the sum of squares χ^2 which can be calculated from the regression matrix considering only those terms from the bank of terms which are part of the current formulation, see Sect. 4.4.5. In the simultaneous optimisation procedure, the corresponding sums of squares χ_j^2 have to be calculated from each of the regression matrices. Equivalent to χ^2 the quality criterion could now be defined as

$$X^2 = \sum_{j=1}^{J} \chi_j^2. \qquad (6.1)$$

In Eq. 6.1, extensive data sets of well measured substances which result in high values for χ_j^2 have an exaggerated influence on the quality criterion even if the representation of the data is satisfactory. To avoid this problem in the simultaneous optimisation, the sums of squares χ_j^2 are reduced by reference sums of squares, $\chi_{o,j}^2$, resulting from equations of state of similar length which are optimised specifically for the corresponding substance. Thus, in the simultaneous optimisation the reduced sum of squares of the substance j,

$$\chi_j^{*2} = \chi_j^2 / \chi_{o,j}^2, \qquad (6.2)$$

becomes one if the current formulation describes the corresponding data set as well as the equation of state which was optimised using the substance specific algorithm. With these reduced sums of squares, the quality criterion of SIMOPT, X_r^{*2} is defined as

$$X^{*2} = \sum_{j=1}^{J} \chi_j^{*2} = \sum_{j=1}^{J} \chi_j^2 / \chi_{o,j}^2. \qquad (6.3)$$

The formulations chosen as first parental generation in O1 are those with the minimum values for X_r^{*2}

At this point, it is important to realise that substance specific reference sums of squares and therefore substance specific optimised equations of state are needed only for the simultaneous optimisation procedure and not for the subsequent nonlinear fits, see Fig. 6.4. Thus, the introduction of reference sums of squares does not make the transfer of simultaneously optimised functional forms to substances which were not considered in the optimisation process more complicated. And since reference equations are needed only to reduce sums of squares it does not matter if they yield unphysical results outside of the regions which are covered by data – reference equations can easily be established for substances with rather restricted data sets, too.

---

[5] Here the expression "formulation" refers to the combination of terms which is considered in the current stadium of the optimisation process. The best formulation becomes the functional form of an equation of state when the results are written to a file after the optimisation procedure is finished.

When the starting population is determined, SIMOPT continues with step O3, the "mutation" of the formulations in the current parental generation, see Sect. 4.4.5. For each "mutated" formulation, the quality criterion is determined in the same way as described above. The old formulation is replaced by its mutant if

$$X^{*2}_{mutant} < X^{*2}_{old} \tag{6.4}$$

holds.

The way in which starting solutions for the repeated regression runs in O5 are determined corresponds exactly to the way described in Sect. 4.4.5, including the use of starting solutions in the original and in the modified form. In order to add the most important term in step R2, the quality criterion has to be calculated for all terms from the bank of terms which have not yet been included in the current formulation. When adding the term k, the sum of squares $\chi^2_{j,k}$ resulting from the j-th regression matrix becomes

$$\chi^2_{j,k} = \chi^2_{j,\text{old}} - \frac{b^2_{Lk,j}}{b_{kk,j}}, \tag{6.5}$$

where b corresponds to an element of the two-dimensional regression matrix and L is the bottom row of the matrix (see Sect. 4.4.2 and Eq. 4.97). If all J values $\chi^2_{j,k}$ are determined, the quality criterion for adding the term k, X^{*2}_k, can be calculated according to Eq. 6.3. The term which yields the smallest value X^{*2}_k is actually inserted into the current formulation. To do so, all J regression matrices have to be transformed according to the instructions given in Table 4.3 with $c = 1$.

After adding a term, the mutation procedure is repeated in R3 if the current formulation consists of 6–21 terms and statistical tests are applied to test the significance of the terms in the formulation (R4, Student t test, see Eqs. 4.98–4.100) and of the entire formulation (R5, simplified Fisher F test, see Eq. 4.104). Although these tests are formally valid only for a single data set, the average values of the statistical probabilities calculated from the J considered regression matrices can be used as criteria for the statistical tests in the simultaneous optimisation procedure.

If the Student t test shows that a term in the current formulation yields no significant contribution, this term is removed by transformation of all J regression matrices according to the instructions given in Table 4.3 with $c = 2$. The precautions used to avoid endless cyclic exchanges are the same as in OPTIM (R4a–R4e), including the modification explained in Sect. 4.4.5, footnote 3. If the formulation as a whole satisfies the conditions of the F test, the next term is again added in R2. If not, it is tested whether the formulation can be improved by exchanging any of the terms in it for any other term from the bank of terms. Therefore, the quality criterion $X^{*2}_{ex,nm}$ has to be determined for every possible exchange of one of the terms in the current formulation (index n) against one of the terms (index m) which are not contained in the current formulation. The required values $\chi^2_{ex,nm,j}$ can be calculated according to Table 4.4. The exchange with the smallest value for $X^{*2}_{ex,nm}$ is carried out by deleting the term n and adding the term m (see above) if one of the exchanges results in a reduction of X^{*2}_*. If the step "exchange of a term" is successful it is repeated until the current formulation cannot be improved further by an exchange of terms. If it is not successful, pairwise exchanges can be tested either for all terms in the equation and in the bank of terms or only for certain kinds of terms. Again, the prescriptions given in Sect. 4.4.3.1 have to be applied to all J matrices to determine X^{*2}_{pex}.

If the exchange of terms was successful, the F test is applied again to test the significance of the formulation as a whole. The next term is added if the formulation now fulfills the criterion of the F test. If it fails again, no further improvement is possible and the procedure continues either with the addition of constraints (R7a) or with the next regression run. When the predetermined number of regression runs has been carried out, the best results replace the poorest formulations in the population (O6) and the optimisation process continues with step O2 for the next generation until either a predetermined number of generations is finished or all formulations in the population are the same. The best formulation in the last generation is regarded as the functional form which yields on average the best representation of the J considered data sets.

6.2 Simultaneously Optimised Equations for Technical Applications

The simultaneous optimisation procedure described in the preceding section was developed as part of a project which aimed to establish a new class of simple equations of state for typical technical applications. The results of this project, which are summarised in this section, are currently being prepared for publication (Span and Wagner, 2000a–c).

The discussion of the performance of typical multiparameter equations of state in Chap. 5 showed that only rather complex "technical equations of state", such as Bender-type equations with 19 terms, are able to satisfy typical technical demands on the accuracy of thermodynamic properties over broad ranges of temperature and pressure. However, it was shown that strong intercorrelations between the terms in such equations of state (see Fig. 6.3) affect both their extrapolation behaviour and the representation of caloric properties within the range where data were available to fit the equation (see Figs. 6.1 and 6.2). These results agree with reports of users who criticise mainly the reliability of such equations but not the accuracy[6] achieved for substances with extensive data sets.

Thus, the development of improved technical equations of state had to focus on functional forms which yield results of comparable accuracy with improved numerical stability. Based on investigations of existing technical multiparameter equations the demands on accuracy, which are summarised in Table 6.1, were formulated for the new class of equations. However, unlike existing equations of state, the new equations should be designed to fulfil these requirements for substances with restricted data sets as well and should extrapolate reasonably well.

---

[6] Except for applications where it is clear from the beginning that the requirements on accuracy are high and cannot be fulfilled by simple technical equations of state, see Sect. 5.

Table 6.1. Demands formulated by Span and Wagner (2000a) on the accuracy of their new class of simple technical equations of state

Uncertainty in	$\rho(p,T)$	$w(p,T)$	$c_p(p,T)$	$p_s(T)$	$\rho'(T)$	$\rho''(T)$
$p \leq 30$ MPa[a]	$\pm 0.2\,\%$[b]	$\pm 1\,\% - \pm 2\,\%$[c]	$\pm 1\,\% - \pm 2\,\%$[c]	$\pm 0.2\,\%$[d]	$\pm 0.2\,\%$	$\pm 0.4\,\%$[d,e]
$p > 30$ MPa[f]	$\pm 0.5\,\%$	$\pm 2\,\%$	$\pm 2\,\%$	–	–	–

[a] larger uncertainties are to be expected in the extended critical region
[b] in the extended critical region $\Delta p/p$ is used instead of $\Delta \rho/\rho$
[c] $\pm 1\,\%$ at gaseous and gas-like supercritical states, $\pm 2\,\%$ at liquid and liquid-like states
[d] larger relative uncertainties have to be tolerated for small vapour pressures and saturated vapour densities
[e] combination of the uncertainties of gas densities and vapour pressures; experimental data of this accuracy are available for only a few substances
[f] states at pressures $p > 100$ MPa are not considered due to their small technical relevance

For the weights assigned to the selected data an important difference results from the requirements formulated above. Reference equations of state are expected to represent available experimental results within their experimental uncertainty, and thus the experimental uncertainties of the measured properties play a key role in the weighting schema explained in Sect. 4.3.4. However, when developing technical equations of state highly accurate data weighted according to these prescriptions are overemphasised. Weighted with the experimental uncertainty a highly accurate $p\rho T$ data point i with $\sigma_{\rho,i}/\rho_i \approx \pm 0.02\,\%$ would contribute $\zeta_i^2/\sigma_{\rho,i}^2 \approx 100$ to the sum of squares for the corresponding substance when being represented just within the demanded $\Delta \rho/\rho = \pm 0.2\,\%$. At higher temperatures for example, a less accurate data point j with $\sigma_{\rho,j}/\rho_j \approx \pm 0.1\,\%$ would contribute only $\zeta_j^2/\sigma_{\rho,j}^2 \approx 4$ to the sum of squares when being represented just within $\Delta \rho/\rho = \pm 0.2\,\%$, too. As a consequence, regions where highly accurate data are available would be overfitted while regions with worse, but still sufficient data situation would not be considered appropriately.

To avoid this problem, selected data have to be weighted with the demanded uncertainty for the corresponding property in general. In this way, the contribution to the sum of squares becomes 1 for data which are represented just within the demanded accuracy; the weighted variance of a fit becomes a measure for the fulfilment of the demands on accuracy. Only where data with experimental uncertainties exceeding the requested uncertainties of the equations have to be considered due to restrictions of the available data set do these data have to be weighted with their experimental uncertainty. This modified weighting schema makes the necessary assessment of a data set much easier and less time-consuming, since the very sophisticated question "how accurate are these data?", which has to be answered when setting up a reference equation of state, is replaced in most cases by the simpler question "are these data accurate enough to fulfil the requests on accuracy?"

6.2 Simultaneously Optimised Equations for Technical Applications

In contrast to older technical equations of state, which were usually formulated in terms of the pressure, the new generation of technical equations is formulated in terms of the reduced Helmholtz energy which is split into an ideal part α^o and a residual part α^r, see Eq. 3.5. Correlations for the ideal gas contribution α^o were used as published in the literature, see Tables 6.3 and 6.5. The general functional form of the equations for the residual part of the reduced Helmholtz energy, α^r, was chosen to correspond to the simple functional form given in Eq. 3.25. This general set-up resulted both from practical requirements on the functional form of technical equations of state and from tests which accompanied the development of the new equations.

Special critical region terms such as Gaussian bell shaped terms (see Sects. 3.1.3.2 and 4.5.3) and nonanalytic terms (see Sect. 3.1.3.2 and 4.5.4) were not used in order to preserve the necessary simplicity of a technical equation of state; in the extended critical region the demanded accuracy of $\Delta p/p \leq \pm 0.2\%$ can easily be achieved without such terms.

Besides the expressions discussed in the preceding sections, we considered two different types of terms as promising candidates for use in technical equations of state. Expressions of the general form

$$A_i = n_i \frac{\delta^{d_i} \tau^{t_i}}{(1-c_i\delta)^2} \tag{6.6}$$

can be interpreted as an empirical simplification of the well known hard sphere terms, see Sects. 3.1.3.1 and 4.4.1.1. The c_i introduced by such "*simplified hard sphere terms*" are additional parameters, which have to be determined during the optimisation process, and due to the quotient the required derivatives become more complicated than derivatives of Eq. 3.25. But compared to the original hard sphere terms the complications involved by Eq. 6.6 were still regarded as acceptable since terms like this were expected to improve the representation of properties at liquid and liquid-like supercritical states.

Tegeler et al. (1997) derived "*square-well terms*" of the form

$$A_i = n_i \, \delta^{d_i} \left(\exp(c_i\tau) - 1\right)^{m_i} \tag{6.7}$$

from an expression for the second virial coefficient which results from an integration of the square-well potential, see Mason and Spurling (1969). With c_i and m_i, Eq. 6.7 introduces two additional parameters and, compared to simple polynomial and exponential terms, derivatives with respect to τ become more complicated. Terms of this form were tested since they were expected to improve the representation of properties at gaseous and gas-like supercritical states.

However, neither simplified hard sphere terms nor square-well terms resulted in a significant improvement of the obtained equations and thus the general set-up of the bank of terms was again restricted to simple polynomial and exponential terms. The final bank of terms contained a total of 583 terms and reads

$$\alpha^r(\tau,\delta) = \sum_{i=1}^{8}\sum_{j=-8}^{12} n_{i,j}\,\delta^i\,\tau^{j/8} + \sum_{i=1}^{5}\sum_{j=-8}^{24} n_{i,j}\,\delta^i\,\tau^{j/8}\exp(-\delta)$$
$$+ \sum_{i=1}^{5}\sum_{j=16}^{56} n_{i,j}\,\delta^i\,\tau^{j/8}\exp(-\delta^2) + \sum_{i=2}^{4}\sum_{j=24}^{38} n_{i,j}\,\delta^i\,\tau^{j/2}\exp(-\delta^3), \quad (6.8)$$

with $\tau = T_c/T$ and $\delta = \rho/\rho_c$. Exponential expressions with density powers up to 6 were used in preliminary banks of terms, but the optimisation algorithm did not select any of the corresponding terms – a restriction to terms up to $\exp(-\delta^3)$ seems appropriate for simple technical equations of state.

With respect to the exponents used, Eq. 6.8 violates two of the recommendations given in Sect. 4.4.1. For pure polynomial terms, the use of density powers up to 8 leads to plots which are too steep in the limit of very high densities and affects the extrapolation to very high pressures, see Sect. 4.6.2. However, for technical equations of state with a restricted number of terms the use of polynomials with high density powers is essential for an accurate representation of liquid properties, while the representation of properties at pressures of several GPa can be regarded as arbitrary under technical aspects.

The use of negative exponents for the inverse reduced temperature is generally expected to improve the representation of properties at low reduced temperatures and thus it was accepted that such terms result in virial coefficients which diverge at very high temperatures. But at this point experiences with simultaneously optimised functional forms disprove common teachings – the final equations do not contain terms with such exponents, although they were contained in the used bank of terms. Obviously, terms with negative exponents for the inverse reduced temperature do not improve the resulting equations of state significantly, at least not if the assessment of the resulting equations does not depend on characteristics of single data sets.

Finally, one common feature of reference equations of state was sacrificed for the requirements of simple technical equations of state – the equations were not constrained to preselected critical points. Preliminary equations which were constrained to preselected critical points represented $p\rho T$ data in the critical region clearly better than the requested $\Delta p/p \leq \pm 0.2\%$ but compared to equations which were not constrained to critical parameters they needed 1 to 2 additional terms to fulfil the requirements outside of the extended critical region. Since an exact representation of critical parameters is usually not important for technical applications this additional numerical expense was regarded as not justified. Where accurate thermodynamic properties are needed in the extended critical region the use of accurate reference equations of state has to be recommended in any rate (see the comparisons in Chap. 5); at best, simple technical equations of state can satisfy average demands on the accuracy of properties in the extended critical region.

For substances for which no reliable data are available in the critical region or at supercritical states at all, estimates for the critical temperature are usually the most accurate information on the location of the critical point (see Sec. 6.2.1.2). To make use of this information, estimated values for one saturated vapour ($\rho'' \approx 0.9\,\rho_{c,\text{est}}$) and one saturated liquid density ($\rho' \approx 1.1\,\rho_{c,\text{est}}$) close to the critical

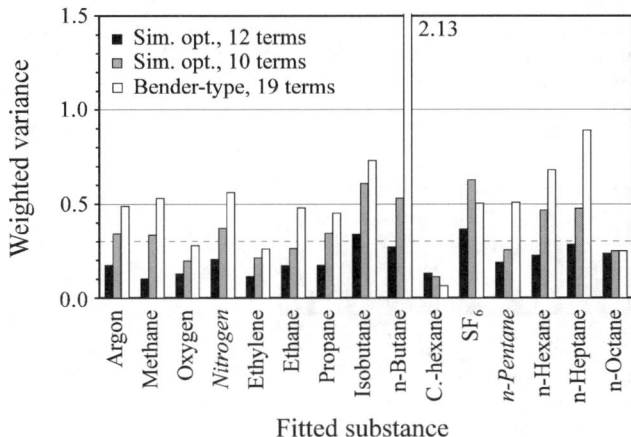

Fig. 6.5. Weighted variances resulting from fits of equations of state with simultaneously optimised functional forms and of Bender-type equations to data sets of 15 non- and weakly polar fluids. The data sets for nitrogen and n-pentane were not used in the simultaneous optimisation.

point ($T \approx 0.9998\ T_{c,est}$) can be introduced into the corresponding data set with very low weights. In this way, the resulting equation of state will yield a reasonable critical point without being strictly constrained to questionable critical parameters.

Based on the foundations explained above and on investigations for 14 nonpolar fluids, 1 weakly polar fluid, and 13 polar fluids Span and Wagner (2000a) found that simultaneously optimised equations of state with 10 to 12 terms are able to fulfil the requirements which were summarised in Table 6.1. The results for the individual substances are discussed briefly in Sects. 6.2.2 and 6.2.3; this section will focus on the discussion of more general features.

For nonpolar fluids Fig. 6.5 compares the performance of equations of state with simultaneously optimised functional forms with the performance of Bender-type equations. The used quality criterion is the *weighted variance* which is defined as

$$\sigma_{wt}^2 = \chi^2/(M-I) \approx \chi^2/M, \qquad (6.9)$$

where M is the number of data used in the fit and I the number of fitted coefficients. The used data sets were weighted with the demanded uncertainties, see above. The data sets of 13 of the 15 shown fluids were used in the simultaneous optimisation process; the data sets for nitrogen and n-pentane were not considered in order to be able to test the applicability of the developed functional forms to other nonpolar fluids. Detailed comparisons show no significant disadvantages for the representation of the data sets for nitrogen and n-pentane – the transferability of the functional forms seems to be ensured.[7] The data set for the weakly polar

---

[7] See the discussion on cyclohexane in Sect. 6.2.2 for information on possible restrictions with regard to molecules with cyclic structure.

232 6 Generalised Functional Forms

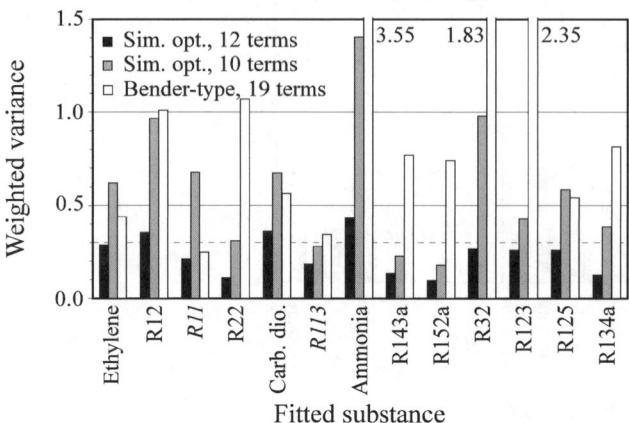

Fig. 6.6. Weighted variances resulting from fits of equations of state with simultaneously optimised functional forms and of Bender-type equations to data sets of 13 polar fluids. The data sets for the refrigerants R11 and R113 were not used in the simultaneous optimisation.

fluid ethylene was included into the optimisation process to make sure that the resulting functional forms also yield acceptable results for weakly polar fluids. The acentric factors of the considered substances range from $\omega = -0.002$ for argon to $\omega = 0.391$ for n-octane.

With a single exception, all of the simultaneously optimised equations presented in Fig. 6.5 yield weighted variances $\sigma_{wt}^2 < 1$; on average the selected data sets are represented within the demanded uncertainties. However, more detailed comparisons show that this criterion is far from sufficient to claim the demanded uncertainties for properties calculated from the corresponding equations of state. In general,[8] equations of state with $\sigma_{wt}^2 > 0.3$ resulted in problems mostly for caloric properties at liquid states, unless the large weighted variance was caused by inconsistencies in the selected data set, see e.g. isobutane and sulphur hexafluoride (SF$_6$). On average, simultaneously optimised equations of state with 10 terms yield better results than Bender-type equations with 19 terms. However, the weighted variances calculated from these equations exceed the limit $\sigma_{wt}^2 = 0.3$ in 9 out of 15 cases; the equations do not fulfil the requirements with regard to accuracy over the whole fluid surface. Simultaneously optimised equations with 11 terms do not result in significant improvements, but equations with 12 terms represent the considered data sets within the demanded uncertainties almost without exception. Generally speaking, these equations are better than required, but 12 terms are needed to describe properties in problematic regions within the uncertainties summarised in Table 6.1 in most cases.

---

[8] To assess the quality of an equation of state based on a single criterion is of course a gross simplification; however this very simple criterion agrees well with results of more detailed comparisons in this case, see Sects. 6.2.2 and 6.2.3.

6.2 Simultaneously Optimised Equations for Technical Applications

Figure 6.6 shows the corresponding results for polar fluids. The data sets for the refrigerants R11 and R113 were not considered in the simultaneous optimisation to test the transferability of the developed functional forms. The data set for ethylene was used as a polar fluid too in order to fill the gap between polar and nonpolar fluids. Ammonia was considered as normal polar fluid, although its properties are clearly influenced by association; at least to a certain degree the functional forms developed for polar fluids should be able to describe associating fluids as well. An extension of the set of used substances to water was tested, but it failed – the properties of the strongly associating fluid water could not be described satisfactorily with functional forms which are suitable for other polar fluids as well. The acentric factors of the considered substances range from $\omega = 0.087$ for ethylene to $\omega = 0.327$ for refrigerant R134a, but it is well known that this property is not sufficient to characterise the similarity of thermodynamic properties of polar fluids.

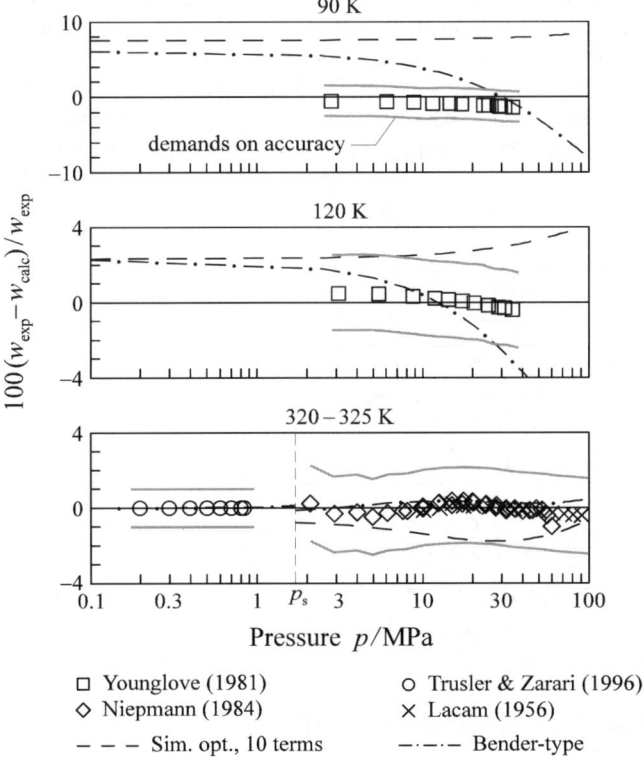

Fig. 6.7. Percentage deviations between experimental data for the speed of sound in propane and values calculated from the simultaneously optimised equation of state with 12 terms. Values calculated from the corresponding equation with 10 terms and from a Bender-type equation are plotted for comparison. The grey lines indicate the demands on accuracy, see Table 6.1.

Bender-type equations describe the properties of polar fluids significantly less accurately than properties of nonpolar fluids. The limit of $\sigma_{wt}^2 = 1$ is exceeded for five substances, for three substances the results are completely unsatisfactory. On average, simultaneously optimised equations with 10 terms yield better results again, but their shortcomings are still more obvious than for nonpolar fluids. Equations with 12 terms describe most of the data sets with $\sigma_{wt}^2 < 0.3$ again and are on the whole able to satisfy the demands summarised in Table 6.1 in problematic regions as well.

The slightly larger weighted variances for ammonia and carbon dioxide are worthy of note. For ammonia the increased variance is clearly caused by the differing thermodynamic behaviour of an associating fluid; in fact it is surprising that the properties of ammonia can still be described almost within the demanded uncertainties using a functional form which was optimised mainly for non- or only weakly associating fluids. The data set for carbon dioxide is characterised by a large number of data in the extended critical region and by $p\rho T$ data which reach up to $T/T_c \approx 3.4$. The simultaneously optimised equation with 12 terms describes properties in the extended critical region well within the expected uncertainties; the observed increase in the weighted variance results from $p\rho T$ data at reduced temperatures above $T/T_c \approx 2$. Data are not available at high reduced temperatures for any of the other polar substances. The simultaneously optimised functional form is based only on carbon dioxide data at this point. The resulting plots are

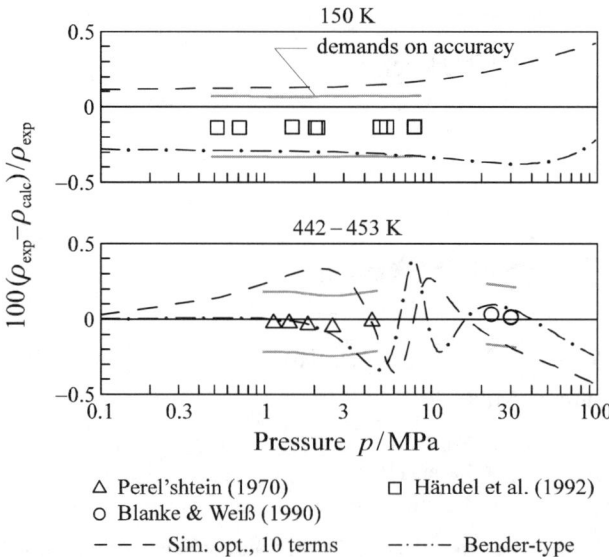

Fig. 6.8. Percentage deviations between experimental data for the density of refrigerant R12 and values calculated from the simultaneously optimised equation of state with 12 terms. Values calculated from the corresponding equation with 10 terms and from a Bender-type equation are plotted for comparison. The grey lines indicate the demands on accuracy, see Table 6.1.

6.2 Simultaneously Optimised Equations for Technical Applications

qualitatively correct even up to much higher temperatures, see Sect. 6.2.1.1, but increased uncertainties up to $\Delta\rho/\rho \approx \pm 1.0\%$ are to be observed for $T/T_c > \approx 2$ and $p > \approx 40$ MPa. Preliminary functional forms which were forced to represent carbon dioxide data at high temperatures more accurately were rejected since they resulted in an unsatisfactory extrapolation behaviour for other substances.

6.2.1 Accuracy Versus Numerical Stability – A Compromise

Figure 6.7 shows a typical example for the superiority of simultaneously optimised functional forms with 12 terms. For propane, the simultaneously optimised equation with 12 terms is able to represent all of the available speed of sound data well within the demanded uncertainty of $\Delta w/w = \pm 1\%$ at gaseous states and $\Delta w/w = \pm 2\%$ at liquid states. In contrast to this, both the simultaneously optimised equation with 10 terms and the Bender-type equation result in deviations up to $\Delta w/w \approx \pm 8\%$ for liquid states at very low temperatures. At higher temperatures and for gaseous states these equations satisfy the demands on accuracy as well.

For thermal properties the disadvantages of 10 term equations are still obvious but less significant. For the density of refrigerant R12 Fig. 6.8 shows deviations

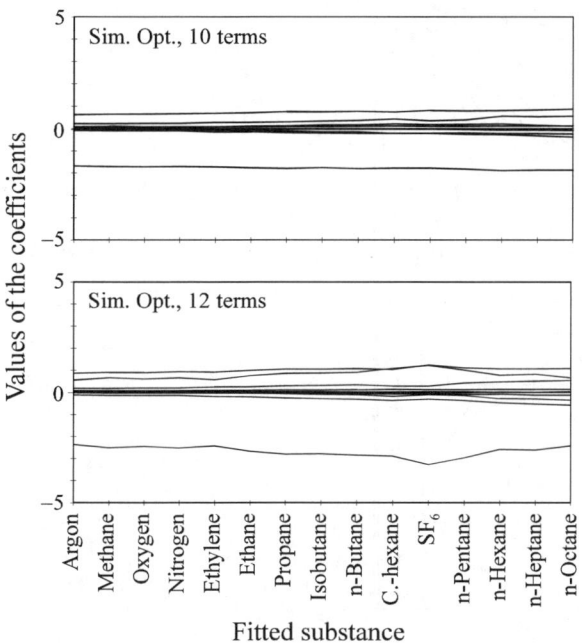

Fig. 6.9. Values of the coefficients of simultaneously optimised equations of state which resulted from fits to data sets of 15 non- and weakly polar fluids. The fluids are ordered by increasing acentric factor, ω.

6 Generalised Functional Forms

between experimental results and values calculated from equations of state with simultaneously optimised functional form and from Bender-type equations. At low temperatures and at supercritical states the uncertainty of values calculated from the simultaneously optimised equation with 10 terms exceeds $\Delta\rho/\rho = \pm 0.2\%$.

Based strictly on the demands formulated in Table 6.1 one has to conclude that simultaneously optimised functional forms with 12 terms form the ideal basis for a new class of technical equations of state. However, compared to equations with 10 terms the resulting equations show one serious disadvantage – the additional terms affect the numerical stability. In Fig. 6.3, the plot of the coefficients of BWR- and Bender-type equations of state resulting from independent fits to data sets of the considered nonpolar substances has been used as an indication for numerical stability; Fig. 6.9 shows the corresponding plots for equations of state which use simultaneously optimised functional forms with 10 and 12 terms.

Compared to the results shown in Fig. 6.3 the plots found for the simultaneously optimised functional form with 10 terms are excellent. On average, the coefficients are small and the coefficients of the dominating terms with relatively large coefficients do not change rapidly from substance to substance. The observed small changes seem to be almost linear in the acentric factor; significant oscillations are

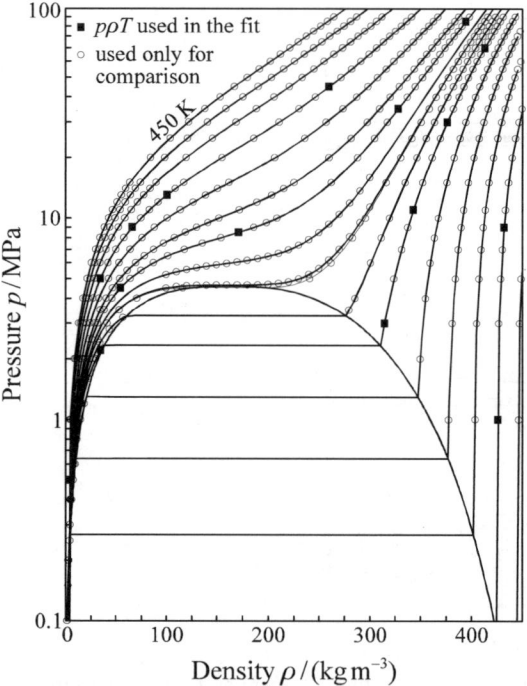

Fig. 6.10. Distribution of the data calculated from the reference equation by Setzmann and Wagner (1991) for methane. The "reduced data set" contains only 21 $p\rho T$ data at the locations indicated by filled circles and 10 data for phase-equilibrium properties.

not observed. The plots which result from the functional form with 12 terms still indicate higher numerical stability than those found for the BWR-equation with only 8 terms and far inferior accuracy, but the observed oscillations and intercorrelations are clearly more pronounced than for the equation with 10 terms.

As discussed above, the major objective of the described project was to establish functional forms for sufficiently accurate equations of state for nonpolar and polar fluids which can be transferred easily to technically important substances with small data sets. Bearing in mind this fact and the results shown in Fig. 6.9 it had to be clarified whether the obvious advantages of equations with 12 terms do not turn into disadvantages when being fitted to small data sets or to substances with uncertain critical parameters.

6.2.1.1 An Investigation of Numerical Stability

The numerical stability of a functional form can be tested systematically by comparisons with well defined data sets calculated from highly accurate reference

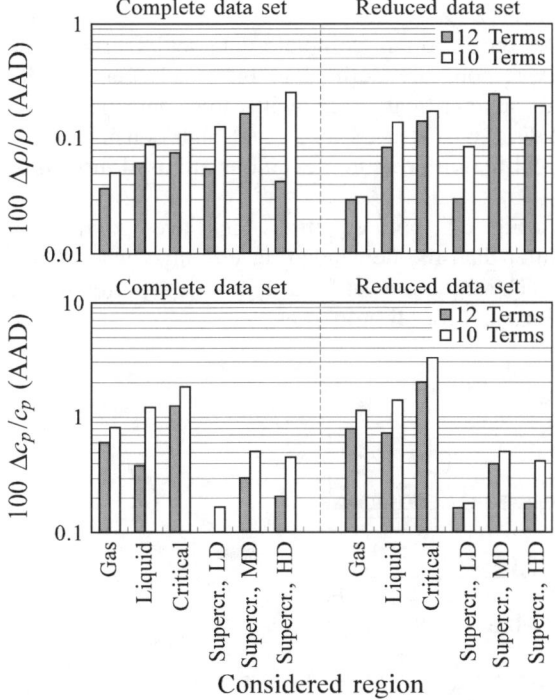

Fig. 6.11. Average absolute deviations between densities and isobaric heat capacities calculated from simultaneously optimised equations with 10 and 12 terms and from the reference equation by Setzmann and Wagner (1991). For an explanation of the different supercritical regions, see the text. In the critical region, pressure deviations are used instead of density deviations.

equations and, less systematically but may be more practice-oriented, by an investigation of equations of state which were fitted to subaverage data sets. The second approach is informative especially if additional data become available in regions which were not covered by data when the corresponding equation was fitted.

For a systematic investigation of numerical stability values for the density, the isobaric heat capacity and the speed of sound of methane were calculated from the highly accurate reference equation by Setzmann and Wagner (1991) at 511 points. The established data set covers temperatures from 95 K to 450 K ($T_t = 90.6941$ K and $T_c = 190.564$ K) and pressures up to 100 MPa at reasonable intervals; Fig. 6.10 illustrates the distribution of the data points in the homogeneous region. Additionally, data for the vapour pressure and the saturated vapour and liquid density were calculated at 20 saturation temperatures.

In a first step, equations of state with simultaneously optimised functional forms with 10 and 12 terms were fitted to the complete data set, considering data of all properties. For the density and the isobaric heat capacity, Fig. 6.11 reports the percentage absolute average deviations

$$100\,\Delta y / y \,(\text{AAD}) = \sum_{m=1}^{M} 100 \left| \Delta y_m / y_m \right| / M \qquad (6.10)$$

which resulted from this fit under the heading "complete data set". To make the comparison more expressive, six regions are defined in Fig. 6.11: Gas, liquid, (extended) critical region, and supercritical fluid with low densities (LD, $\rho \leq 0.6\,\rho_c$), medium densities (MD, $0.6\,\rho_c < \rho < 1.5\,\rho_c$), and high densities (HD, $\rho \geq 1.5\,\rho_c$). In the extended critical region pressure deviations are used instead of density deviations.

As it was to be expected from Fig. 6.5, both equations yield average absolute deviations which are clearly smaller than the demanded uncertainties when being fitted to the complete data sets. The equation with 12 terms is superior to the equation with 10 terms in all regions and for all properties.

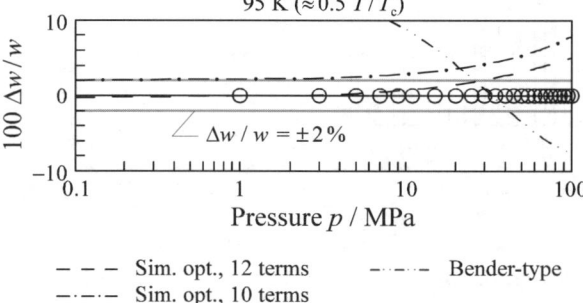

Fig. 6.12. Percentage deviations $100\,(w_{\text{cal}} - w_{\text{SW}}) / w_{\text{SW}}$ between speeds of sound calculated from the reference equation by Setzmann and Wagner (1991), w_{SW}, and from simultaneously optimised equations with 10 and 12 terms and a Bender-type equation which were fitted to the "reduced data set".

6.2 Simultaneously Optimised Equations for Technical Applications

In a second step, both equations were fitted to a reduced data set which consisted of 21 $p\rho T$ data in the range $100\text{ K} \leq T \leq 300\text{ K}$ and $p \leq 95$ MPa, 4 vapour pressures and 4 saturated liquid densities in the range $100\text{ K} \leq T \leq 185\text{ K}$ and 2 saturated vapour densities at temperatures of 160 K and 185 K. To imitate a small but accurate set of experimental data the calculated values were scattered with a standard deviation of $\sigma_y = \pm 0.02\,\%$. The location of the $p\rho T$ data used in the reduced data sets is indicated by filled symbols in Fig. 6.10. Under the heading "reduced data set" Fig. 6.11 summarises the results of comparisons between the data *which were not used in the fit* and values predicted from the simultaneously optimised equations of state.

Even for equations which were fitted to such an extremely small data set consisting only of 31 data for thermal properties the observed average absolute deviations exceed the demanded uncertainties only slightly in a single region, namely for the representation of $p\rho T$ data at supercritical states with medium densities. On average, the advantages of the equation with 12 terms are less pronounced than before, but still the longer equation yields better results. To give an example of the performance of equations fitted to the reduced data set, Fig. 6.12 shows deviations between speeds of sound calculated from the reference equation by Setzmann and Wagner (1991), from simultaneously optimised equations with 10 and 12 terms, and from a Bender-type equation; the circles indicate the locations where data were used in the initial fit and in the assessment of equations fitted to the reduced

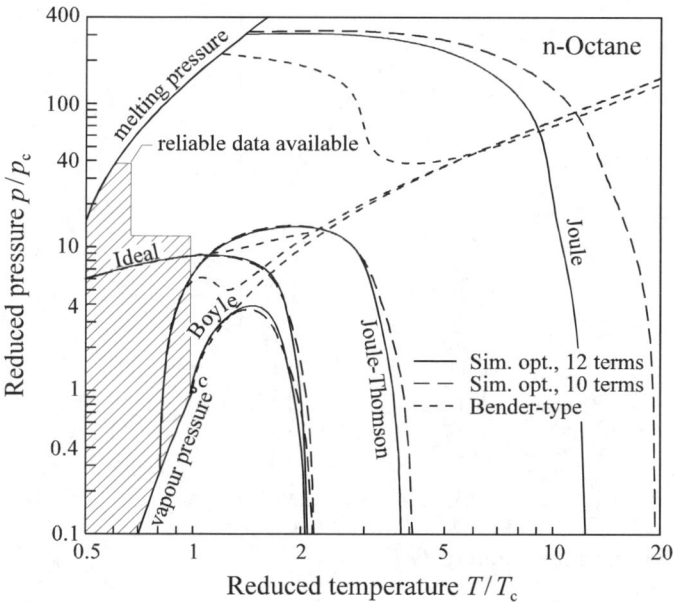

Fig. 6.13. Ideal curves of n-octane calculated from simultaneously optimised equations of state with 10 and 12 terms and from a Bender-type equation. The hatched area corresponds to the region where reliable data were available to fit the equations.

data set. Even at these low temperature liquid states, the simultaneously optimised equations result in very reasonable speeds of sound. The values calculated from the equation with 12 terms deviate by more than ±2% only for pressures above about 30 MPa. The equation with 10 terms results in slightly larger deviations, but this behaviour corresponds exactly to the problems observed for equations fitted to larger data sets too, see Fig. 6.7. Multiparameter equations of state without optimised functional form such as Bender-type equations yield completely unreliable results for derived caloric properties when being fitted to such small data sets.

Systematic investigations like the one described above prove the superior numerical stability of equations of state with simultaneously optimised functional form. No clear indications for disadvantages of functional forms with 12 terms became obvious.

When setting up the new class of technical equations for nonpolar and polar fluids simultaneously optimised functional forms had to be fitted to very restricted data sets especially for the higher alkanes. For n-heptane and n-octane reliable data were available only for liquid states, see the brief discussion of the data sets in Sect. 6.2.2. For the representation of properties in the gas phase and at supercritical states we could rely only on some questionable data for the second virial coefficient and on the numerical stability of the used functional forms. As a very sensi-

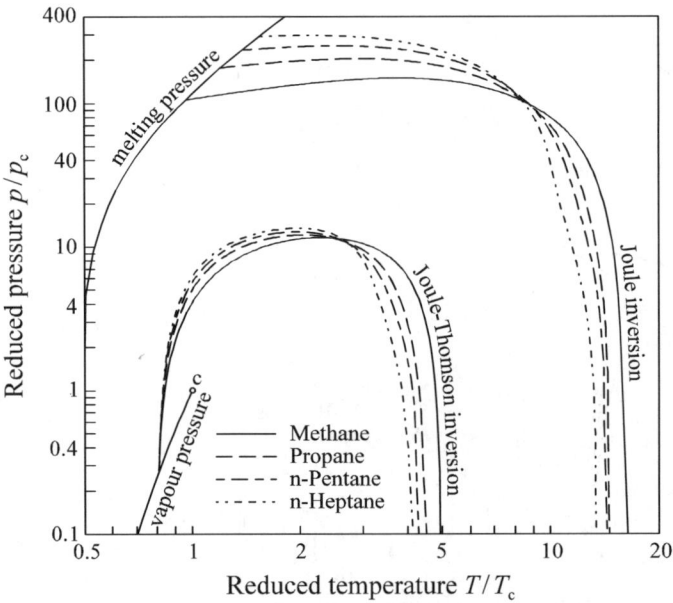

Fig. 6.14. Joule-Thomson and Joule inversion curves of methane, propane, n-pentane, and n-heptane calculated from simultaneously optimised equations of state with 12 terms. The plotted phase boundaries correspond to those of methane.

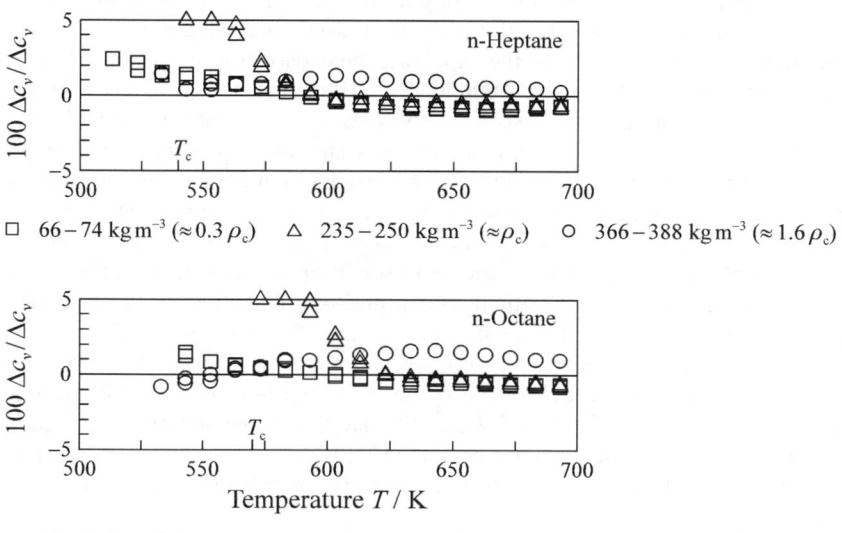

Fig. 6.15. Percentage deviations $100\,(c_{v,\text{calc}} - c_{v,exp})/c_{v,exp}$ between experimental data for the isochoric heat capacity of n-heptane and n-octane by Abdulagatov (1998) and values predicted by the corresponding simultaneously optimised equations of state with 12 terms.

tive test for the extrapolation behaviour of the resulting equations, Fig. 6.13 shows plots of the ideal curves[9] of n-octane calculated from simultaneously optimised equations of state with 10 and 12 terms and from a Bender-type equation. The equations with simultaneously optimised functional forms show qualitatively correct plots even for the outermost curve, the Joule inversion curve. However, deformations become obvious mainly for the Joule-Thomson and Joule inversion curves calculated from the equation with 10 terms;[10] the functional form with 12 terms is even superior with regard to its extrapolation behaviour. This observation agrees with very reasonable plots found for thermal properties as well as for derived caloric properties – the "theoretically expected" plots in the p,ρ- and c_p,T-diagrams for n-octane, Figs. 6.1 and 6.2, were calculated from the simultaneously optimised equation with 12 terms. The ideal curves calculated from the Bender-type equation show unreasonable plots right outside of the region where the equations were fitted to data.

Similar results can be found for other polar and nonpolar substances as well. Fig. 6.14 shows Joule-Thomson and Joule inversion curves calculated from simultaneously optimised equations of state with 12 terms for methane, propane, n-pentane, and n-heptane. Slight deformations become obvious only for n-heptane.

---

[9] See Sect. 4.6.3 for a detailed discussion on these curves.
[10] For an assessment of the performance of simultaneously optimised equations one has to be aware that only very few reference equations for substances with extensive data sets have so far been able to predict Joule inversion curves at least qualitatively correctly, see Sect. 4.6.3.

Although the corresponding equation is based almost exclusively on liquid data it still predicts a qualitatively correct plot even of the Joule inversion curve.

Experimental investigations of the isochoric heat capacity at gaseous, liquid, critical, and supercritical states by Abdulagatov (1998) became available both for n-heptane and n-octane after the work on the corresponding equations had been finished. For gaseous, critical, and supercritical states the representation of these data can be regarded as purely predictive since data had not been available in these regions either for thermal or for caloric properties. At liquid states, data for thermal and caloric properties, but not for the isochoric heat capacity, had been previously available. Figure 6.15 shows the representation of representative data on isochores with gas-like, almost critical, and liquid-like densities. In the extended critical region the deviations between experimental data and predicted heat capacities exceed $\Delta c_v / c_v = \pm 5\,\%$; this result was to be expected since simple technical equations cannot be expected to follow the steep increase of the isochoric heat capacity in this region, see Sects. 4.5.2 and 5.2. However, outside of the extended critical region the heat capacities are predicted within $\Delta c_v / c_v = \pm 2\,\% - \pm 3\,\%$, and thus within the uncertainty claimed for this property for substances with far better data sets.[11]

The numerical stability of equations of state which use simultaneously optimised functional forms with 12 terms is obviously sufficient for the applications they were developed for. Functional forms with 10 terms are numerically more stable, but they yield worse results for the considered thermodynamic properties even when fitted to very restricted data sets.

6.2.1.2 The Influence of Uncertain Critical Parameters

When setting up reference equations of state, the accuracy of the results which are available for the critical parameters T_c, p_c, and ρ_c is an essential point, since such equations are usually constrained to preselected critical parameters. Faulty values of critical parameters are likely to distort the representation of properties in the extended critical region. For reasons which were discussed in Sect. 6.2 the new equations for technical applications were not constrained to certain critical parameters[12] and thus the influence of uncertain values for T_c, p_c, and ρ_c is reduced drastically. However, values for T_c and ρ_c are still needed as reducing parameters in Eq. 6.8 and in the resulting equations.

When using only simple polynomial and exponential terms, see Eq. 6.8, faulty values for T_c do not result in changes regarding the representation of thermodynamic properties. If τ is reduced with a critical temperature T_c^* instead of the "true" critical temperature T_c the coefficients of the equation become

---

[11] At gaseous and supercritical states relative uncertainties for isochoric heat capacities are always slightly larger than those for isobaric heat capacities since the usually well known contribution from derivatives of thermal properties, see 4.140, is missing for the isochoric heat capacity.

[12] In the way discussed in Sect. 6.2, reasonable critical temperatures were enforced for cyclohexane, n-heptane, and n-octane without constraining the equations strictly to the corresponding critical parameters.

6.2 Simultaneously Optimised Equations for Technical Applications

Table 6.2. Average and maximum deviations between experimentaly determined and predicted values for critical parameters

Group of substances	Reference	average deviations (RMS[a])			maximum deviations		
		$T_c/\%$	$p_c/\%$	$\rho_c/\%$	$T_c/\%$	$p_c/\%$	$\rho_c/\%$
n-Alkanes	Teja et al. (1990)[b]	0.07	0.72	0.50	0.15	−2.15	−1.20
	Riedel (1963)[c]	0.23	3.01	7.04[d]	0.24	−3.67	−10.4[d]
Isomers	Riedel (1963)[c]	0.24	2.00	1.36	0.73	−3.42	−2.64
Alcohols	Teja et al. (1990)[b]	0.09	0.39	1.36	−0.16	−0.79	2.84
	Somayajulu (1989)[c]	1.06	2.96	2.78	−4.45	6.54	−6.59
Others	Somayajulu (1989)[c]	0.30	1.59	2.42	0.74	−3.14	7.09

[a] root mean square deviation, RMS $100 \Delta y / y = [\Sigma (100 (y_{calc} - y_{exp}) / y_{exp})^2 / M^2]^{0.5}$
[b] based on the number of carbon atoms in a molecule
[c] based on a group contribution method
[d] large deviations due to more recent findings regarding the critical density of higher alkanes

$$n_i^* = n_i \left(\frac{T_c}{T_c^*}\right)^{t_i} ; \qquad (6.11)$$

calculations are not affected at all. Due to the set-up of the exponential functions in Eq. 6.8 the same does not hold for faulty critical densities ρ_c^*, but most of the influence can still be compensated by changed coefficients.

When setting up equations of state for technically important substances with restricted data sets, one usually has to rely either on questionable experimental results for the critical properties or on predicted values. To estimate typical uncertainties of predicted critical properties 26 algorithms for the determination of critical parameters of n-alkanes, 12 algorithms for isomers of alkanes, 17 algorithms for alcohols, and 13 algorithms for other organic substances were tested by Span and Wagner (2000a). In general, algorithms based on group contribution methods were considered; for n-alkanes and alcohols simple methods based on the number of carbon atoms or on the molar mass were considered as well. For the 18 n-alkanes from methane to octadecane, 15 isomers with 4 to 8 carbon atoms, 20 alcohols ranging from methanol to decanol, and 22 organic substances belonging to the groups of cyclo-alkanes, of aromatic substances, and of unsaturated hydrocarbons deviations between experimental and predicted results for the critical parameters were investigated. For the algorithms which led to the smallest deviations the results are summarised in Table 6.2 for each of the groups. If the normal boiling temperature and the saturated liquid density at this state are accepted as additional parameters, the algorithm proposed by Vetere (1995) yields slightly better results in many cases.

With the results of these comparisons in mind one has to be aware of typical uncertainties of ±3% for critical densities of substances with restricted data set. Errors up to ±10% may be encountered in single cases. To assess the effect of such uncertainties simultaneously optimised equations of state with 10 and 12

244 6 Generalised Functional Forms

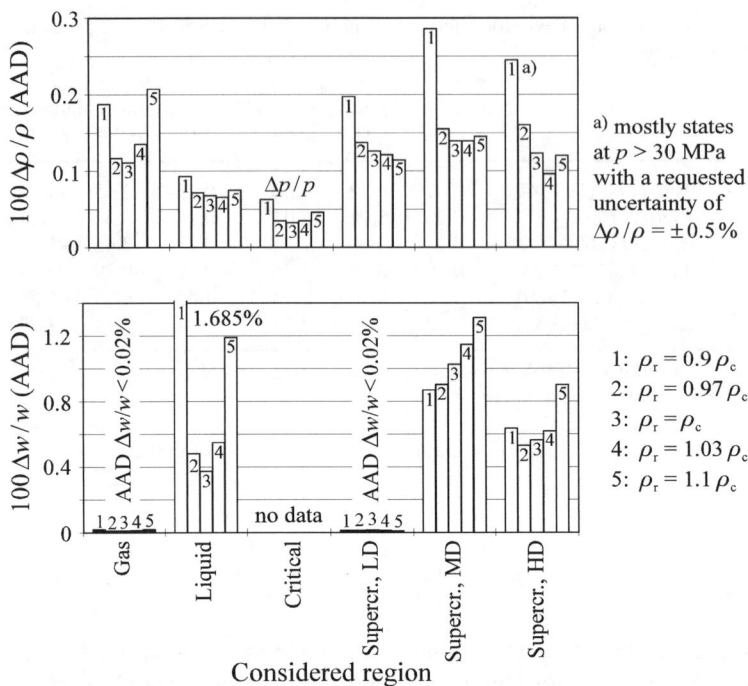

Fig. 6.16. Percentage average absolute deviations resulting from fits to propane data for simultaneously optimised equations of state with 12 terms and different reducing densities.

terms have been fitted to the complete data set for propane, whereby the density has been reduced with $\rho_r = 0.9\rho_c$, $\rho_r = 0.97\rho_c$, $\rho_r = \rho_c$, $\rho_r = 1.03\rho_c$, and $\rho_r = 1.1\rho_c$. For the functional form with 12 terms the results of this test are summarised in Fig. 6.16. The indicated regions correspond to those defined in Sect. 6.2.1.1, see Fig. 6.11.

As was to be expected, the best results were found for $\rho_r = \rho_c$, the case which was assumed when optimising the functional form. However, changes of $\Delta\rho_r/\rho_c = \pm 3\%$ hardly affect the accuracy of the resulting equations. The observed average deviations are still far smaller than the demanded uncertainties. Changes of $\Delta\rho_r/\rho_c = \pm 10\%$ result in significantly worse equations of state. The average deviations are still smaller than the demanded uncertainties, except for $p\rho T$ data in the gas phase for $\rho_r = 1.1\rho_c$ and at supercritical states with medium densities for $\rho_r = 0.9\rho_c$, but the maximum deviations exceed the demanded uncertainties in other regions as well.

However, an uncertainty of $\pm 10\%$ in the critical density has to be assumed only for uncommon substances with very restricted data sets. In this sense, the results summarised in Fig. 6.16 prove that typical uncertainties of critical parameters are no threat for technical equations of state with simultaneously optimised functional forms. Simultaneously optimised equations with 10 terms are slightly more sensitive to changes in ρ_r; thus, the uncertainty of critical parameters does not speak against functional forms with 12 terms in any way.

6.2.1.3 Conclusions Regarding Required Data Sets

When discussing requirements on the data set necessary to establish technical equations of state with simultaneously optimised functional forms one has to distinguish between two very different situations.

Where equations are based on data published in scientific literature there is hardly a chance to exercise an influence on the number of available data, on the properties which have been measured, or on the distribution of the data. The comparisons of Sect. 6.2.1.1 have shown that surprisingly small data sets with a very unbalanced distribution of the data may still be sufficient to establish equations of state which yield reliable results even far outside of the region where data were available. However, this has to be verified in every single case. Absolute plots of derived caloric properties (see Fig. 6.2) or plots of ideal curves (see Figs. 6.13 and 6.14 and Sect. 4.6.3) calculated from the resulting equation are sensitive tools to detect unreasonable behaviour. Equations which were not fitted at least to vapour pressures and to data for thermal properties at liquid states cannot be expected to describe the whole range of fluid states. Critical parameters may be obtained from predictive algorithms with sufficient accuracy, see Sect. 6.2.1.2. Information on the heat capacity of the ideal gas is required to describe caloric properties as well, see Sects. 3.1.1.1 and 4.2.

Due to the numerical stability of simultaneously optimised functional forms the number of substances, for which accurate and reliable equations of state can be established based only on literature data, will be increased significantly. But still this group of substances is only a small subset of the group of technically important substances. Thus, additional measurements will become necessary either to supplement published data or as only source of experimental information. At this point it becomes much more important to discuss requirements on the available data sets.

For substances with simple molecular set-up the required information on caloric properties of the ideal gas can usually be derived from published spectroscopic data with sufficiently high accuracy, see Sect. 4.2. Where such information is not available, where the molecular set-up becomes too complicated, or where features like hindered internal rotations make an accurate theoretically founded prediction of ideal gas heat capacities impossible, gas phase speeds of sound measured with spherical resonators are the most reliable source of information on caloric properties of the ideal gas. In this case, such data are needed over a range of temperatures which is as broad as possible, but 10–20 data points will usually be sufficient, since the temperature dependence of the ideal gas heat capacity is easy to fit.

At reduced temperature below $T/T_c \approx 0.9$, information which constrains the second virial coefficients calculated from the equation of state is sufficient to guarantee the required accuracy in the gas phase. This information can also be obtained most easily from speed of sound measurements with spherical resonators. Usually about three isotherms with five data points each will be sufficient to enable an accurate representation of gas phase properties. Due to experimental limitations, the data should not approach the phase boundary. However, since the residual contribution to the speed of sound is very small in the gas phase, see Fig. 5.19, the speeds of sound need to be fitted to far less than the demanded uncertainty of $\Delta w/w = \pm 1\%$ to guarantee a sufficiently accurate description of $p\rho T$ data, too. Where the resulting equation does not represent gas phase speeds of sound clearly within $\Delta w/w < \pm 0.1\%$ for pressures below about 1 MPa it needs to be checked whether the second acoustic virial coefficient is represented within $\Delta \beta_a/\beta_a \leq \approx \pm 2\%$. Larger deviations in β_a and B are common at $T/T_c < \approx 0.7$ where the plot of the second virial coefficients becomes very steep.

At liquid and supercritical states it is most advantageous to rely on measured $p\rho T$ data. The data set does not need to be large – about 30–50 data points should be absolutely sufficient, as shown by the results with even smaller data sets discussed in Sect. 6.2.1.1. But the data should cover a sufficiently broad range of states and they should be clearly more accurate than the resulting equation needs to be; significant systematic errors in parts of the data set are likely to distort the prediction of derived properties. To enable the resulting equation to fulfil the demands on accuracy summarised in Table 6.1, an experimental uncertainty of at least $\Delta \rho/\rho \leq \pm 0.05\%$ should be aimed for at $p \leq 30$ MPa; at $p > 30$ MPa uncertainties of $\Delta \rho/\rho \leq \pm 0.1\% - \pm 0.2\%$ can be regarded as sufficient.

Information on thermal properties at phase equilibrium is essential to describe the phase equilibrium itself and to achieve a consistent description of caloric properties in the liquid phase. However, since the temperature dependence of these properties is predicted rather accurately by equations with simultaneously optimised functional form, data sets with 5–10 accurate vapour pressures and saturated liquid densities should be sufficient. If possible, the data should cover the range $T_t \leq T \leq \approx 0.99 T_c$. 3–5 data for the saturated vapour density are helpful mainly at temperatures in the range $\approx 0.9 T_c \leq T \leq \approx 0.99 T_c$. At higher temperatures vapour pressures can still be represented accurately, but the uncertainty of calculated values for the saturated vapour and liquid density increases due to fundamental restrictions of simple technical equations of state. Where such data are used in fits with high weights they may distort the representation of properties in other regions. At low reduced temperatures the relative uncertainty of results on the vapour pressure increases due to experimental difficulties regarding the measurement of small pressures and due to an increasing influence of impurities. If such data are overfitted they may distort the representation of other properties as well. Where an accurate description of phase equilibria at low reduced temperatures is regarded as important it may be useful to include some data on the enthalpy of evaporation, Δh_v, at these conditions.

Some data on derived caloric properties such as the isochoric heat capacity or the speed of sound at liquid and liquid-like supercritical states are helpful to verify a certain level of accuracy for these properties. But in general reasonable predictions can be expected if the equation was fitted to a consistent set of thermal properties, see Fig. 6.11. Increased uncertainties may be encountered most likely at low reduced temperatures.

Table 6.3. Characteristic parameters of the considered non- and weakly polar substances and references for the used correlations for caloric properties of the corresponding ideal gases

Substance	Correlation used for α^{o}	$\dfrac{M}{\text{g mol}^{-1}}$	$\dfrac{T_c}{\text{K}}$	$\dfrac{p_c}{\text{MPa}}$	$\dfrac{\rho_c}{\text{kg m}^{-3}}$	$\dfrac{T_c}{T_t}$	ω
	n-Alkanes						
Methane	Setzmann, Wagner (1991)	16.043	190.564	4.599	162.66	0.476	0.011
Ethane	Jaeschke, Schley (1995)	30.070	305.322	4.872	206.60	0.296	0.099
Propane	Jaeschke, Schley (1995)	44.097	369.825	4.248	220.48	0.231	0.153
n-Butane	Jaeschke, Schley (1995)	58.123	425.125	3.796	227.84	0.317	0.200
n-Pentane	Jaeschke, Schley (1995)	72.150	469.70	3.370	232.00	0.305	0.251
n-Hexane	Jaeschke, Schley (1995)	86.177	507.82	3.034	233.18	0.350	0.308
n-Heptane	Jaeschke, Schley (1995)	100.204	540.13	2.736	232.00	0.338	0.350
n-Octane	Jaeschke, Schley (1995)	114.231	569.32	2.497	234.90	0.380	0.391
	Other non- or weakly polar substances						
Argon	Tegeler et al. (1997)	39.948	150.687	4.863	535.60	0.556	−0.002
Oxygen	Schmidt, Wagner (1985)	31.999	154.595	5.043	436.14	0.352	0.022
Nitrogen	Span et al. (1998b)	28.013	126.192	3.396	313.30	0.500	0.037
Ethylene	Smukala et al. (1999)	28.054	282.35	5.042	214.24	0.368	0.087
Isobutane	Jaeschke, Schley (1995)	58.123	407.817	3.640	224.36	0.278	0.185
Cyclohexane	Penoncello et al. (1995)	84.161	553.60	4.078	273.02	0.505	0.209
SF_6	Cole, de Reuck (1989)	146.054	318.7232	3.755	742.15	0.701	0.218

Predicted values for the critical temperature and density are sufficient to be used as reducing parameters for τ and δ. Where the accuracy of properties in the extended critical region is regarded as only of minor importance, only few $p\rho T$ data are required in this region. To improve the representation of properties in the extended critical region, it is most useful to measure additional $p\rho T$ data on 2–3 supercritical isotherms (e.g. $T/T_c = 1.005, 1.02, 1.05$) which cover a density range of approximately $0.6\rho_c \leq \rho \leq 1.4\rho_c$. But even when using such data one has to be aware of the fundamental restrictions regarding the representation of properties in the extended critical region by simple technical equations of state. For an accurate description of properties in the critical region far more sophisticated equations of state are required, see Sects. 4.5.2–4.5.4 and Chap. 5.

To sum up, about 100 data points are sufficient to guarantee an accurate description of the whole range of fluid states. Compared to multiparameter equations without simultaneously optimised functional form this is a major advantage, but the requirements on the experimental equipment are still high since data for different properties are needed over broad ranges of temperature and pressure and with high accuracy.

6.2.2 Results for Non- and Weakly Polar Fluids

Data sets for a total of 15 non- and weakly polar substances were considered by Span and Wagner (2000a/b) when working on a simultaneously optimised functional form for nonpolar fluids. Table 6.3 summarises information on relevant characteristic parameters of these substances and gives references for the correlations which were used to describe caloric properties of the corresponding ideal gas. The reported molar masses are rounded to $1~\text{mg}\,\text{mol}^{-1}$ corresponding to a relative uncertainty of $\pm 0.003\%$ in the worst case; for highly accurate values see Coplen (1997). The used critical parameters and triple point temperatures are summarised in Table 6.3 without giving references or uncertainties; due to the limited requirements on the accuracy of these parameters this set-up is justifiable to avoid a further extension of the table. This section focuses on the development

Table 6.4. Coefficients n_i of the simultaneously optimised equations of state for nonpolar fluids, Eq. 6.12

i	Methane n_i	Ethane n_i	Propane n_i	n–Butane n_i	n–Pentane n_i
1	$0.89269676 \cdot 10^{+00}$	$0.97628068 \cdot 10^{+00}$	$0.10403973 \cdot 10^{+01}$	$0.10626277 \cdot 10^{+01}$	$0.10968643 \cdot 10^{+01}$
2	$-0.25438282 \cdot 10^{+01}$	$-0.26905251 \cdot 10^{+01}$	$-0.28318404 \cdot 10^{+01}$	$-0.28620952 \cdot 10^{+01}$	$-0.29988888 \cdot 10^{+01}$
3	$0.64980978 \cdot 10^{+00}$	$0.73498222 \cdot 10^{+00}$	$0.84393810 \cdot 10^{+00}$	$0.88738233 \cdot 10^{+00}$	$0.99516887 \cdot 10^{+00}$
4	$0.20793471 \cdot 10^{-01}$	$-0.35366206 \cdot 10^{-01}$	$-0.76559592 \cdot 10^{-01}$	$-0.12570581 \cdot 10^{+00}$	$-0.16170709 \cdot 10^{+00}$
5	$0.70189104 \cdot 10^{-01}$	$0.84692031 \cdot 10^{-01}$	$0.94697373 \cdot 10^{-01}$	$0.10286309 \cdot 10^{+00}$	$0.11334460 \cdot 10^{+00}$
6	$0.23700378 \cdot 10^{-03}$	$0.24154594 \cdot 10^{-03}$	$0.24796475 \cdot 10^{-03}$	$0.25358041 \cdot 10^{-03}$	$0.26760595 \cdot 10^{-03}$
7	$0.16653334 \cdot 10^{+00}$	$0.23964954 \cdot 10^{+00}$	$0.27743760 \cdot 10^{+00}$	$0.32325200 \cdot 10^{+00}$	$0.40979882 \cdot 10^{+00}$
8	$-0.43855669 \cdot 10^{-01}$	$-0.42780093 \cdot 10^{-01}$	$-0.43846001 \cdot 10^{-01}$	$-0.37950761 \cdot 10^{-01}$	$-0.40876423 \cdot 10^{-01}$
9	$-0.15726780 \cdot 10^{+00}$	$-0.22308832 \cdot 10^{+00}$	$-0.26991065 \cdot 10^{+00}$	$-0.32534802 \cdot 10^{+00}$	$-0.38169482 \cdot 10^{+00}$
10	$-0.35311675 \cdot 10^{-01}$	$-0.51799954 \cdot 10^{-01}$	$-0.69313413 \cdot 10^{-01}$	$-0.79050969 \cdot 10^{-01}$	$-0.10931957 \cdot 10^{+00}$
11	$-0.29570024 \cdot 10^{-01}$	$-0.27178426 \cdot 10^{-01}$	$-0.29632146 \cdot 10^{-01}$	$-0.20636721 \cdot 10^{-01}$	$-0.32073223 \cdot 10^{-01}$
12	$0.14019842 \cdot 10^{-01}$	$0.11246305 \cdot 10^{-01}$	$0.14040127 \cdot 10^{-01}$	$0.57053809 \cdot 10^{-02}$	$0.16877016 \cdot 10^{-01}$

i	n-Hexane n_i	n-Heptane n_i	n-Octane n_i	Argon n_i	Oxygen n_i
1	$0.10553238 \cdot 10^{+01}$	$0.10543748 \cdot 10^{+01}$	$0.10722545 \cdot 10^{+01}$	$0.85095715 \cdot 10^{+00}$	$0.88878286 \cdot 10^{+00}$
2	$-0.26120616 \cdot 10^{+01}$	$-0.26500682 \cdot 10^{+01}$	$-0.24632951 \cdot 10^{+01}$	$-0.24003223 \cdot 10^{+01}$	$-0.24879433 \cdot 10^{+01}$
3	$0.76613883 \cdot 10^{+00}$	$0.81730048 \cdot 10^{+00}$	$0.65386674 \cdot 10^{+00}$	$0.54127841 \cdot 10^{+00}$	$0.59750191 \cdot 10^{+00}$
4	$-0.29770321 \cdot 10^{+00}$	$-0.30451391 \cdot 10^{+00}$	$-0.36324974 \cdot 10^{+00}$	$0.16919771 \cdot 10^{-01}$	$0.96501817 \cdot 10^{-02}$
5	$0.11879908 \cdot 10^{+00}$	$0.12253869 \cdot 10^{+00}$	$0.12713270 \cdot 10^{+00}$	$0.68825965 \cdot 10^{-01}$	$0.71970429 \cdot 10^{-01}$
6	$0.27922861 \cdot 10^{-03}$	$0.27266473 \cdot 10^{-03}$	$0.30713573 \cdot 10^{-03}$	$0.21428033 \cdot 10^{-03}$	$0.22337443 \cdot 10^{-03}$
7	$0.46347590 \cdot 10^{+00}$	$0.49865826 \cdot 10^{+00}$	$0.52656857 \cdot 10^{+00}$	$0.17429895 \cdot 10^{+00}$	$0.18558686 \cdot 10^{+00}$
8	$0.11433197 \cdot 10^{-01}$	$-0.71432815 \cdot 10^{-03}$	$0.19362863 \cdot 10^{-01}$	$-0.33654496 \cdot 10^{-01}$	$-0.38129368 \cdot 10^{-01}$
9	$-0.48256969 \cdot 10^{+00}$	$-0.54236896 \cdot 10^{+00}$	$-0.58939427 \cdot 10^{+00}$	$-0.13526800 \cdot 10^{+00}$	$-0.15352245 \cdot 10^{+00}$
10	$-0.93750559 \cdot 10^{-01}$	$-0.13801822 \cdot 10^{+00}$	$-0.14069964 \cdot 10^{+00}$	$-0.16387351 \cdot 10^{-01}$	$-0.26726815 \cdot 10^{-01}$
11	$-0.67273247 \cdot 10^{-02}$	$-0.61595287 \cdot 10^{-02}$	$-0.78966331 \cdot 10^{-02}$	$-0.24987667 \cdot 10^{-01}$	$-0.25675299 \cdot 10^{-01}$
12	$-0.51141584 \cdot 10^{-02}$	$0.48602510 \cdot 10^{-03}$	$0.33036598 \cdot 10^{-02}$	$0.88769205 \cdot 10^{-02}$	$0.95714302 \cdot 10^{-02}$

6.2 Simultaneously Optimised Equations for Technical Applications

of relations for the residual part of the reduced Helmholtz energy α. The resulting equations are discussed very briefly in combination with a characterisation of the available data sets; the information on the available data sets may be helpful especially for readers who are looking for suitable fluids for their own applications.

Based on selected data sets for methane, ethane, propane, n-butane, n-hexane, n-heptane, n-octane, argon, oxygen, ethylene, isobutane, cyclohexane, and sulphur hexafluoride which were weighted with the demanded uncertainties summarised in Table 6.1 the simultaneous optimisation algorithm resulted in the following functional form for the reduced Helmholtz energy of non- or weakly polar fluids:

$$\alpha(\tau,\delta) = \alpha^\circ(\tau,\delta) + n_1 \delta \tau^{0.250} + n_2 \delta \tau^{1.125} + n_3 \delta \tau^{1.500}$$
$$+ n_4 \delta^2 \tau^{1.375} + n_5 \delta^3 \tau^{0.250} + n_6 \delta^7 \tau^{0.875}$$
$$+ n_7 \delta^2 \tau^{0.625} e^{-\delta} + n_8 \delta^5 \tau^{1.750} e^{-\delta} + n_9 \delta \tau^{3.625} e^{-\delta^2}$$
$$+ n_{10} \delta^4 \tau^{3.625} e^{-\delta^2} + n_{11} \delta^3 \tau^{14.5} e^{-\delta^3} + n_{12} \delta^4 \tau^{12.0} e^{-\delta^3}, \quad (6.12)$$

with $\alpha^r = a^r/(RT)$ and $\tau = T_c/T$, $\delta = \rho/\rho_c$. For the considered substances both the used critical parameters and the molar masses which are required for conversions to specific properties are summarised in Table 6.3. Data sets for n-pentane and nitrogen were used to test the transferability of the functional form — no significant disadvantages were found for the representation of the data which are available for these substances. For all 15 substances Table 6.4 summarises the coefficients n_i of Eq. 6.12. When working with these coefficients, $R = 8.314510$ J mol^{-1} K^{-1} has to be used for the gas constant.

Results of this equation were already discussed in the preceding sections, whereby Eq. 6.12 was referred to as the "simultaneously optimised equation of state with 12 terms" where nonpolar fluids were discussed. As a very global qual-

Table 6.4. Coefficients n_i of the simultaneously optimised equations of state for nonpolar fluids, Eq. 6.12 – Continued

i	Nitrogen n_i	Ethylene n_i	Isobutane n_i	Cyclohexane n_i	SF$_6$ n_i
1	$0.92296567 \cdot 10^{+00}$	$0.90962230 \cdot 10^{+00}$	$0.10429332 \cdot 10^{+01}$	$0.10232354 \cdot 10^{+01}$	$0.12279403 \cdot 10^{+01}$
2	$-0.25575012 \cdot 10^{+01}$	$-0.24641015 \cdot 10^{+01}$	$-0.28184273 \cdot 10^{+01}$	$-0.29204964 \cdot 10^{+01}$	$-0.33035623 \cdot 10^{+01}$
3	$0.64482463 \cdot 10^{+00}$	$0.56175311 \cdot 10^{+00}$	$0.86176232 \cdot 10^{+00}$	$0.10736630 \cdot 10^{+01}$	$0.12094019 \cdot 10^{+01}$
4	$0.10831020 \cdot 10^{-01}$	$-0.19688013 \cdot 10^{-01}$	$-0.10613619 \cdot 10^{+00}$	$-0.19573985 \cdot 10^{+00}$	$-0.12316000 \cdot 10^{+00}$
5	$0.73924167 \cdot 10^{-01}$	$0.78831145 \cdot 10^{-01}$	$0.98615749 \cdot 10^{-01}$	$0.12228111 \cdot 10^{+00}$	$0.11044657 \cdot 10^{+00}$
6	$0.23532962 \cdot 10^{-03}$	$0.21478776 \cdot 10^{-03}$	$0.23948209 \cdot 10^{-03}$	$0.28943321 \cdot 10^{-03}$	$0.32952153 \cdot 10^{-03}$
7	$0.18024854 \cdot 10^{+00}$	$0.23151337 \cdot 10^{+00}$	$0.30330005 \cdot 10^{+00}$	$0.27231767 \cdot 10^{+00}$	$0.27017629 \cdot 10^{+00}$
8	$-0.45660299 \cdot 10^{-01}$	$-0.37804454 \cdot 10^{-01}$	$-0.41598156 \cdot 10^{-01}$	$-0.44833320 \cdot 10^{-01}$	$-0.62910351 \cdot 10^{-01}$
9	$-0.15521060 \cdot 10^{+00}$	$-0.20122739 \cdot 10^{+00}$	$-0.29991937 \cdot 10^{+00}$	$-0.38253334 \cdot 10^{+00}$	$-0.31828890 \cdot 10^{+00}$
10	$-0.38111490 \cdot 10^{-01}$	$-0.44960157 \cdot 10^{-01}$	$-0.80369343 \cdot 10^{-01}$	$-0.89835333 \cdot 10^{-01}$	$-0.99557419 \cdot 10^{-01}$
11	$-0.31962422 \cdot 10^{-01}$	$-0.28342960 \cdot 10^{-01}$	$-0.29761373 \cdot 10^{-01}$	$-0.24874965 \cdot 10^{-01}$	$-0.36909694 \cdot 10^{-01}$
12	$0.15513532 \cdot 10^{-01}$	$0.12652824 \cdot 10^{-01}$	$0.13059630 \cdot 10^{-01}$	$0.10836132 \cdot 10^{-01}$	$0.19136427 \cdot 10^{-01}$

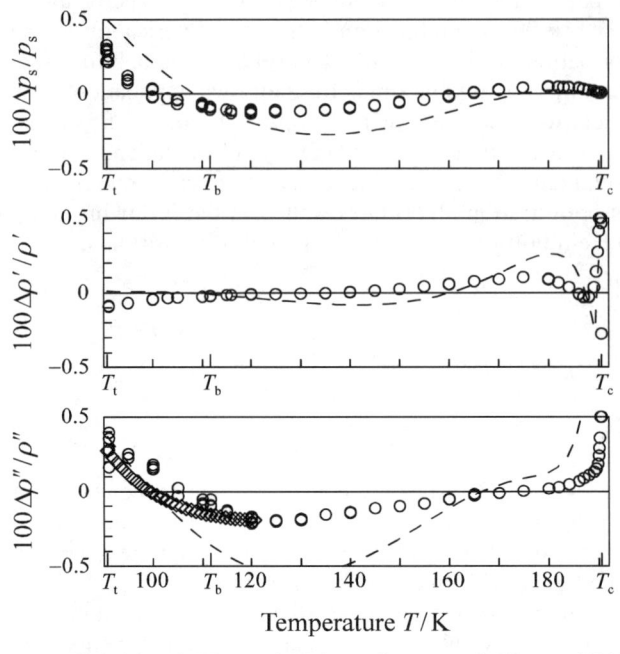

○ Kleinrahm & Wagner (1986)   ◇ Setzmann & Wagner (1991)
– – – Bender-type

Fig. 6.17. Percentage deviations $100\,\Delta y/y = 100\,(y_{exp} - y_{calc})/y_{exp}$ with $y = p_s,\, \rho',\, \rho''$ between accurate experimental results for methane and values calculated from Eq. 6.12. For comparison, values calculated from the corresponding Bender-type equation are plotted as dashed lines.

ity criterion, Fig. 6.5 shows weighted variances which resulted from fits of Eq. 6.12. In general, Eq. 6.12 describes the available data sets with $\sigma_{wt}^2 < 0.3$; the only exceptions, isobutane ($\sigma_{wt}^2 = 0.34$) and SF_6 ($\sigma_{wt}^2 = 0.38$), will be discussed below. The numerical stability of Eq. 6.12 and its extrapolation behaviour were analysed in Sect. 6.2.1.1. Plots of ideal curves were presented in Figs. 6.13 and 6.14. In this section, the performance of Eq. 6.12 is illustrated by exemplary comparisons with selected data. For more complete comparisons see Span and Wagner (2000b).

The data set which is available for *methane* was recently described in detail by Setzmann and Wagner (1991) and by Wagner and de Reuck (1996). The used data set covers fluid states from the triple point temperature to a temperature of $T_{max} \approx 623$ K ($T_{max}/T_c \approx 3.27$) at pressures up to $p_{max} = 100$ MPa. Data which describe thermal, including the saturated vapour density, and caloric properties on the vapour-liquid phase boundary are available from the triple point temperature to the critical temperature. The simultaneously optimised equation of state, Eq. 6.12 with the parameters summarised in Tables 6.3 and 6.4 and in combination with the correlation for α^o published by Setzmann and Wagner (1991), describes the selected data set well within the uncertainties claimed in Table 6.1. Larger deviations between values calculated from the simultaneously optimised equation and from the group 1 reference equation[13] published by Setzmann and Wagner (1991) were observed only for speeds of sound at low temperature liquid states where the limit of $\Delta w/w = \pm 2\%$ is exceeded at $p \approx 30$ MPa with maximum deviations of $\Delta w/w \approx -4\%$ at $p = 100$ MPa. The observed plot of the speed of sound corresponds almost exactly to the one shown for the simultaneously optimised equation with 12 terms which was fitted to the "reduced data set", see Fig. 6.12.

As an example for the representation of thermal properties on the phase boundary Fig. 6.17 shows deviations between accurate experimental data for the vapour pressure and the saturated vapour and liquid density of methane and values calculated from Eq. 6.12. For comparison, values calculated from the corresponding Bender-type equation are plotted as dashed lines. The simultaneously optimised

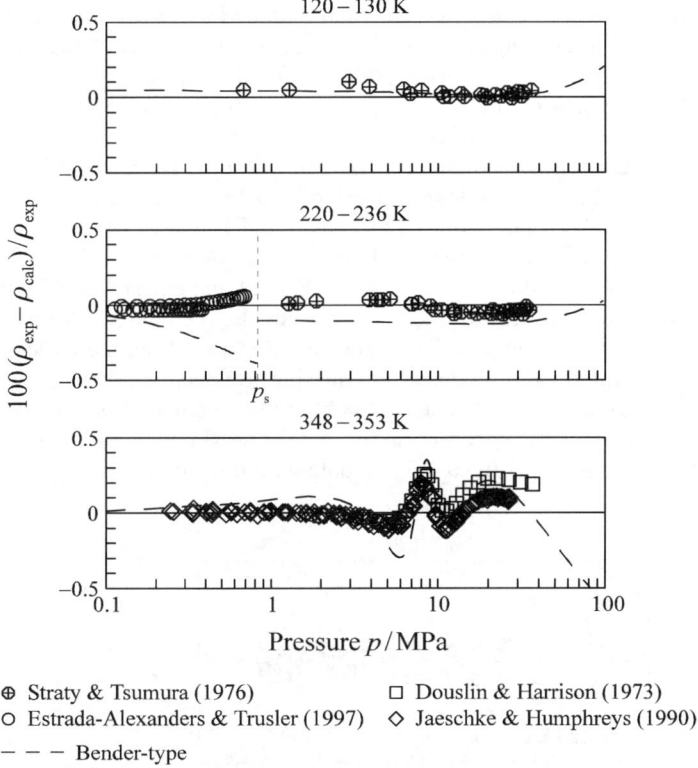

⊕ Straty & Tsumura (1976)   □ Douslin & Harrison (1973)
○ Estrada-Alexanders & Trusler (1997)   ◇ Jaeschke & Humphreys (1990)
– – – Bender-type

Fig. 6.18. Percentage deviations between experimental results for the density of ethane and values calculated from Eq. 6.12. For comparison, values calculated from the corresponding Bender-type equation are plotted as dashed lines.

---

[13] See Sect. 5 for an explanation of the characterisation as "group 1 reference equation".

equation represents the data far within the claimed uncertainty for $T_t \leq T \leq \approx 0.995\,T_c$. Vapour pressures are represented accurately up to the critical temperature; for saturated vapour and liquid densities, the theoretically expected increased uncertainties are observed in the critical region.

For *ethane*, reliable experimental results are available for thermodynamic properties from the triple point temperature to $T_{max} \approx 623$ K ($T_{max}/T_c \approx 2.04$) at pressures up to $p_{max} \approx 52$ MPa. Detailed information on the available data was published by Friend et al. (1991). Both for thermal and caloric properties the published data are sufficiently accurate for the development of technical equations of state. On the phase boundary, reliable data are available for the vapour pressure and the saturated liquid density from the triple point temperature to the critical temperature. For the saturated vapour density reliable data are available only for temperatures above 248 K ($T/T_c \approx 0.81$). The simultaneously optimised equation of state represents the available data within the demanded uncertainties without restrictions. As an example for the representation of $p\rho T$ data Fig. 6.18 shows percentage deviations between selected experimental data and densities calculated from Eq. 6.12. On the supercritical isotherm 348–353 K ($T/T_c \approx 1.15$) the observed density deviations exceed the limit of $\Delta \rho/\rho = \pm 0.2\,\%$ slightly; however, in terms of pressure these deviations are still very small ($\Delta p/p \approx \pm 0.1\,\%$) and thus unavoidable for technical equations of state.

For *propane,* reliable experimental data are available from the triple point temperature to $T_{max} \approx 623$ K ($T_{max}/T_c \approx 1.68$); the considered data set was restricted to pressures up to $p_{max} \approx 100$ MPa. The work of Goodwin and Haynes (1982a) contains the most recent detailed review of the available data set. At liquid and liquid-like supercritical states reliable data are available for thermal and caloric data from different authors. However, compared to methane and ethane, the data situation is quite unsatisfactory for gaseous and gas-like supercritical states. Accurate experimental results for the saturated vapour density are missing. An accurate description of caloric properties at gaseous states became possible only recently when Trusler and Tsarari (1996) published their speed of sound data. The simultaneously optimised equation, Eq. 6.12, represents the available data within the demanded uncer-

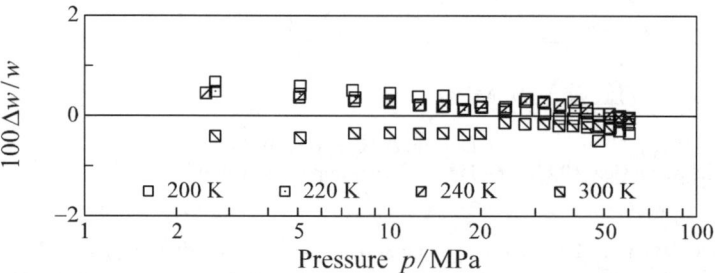

Fig. 6.19. Percentage deviations $100\,\Delta w/w = 100\,(w_{exp} - w_{calc})/w_{exp}$ between experimental data for the speed of sound in the liquid phase of normal butane measured by Niepmann (1984) and values calculated from the corresponding simultaneously optimised equation of state, Eq. 6.12.

tainties without significant restrictions. As an example of the accuracy of calculated speeds of sound a comparison for propane has already been shown in Fig. 6.7.

For *n-butane,* reliable experimental data are available from the triple point temperature to $T_{max} \approx 693$ K ($T_{max}/T_c \approx 1.63$) and up to pressures of $p_{max} \approx 69$ MPa. For liquid and liquid-like supercritical states, the data situation is satisfactory again, while accurate data for gaseous states are rare. For the development of technical equations of state vapour pressure and saturated liquid density are described by experimental data with sufficient accuracy; reliable data for the saturated vapour density are not available. An extensive review was published by Goodwin and Haynes (1982b). Important new data sets for properties of gaseous n-butane include those of Gupta and Eubank (1997) and Ewing et al. (1988). In the liquid region, measurements of the speed of sound published by Niepmann (1984) and of the isochoric heat capacity published recently by Magee and Lüddecke (1998) have improved the situation with regard to caloric properties. The simultaneously optimised equation of state describes the available data within the demanded uncertainties. The representation of speeds of sound in the liquid phase of n-butane was used in Sect. 5.2 to point out typical shortcomings of MBWR-type equations of state, see Fig. 5.24. Figure 6.19 compares the same data set with values calculated from Eq. 6.12. Obviously, this new class of technical equations of state is even superior to subaverage reference equations with regard to certain features.

From all the considered nonpolar and polar substances, the data sets which are available for n-pentane and n-hexane are the most restricted ones. For *n-pentane,* experimental data are available from the triple point temperature to $T_{max} \approx 573$ K ($T_{max}/T_c \approx 1.22$) and up to pressures of $p_{max} \approx 69$ MPa. At liquid and liquid-like states, only the $p\rho T$ data published by Kratzke et al. (1985) build a sufficiently accurate foundation for the development of an equation of state; basically, the same is true for vapour pressures and saturated liquid densities. Data for the saturated vapour density are not available. The only reliable data for caloric properties are those published by Ewing et al. (1989) for the speed of sound in the gas phase

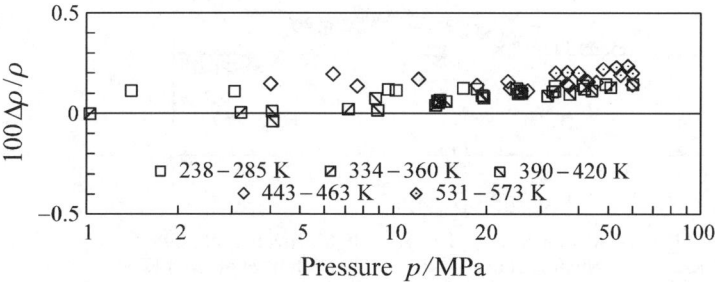

Fig. 6.20. Percentage deviations $100 \, \Delta\rho/\rho = 100 \, (\rho_{exp} - \rho_{calc})/\rho_{exp}$ between experimental data for the density of n-pentane measured by Kratzke et al. (1985) and values calculated from the corresponding simultaneously optimised equation of state, Eq. 6.12.

at very low pressures; no other reliable data are available in the gas phase. No complete reviews have been published for the data sets available for n-pentane and for the higher n-alkanes; recently Cibulka and Hnedkovsky (1996) have published an extensive compilation which summarises at least the available data on liquid densities considering n-alkanes from pentane to hexadecane. Based on the available data, it is remarkable that the simultaneously optimised equation of state describes not only the data within the demanded uncertainties, but that it also extrapolates well to very high pressures and temperatures, see the ideal curves plotted in Fig. 6.14. The representation of $p\rho T$ data by Kratzke et al. (1985) is illustrated in Fig. 6.20.

For *n-hexane,* reliable experimental data are available from $T_{\min} = 263$ K to $T_{\max} \approx 548$ K ($T_{\max}/T_c \approx 1.08$) and up to pressures of $p_{\max} \approx 92$ MPa. The description of the gaseous phase has to rely mainly on questionable data published for second virial coefficients. At liquid states the $p\rho T$ data set published by Sauermann et al. (1995) is the most reliable foundation for an equation of state; this reference contains reliable vapour pressures and saturated liquid densities as well. Data for the saturated vapour density are not available. The simultaneously optimised equation of state describes the selected data sets well within the demanded uncertainties, respectively within their experimental uncertainties, and extrapolates well to very high temperatures and pressures.

For *n-heptane,* reliable data on thermal and caloric properties are available from different authors at least in the liquid region from the triple point temperature to 310 K; the data set used was restricted to pressures up to $p_{\max} = 100$ MPa. At higher temperatures caloric data are missing and larger inconsistencies are encountered for $p\rho T$ data. No reliable data are available above $T_{\max} \approx 523$ K ($T_{\max}/T_c \approx 0.97$). The critical temperature of the new equation had to be constrained to a reasonable value in the way described in Sect. 6.2. The description of the gaseous phase has to rely on questionable data published for the second virial coefficients again. Data for the vapour pressure are available up to the critical

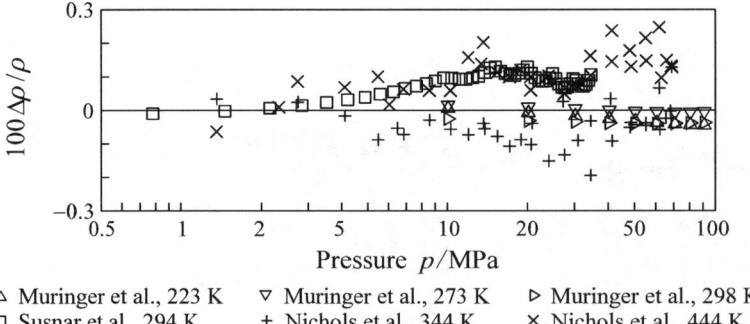

Fig. 6.21. Percentage deviations $100 \, \Delta\rho/\rho = 100 \, (\rho_{\exp} - \rho_{\text{calc}})/\rho_{\exp}$ between experimental data for the density of n-heptane measured by Muringer et al. (1985), Susnar et al. (1992), and Nichols et al. (1955) and values calculated from the corresponding simultaneously optimised equation of state, Eq. 6.12.

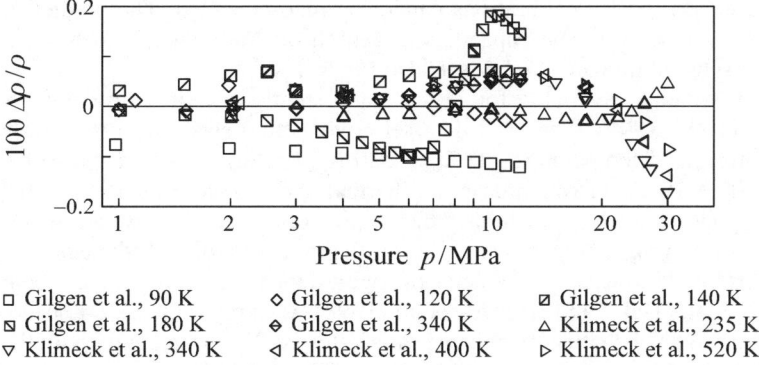

Fig. 6.22. Percentage deviations $100\,\Delta\rho/\rho = 100\,(\rho_{\text{exp}}-\rho_{\text{calc}})/\rho_{\text{exp}}$ between experimental results for the density of argon by Gilgen et al. (1994) and Klimeck et al. (1998) and values calculated from the simultaneously optimised equation of state, Eq. 6.12.

temperature but reliable data for the saturated liquid density are available only at ambient temperatures. Saturated vapour densities are not available at all. The simultaneously optimised equation of state describes the reliable data sets within the demanded uncertainties. After the development of the simultaneously optimised equations had been finished isochoric heat capacities became available for n-heptane at gaseous, supercritical and liquid states. The predictive description of these properties has already been discussed in Sect. 6.2.1.1, see also Fig. 6.15.

In calibration applications, n-heptane is frequently recommended as reference fluid for liquid densities around 700 kg m^{-3}. This recommendation is justified only for temperatures up to 310 K, and even below this limit the uncertainty of the available data is significantly larger than for typical reference fluids like water, nitrogen, or argon. However, where an uncertainty of $\Delta\rho/\rho \leq \pm 0.1\,\%$ is regarded as sufficient, Eq. 6.12 can even be used as reference for calibrations in this region. Figure 6.21 shows deviations between selected experimental results for the density of n-heptane and values calculated from Eq. 6.12. At temperatures above 310 K a considerable scatter and inconsistencies become obvious even for the most accurate available data sets.

In the liquid region, the data situation is satisfactory for *n-octane* at temperatures from 258 K to 373 K. At higher temperatures, no data for caloric properties are available and the only set of at least reasonable $p\rho T$ data scatters by more than $\Delta\rho/\rho = \pm 0.2\,\%$. Reliable data for the vapour pressure are available at temperatures above 350 K and the saturated liquid density is described with sufficient accuracy between 245 K and 475 K. The highest temperature for which data could be used was $T_{\max} \approx 548$ K ($T_{\max}/T_{\text{c}} \approx 0.96$) and the highest pressure $p_{\max} \approx 96$ MPa. The critical temperature of the new equation had to be constrained to a reasonable value in the way described in Sect. 6.2. The reliable data sets are described within the demanded uncertainties, but obvious inconsistencies in the data set result in deviations which are larger than those observed for the data set of Muringer et al. (1985) in Fig. 6.21. Based on the data set which is available today the use of n-octane as reference fluid for calibrations cannot be recommended. The predictive description of the isochoric heat capacities measured by Abdulagatov (1998) has already been discussed in Sect. 6.2.1.1, see also Fig. 6.15.

The data set which is available for *argon* was recently described in detail by Tegeler et al. (1997/1999). The used data set covers fluid states from the triple point temperature to a temperature of $T_{max} \approx 520$ K ($T_{max}/T_c \approx 3.45$) at pressures up to $p_{max} = 100$ MPa. Data which describe thermal and caloric properties on the vapour-liquid phase boundary, including the saturated vapour density, are available from the triple point temperature to the critical temperature. With regard to thermal properties, the reference character of argon depends mainly on the recent data of Gilgen et al. (1994) and Klimeck et al. (1998); with regard to caloric properties at gaseous states the situation became satisfactory only very recently with the publications of Estrada-Alexanders and Trusler (1995/1996). In general, the simultaneously optimised equation of state describes the selected data set within the uncertainties claimed in Table 6.1. Larger deviations are again observed only for speeds of sound at low temperature liquid states, where the limit of $\Delta w/w = \pm 2\%$ is exceeded at $p \approx 60$ MPa with maximum deviations of $\Delta w/w \approx -3\%$ at $p = 100$ MPa. To illustrate the quality of the simultaneously optimised equation of state Fig. 6.22 uses the set of highly accurate $p\rho T$ data which was already shown in Fig. 5.4. The observed deviations compare well with results of typical group 2 reference equations of state again, see for instance Fig. 5.5 for carbon dioxide.

At pressures up to $p_{max} \approx 92$ MPa the data set which is available for *oxygen* covers a temperature range from close to the triple point temperature to $T_{max} \approx 303$ K ($T_{max}/T_c \approx 1.96$). The available experimental data are at least 20 years old and do not fulfil recent demands on reference data. However, for the development of an equation of state for technical applications the data situation is satisfactory both with regard to thermal and caloric properties in the homogeneous region and on the vapour-liquid phase boundary. The IUPAC book by Wagner and

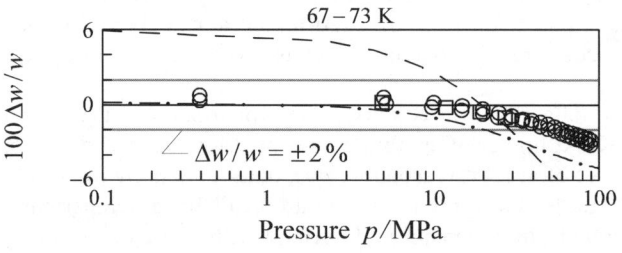

○ Van Itterbeek & Van Dael (1962)   □ Straty & Younglove (1973)
— · — Schmidt & Wagner (1985)   – – – Bender-type

Fig. 6.23. Percentage deviations $100\,\Delta w/w = 100\,(w_{exp} - w_{calc})/w_{exp}$ between experimental data for the speed of sound in the liquid phase of oxygen and values calculated from the corresponding simultaneously optimised equation of state, Eq. 6.12. Values calculated from the reference equation by Schmidt and Wagner (1985) and from a Bender-type equation are plotted as dash-dotted lines for comparison.

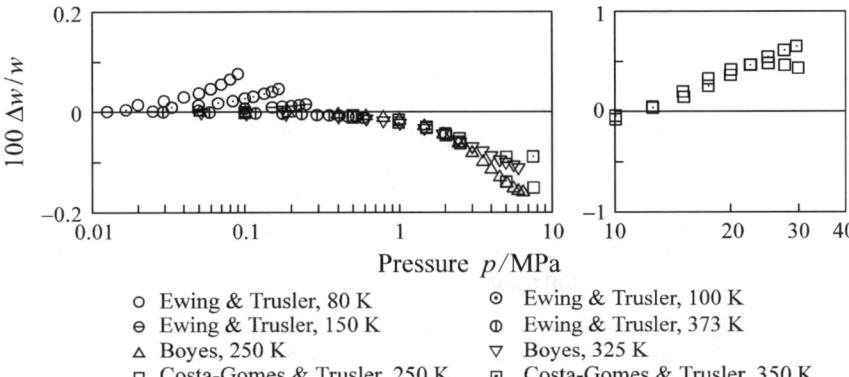

Fig. 6.24. Percentage deviations $100 \, \Delta w/w = 100 \, (w_{exp} - w_{calc})/w_{exp}$ between experimental data for the speed of sound in nitrogen by Ewing and Trusler (1992), Boyes (1992), and Costa Gomes and Trusler (1998a) and values calculated from the simultaneously optimised equation of state, Eq. 6.12.

de Reuck (1987) contains a comprehensive review of the available data sets. The simultaneously optimised equation of state describes the selected data set well within the claimed uncertainties. Again, larger deviations occur only for speeds of sound at low temperature liquid states, where the limit of $\Delta w/w = \pm 2\%$ is exceeded at $p \approx 60$ MPa with maximum deviations of $\Delta w/w \approx -3\%$ at $p \approx 90$ MPa. For a proper assessment of this shortcoming, one has to be aware that the recent IUPAC standard for oxygen, the group 2 reference equation by Schmidt and Wagner (1985), results in deviations of almost the same magnitude but with the opposite sign, while the best known technical equations of state without an optimised functional form yield significantly larger deviations. These facts are illustrated in Fig. 6.23.

In terms of reduced properties the data set which is available for *nitrogen* covers by far the broadest range of temperatures. At pressures up to $p_{max} = 100$ MPa data were used for temperatures from the triple point temperature to $T_{max} \approx 1074$ K ($T_{max}/T_c \approx 8.51$). Most of the available data sets describe far supercritical states, but for gaseous, critical, and liquid states highly accurate data are available as well. Both thermal and caloric properties on the vapour-liquid phase boundary are well known. A recent review of the available data set was published by Span et al. (1998b) and, in more detail, by Span et al. (1999). The simultaneously optimised equation of state represents the available data within the demanded uncertainty with just two exceptions. Larger deviations occur for speeds of sound at low temperature liquid states, where the limit of $\Delta w/w = \pm 2\%$ is exceeded at $p \approx 30$ MPa with maximum deviations of $\Delta w/w \approx -4\%$ at $p = 100$ MPa. At temperatures of $T > \approx 500$ K ($T/T_c > 4$) the deviations observed for $p\rho T$ data exceed $\Delta \rho/\rho = \pm 0.5\%$ at $p \approx 50$ MPa with maximum deviations up to $\Delta \rho/\rho \approx \pm 1\%$ at $p = 100$ MPa. For none of the other substances data have been used at such high reduced temperatures and, based only on the data set for nitrogen, the representa-

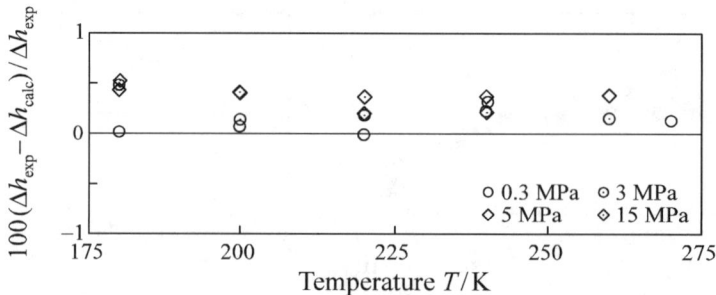

Fig. 6.25. Percentage deviations between experimental data for enthalpy differences in nitrogen measured by Grini and Owren (1997) at supercritical states and values calculated from the corresponding simultaneously optimised equation of state, Eq. 6.12.

tion of data in this region could not be improved in the simultaneous optimisation process without affecting the overall performance of the resulting functional form. Finally, nitrogen was excluded from the group of substances used in the simultaneous optimisation for use as an ambitious test for the transferability of the simultaneously optimised functional forms.

However, problems at extreme states should not obscure the fact that the simultaneously optimised equation of state yields excellent results in the technically most important regions. Figure 6.24 shows the representation of highly accurate speed of sound data at gaseous and supercritical states, see also Figs. 5.16, 5.18, and 5.21 for a comparison with other equations of state. The observed deviations stay well within the demanded uncertainty of $\Delta w/w = \pm 1\,\%$; at pressures up to 10 MPa the uncertainty of calculated speeds of sound is even $\Delta w/w < \pm 0.2\,\%$. In Fig. 6.25 deviations between experimental results for isobaric enthalpy differences and corresponding values calculated from the simultaneously optimised equation of state are plotted, see also Fig. 5.28 for comparison. The observed deviations stay within $\Delta(\Delta h)/\Delta h \leq \approx \pm 0.5\,\%$ and are thus about four times smaller than the maximum deviations observed for the best technical equations of state without optimised functional form.

For *ethylene*, reliable data are available from the triple point temperature to $T_{\max} \approx 473$ K ($T_{\max}/T_c \approx 1.68$) whereby the used data set has been restricted to $p_{\max} = 100$ MPa again. Highly accurate data are available for thermal properties in the homogeneous regions as well as at vapour-liquid equilibrium states. The experimental results for caloric properties show larger uncertainties and inconsistencies, but on the level of accuracy required here the data situation is satisfactory as well. An extensive review of the data set which is available for ethylene was published by Jacobsen et al. (1988). Important recent publications on experimental results are especially those of Nowak et al. (1996a/b) and a recent review on the data set can be found in Smukala et al. (1999). The simultaneously optimised equation of state represents the available data within the demanded uncertainties.

For ethylene, accurate data for properties in the region of ambient temperatures and typical pipeline pressures are frequently needed in technical applications;

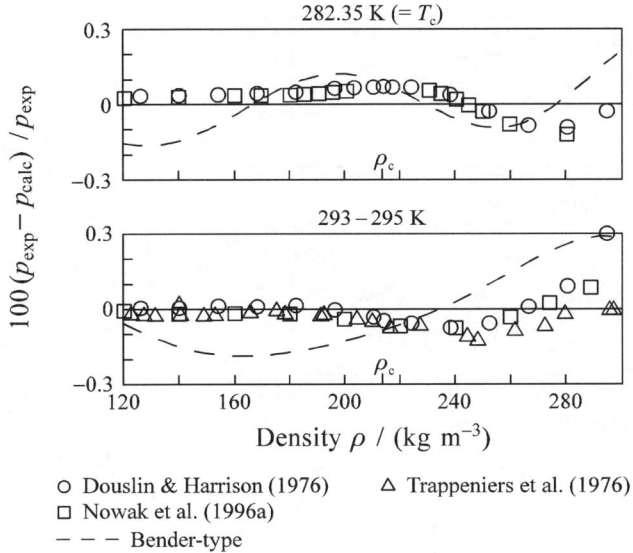

Fig. 6.26. Percentage pressure deviations between experimental $p\rho T$ data in the critical region of ethylene and values calculated from the corresponding simultaneously optimised equation of state, Eq. 6.12. Values calculated from a Bender-type equation are plotted as dashed lines for comparison.

typical technical equations of state fail at this point, since these conditions correspond to states in the extended critical region. In terms of pressure Fig. 6.26 shows deviations between selected experimental $p\rho T$ data in this region and values calculated from the corresponding simultaneously optimised equation of state. In general, the observed deviations stay within $\Delta p/p \approx \pm 0.1\%$ and are thus in an order which is typical for group 2 reference equations of state, but not for technical equations of state. However, for calculations with high demands on accuracy the use of more sophisticated reference equations has to be recommended, see the comparisons shown in Figs. 5.12 and 5.13.

For *isobutane,* experimental data are available in a very broad range of temperatures, namely from the triple point temperature to $T_{max} \approx 1501$ K ($T_{max}/T_c \approx 3.68$) and up to pressures of $p_{max} \approx 106$ MPa. However, the accuracy of the data for thermal properties is at least partly questionable and reliable data for caloric properties are available only up to pressures of 1 MPa; at liquid states only obsolete data for the heat capacity along the saturated liquid line have been published. No reliable data are available for the properties of the saturated vapour. A review of the available data was published by Goodwin and Haynes (1982c).

The simultaneously optimised equation of state represents the reliable experimental results well within the demanded uncertainties, but large inconsistencies between the available sets of $p\rho T$ data result in the large weighted variance which was reported in Fig. 6.5. Figure 6.27 shows two typical examples for such incon-

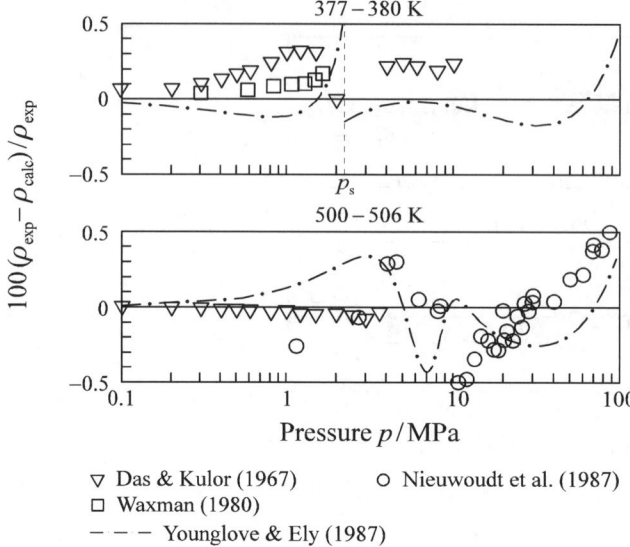

Fig. 6.27. Percentage density deviations between experimental data for isobutane and values calculated from the corresponding simultaneously optimised equation of state, Eq. 6.12. Values calculated from the MBWR-type equation of state by Younglove and Ely (1987) are plotted as dash-dotted lines for comparison.

sistencies. A comparison with the MBWR-type equation by Younglove and Ely (1987) is not able to clarify the situation – at least at supercritical temperatures the plot of this equation is as questionable as the plots implied by the data sets.

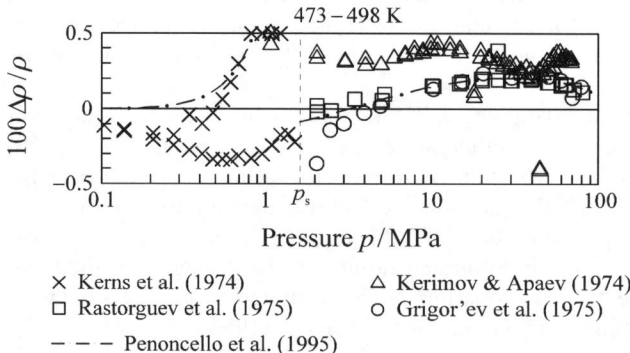

Fig. 6.28. Percentage deviations $100\,\Delta\rho/\rho = 100\,(\rho_{\mathrm{exp}} - \rho_{\mathrm{calc}})/\rho_{\mathrm{exp}}$ between experimental results for the density of cyclohexane and values calculated from the corresponding simultaneously optimised equation of state, Eq. 6.12. For comparison, values calculated from the equation of Penoncello et al. (1995) are plotted as a dash-dotted line.

For *cyclohexane,* a large data set of questionable quality is available. For temperatures from the triple point to 473 K ($T/T_c \approx 0.85$) reliable data are available both for thermal and caloric properties at liquid states. Above 350 K, the data sets which are available for the vapour pressure and the saturated liquid density show inconsistencies which clearly exceed the demanded value of $\Delta y/y \approx \pm 0.2\%$ and no vapour pressures are available above 423 K at all. The reliable data sets at temperatures up to 473 K are represented by the simultaneously optimised equation of state well within the demanded uncertainties. At temperatures above 473 K experimental results are available for the density and the isobaric heat capacity, but these data cannot be described consistently with the accurate data at lower temperatures. The reference equation published by Penoncello et al. (1995) cannot clarify the obvious inconsistencies; within the large scatter of the data it follows the plot implied by the $p\rho T$ data up to $T_{max} \approx 748$ K ($T_{max}/T_c \approx 1.35$), but it fails to represent isobaric heat capacities at high temperatures and accurate liquid speeds of sound at lower temperatures. As examples of this unsatisfactory situation, Figs. 6.28 and 6.29 show the representation of $p\rho T$ data and of isobaric heat capacities at high temperatures. The simultaneously optimised equation had to be constrained to a reasonable critical temperature in the way described in Sect. 6.2.

Without additional experimental data, it cannot be clarified whether the problems observed for cyclohexane depend solely on inconsistencies in the data set or whether the simultaneously optimised functional form for nonpolar fluids is unsuitable for the description of cyclic alkanes, at least at high temperatures. How-

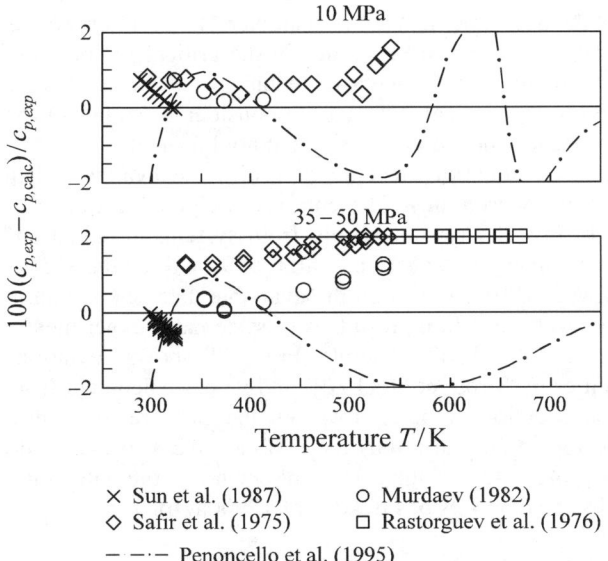

× Sun et al. (1987) ○ Murdaev (1982)
◇ Safir et al. (1975) □ Rastorguev et al. (1976)
— · — Penoncello et al. (1995)

Fig. 6.29. Percentage deviations between experimental results for the isobaric heat capacity of cyclohexane and values calculated from the corresponding simultaneously optimised equation of state, Eq. 6.12. For comparison, values calculated from the equation of Penoncello et al. (1995) are plotted as dash-dotted lines.

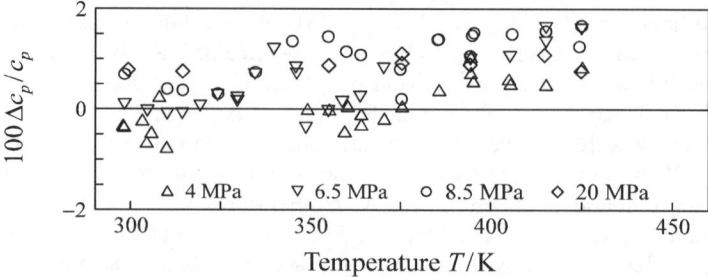

Fig. 6.30. Percentage deviations $100 \, \Delta c_p/c_p = 100 \, (c_{p,\mathrm{exp}} - c_{p,\mathrm{calc}})/c_{p,\mathrm{exp}}$ between experimental data for the isobaric heat capacity of sulphur hexafluoride measured by Sirota et al. (1979) and values calculated from the simultaneously optimised equation of state, Eq. 6.12.

ever, based on experiences made with generalised equations of state it seems likely that the observed problems are caused by systematic errors of experimental results for the saturated liquid density at high temperatures, see Sect. 7.2.2.3.

The data set which was used for *sulphur hexafluoride (SF$_6$)* covers the range from the triple point temperature to $T_{\mathrm{max}} \approx 521$ K ($T_{\mathrm{max}}/T_\mathrm{c} \approx 1.73$) at pressures up to $p_{\mathrm{max}} \approx 59$ MPa. Data are available for thermal and caloric properties at homogeneous states and for the thermal phase equilibrium properties. A review of the available data set was published by Cole and de Reuck (1989). Sulphur hexafluoride is regarded as the most important reference fluids for the investigation of critical phenomena and thus the available data set is characterised by an enormous amount of data in the critical region. Since properties in the critical region were considered as less important for the development of technical equations of state data sets which are restricted to the critical region were considered neither in the simultaneous optimisation procedure nor in the subsequent nonlinear fit.

The simultaneously optimised equation of state describes the available data generally within the demanded uncertainties. The enlarged weighted variance of the fit which was reported in Fig. 6.5 results mainly from systematic deviations observed for the isobaric heat capacities by Sirota et al. (1979) and for the speeds of sound by Vacek and Zollweg (1993). Based on the available data sets it cannot be clarified whether the observed deviations result from systematic experimental errors or from the equation of state. As an example, Fig. 6.30 shows deviations between isobaric heat capacities by Sirota et al. (1979) and values calculated from the simultaneously optimised equation of state. However, the data are still represented within the demanded uncertainty and thus the assumed inconsistencies do not affect the results of this project. On a higher level of accuracy, the data situation will soon be improved by the data sets of Funke et al. (2000a/b).

6.2.3 Results for Polar Fluids

Data sets for a total of 13 polar substances were considered by Span and Wagner (2000a/c) when working on a simultaneously optimised functional form for polar

Table 6.5. Characteristic parameters of the considered polar substances and references for the used correlations for caloric properties of the corresponding ideal gases

Substance	Correlation used for α°	$\dfrac{M}{\text{g mol}^{-1}}$	$\dfrac{T_c}{\text{K}}$	$\dfrac{p_c}{\text{MPa}}$	$\dfrac{\rho_c}{\text{kg m}^{-3}}$	$\dfrac{T_c}{T_t}$	ω
	Halogenated Hydrocarbons						
R11	Marx et al. (1992)	137.368	471.06	4.3935	565.0	0.345	0.187
R12	Marx et al. (1992)	120.914	385.12	4.1361	565.0	0.301	0.179
R22	Wagner et al. (1993)	86.469	369.28	4.9885	520.0	0.313	0.221
R32	Tillner-Roth, Yokozeki (1997)	52.024	351.35	5.795	427.0	0.388	0.277
R113	Marx et al. (1992)	187.376	487.21	3.3922	560.0	0.486	0.252
R123	Younglove, McLinden (1994)	152.931	456.82	3.672	553.0	0.363	0.283
R125	Outcalt, McLinden (1995)	120.022	339.33	3.629	571.3	0.508	0.304
R134a	Tillner-Roth, Baehr (1994)	102.032	374.18	4.0563	508.0	0.454	0.327
R143a	Li et al. (1997)	84.040	345.86	3.764	434.1	0.466	0.262
R152a	Tillner-Roth (1995)	66.051	386.41	4.5198	368.0	0.400	0.275
	Other polar substances						
Ethylene	Smukala et al. (1999)	28.054	282.35	5.042	214.24	0.368	0.087
CO_2	Span, Wagner (1996)	44.010	304.1282	7.3773	467.6	0.712	0.225
Ammonia	Tillner-Roth et al. (1993)	17.031	405.40	11.3393	225.0	0.482	0.256

fluids. Due to intense investigations of thermodynamic properties of working fluids for refrigeration applications, above average data sets are available mainly for halogenated hydrocarbons. Although these substances are rather similar from a chemical point of view the set-up of their molecules results in very different polarities – halogenated hydrocarbons are an ideal reference group to investigate thermodynamic properties of polar substances. Table 6.5 summarises information on characteristic parameters of the substances used and gives references for the correlations which were used to describe caloric properties of the corresponding ideal gas. The reported molar masses are rounded to 1 mg mol^{-1} corresponding to a relative uncertainty of $\pm 0.003\%$ in the worst case; for highly accurate values see Coplen (1997). The used critical parameters and triple point temperatures are summarised in Table 6.5 without giving references again. Just like Sect. 6.2.2 for nonpolar fluids, this section focuses on an assessment of the equations developed for the residual part of the reduced Helmholtz energy α and on a brief description of the available data sets.

Based on selected data sets for the halogenated hydrocarbons R12 (dichlorodifluoromethane), R22 (chlorodifluoromethane), R32 (trifluoromethane), R123 (2,2-dichloro-1,1,1-trifluoroethane), R125 (pentafluoroethane), R134a (1,1,1,2-tetrafluoroethane), R143a (1,1,1-trifluoroethane), R152a (1,1-difluoroethane), ethylene, carbon dioxide, and ammonia which were weighted with the demanded uncertainties summarised in Table 6.1 the simultaneous optimisation

Table 6.6. Coefficients n_i of the simultaneously optimised equations of state for polar fluids, Eq. 6.13

i	R11 n_i	R12 n_i	R22 n_i	R32 n_i
1	$0.10656383 \cdot 10^{+01}$	$0.10557228 \cdot 10^{+01}$	$0.96268924 \cdot 10^{+00}$	$0.92876414 \cdot 10^{+00}$
2	$-0.32495206 \cdot 10^{+01}$	$-0.33312001 \cdot 10^{+01}$	$-0.25275103 \cdot 10^{+01}$	$-0.24673952 \cdot 10^{+01}$
3	$0.87823894 \cdot 10^{+00}$	$0.10197244 \cdot 10^{+01}$	$0.31308745 \cdot 10^{+00}$	$0.40129043 \cdot 10^{+00}$
4	$0.87611569 \cdot 10^{-01}$	$0.84155115 \cdot 10^{-01}$	$0.72432837 \cdot 10^{-01}$	$0.55101049 \cdot 10^{-01}$
5	$0.29950049 \cdot 10^{-03}$	$0.28520742 \cdot 10^{-03}$	$0.21930233 \cdot 10^{-03}$	$0.11559754 \cdot 10^{-03}$
6	$0.42896949 \cdot 10^{+00}$	$0.39625057 \cdot 10^{+00}$	$0.33294864 \cdot 10^{+00}$	$-0.25209758 \cdot 10^{+00}$
7	$0.70828452 \cdot 10^{+00}$	$0.63995721 \cdot 10^{+00}$	$0.63201229 \cdot 10^{+00}$	$0.42091879 \cdot 10^{+00}$
8	$-0.17391823 \cdot 10^{-01}$	$-0.21423411 \cdot 10^{-01}$	$-0.32787841 \cdot 10^{-02}$	$0.37071833 \cdot 10^{-02}$
9	$-0.37626522 \cdot 10^{+00}$	$-0.36249173 \cdot 10^{+00}$	$-0.33680834 \cdot 10^{+00}$	$-0.10308607 \cdot 10^{+00}$
10	$0.11605284 \cdot 10^{-01}$	$0.19341990 \cdot 10^{-02}$	$-0.22749022 \cdot 10^{-01}$	$-0.11592089 \cdot 10^{+00}$
11	$-0.89550567 \cdot 10^{-01}$	$-0.92993833 \cdot 10^{-01}$	$-0.87867308 \cdot 10^{-01}$	$-0.44350855 \cdot 10^{-01}$
12	$-0.30063991 \cdot 10^{-01}$	$-0.24876461 \cdot 10^{-01}$	$-0.21108145 \cdot 10^{-01}$	$-0.12788805 \cdot 10^{-01}$

i	R113 n_i	R123 n_i	R125 n_i	R134a n_i
1	$0.10519071 \cdot 10^{+01}$	$0.11169730 \cdot 10^{+01}$	$0.11290996 \cdot 10^{+01}$	$0.10663189 \cdot 10^{+01}$
2	$-0.28724742 \cdot 10^{+01}$	$-0.30745930 \cdot 10^{+01}$	$-0.28349269 \cdot 10^{+01}$	$-0.24495970 \cdot 10^{+01}$
3	$0.41983153 \cdot 10^{+00}$	$0.51063873 \cdot 10^{+00}$	$0.29968733 \cdot 10^{+00}$	$0.44645718 \cdot 10^{-01}$
4	$0.87107788 \cdot 10^{-01}$	$0.94478812 \cdot 10^{-01}$	$0.87282204 \cdot 10^{-01}$	$0.75656884 \cdot 10^{-01}$
5	$0.24105194 \cdot 10^{-03}$	$0.29532752 \cdot 10^{-03}$	$0.26347747 \cdot 10^{-03}$	$0.20652089 \cdot 10^{-03}$
6	$0.70738262 \cdot 10^{+00}$	$0.66974438 \cdot 10^{+00}$	$0.61056963 \cdot 10^{+00}$	$0.42006912 \cdot 10^{+00}$
7	$0.93513411 \cdot 10^{+00}$	$0.96438575 \cdot 10^{+00}$	$0.90073581 \cdot 10^{+00}$	$0.76739111 \cdot 10^{+00}$
8	$-0.96713512 \cdot 10^{-02}$	$-0.14865424 \cdot 10^{-01}$	$-0.68788457 \cdot 10^{-02}$	$0.17897427 \cdot 10^{-02}$
9	$-0.52595315 \cdot 10^{+00}$	$-0.49221959 \cdot 10^{+00}$	$-0.44211186 \cdot 10^{+00}$	$-0.36219746 \cdot 10^{+00}$
10	$0.22691984 \cdot 10^{-01}$	$-0.22831038 \cdot 10^{-01}$	$-0.35041493 \cdot 10^{-01}$	$-0.67809370 \cdot 10^{-01}$
11	$-0.14556325 \cdot 10^{+00}$	$-0.14074860 \cdot 10^{+00}$	$-0.12698630 \cdot 10^{+00}$	$-0.10616419 \cdot 10^{+00}$
12	$-0.27419950 \cdot 10^{-01}$	$-0.25117301 \cdot 10^{-01}$	$-0.25185874 \cdot 10^{-01}$	$-0.18185791 \cdot 10^{-01}$

algorithm resulted in the following functional form for the reduced Helmholtz energy of polar fluids:

$$\alpha(\tau,\delta) = \alpha^\circ(\tau,\delta) + n_1 \delta\tau^{0.250} + n_2 \delta\tau^{1.250} + n_3 \delta\tau^{1.500}$$
$$+ n_4 \delta^3 \tau^{0.250} + n_5 \delta^7 \tau^{0.875} + n_6 \delta\tau^{2.375} e^{-\delta}$$
$$+ n_7 \delta^2 \tau^{2.000} e^{-\delta} + n_8 \delta^5 \tau^{2.125} e^{-\delta} + n_9 \delta\tau^{3.500} e^{-\delta^2}$$
$$+ n_{10} \delta\tau^{6.50} e^{-\delta^2} + n_{11} \delta^4 \tau^{4.75} e^{-\delta^2} + n_{12} \delta^2 \tau^{12.5} e^{-\delta^3}, \quad (6.13)$$

with $\alpha^r = a^r/(RT)$, $\tau = T_c/T$, $\delta = \rho/\rho_c$. For the considered substances both the used critical parameters and the molar masses which are required for conversions to specific properties are summarised in Table 6.5. Data sets for R11 (trichlorofluoromethane) and R113 (1,1,2-trichlorotrifluoroethane) were used to

6.2 Simultaneously Optimised Equations for Technical Applications

Table 6.6. Coefficients n_i of the simultaneously optimised equations of state for polar fluids, Eq. 6.13 – Continued

i	R143a n_i	R152a n_i	Carbon dioxide n_i	Ammonia n_i
1	$0.10306886 \cdot 10^{+01}$	$0.95702326 \cdot 10^{+00}$	$0.89875108 \cdot 10^{+00}$	$0.73022720 \cdot 10^{+00}$
2	$-0.29497307 \cdot 10^{+01}$	$-0.23707196 \cdot 10^{+01}$	$-0.21281985 \cdot 10^{+01}$	$-0.11879116 \cdot 10^{+01}$
3	$0.69435230 \cdot 10^{+00}$	$0.18748463 \cdot 10^{+00}$	$-0.68190320 \cdot 10^{-01}$	$-0.68319136 \cdot 10^{+00}$
4	$0.71552102 \cdot 10^{-01}$	$0.63800843 \cdot 10^{-01}$	$0.76355306 \cdot 10^{-01}$	$0.40028683 \cdot 10^{-01}$
5	$0.19155982 \cdot 10^{-03}$	$0.16625977 \cdot 10^{-03}$	$0.22053253 \cdot 10^{-03}$	$0.90801215 \cdot 10^{-04}$
6	$0.79764936 \cdot 10^{-01}$	$0.82208165 \cdot 10^{-01}$	$0.41541823 \cdot 10^{+00}$	$-0.56216175 \cdot 10^{-01}$
7	$0.56859424 \cdot 10^{+00}$	$0.57243518 \cdot 10^{+00}$	$0.71335657 \cdot 10^{+00}$	$0.44935601 \cdot 10^{+00}$
8	$-0.90946566 \cdot 10^{-02}$	$0.39476701 \cdot 10^{-02}$	$0.30354234 \cdot 10^{-03}$	$0.29897121 \cdot 10^{-01}$
9	$-0.24199452 \cdot 10^{+00}$	$-0.23848654 \cdot 10^{+00}$	$-0.36643143 \cdot 10^{+00}$	$-0.18181684 \cdot 10^{+00}$
10	$-0.70610813 \cdot 10^{-01}$	$-0.80711618 \cdot 10^{-01}$	$-0.14407781 \cdot 10^{-02}$	$-0.98416660 \cdot 10^{-01}$
11	$-0.75041709 \cdot 10^{-01}$	$-0.73103558 \cdot 10^{-01}$	$-0.89166707 \cdot 10^{-01}$	$-0.55083744 \cdot 10^{-01}$
12	$-0.16411241 \cdot 10^{-01}$	$-0.15538724 \cdot 10^{-01}$	$-0.23699887 \cdot 10^{-01}$	$-0.88983219 \cdot 10^{-02}$

test the transferability of the functional form – again, no significant disadvantages were found for these substances. For 12 of the considered substances Table 6.6 summarises the coefficients n_i of Eq. 6.13. When working with these coefficients, $R = 8.314510 \text{ J mol}^{-1} \text{K}^{-1}$ has to be used for the gas constant. For ethylene the equation of state which uses the simultaneously optimised functional form for nonpolar fluids (see Eq. 6.12 and Table 4.4) yields slightly better results and thus the corresponding coefficients for Eq. 6.13 are not presented here.

Results of Eq. 6.13 were already discussed in the preceding sections, whereby the equation was referred to as the "simultaneously optimised equation of state with 12 terms" when discussing polar fluids. As a very global quality criterion, Fig. 6.6 shows weighted variances which resulted from fits of Eq. 6.13. In general, Eq. 6.13 describes the available data sets with $\sigma_{wt}^2 < 0.3$; the only exceptions, R12 ($\sigma_{wt}^2 = 0.35$), carbon dioxide ($\sigma_{wt}^2 = 0.36$), and ammonia ($\sigma_{wt}^2 = 0.43$), will be discussed below.

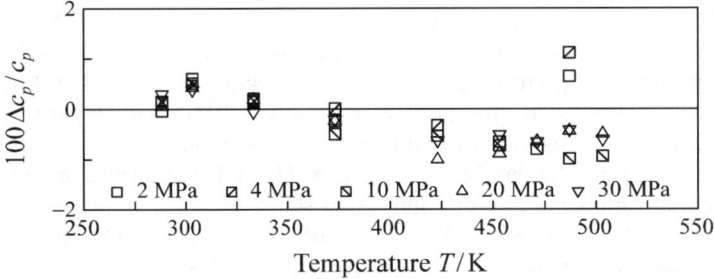

Fig. 6.31. Percentage deviations $100 \, \Delta c_p/c_p = 100 \, (c_{p,\text{exp}} - c_{p,\text{calc}})/c_{p,\text{exp}}$ between experimental data for the isobaric heat capacity of R11 measured by Wirbser et al. (1992) and values calculated from the corresponding simultaneously optimised equation of state, Eq. 6.13.

266   6 Generalised Functional Forms

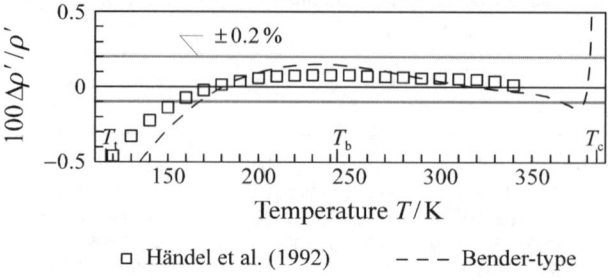

□ Händel et al. (1992)   − − − Bender-type

Fig. 6.32. Percentage deviations $100\,\Delta\rho'/\rho' = 100\,(\rho'_{\text{exp}}-\rho'_{\text{calc}})/\rho'_{\text{exp}}$ between highly accurate experimental data for the saturated liquid density of R12 and values calculated from the corresponding simultaneously optimised equation of state, Eq. 6.13. For comparison, values calculated from a Bender-type equation are plotted as a dashed line.

The data set which is available for the refrigerant *R11* (*trichlorofluoromethane*) covers the temperature range from the triple point temperature to $T_{\max} \approx 595$ K ($T_{\max}/T_{\text{c}} \approx 1.26$); the used data set was restricted to pressures of $p_{\max} = 100$ MPa. However, reliable $p\rho T$ data are available only at temperatures above 254 K ($T_{\text{t}} = 162.7$ K) and only up to $T/T_{\text{c}} \approx 1.01$ and up to pressures of $p = 31$ MPa. At higher temperatures and pressures data are available only for speeds of sound and isobaric heat capacities. Thus, for R11 the available data set is limited especially for thermal properties. With regard to vapour-liquid equilibrium states, reliable experimental results are available for the vapour pressure, for the saturated liquid density, and for the isobaric heat capacity and the speed of sound of the saturated liquid. Reliable data are lacking for the saturated vapour density. Reviews on the available data set were published by Marx et al. (1992) and Jacobsen et al. (1992). The simultaneously optimised equation of state, Eq. 6.13 with the set of coefficients presented in Table 6.6, represents the available data within the demanded uncertainties. Enlarged deviations are observed only for speeds of sound in the saturated liquid, where the limit of $\Delta w/w \leq \pm 2\,\%$ is exceeded for temperatures below ≈ 180 K with maximum deviations of $\Delta w/w \approx -3.5\,\%$. As an example of the representation of isobaric heat capacities, Fig. 6.31 shows deviations between experimental results published by Wirbser et al. (1992) and values calculated from the simultaneously optimised equation of state.

The data set for the refrigerant *R12* (*dichlorodifluoromethane*) covers the temperature range from the triple point temperature to $T_{\max} \approx 475$ K ($T_{\max}/T_{\text{c}} \approx 1.23$); the used data set was restricted to pressures of $p_{\max} = 100$ MPa. With regard to thermal properties, the data situation is satisfactory with shortcomings only in the extended critical region. With the data set published by Händel et al. (1992) highly accurate $p\rho T$ data are available at gaseous and liquid states. Data for caloric properties are available only up to pressures of $p = 6.4$ MPa. At vapour-liquid equilibrium states, reliable data are again available for the vapour pressure, for the saturated liquid density, and for the isobaric heat capacity and the speed of sound of the saturated liquid. At just three temperatures accurate experimental results were published for the saturated vapour density. More detailed reviews on the available

6.2 Simultaneously Optimised Equations for Technical Applications

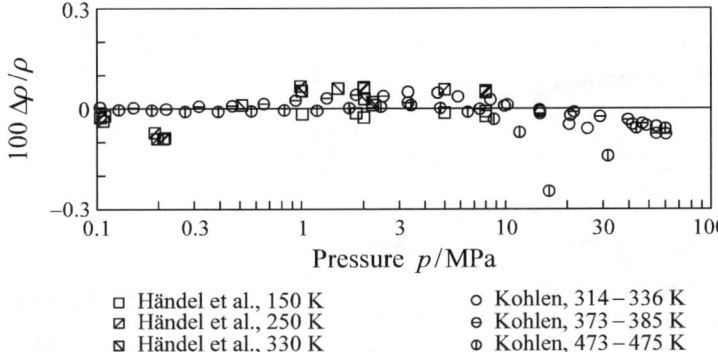

Fig. 6.33. Percentage deviations $100 \, \Delta\rho/\rho = 100 \, (\rho_{\text{exp}} - \rho_{\text{calc}})/\rho_{\text{exp}}$ between accurate experimental data for the density of R22 published by Händel et al. (1992) and Kohlen (1987) and values calculated from the corresponding simultaneously optimised equation of state, Eq. 6.13.

data sets were published by Marx et al. (1992) and Penoncello et al. (1992). A comparison with $p\rho T$ data for R12 has already been shown in Fig. 6.12.

In general, the simultaneously optimised equation of state, Eq. 6.13, represents the available data sets well within the demanded uncertainties. However, among the used polar substances R12 is the one with the smallest reduced triple point temperature, see Table 6.5. Typical problems which have been reported for low temperature liquid states become more pronounced for R12; at very low temperatures deviations which exceed the demanded uncertainties are observed for ρ', c_p', and w'. As an example for this problem, Fig. 6.32 shows deviations between experimental results for the saturated liquid density and values calculated from the simultaneously optimised equation of state, Eq. 6.13. These enlarged deviations result in the above average value of the weighted variance of the substance specific fit, see Fig. 6.6.

The data set which was used for the refrigerant R22 (*chlorodifluoromethane*) covers the temperature range from the triple point temperature to $T_{\text{max}} \approx 524$ K ($T_{\text{max}}/T_c \approx 1.42$) at pressures up to $p_{\text{max}} \approx 85$ MPa. Both with regard to thermal and caloric properties the data situation is clearly better than for the other "old" refrigerants. State-of-the-art data are available especially for the $p\rho T$-relation, for the isobaric heat capacity, for the speed of sound at liquid and gaseous states, and for thermal properties on the phase boundary including some accurate results for the saturated vapour density. Detailed reviews on the available data set were published by Wagner et al. (1993) and Kamei et al. (1995). The simultaneously optimised equation of state for polar fluids, Eq. 6.13, generally represents the available data well within the demanded uncertainty; slightly enlarged deviations are to be observed for liquid states close to the triple point temperature and for $p\rho T$ data at supercritical states close to T_{max} and at $p > \approx 20$ MPa. Figure 6.33 shows deviations between a representative subset of the accurate experimental results for the density of R22 published by Händel et al. (1992) and Kohlen (1987) and values calculated from the simultaneously optimised equation of state, Eq. 6.13.

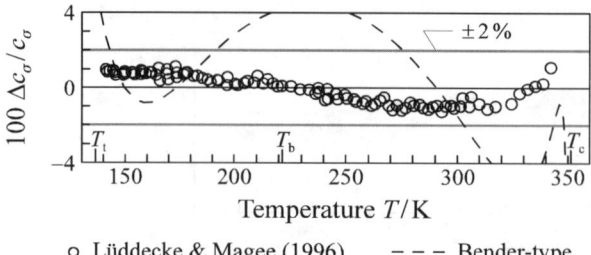

o   Lüddecke & Magee (1996)   – – – Bender-type

Fig. 6.34. Percentage deviations $100\,\Delta c_\sigma/c_\sigma = 100\,(c_{\sigma,\mathrm{exp}} - c_{\sigma,\mathrm{calc}})/c_{\sigma,\mathrm{exp}}$ between recent experimental results for the heat capacity along the saturated liquid line of R32 and values calculated from the corresponding simultaneously optimised equation of state, Eq. 6.13. For comparison, values calculated from a Bender-type equation are plotted as a dashed line.

The data set for the refrigerant *R32* (*trifluoromethane*) covers the temperature range from the triple point temperature to $T_{\max} \approx 420$ K ($T_{\max}/T_c \approx 1.20$) at pressures up to $p_{\max} \approx 72$ MPa. The investigation of the thermodynamic properties of the "alternative", non ozone depleting refrigerant R32 started only in the early 1990s. The data set which is available today is smaller than the data sets for refrigerants such as R12 or R22, but most of the data were measured with modern experimental equipment and thus the data set is more consistent than average data sets for the "old" refrigerants – the number of "selected" data is larger than that for R22. Basically, similar situations were found for all of the "new" refrigerants. Data for thermal properties are available both at homogeneous and at vapour-liquid equilibrium states, with restrictions for the supercritical fluid at medium densities. Even for the saturated vapour density reasonable data are available from about the normal boiling temperature to the critical temperature. Accurate experimental results for caloric properties are available essentially for liquid states. A detailed review of the available data sets was published by Tillner-Roth and Yokozeki (1997). In general, the simultaneously optimised equation of state represents the available data within the demanded uncertainties; slightly enlarged deviations up to $\Delta p/p \approx \pm 0.3\,\%$ are to be observed for states in the extended critical region close to $T_{\max}$, where the data situation is scarce. As an example for the representation of caloric properties at liquid states, Fig. 6.34 shows deviations between experimental results for the heat capacity along the saturated liquid line, c_σ, and values calculated from the simultaneously optimised equation of state.

For the refrigerant *R113* (*1,1,2-trichlorotrifluoroethane*), only a very limited data set is available which covers the temperature range from the triple point temperature to $T_{\max} \approx 538$ K ($T_{\max}/T_c \approx 1.10$) at pressures up to $p_{\max} \approx 30$ MPa. Thermal properties of the homogeneous and saturated liquid are described with sufficient accuracy, but at gaseous and supercritical states the data situation is unsatisfactory. The representation of caloric properties is based mainly on experimental results for the isobaric heat capacity at liquid and gaseous states. No data for the saturated vapour density and for caloric properties on the vapour-liquid phase boundary could be incorporated in the set of selected experimental data. A detailed

6.2 Simultaneously Optimised Equations for Technical Applications

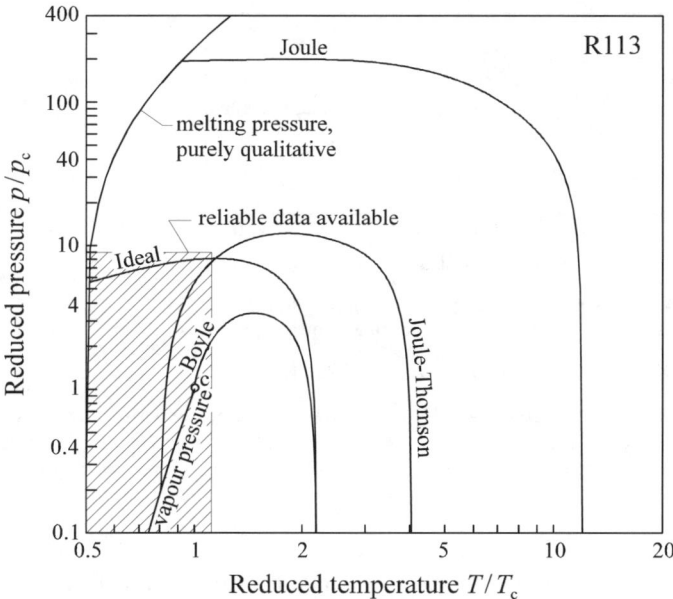

Fig. 6.35. Ideal curves of R113 calculated from the corresponding simultaneously optimised equation of state, Eq. 6.13. The hatched area corresponds to the region where reliable data were available to fit the equation.

review of the available data set was given by Marx et al. (1992). The simultaneously optimised equation of state, Eq. 6.13, describes the available data well within the demanded uncertainties.

Among the polar fluids considered by Span and Wagner (2000a/c), the data set for R113 is the most restricted one of all. Therefore, it is appropriate to highlight the extrapolation behaviour of the new equations of state for polar fluids using R113 as an example. Just like the simultaneously optimised equations for nonpolar fluids, the equations for polar fluids are numerically very stable. However, plots of the coefficients as they were given in Figs. 6.3 and 6.9 cannot prove this thesis since a proper alignment of the considered substances becomes difficult – the similarity of polar fluids cannot be expressed in terms of a single order parameter like the acentric factor ω. Under these circumstances, the plot of the ideal curves calculated from the simultaneously optimised equations becomes the most expressive test of reasonable extrapolation behaviour, see also Sects. 4.6.3. and 6.2.1.1. Figure 6.35 shows ideal curves which were calculated from the simultaneously optimised equation of state for R113. Similar to the functional form for nonpolar fluids, the functional form introduced for polar fluids results in reasonable plots of the ideal curves even when being fitted to restricted data sets. The group 2 reference equation for R113 by Marx et al. (1992) yields similar results for the Boyle, Ideal, and Joule-Thomson inversion curves, but it cannot predict a reasonable plot of the Joule inversion curve.

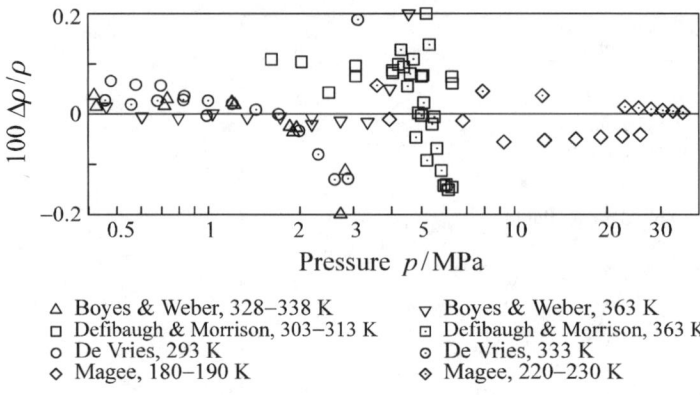

△ Boyes & Weber, 328–338 K
□ Defibaugh & Morrison, 303–313 K
○ De Vries, 293 K
◇ Magee, 180–190 K

▽ Boyes & Weber, 363 K
▣ Defibaugh & Morrison, 363 K
◉ De Vries, 333 K
◈ Magee, 220–230 K

Fig. 6.36. Percentage deviations $100 \, \Delta\rho/\rho = 100 \, (\rho_{\mathrm{exp}} - \rho_{\mathrm{calc}})/\rho_{\mathrm{exp}}$ between experimental results for the density of R125 by Boyes and Weber (1995), Defibaugh and Morrison (1992), de Vries (1997), and Magee (1996) and values calculated from the corresponding simultaneously optimised equation of state, Eq. 6.13.

At pressures up to $p_{\max} \approx 76$ MPa, the data set which is available for the refrigerant *R123 (2,2-dichloro-1,1,1-trifluoroethane)* covers the temperature range from the triple point temperature to $T_{\max} \approx 523$ K ($T_{\max}/T_{\mathrm{c}} \approx 1.14$); however, at pressures of $p > \approx 39$ MPa experimental results are available only for the speed of sound at liquid states. At gaseous and liquid states a sufficient number of $p\rho T$ data is available, but significant inconsistencies between the different data sets are to be observed. At supercritical temperatures, the data situation becomes questionable even for the development of a technical equation of state. Accurate experimental data are available for the vapour pressure and the saturated liquid density; data sets of limited accuracy have been published for the saturated vapour density. Reliable caloric data are essentially available only for liquid states. A recent review on the available set of experimental data was published by Younglove and McLinden (1994). In general, the simultaneously optimised equation of state represents the available data within the demanded uncertainties. Enlarged deviations are observed only for the heat capacity along the saturated liquid line, c_σ, at temperatures close to the triple point temperature. The limit of $\Delta c_\sigma/c_\sigma \leq \pm 2\%$ is exceeded for $T < \approx 185$ K with deviations increasing to $\Delta c_\sigma/c_\sigma \approx -4\%$ at $T_{\mathrm{t}} \approx 166$ K.

The data set which is available for the refrigerant *R125 (pentafluoroethane)* covers the temperature range from the triple point temperature to $T_{\max} \approx 400$ K ($T_{\max}/T_{\mathrm{c}} \approx 1.18$) at pressures up to $p_{\max} \approx 68$ MPa. Although all of the data result from recent experimental investigations, significant inconsistencies are observed among the available data sets especially for thermal properties. In the extended critical region, the data situation is scarce. At vapour-liquid equilibrium states, accurate data are available for the vapour pressure, the saturated liquid density, the saturated liquid heat capacity, and the speed of sound in the saturated liquid and vapour phase at temperatures of $T > 293$ K ($T/T_{\mathrm{c}} > 0.86$). Reliable experimental results for the saturated vapour density are lacking. At homogeneous states, accu-

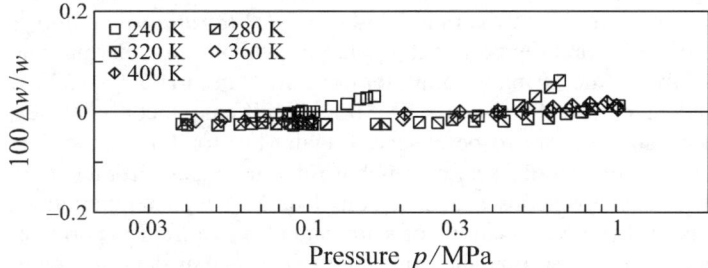

Fig. 6.37. Percentage deviations $100\,\Delta w/w = 100\,(w_{\text{exp}} - w_{\text{calc}})/w_{\text{exp}}$ between experimental data for the speed of sound in R125 by Gillis (1997) and values calculated from the corresponding simultaneously optimised equation of state, Eq. 6.13.

rate caloric data are available for both the gas and liquid phase. A review of the available data set has been published recently by Piao and Noguchi (1998). The simultaneously optimised equation of state, Eq. 6.13, represents the available data well within the demanded uncertainties.

In Chap. 5, data for R125 were compared with results from the equation by Piao and Noguchi (1998) as an example for the performance of subaverage group 2 reference equations of state, see Figs. 5.7 and 5.20. For the same data sets, Figs. 6.36 and 6.37 show deviations between densities and gas phase speeds of sound and values calculated from the simultaneously optimised equation of state for R125. In both cases, the simultaneously optimised equation of state is superior to the reference equation of state – although this new generation of equations of state was clearly designed for technical applications it compares well with typical group 2 reference equations in many cases.

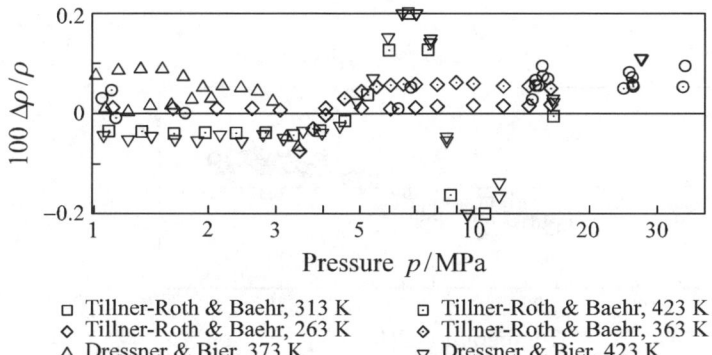

Fig. 6.38. Percentage deviations $100\,\Delta\rho/\rho = 100\,(\rho_{\text{exp}} - \rho_{\text{calc}})/\rho_{\text{exp}}$ between experimental results for the density of R134a by Tillner-Roth and Baehr (1992/1993), Dressner and Bier (1993) and Klomfar et al. (1993) and values calculated from the corresponding simultaneously optimised equation of state, Eq. 6.13.

Among the halogenated methanes and ethanes the data set which is available for the thermodynamic properties of refrigerant *R134a* (*1,1,1,2-tetrafluoroethane*) is the most extensive one. Accurate experimental data are available for thermal and caloric properties from the triple point temperature up to $T_{max} \approx 523$ K ($T_{max}/T_c \approx 1.40$) and at pressures to $p_{max} \approx 75$ MPa. Significant inconsistencies, even between recent data sets, are to be observed again, but for R134a the data situation is good enough to identify sets of reliable reference data. Problems are caused only by obviously erroneous speed of sound data at high temperatures, where similar results of different equations of state cannot be verified experimentally. On the vapour-liquid phase boundary accurate experimental data are available for the vapour pressure, for the saturated liquid density and for the heat capacity of the saturated liquid; data for the saturated vapour density are available, but the accuracy of these data is unsatisfactory. The simultaneously optimised equation of state represents the available data in general within the demanded uncertainties. In Sect. 5.1, $p\rho T$ data for R134a were used to illustrate the performance of an above average group 2 reference equation of state, see Fig. 5.6. Figure 6.38 compares the same set of experimental data with the simultaneously optimised equation of state for R134a. The reference equation by Tillner-Roth and Baehr (1994) yields better results, especially in the range of medium densities at supercritical states, but the results of the simultaneously optimised equation again come close to the results of average group 2 reference equations of state.

The data set for the refrigerant *R143a* (*1,1,1-trifluoroethane*) covers the temperature range from the triple point temperature to $T_{max} \approx 433$ K ($T_{max}/T_c \approx 1.25$) at pressures up to $p_{max} \approx 35$ MPa. Within these limits a sufficient number of accurate $p\rho T$ data is available for all homogeneous states and reliable experimental results for the vapour pressure and the saturated liquid density are available at least for temperatures above about 200 K and 245 K, respectively. No sufficiently accurate data are available for the saturated vapour density. Significant inconsistencies can again be observed between different sets of recent $p\rho T$ data, but based on the demands on the accuracy of technical equations of state these inconsistencies can

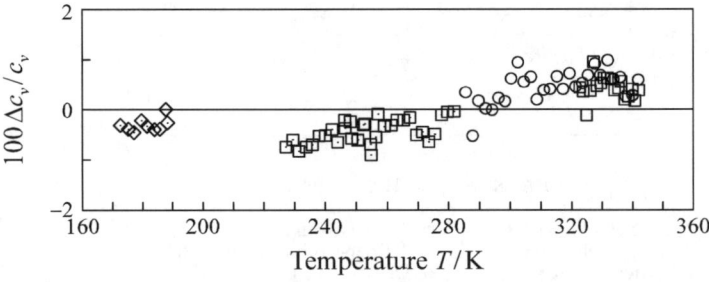

Fig. 6.39. Percentage deviations $100\,\Delta c_v/c_v = 100\,(c_{v,\mathrm{exp}} - c_{v,\mathrm{calc}})/c_{v,\mathrm{exp}}$ between experimental results for the isobaric heat capacity of R143a published by Magee (1998) and values calculated from the corresponding simultaneously optimised equation of state, Eq. 6.13.

be tolerated. State-of-the-art experimental results for caloric properties are published for gaseous and liquid states. A review of the data sets was given by Li et al. (1997). The simultaneously optimised equation of state for R143a represents the available data within the demanded uncertainties. As an example of the representation of isochoric heat capacities at liquid states, Fig. 6.39 shows deviations between accurate experimental results published by Magee (1996) and values calculated from the simultaneously optimised equation.

The data set which is available for thermal properties of the refrigerant *R152a* (*1,1-difluoroethane*) covers the range from the triple point temperature to $T_{\max} \approx 471$ K ($T_{\max}/T_c \approx 1.22$) at pressures up to $p_{\max} \approx 58$ MPa. Reliable experimental results are available for all homogeneous states, for the vapour pressure, and for the saturated liquid density. Data of questionable accuracy have been published for the saturated vapour density at temperatures above ≈ 310 K. However, accurate experimental information on caloric properties of R152a is scarce; gas

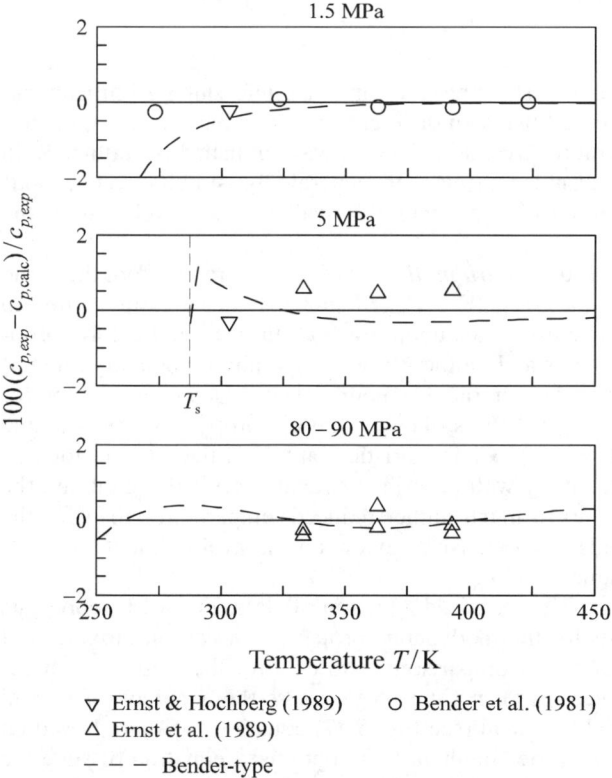

▽ Ernst & Hochberg (1989)   ○ Bender et al. (1981)
△ Ernst et al. (1989)
– – – Bender-type

Fig. 6.40. Percentage deviations between experimental data for the isobaric heat capacity and values calculated from the simultaneously optimised equation of state for carbon dioxide, Eq. 6.13. Values calculated from a Bender-type equation are plotted as dashed lines.

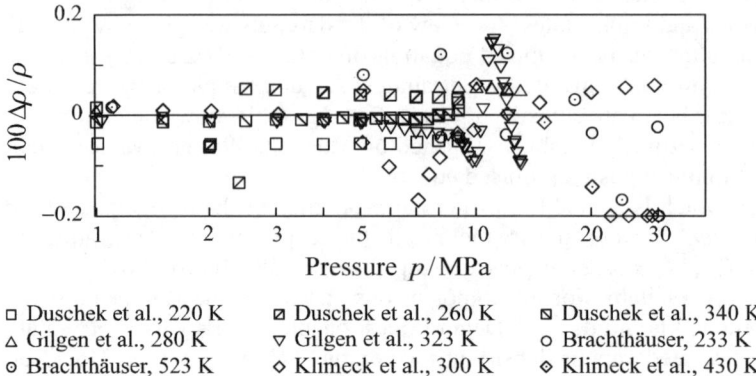

Fig. 6.41. Percentage deviations $100 \, \Delta\rho/\rho = 100 \, (\rho_{\text{exp}} - \rho_{\text{calc}})/\rho_{\text{exp}}$ between experimental results for the density of carbon dioxide by Duschek et al. (1990), Gilgen et al. (1992), Brachthäuser (1993), and Klimeck et al. (2000) and values calculated from the corresponding simultaneously optimised equation of state, Eq. 6.13.

phase data are restricted to very low pressures and at liquid states significant inconsistencies are to be observed between different sets of isobaric heat capacities. More detailed information on the available data set was published by Tillner-Roth (1995) and Outcalt and McLinden (1996). In general, the simultaneously optimised equation of state for R152a describes the available data well within the demanded uncertainties.

The data set which was used for *carbon dioxide* covers the range from the triple point temperature to $T_{\text{max}} \approx 1024$ K ($T_{\text{max}}/T_c \approx 3.38$) at pressures up to $p_{\text{max}} = 100$ MPa. Accurate results are available for both thermal and caloric properties for all homogeneous states and on the vapour-liquid phase boundary, including highly accurate data for the saturated vapour density. However, except for isobaric heat capacities at low pressures, data for caloric properties are available only up to $T \approx 393$ K ($T/T_c \approx 1.29$). Characteristic features of the data set for carbon dioxide are an experimentally well described extended critical region and the fact that data are available up to much higher reduced temperatures than for the other considered polar fluids. An extensive review on the available data set was published by Span and Wagner (1996).

In general, the simultaneously optimised equation of state, Eq. 6.13, represents reliable experimental results for thermodynamic properties of carbon dioxide well with the demanded uncertainties. Comparisons with data for the isobaric heat capacity of carbon dioxide were used in Chap. 5 to assess the accuracy of typical reference (see Fig. 5.26) and technical (see Fig. 5.27) equations of state. Based on the same data set the results of the simultaneously optimised equation of state are illustrated in Fig. 6.40. Although designed as a technical equation of state, Eq. 6.13 describes the data at least as accurately as the group 2 reference equations of state shown in Fig. 5.26. Up to temperatures of about 400 K, the same is true for the representation of experimental results for the density of carbon dioxide, see

Fig. 6.41 for the simultaneously optimised equation of state, Fig. 5.5 for typical reference equations of state, and Fig. 5.11 for typical technical equations of state.

At temperatures above $T \approx 400$ K and pressures above $p \approx 20$ MPa increased deviations are observed between experimental and calculated results for the density, see the data of Klimeck et al. (2000) at 430 K and of Brachthäuser (1993) at 523 K. Up to $T \approx 600$ K the deviations stay within $\Delta\rho/\rho \leq \approx \pm 0.4\%$, but at higher temperatures they increase even up to $\Delta\rho/\rho \approx \pm 1\%$. The reason for these enlarged deviations was already discussed in Sect. 6.2 – without additional experimental results for thermodynamic properties of other polar fluids at high reduced temperatures, improvements regarding this problem are not to be expected.

The data set for the refrigerant *ammonia* covers the temperature range from the triple point temperature to $T_{\max} \approx 574$ K ($T_{\max}/T_c \approx 1.42$) at pressures up to $p_{\max} \approx 85$ MPa. For thermal properties, the available data set is satisfactory in general, but experimental results for caloric properties are available only at pressures up to $p \approx 2$ MPa. A brief review of the available data set was published recently by Tillner-Roth et al. (1993). In general, the simultaneously optimised equation of state represents the available data within the demanded uncertainties up to $T \approx 430$ K ($T/T_c \approx 1.06$). At higher temperatures and pressures above ≈ 10 MPa, enlarged deviations on the order of $\Delta\rho/\rho \approx \pm 0.5\%$ are to be observed between calculated densities and the $p\rho T$ data published by Garnjost (1974). Keeping in mind that the properties of ammonia are clearly influenced by association, this result shows both the broad applicability of the simultaneously optimised functional form for polar fluids and its limitations; as already discussed, properties of the strongly associating fluid water could not be described accurately with equations of state which use this functional form.

6.3 Simultaneously Optimised Reference Equations of State

As shown above, the technical equations of state with simultaneously optimised functional forms presented by Span and Wagner (2000a–c) come close to the performance of typical group 2 reference equations of state in most cases. However, the objectives of this project clearly aimed at demands on technical equations of state – the transferability of the functional form was regarded as more important than improved accuracy, the considered pressure range was restricted to technically relevant pressures, and the equations were not constrained to preselected critical data.

When working on reference equations of state, advantages from simultaneously optimised functional forms are expected due to a slightly different approach. A transfer of functional forms is not intended for reference equations of state. However, most of the available group 2 reference equations of state suffer from the fact that they are based on data sets which are restricted either with regard to the covered temperature and pressure range or with regard to the availability of data in certain regions; missing experimental results for thermal and caloric properties at gaseous states or for caloric properties at liquid states are typical problems. In such cases, it can be advantageous to optimise the functional form considering both data for the substance under investigation and data for one or more thermodynamically similar substances with better data sets. In this way, the resulting functional form is constrained to information from better data sets in regions where the data set of the substance under investigation has certain shortcomings. Better predictive capabilities can be expected for the resulting equation of state.

This approach has been tested by Klimeck et al. (1999a) recently. As pure component basis for a thermodynamic property model for natural gases group 2 reference equations with 23 terms were developed for methane, nitrogen, and ethane using the simultaneous optimisation algorithm described in Sect. 6.1. As a result, especially the extrapolation behaviour of the ethane equation could be improved. Since reduced temperatures T/T_c for natural gas like mixtures may very well exceed the value $T_{max}/T_c \approx 2.04$ found for ethane[14] in technical applications, a reasonable extrapolation behaviour becomes an essential feature for pure component equations of state which are to be incorporated into accurate mixture approaches.

A long-term project has been started between the *Thermophysics Division* of the *National Institute of Standards and Technology* in Boulder, Colorado, and the *Lehrstuhl für Thermodynamik* of the *Ruhr-Universität Bochum* in Bochum intended to develop group 1 reference equations for ethane and propane which will be based on new experimental results for these substances. As a by-product, improved group 2 reference equations of state shall be developed for n-butane and isobutane by simultaneous optimisation considering the data sets of all four substances – the simultaneous optimisation approach will have an impact on the future development of reference equations of state as well, especially with regard to the description of substances with limited data sets.

---

[14] See Sect. 6.2.2 for a description of the data set which is available for ethane and Sect. 8.3 for a description of the used mixture approaches.

7 Generalised Equations of State

In the preceding section it was shown that state-of-the-art technical equations of state with simultaneously optimised functional form are numerically very stable and that they have surprising predictive capabilities. Such equations disprove common teachings which say that empirical equations of state should only be used in regions where they were fitted to a sufficiently dense set of accurate experimental data. However, even equations of state with simultaneously optimised functional form are bound to fail when fitted either to extremely small data sets or to small data sets with large inconsistencies. Thus, thermodynamic property models with further restricted numerical flexibility and with increased predictive capabilities are still needed to describe the broad variety of substances which is relevant for instance for applications in the chemical or petrochemical industry.[1] An immeasurable multitude of models and equations of state has been developed either to fulfil the corresponding technical demands, or, more often, to advance the scientific search for the physically "true" description of the thermodynamic properties of fluids.

For a brief overview, the generalised equations of state for pure substances which are in use today may be subdivided into five groups:

- *Cubic equations of state* are direct descendants of the famous van der Waals equation of state, which was able to correctly describe the typical behaviour of fluids for the very first time. Classical representatives of this group are the well known equations of state by Redlich and Kwong (1949), Soave (1972), and Peng and Robinson (1976). Information on at least some of these equations can be found in textbooks on thermodynamics today, see, e.g., Baehr (1992) or Reid et al. (1987). However, the work on cubic equations of state continues. Examples for recent developments can be found in the articles by Trebble and Bishnoi (1987), Schwarzentruber et al. (1989), Zhang et al. (1997), Ji and Lempe (1997/1998), and Nagel and Bier (1998). In particular, the introduction of "volume translations" has improved cubic equations of state with respect to some of their traditional shortcomings, but it has also sacrificed their simplicity in part.
- *Semiempiric equations of state* have tried to improve the description of the contribution of repulsive forces between molecules which are overly simplified in cubic equations. In most cases, these forces were modelled by some kind of hard sphere term. A well known example for this kind of equation of state is

---

[1] The need for generalised relations becomes even more pressing when considering mixtures, see Chap. 8.

the formulation by Carnahan and Starling (1969). Recent work focuses on an improved description of the effects of attractive forces as well. Articles like those of Haar and Shenker (1971), Boublik (1975), Boublik and Nezbeda (1977), and Kohler and Haar (1981) were formulated in a very theoretical way – the corresponding equations of state were used mainly by members of a small scientific community. Only the work of Deiters (1981a/b/1982) resulted in a certain acceptance of advanced semiempiric equations of state in technical applications. A recent and promising development in the tradition of semiempiric equations of state is the "*BACKONE* family of equations of state", see Saager et al. (1992), Saager and Fischer (1992), Müller et al. (1996), and Calero et al. (1998).

- *Generalised empirical equations* are based on an extended corresponding states approach in general. The acentric factor ω and / or similar parameters are either used to interpolate between properties of reference fluids, see Lee and Kessler (1975) and Plöcker et al. (1978), or to calculate values of coefficients $n_i(\omega,...)$, see Starling (1973), Yamada (1973), Brule et al. (1982), Platzer and Maurer (1989), and Soave (1995).

- *Shape factor models* are based on empirical equations for reference fluids and corresponding states approaches as well. However, the formulations which describe similarity – the shape functions – are much more complex in these models; yet no general temperature and density dependent formulation for these functions has been found. The work on shape factor models was initiated by Leach et al. (1968) and Fisher and Leland (1970) and was continued by the group around Ely for a long time, see Ely (1990). Recent developments were reported by Huber and Ely (1994) and Estela-Uribe and Trusler (1998).

- *Group contribution methods* are common especially to describe phase equilibria of mixtures with complex compounds. However, more recently the use of group contribution based cubic equations of state has become common besides the well known excess-free-enthalpy (g^E) models; for an overview see Gmehling et al. (1993) and Gmehling (1995). For the first time, Tillner-Roth (1998) discussed a group contribution concept which is based on accurate multiparameter equations of state. However, a broad application of this concept is yet not in sight.[2]

Besides this multitude of generalised models for the calculation of thermodynamic properties, a variety of articles have been published which assess the performance

---

[2] The numerical efforts which are caused by multiparameter equations of state are justified only if the results are significantly more accurate than those from simple equations of state. This is true for the use of multiparameter equations in group contribution approaches as well. Extensive sets of accurate data or accurate equations of state are required for a broad variety of substances to identify group contributions for different characteristic groups with the required accuracy – this necessary basis is lacking today. A significant extension of the group of described pure substances, which is stil restricted to n-alkanes, is not to be expected until a suitable foundation has been set up by long lasting systematic efforts. However, at such time it will be interesting to see whether the simplifications involved by group contribution approaches allow a sufficiently accurate description of thermodynamic properties. If the corresponding problems can be solved, the long term potential of such approaches is tremendous especially for the description of mixtures.

of different approaches more or less systematically, see e.g. Toledo and Reich (1988), Sharif and Groves (1990), Wong et al. (1990), Ye et al. (1992), Garipis and Stamatoudis (1992), Karkaris et al. (1992), and Hemptinne and Ungerer (1995). The results of these comparisons coincide perfectly with the author's experiences – the uncertainties of generalised equations of state and thermodynamic property models exceed the demands formulated for advanced technical equations of state in Sect. 6.2 far, in most cases by more than one order of magnitude. Correspondingly, comparisons to the data sets discussed in Chap. 6 result usually in weighted variances $\sigma_{wt}^2 \geq \approx 30-100$.

Since this work focuses on accurate empirical equations of state, the intention of this section is not to discuss all of the approaches mentioned above in detail; to do so would require an additional volume. Instead of this, three approaches are described which belong to the most accurate generalised models, which involve empirical multiparameter equations, and which are regarded as promising starting points for the development of more accurate[3] future models.

7.1 BACKONE Equations of State

The general set-up of BACKONE equations of state follows the set-up which was given for equations of state with hard sphere terms in Eq. 3.35, with the only exception that the residual contribution α^{r*} which describes the effect of attractive intermolecular forces is subdivided even further for polar fluids. For nonpolar fluids the SIMBACKONE equation of state can be written as

$$\frac{a(T,\rho)}{RT} = \frac{a^\circ(T,\rho) + a^h(T,\rho) + a^a(T,\rho)}{RT} = \alpha^\circ(\tau,\delta) + \alpha^h(\tau,\delta) + \alpha^a(\tau,\delta), \quad (7.1)$$

with $\tau = T_0/T$ and $\delta = \rho/\rho_0$, where T_0 and ρ_0 are the corresponding parameters of the hard sphere term. The repulsive term α^h corresponds to the contribution of the hard sphere term presented in Sect. 3.1.3.1, see Eq. 3.37. The term α^a describes the attractive dispersion forces between nonpolar molecules.

The functional form used for α^a was established by Müller et al. (1996) using data for methane, oxygen, and ethane. Data for all three substances could be used simultaneously due to a combination of a typical empirical multiparameter form with 28 terms with the substance specific anisotropy parameter φ, which is used in the hard sphere term as well. The resulting form reads

$$\alpha^a(\tau,\delta) = \sum_{i=1}^{19} c_i^a \tau^{n_i^a/2} \delta^{m_i^a} \varphi^{j_i} + \sum_{i=20}^{28} c_i^a \tau^{n_i^a/2} \delta^{m_i^a} \varphi^{j_i} \exp(-\delta^{l_i^a}) \quad (7.2)$$

---

[3] Based on recent knowledege, it is not to be expected that generalised equations of state can strictly fulfil the demands which were given in Table 6.1, except for very simple substances. When working on more accurate future models, it is useful to strive for formulations which fulfil the corresponding demands for most states and which do not fail completely in regions which are difficult to describe. Such equations would result in typical weighted variances of $\sigma_{wt}^2 \approx 1$, see also the discussion in Chap. 6.

and was established using the optimisation algorithm by Setzmann and Wagner (1989). The required values for c_i^a, j_i, l_i^a, m_i^a, and n_i^a were given by Müller et al. (1996) and do not depend on the considered substance. Thus, the use of $\alpha^a(\tau,\delta)$ does not increase the number of substance specific parameters; just as in the hard sphere terms the adjustable parameters are T_0, ρ_0, and φ. To establish Eq. 7.2 ppT data, linearised Maxwell data (see Sect. 4.3.2), enthalpies of evaporation, isochoric heat capacities, and data for the second virial coefficient were used. Data for properties which result in nonlinear residua were avoided.

For polar substances, Eq. 7.1 is extended by an additional term which considers the effects of the corresponding polar interactions. For dipolar substances the so called DIBACKONE equation becomes

$$\alpha(\tau,\delta) = \alpha^o(\tau,\delta) + \alpha^h(\tau,\delta) + \alpha^a(\tau,\delta) + \alpha^d(\tau,\delta). \tag{7.3}$$

Both the functional form and the not substance dependent coefficients of the dipolar contribution α^d were determined by Saager and Fischer (1992) based on fits to molecular simulation results by Saager et al. (1991). The functional form of α^d reads

$$\alpha^d(\tau,\delta) = \sum_{i=11}^{28} c_i^d \hat{\tau}^{n_i^d/2} \delta^{m_i^d/2} \left(\mu^{*2}\right)^{k_i^d/4} + \sum_{i=1}^{10} c_i^d \hat{\tau}^{n_i^d/2} \delta^{m_i^d/2} \left(\mu^{*2}\right)^{k_i^d/4} \exp\left(-\delta^{l_i^d}\right), \tag{7.4}$$

with the reduced dipole moment μ^{*2} as the only substance specific parameter. For the polar contribution Saager and Fischer (1992) replaced the inverse reduced temperature τ by $\hat{\tau} = 1.15\,\tau$. More recently, Müller et al. (1996) used $\hat{\tau} = 1.13\,\tau$ without changing the coefficients c_i of Eqs. 7.4 and 7.6; for the results presented in Sect. 7.1.2 this change was adopted. The required values for c_i^d, k_i^d, l_i^d, m_i^d, and n_i^d are given by Saager and Fischer (1992) or Müller et al. (1996). The use of non-integer exponents $m_i/2$ is unsatisfactory from a theoretical point of view. The smallest uneven value of m_i is $m_i = 3$ and thus in a Z,ρ-diagram the DIBACKONE equation results in an infinite curvature of isotherms in the limit of vanishing density and it cannot be evaluated for third or higher virial coefficients, see also Sect. 4.4.1.

For quadrupolar fluids the so called QUABACKONE equation becomes

$$\alpha(\tau,\delta) = \alpha^o(\tau,\delta) + \alpha^h(\tau,\delta) + \alpha^a(\tau,\delta) + \alpha^q(\tau,\delta). \tag{7.5}$$

The functional form and the coefficients of the quadrupolar contribution α^q were determined by Saager and Fischer (1992) again, based on molecular dynamic simulations published in the same article. The functional form of α^q reads

$$\alpha^q(\tau,\delta) = \sum_{i=7}^{17} c_i^q \hat{\tau}^{n_i^q/2} \delta^{m_i^q/2} \left(Q^{*2}\right)^{k_i^q/4} + \sum_{i=1}^{6} c_i^q \hat{\tau}^{n_i^q/2} \delta^{m_i^q/2} \left(Q^{*2}\right)^{k_i^q/4} \exp\left(-\delta^2\right), \tag{7.6}$$

with the reduced quadrupol moment Q^{*2} as the only substance specific parameter. The required values for c_i^q, k_i^q, m_i^q, and n_i^q are given by Saager and Fischer (1992) and Müller et al. (1996) again. The use of non-integer exponents $m_i/2$ with a smallest uneven value of $m_i = 5$ allows calculations of third but not of fourth and higher virial coefficients and is still unsatisfactory from a theoretical point of view.

Calero et al. (1998) found that most of the H-FKW and H-FCKW refrigerants which are either used or discussed today are not represented satisfactorily either by Eq. 7.3 or by Eq. 7.5. Based on this observation the hypothesis was formulated, that the interactions between the molecules of these substances are influenced both by dipolar and quadrupolar moments. Since no explicit formulation for the consideration of such mixed interactions is available yet, it was assumed that the extended formulation

$$\alpha(\tau,\delta) = \alpha^{\text{o}}(\tau,\delta) + \alpha^{\text{h}}(\tau,\delta) + \alpha^{\text{a}}(\tau,\delta) + \alpha^{\text{d}}(\tau,\delta) + \alpha^{\text{q}}(\tau,\delta) \tag{7.7}$$

yields a good approximation for these effects. This D+QBACKONE called five parameter equation could be used to describe the properties of eight refrigerants with some success.

BACKONE equations of state can be used like typical multiparameter equations of state in terms of the reduced Helmholtz energy if the corresponding algorithms are designed for the use of hard sphere terms, see Sect. 3.1.3.1, and non-integer density exponents. Based on the generalised coefficients c_i^{a}, c_i^{d}, and c_i^{q} and on the substance specific parameters φ, μ^{*2}, and Q^{*2} BACKONE equations can be reformulated as

$$\alpha(\tau,\delta) = \alpha^{\text{o}}(\tau,\delta) + \alpha^{\text{h}}(\tau,\delta) + \sum_{j=1}^{J_{\text{pol}}} c_j \, \tau^{t_j} \, \delta^{d_j} + \sum_{j=J_{\text{pol}}+1}^{J_{\text{pol}}+J_{\text{exp}}} c_j \, \tau^{t_j} \, \delta^{d_j} \exp(-\delta^{l_j}) \tag{7.8}$$

with
$$c_j = c_i^{\text{a}} \, \varphi^{j_i},$$

$$c_j = c_i^{\text{d}} \, 1.13^{n_i^{\text{d}}/2} \left(\mu^{*2}\right)^{k_i^{\text{d}}/4},$$

and
$$c_j = c_i^{\text{q}} \, 1.13^{n_i^{\text{q}}/2} \left(Q^{*2}\right)^{k_i^{\text{q}}/4},$$

respectively. Written in this form, BACKONE equations correspond to multiparameter equations of state with a hard sphere term and 28 polynomial and exponential terms for nonpolar fluids, with 56 polynomial and exponential terms for dipolar fluids, with 45 polynomial and exponential terms for quadrupolar fluids, and with 73 polynomial and exponential terms for fluids where both polar contributions have to be considered. Thus, BACKONE equations are in fact rather complex equations of state, but they require only 3 to 5 substance specific parameters. These parameters have been determined successfully by fitting equations just to some vapour pressure and saturated liquid density data, see Müller et al. (1996).

7.1.1 Fitting BACKONE Equations of State to Data

Compared to multiparameter equations of state, a typical disadvantage of semi-empiric and generalised equations of state is that the derivatives of α^{r} (and of the derivatives of α^r required for the calculation of thermodynamic properties) with respect to the adjustable parameters become nonlinear. This is true for BACKONE equations of state as well. A systematic description of the methods which are used to fit equations of state to data was given in Sects. 4.1 and 4.3. This section gives just a simple example to illustrate the complications caused by the set-up of the BACKONE equation. To deduce all the required residua and derivatives would go beyond the scope of this section.

For a typical multiparameter equation of state the residuum of a $p\rho T$ data point was given as

$$\zeta_{p\rho T} = \left(\frac{p-\rho RT}{\rho^2 RT}\right) - \sum_{i=1}^{I} n_i \left(\frac{1}{\rho_r}\left(\frac{\partial A_i}{\partial \delta}\right)_\tau\right), \qquad (7.9)$$

in Sect. 4.3.1, see also Eq. 4.45. With this residuum, the derivatives $(\partial \zeta_{p\rho T}/\partial n_j)_{n_i \neq n_j,\tau,\delta}$ which are required to determine the coefficients n_i become

$$\left(\frac{\partial \zeta_{p\rho T}}{\partial n_j}\right)_{n_i \neq n_j,\tau,\delta} = -\frac{1}{\rho_r} d_j \, \delta^{d_j-1} \, \tau^{t_j}, \qquad (7.10)$$

if A_j corresponds to a polynomial term and

$$\left(\frac{\partial \zeta_{p\rho T}}{\partial n_j}\right)_{n_i \neq n_j,\tau,\delta} = -\frac{1}{\rho_r}\left(d_j - \gamma_j p_j \delta^{p_j}\right)\delta^{d_j-1} \tau^{t_j} \exp\left(-\gamma_j \delta^{p_j}\right), \qquad (7.11)$$

if A_j corresponds to an exponential term; for the underlying functional form see e.g. Eq. 3.25. These derivatives are completely decoupled, since they contain neither n_j itself nor any other of the fitted coefficients n_i, and can be used both in linear and nonlinear fits. Starting values for the coefficients n_j, which are required in nonlinear fits, can be determined in preceding linear fits.

When fitting parameters of BACKONE equations of state, Eq. 7.9 has to be rewritten as

$$\zeta_{p\rho T} = \left(\frac{p-\rho RT}{\rho^2 RT}\right) - \left(\frac{\partial \alpha^h}{\partial \rho}\right)_T - \left(\frac{\partial \alpha^a}{\partial \rho}\right)_T - \left(\frac{\partial \alpha^d}{\partial \rho}\right)_T - \left(\frac{\partial \alpha^q}{\partial \rho}\right)_T, \qquad (7.12)$$

whereby the polar contributions are optional, see above. Derivatives with respect to reduced variables are avoided in Eq. 7.12. When fitting the reducing parameters ρ_0 and T_0 derivatives at constant T and ρ are not identical with derivatives at constant δ and τ and fits which are based on residua formulated in terms of reduced variables become numerically instable. With this residuum the derivatives with respect to the adjustable parameters read

$$\left(\frac{\partial \zeta_{p\rho T}}{\partial Q^{*2}}\right)_{\rho_0,T_0} = -\sum_{i=7}^{17} c_i^q \frac{m_i^q k_i^q}{8\rho_0} \left(\frac{\rho}{\rho_0}\right)^{\frac{m_i^q}{2}-1} \left(\frac{1.13 T_0}{T}\right)^{\frac{n_i^q}{2}} (Q^{*2})^{\frac{k_i^q}{4}-1}$$

7.1 BACKONE Equations of State

$$-\sum_{i=1}^{6} c_i^q \frac{k_i^q}{4\rho_0} \left(\frac{m_i^q}{2} - 2\left(\frac{\rho}{\rho_0}\right)^2\right) \left(\frac{\rho}{\rho_0}\right)^{\frac{m_i^q}{2}-1} \left(\frac{1.13 T_0}{T}\right)^{\frac{n_i^q}{2}} (Q*2)^{\frac{k_i^q}{4}-1}$$

$$\cdot \exp\left(-\left(\frac{\rho}{\rho_0}\right)^2\right),$$
(7.13)

$$\left(\frac{\partial \zeta_{p\rho T}}{\partial \mu*2}\right)_{\rho_0, T_0} = -\sum_{i=11}^{28} c_i^d \frac{m_i^d k_i^d}{8\rho_0} \left(\frac{\rho}{\rho_0}\right)^{\frac{m_i^d}{2}-1} \left(\frac{1.13 T_0}{T}\right)^{\frac{n_i^d}{2}} (\mu*2)^{\frac{k_i^d}{4}-1}$$

$$-\sum_{i=1}^{10} c_i^d \frac{k_i^d}{4\rho_0} \left(\frac{m_i^d}{2} - l_i^d \left(\frac{\rho}{\rho_0}\right)^{l_i^d}\right) \left(\frac{\rho}{\rho_0}\right)^{\frac{m_i^d}{2}-1} \left(\frac{1.13 T_0}{T}\right)^{\frac{n_i^d}{2}} (\mu*2)^{\frac{k_i^d}{4}-1}$$

$$\cdot \exp\left(-\left(\frac{\rho}{\rho_0}\right)^{l_i^d}\right),$$
(7.14)

$$\left(\frac{\partial \zeta_{p\rho T}}{\partial \varphi}\right)_{\rho_0, T_0} = -\left[\frac{2\varphi}{1-\xi} + \frac{2\varphi + 3 + (2\varphi - 3)\xi}{(1-\xi)^3}\right] \cdot \left(\frac{\partial \xi}{\partial \rho}\right)_{T, T_0, \rho_0}$$

$$-\sum_{i=1}^{19} c_i^a \frac{m_i^a j_i}{\rho_0} \left(\frac{\rho}{\rho_0}\right)^{m_i^a - 1} \left(\frac{T_0}{T}\right)^{\frac{n_i^a}{2}} \varphi^{j_i - 1}$$

$$-\sum_{i=20}^{28} c_i^a \frac{j_i}{\rho_0} \left(m_i^a - l_i^a \left(\frac{\rho}{\rho_0}\right)^{l_i^a}\right) \left(\frac{\rho}{\rho_0}\right)^{m_i^a - 1} \left(\frac{T_0}{T}\right)^{\frac{n_i^a}{2}} \varphi^{j_i - 1} \exp\left(-\left(\frac{\rho}{\rho_0}\right)^{l_i^a}\right),$$
(7.15)

$$\left(\frac{\partial \zeta_{p\rho T}}{\partial T_0}\right)_{\rho_0, \varphi, \mu, Q} = -\left\{\left[-\frac{\varphi^2 - 1}{(1-\xi)^2} + \frac{3(\varphi^2 + 3\varphi) + (\varphi^2 - 3\varphi)(1+2\xi)}{(1-\xi)^4}\right]\right.$$

$$\cdot \rho \left(\frac{\partial \xi}{\partial \rho}\right)_{T, T_0, \rho_0} + \left[-\frac{\varphi^2 - 1}{1-\xi} + \frac{\varphi^2 + 3\varphi + (\varphi^2 - 3\varphi)\xi}{(1-\xi)^3}\right]\right\} \frac{n}{\rho_0} \frac{\gamma(1-a)\left(\frac{T_0}{T}\right)^{-\gamma - 1}}{T \left(a + (1-a)\left(\frac{T_0}{T}\right)^{-\gamma}\right)^2}$$

$$-\sum_{i=1}^{19} c_i^a \frac{m_i^a}{\rho} \left(\frac{\rho}{\rho_0}\right)^{m_i^a} \frac{n_i^a}{2T} \left(\frac{T_0}{T}\right)^{\frac{n_i^a}{2}-1} \varphi^{j_i}$$

$$-\sum_{i=20}^{28} c_i^a \left(\frac{m_i^a}{\rho} - \frac{l_i^a}{\rho}\left(\frac{\rho}{\rho_0}\right)^{l_i^a}\right) \left(\frac{\rho}{\rho_0}\right)^{m_i^a} \cdot \frac{n_i^a}{2T} \left(\frac{T_0}{T}\right)^{\frac{n_i^a}{2}-1} \varphi^{j_i} \exp\left(-\left(\frac{\rho}{\rho_0}\right)^{l_i^a}\right)$$

$$-\sum_{i=11}^{28} c_i^{\mathrm{d}} \frac{m_i^{\mathrm{d}}}{2\rho} \left(\frac{\rho}{\rho_0}\right)^{\frac{m_i^{\mathrm{d}}}{2}} \frac{1.13\, n_i^{\mathrm{d}}}{2T} \left(\frac{1.13\, T_0}{T}\right)^{\frac{n_i^{\mathrm{d}}}{2}-1} \left(\mu*^2\right)^{\frac{k_i^{\mathrm{d}}}{4}}$$

$$-\sum_{i=1}^{10} c_i^{\mathrm{d}} \left(\frac{m_i^{\mathrm{d}}}{2\rho} - \frac{l_i^{\mathrm{d}}}{\rho} \left(\frac{\rho}{\rho_0}\right)^{l_i^{\mathrm{d}}} \right) \left(\frac{\rho}{\rho_0}\right)^{\frac{m_i^{\mathrm{d}}}{2}}$$

$$\cdot \frac{1.13\, n_i^{\mathrm{d}}}{2T} \left(\frac{1.13\, T_0}{T}\right)^{\frac{n_i^{\mathrm{d}}}{2}-1} \left(\mu*^2\right)^{\frac{k_i^{\mathrm{d}}}{4}} \exp\left(-\left(\frac{\rho}{\rho_0}\right)^{l_i^{\mathrm{d}}}\right)$$

$$-\sum_{i=7}^{17} c_i^{\mathrm{q}} \frac{m_i^{\mathrm{q}}}{2\rho} \left(\frac{\rho}{\rho_0}\right)^{\frac{m_i^{\mathrm{q}}}{2}} \frac{1.13\, n_i^{\mathrm{q}}}{2T} \left(\frac{1.13\, T_0}{T}\right)^{\frac{n_i^{\mathrm{q}}}{2}-1} \left(Q*^2\right)^{\frac{k_i^{\mathrm{q}}}{4}}$$

$$-\sum_{i=1}^{6} c_i^{\mathrm{q}} \left(\frac{m_i^{\mathrm{q}}}{2\rho} + \frac{l_i^{\mathrm{q}}}{\rho} \left(\frac{\rho}{\rho_0}\right)^{l_i^{\mathrm{q}}} \right) \left(\frac{\rho}{\rho_0}\right)^{\frac{m_i^{\mathrm{q}}}{2}}$$

$$\cdot \frac{1.13\, n_i^{\mathrm{q}}}{2T} \left(\frac{1.13\, T_0}{T}\right)^{\frac{n_i^{\mathrm{q}}}{2}-1} \left(Q*^2\right)^{\frac{k_i^{\mathrm{q}}}{4}} \exp\left(-\left(\frac{\rho}{\rho_0}\right)^{l_i^{\mathrm{q}}}\right), \quad \text{and} \qquad (7.16)$$

$$\left(\frac{\partial \zeta_{p\rho T}}{\partial \rho_0}\right)_{T_0,\varphi,\mu,Q} = -\left\{ \left[-\frac{\varphi^2-1}{(1-\xi)^2} + \frac{3(\varphi^2+3\varphi)+(\varphi^2-3\varphi)(1+2\xi)}{(1-\xi)^4} \right] \right.$$

$$\left. \cdot \rho \left(\frac{\partial \xi}{\partial \rho}\right)_{T,T_0,\rho_0} + \left[-\frac{\varphi^2-1}{1-\xi} + \frac{\varphi^2+3\varphi+(\varphi^2-3\varphi)\xi}{(1-\xi)^3} \right] \right\}$$

$$\cdot \left(-\frac{n}{\rho_0^2}\right) \left[a + (1-a)\left(\frac{T_0}{T}\right)^{-\gamma} \right]^{-1}$$

$$-\sum_{i=1}^{19} c_i^{\mathrm{a}} \frac{m_i^{\mathrm{a}2}}{\rho^2} \left(\frac{\rho}{\rho_0}\right)^{m_i^{\mathrm{a}}+1} \left(\frac{T_0}{T}\right)^{\frac{n_i^{\mathrm{a}}}{2}} \varphi^{j_i}$$

$$-\sum_{i=20}^{28} c_i^{\mathrm{a}} \left(-\frac{m_i^{\mathrm{a}2}}{\rho^2} + \frac{l_i^{\mathrm{a}}(l_i^{\mathrm{a}}+2m_i^{\mathrm{a}})}{\rho^2}\left(\frac{\rho}{\rho_0}\right)^{l_i^{\mathrm{a}}} - \frac{l_i^{\mathrm{a}2}}{\rho^2}\left(\frac{\rho}{\rho_0}\right)^{2l_i^{\mathrm{a}}} \right) \left(\frac{\rho}{\rho_0}\right)^{m_i^{\mathrm{a}}+1}$$

$$\cdot \left(\frac{T_0}{T}\right)^{\frac{n_i^{\mathrm{a}}}{2}} \varphi^{j_i} \exp\left(-\left(\frac{\rho}{\rho_0}\right)^{l_i^{\mathrm{a}}}\right)$$

$$-\sum_{i=11}^{28} c_i^{\mathrm{d}} \frac{m_i^{\mathrm{d}2}}{4\rho^2} \left(\frac{\rho}{\rho_0}\right)^{m_i^{\mathrm{d}}+1} \left(\frac{1.13\, T_0}{T}\right)^{\frac{n_i^{\mathrm{d}}}{2}} \left(\mu*^2\right)^{\frac{k_i^{\mathrm{d}}}{4}}$$

$$-\sum_{i=1}^{10} c_i^{\mathrm{d}} \left(-\frac{m_i^{\mathrm{d}2}}{4\rho^2} + \frac{l_i^{\mathrm{d}}\left(l_i^{\mathrm{d}} + m_i^{\mathrm{d}}\right)}{\rho^2} \left(\frac{\rho}{\rho_0}\right)^{l_i^{\mathrm{d}}} - \frac{l_i^{\mathrm{d}2}}{\rho^2}\left(\frac{\rho}{\rho_0}\right)^{2l_i^{\mathrm{d}}} \right) \left(\frac{\rho}{\rho_0}\right)^{\frac{m_i^{\mathrm{d}}}{2}+1}$$

$$\cdot \left(\frac{1.13 T_0}{T}\right)^{\frac{n_i^{\mathrm{d}}}{2}} (\mu*2)^{\frac{k_i^{\mathrm{d}}}{4}} \exp\left(-\left(\frac{\rho}{\rho_0}\right)^{l_i^{\mathrm{d}}}\right)$$

$$-\sum_{i=7}^{17} -c_i^{\mathrm{q}} \frac{m_i^{\mathrm{q}2}}{4\rho^2}\left(\frac{\rho}{\rho_0}\right)^{\frac{m_i^{\mathrm{q}}}{2}+1}\left(\frac{1.13 T_0}{T}\right)^{\frac{n_i^{\mathrm{q}}}{2}} (Q*2)^{\frac{k_i^{\mathrm{q}}}{4}}$$

$$-\sum_{i=1}^{6} c_i^{\mathrm{q}} \left(-\frac{m_i^{\mathrm{q}2}}{4\rho^2} + \frac{l_i^{\mathrm{q}}\left(l_i^{\mathrm{q}} + m_i^{\mathrm{q}}\right)}{\rho^2} \left(\frac{\rho}{\rho_0}\right)^{l_i^{\mathrm{q}}} - \frac{l_i^{\mathrm{q}2}}{\rho^2}\left(\frac{\rho}{\rho_0}\right)^{2l_i^{\mathrm{q}}} \right) \left(\frac{\rho}{\rho_0}\right)^{\frac{m_i^{\mathrm{q}}}{2}+1}$$

$$\cdot \left(\frac{1.13 T_0}{T}\right)^{\frac{n_i^{\mathrm{q}}}{2}} (Q*2)^{\frac{k_i^{\mathrm{q}}}{4}} \exp\left(-\left(\frac{\rho}{\rho_0}\right)^{l_i^{\mathrm{q}}}\right), \tag{7.17}$$

with
$$\left(\frac{\partial \xi}{\partial \rho}\right)_{T, T_0, \rho_0} = \frac{n}{\rho_0}\left[a + (1-a)\tau^{-\gamma}\right]^{-1},$$

$n = 0.1617$, $a = 0.689$, and $\gamma = 0.3674$. The derivatives of $\zeta_{\rho\rho T}$, Eqs. 7.12–7.17, are highly intercorrelated and cannot be used in linear fits. Starting values for the adjustable parameters have to be estimated. The derivatives of residua for properties which involve higher derivatives of α with respect to ρ or T become correspondingly more complex, see Sects. 4.3.2 and 4.3.3.[4]

A comparison of Eqs. 7.9–7.11 with Eqs. 7.11–7.17 illustrates that the problems which have to be solved when fitting an equation of state to experimental data depend mainly on the set-up of the equation and only in the second place on the number of fitted parameters. The number of fitted parameters and intercorrelations between these parameters become relevant only where fits to very small data sets require a high degree of numerical stability, see Sects. 6.2.1 and 7.2.2.3.

7.1.2 Some Results

To assess the quality of BACKONE equations of state, published equations were compared with data sets described in Sects. 6.2.2 and 6.2.3 for four nonpolar fluids and nine polar fluids. However, these comparisons may be considered as unfair to a certain degree, since both Müller et al. (1996) and Calero et al. (1998)[5] fitted

---

[4] In Sects. 4.3.2 and 4.3.3, the corresponding residua are formulated based on derivatives of α with respect to τ and δ. These derivatives can easily be transformed into derivatives with respect to T and ρ using Eqs. 3.42–3.47.

[5] Calero et al. (1998) fitted their equations only to four vapour pressure and saturated liquid density data to demonstrate the predictive capabilities of BACKONE equations of state.

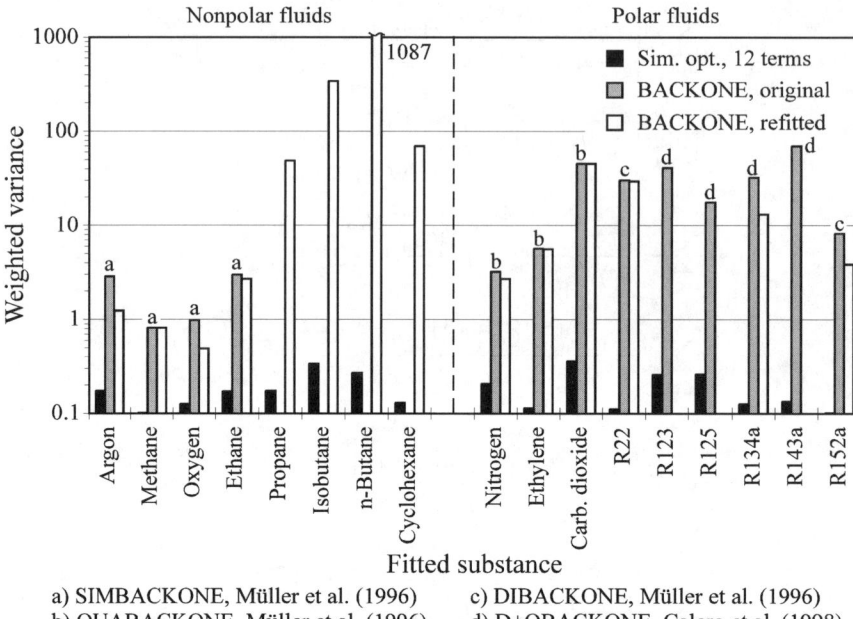

Fig. 7.1. Weighted variances which result from comparisons and fits of BACKONE equations to data sets of 8 nonpolar and 9 polar fluids. Results of the simultaneously optimised equations presented in Sects. 6.2.2 and 6.2.3 are given for comparison.

their equations only to thermal properties while the data sets which were used for comparison contained caloric properties as well. Thus, Span (1998b) refitted published equations to the complete data sets with some success. For some of the considered substances, the performance of the equations could be improved significantly (e.g. for argon, oxygen, R134a, and R152a), while no significant improvements could be achieved for other substances. Furthermore, BACKONE equations were established for four more complex nonpolar fluids, namely for propane, isobutane, n-butane, and cyclohexane. In terms of the weighted variance, see Eq. 6.9, the results of these comparisons and fits are summarised in Fig. 7.1. The considered data sets contained both thermal and caloric properties and were weighted according to the demands summarised in Table 6.1. For information on the correlations which were used to describe the ideal part of the reduced Helmholtz energy, α_i^0, see Tables 6.3 and 6.5.

For methane and oxygen the SIMBACKONE equation yields very good results. When using the weighted variance as a criterion, one has to conclude that the equations even overfulfil the demands formulated for future generalised equations of state, $\sigma_{wt}^2 \approx 1$ (see footnote 3), for these substances. However, this result is no surprise since data sets for methane and oxygen were used when setting up the rather complex functional form of α_i^a, see Eq. 7.2.

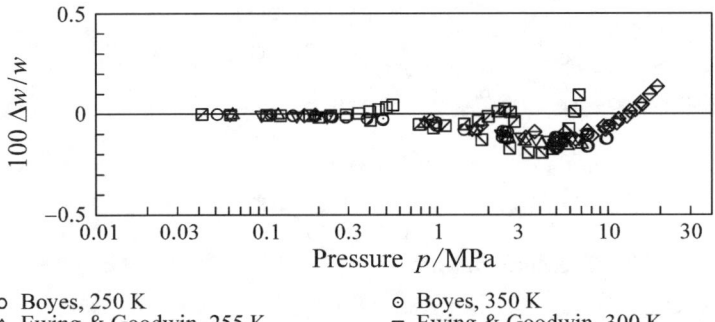

○ Boyes, 250 K
△ Ewing & Goodwin, 255 K
□ Estrada-Alexanders & Trusler, 110 K
◙ Estrada-Alexanders & Trusler, 180 K
✧ Estrada-Alexanders & Trusler, 450 K
○ Boyes, 350 K
▽ Ewing & Goodwin, 300 K
◪ Estrada-Alexanders & Trusler, 140 K
◇ Estrada-Alexanders & Trusler, 300 K

Fig. 7.2. Percentage deviations $100 \Delta w/w = 100 (w_{exp} - w_{calc})/w_{exp}$ between experimental data for the speed of sound in argon by Ewing and Goodwin (1992), Boyes (1992), and Estrada-Alexanders and Trusler (1995) and values calculated from the SIMBACKONE equation by Müller et al. (1996).

Transferred to other simple substances like argon, the SIMBACKONE equation still yields results of comparable accuracy. At gaseous and gas-like supercritical states, both thermal and caloric properties are represented well within the demanded uncertainties. As an example, Fig. 7.2 shows deviations between highly accurate experimental results for the speed of sound in argon and values calculated from the corresponding SIMBACKONE equation by Müller et al. (1996); for comparison see also Fig. 5.17. At liquid states, liquid-like supercritical states, and in the extended critical region the observed uncertainties exceed the demands formulated in Table 6.1, but for a generalised equation of state with just three

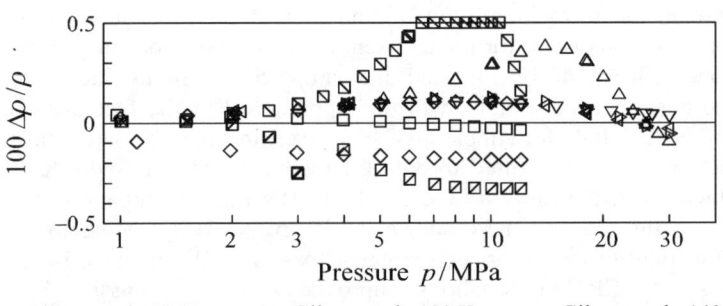

□ Gilgen et al., 90 K
◙ Gilgen et al., 180 K
▽ Klimeck et al., 340 K
◇ Gilgen et al., 120 K
✧ Gilgen et al., 340 K
◁ Klimeck et al., 400 K
◪ Gilgen et al., 140 K
△ Klimeck et al., 235 K
▷ Klimeck et al., 520 K

Fig. 7.3. Percentage deviations $100 \Delta\rho/\rho = 100 (\rho_{exp} - \rho_{calc})/\rho_{exp}$ between experimental results for the density of argon by Gilgen et al. (1997a) and Klimeck et al. (1998) and values calculated from the SIMBACKONE equation by Müller et al. (1996).

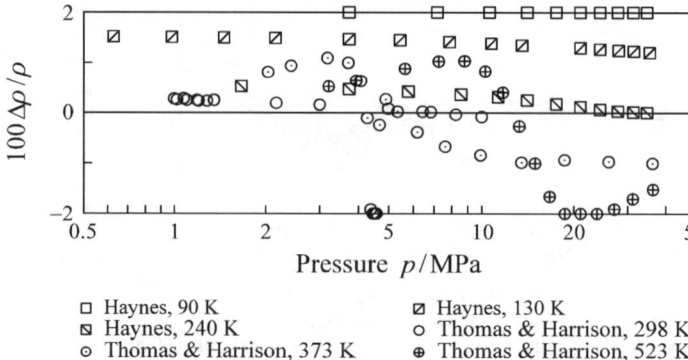

Fig. 7.4. Percentage deviations $100 \, \Delta\rho / \rho = 100 \, (\rho_{exp} - \rho_{calc}) / \rho_{exp}$ between experimental results for the density of propane by Haynes (1983) and Thomas and Harrison (1982) and values calculated from the SIMBACKONE equation by Span (1998b).

adjustable parameters, the results are still surprisingly good. For a broad range of states, the representation of highly accurate $p\rho T$ data is illustrated in Fig. 7.3, see also Figs. 5.4 and 6.22 for comparison.

However, when dealing with more complex nonpolar molecules, the performance of SIMBACKONE equations worsens drastically. As an example, Fig. 7.4 shows deviations between experimental results for the density of propane and values calculated from the SIMBACKONE equation by Span (1998b). In the liquid phase and at high temperatures the observed density deviations increase up to $\Delta\rho / \rho \approx \pm 2\%$ and thus they exceed the demanded value by a whole order of magnitude. For speeds of sound at liquid states deviations up to $\Delta w / w \approx +15\%$ are to be observed for propane. For isobutane and n-butane the results become significantly worse.

In the BACKONE family of equations of state, the anisotropy parameter φ, which is contained in the formulations for α^h and α_3^a is used to account for the different shape or, more precisely, for the different elongation of molecules. From the underlying theory, it was to be expected, that properties of fluids like cyclohexane cannot be described accurately with this approach. But the fact that the SIMBACKONE equation fails for simple alkanes as well indicates a major shortcoming. The relation for α^a was fitted to data for methane, oxygen, and ethane, thus to fluids which cover the range $1.00 \leq \varphi \leq 1.21$. The fits for propane and n-butane resulted in values of $\varphi \approx 1.31$ and $\varphi \approx 1.35$, respectively. Most likely, Eq. 7.2 does not account properly for such larger values of φ. Without sacrificing the general set-up of BACKONE equations, improvements may be possible by a reformulation of the equation for α_3^a Eq. 7.2, based on data for a broader set of substances including substances with more pronounced elongation.

For weakly polar fluids, such as nitrogen and ethylene, the results found for BACKONE equations are comparable to those found for simple nonpolar fluids, but the obtained weighted variances already exceed the demanded value of $\sigma_{wt}^2 \approx 1$

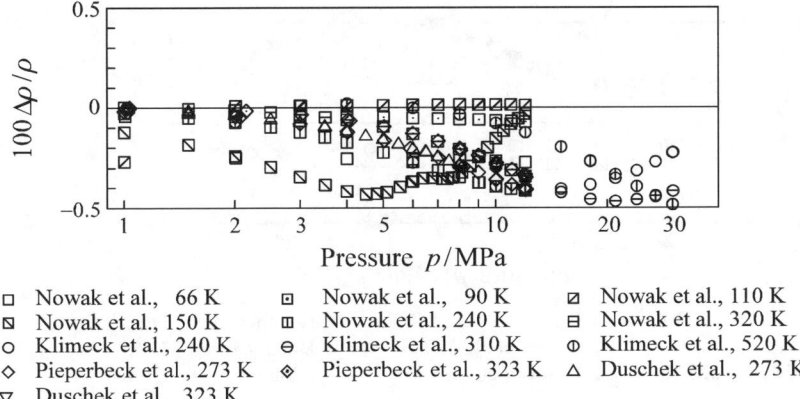

Fig. 7.5. Percentage deviations $100\, \Delta\rho/\rho = 100\,(\rho_{\mathrm{exp}}-\rho_{\mathrm{calc}})/\rho_{\mathrm{exp}}$ between experimental data for the density of nitrogen by Nowak et al. (1997a), Klimeck et al. (1998), Pieperbeck et al. (1991), and Duschek et al. (1988) and values calculated from the QUABACKONE equation by Müller et al. (1996).

considerably. As an example, Fig. 7.5 shows deviations between highly accurate experimental results for the density of nitrogen and values calculated from the corresponding QUABACKONE equation by Müller et al. (1996); for comparison see also Figs. 5.3 and 5.10. At low temperatures liquid densities are described surprisingly well, but at higher temperatures the uncertainty of calculated densities increases to $\Delta\rho/\rho \leq \pm 0.5\,\%$, exceeding the demands formulated for advanced technical equations of state by a factor of 2.5. More severe shortcomings are to be observed for $p\rho T$ data at high density gaseous states, in the extended critical region, and at temperatures above 500 K and pressures above 30 MPa. However, the representation of caloric properties satisfies the demands formulated in Table 6.1 in general; speeds of sound are represented with surprisingly high accuracy both at liquid and supercritical states. Similar results were found for ethylene as well.

When considering substances with more pronounced polarity, the results of BACKONE equations again become worse. In general, the obtained weighted variances exceed the limit of $\sigma^2_{\mathrm{wt}} = 10$. Figure 7.6 illustrates the representation of $p\rho T$ data for carbon dioxide by the QUABACKONE equation of state by Müller et al. (1996); for comparison see Figs. 5.5, 5.11, and 6.41. Just as for propane, the observed density deviations increase up to $\Delta\rho/\rho \approx \pm 2\,\%$ and results for caloric properties become even worse, especially at liquid states.

The D+QBACKONE concept introduced by Calero et al. (1998) improves the representation of properties of the haloginated ethane derivatives, but it still does not yield satisfactory results. Figure 7.7 shows deviations between experimental results for the density of refrigerant R134a and corresponding values calculated from the QUABACKONE equation and from the D+QBACKONE equation, both published by Calero et al. (1998). The D+QBACKONE equation was refitted by

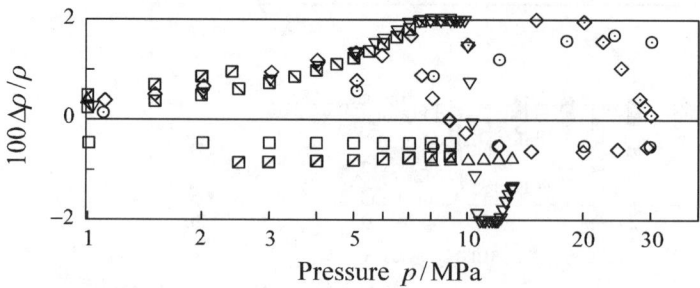

□ Duschek et al., 220 K	▨ Duschek et al., 260 K	▨ Duschek et al., 340 K
△ Gilgen et al., 280 K	▽ Gilgen et al., 323 K	○ Brachthäuser, 233 K
○ Brachthäuser, 523 K	◇ Klimeck et al., 300 K	◈ Klimeck et al., 430 K

Fig. 7.6. Percentage deviations $100 \, \Delta\rho/\rho = 100\,(\rho_{\text{exp}} - \rho_{\text{calc}})/\rho_{\text{exp}}$ between experimental results for the density of carbon dioxide by Duschek et al. (1990), Gilgen et al. (1992), Brachthäuser (1993), and Klimeck et al. (2000) and values calculated from the corresponding QUABACKONE equation by Müller et al. (1996).

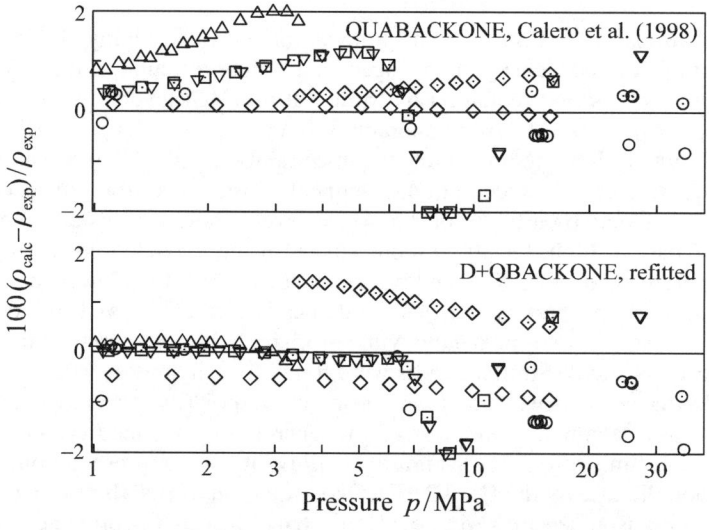

□ Tillner-Roth & Baehr, 313 K	▣ Tillner-Roth & Baehr, 423 K
◇ Tillner-Roth & Baehr, 263 K	◈ Tillner-Roth & Baehr, 363 K
△ Dressner & Bier, 373 K	▽ Dressner & Bier, 423 K
○ Klomfar et al., 236 K	○ Klomfar et al., 298 K

Fig. 7.7. Percentage deviations between experimental results for the density of R134a by Tillner-Roth and Baehr (1992/1993), Dressner and Bier (1993) and Klomfar et al. (1993) and values calculated from the corresponding QUABACKONE equation by Calero et al. (1998) and from the refitted D+QBACKONE equation by Span (1998b).

Span (1998b). The five parameter D+QBACKONE equation is superior especially with respect to the representation of properties at gaseous- and gas-like supercritical states. At these states, the accuracy of the formulation can be regarded as satisfactory. But at the same time, the description of properties at liquid states becomes worse. The largest observed deviations are still in an order of $\Delta \rho / \rho \approx \pm 2\%$. Both equations fail to fulfil the demands formulated for advanced generalised equations above.

However, the shortcomings observed for the polar fluids do not necessarily mean, that the correlations for the polar contributions to α, Eqs. 7.4 and 7.6, need to be reformulated as well. For all of the haloginated ethanes the fits result in anisotropy factors $\varphi \geq 1.38$; the observed problems may be related to the assumed erroneous behaviour of α^a as well. The question, whether it is necessary to consider dipole / quadrupole interactions as an additional independent contribution to α, as it was discussed briefly by Calero et al. (1998), cannot be answered before the problems regarding α^a are overcome.

To sum up, the group of substances for which BACKONE equations satisfy the demands formulated for advanced technical applications is still restricted to simple non- and weakly polar substances without or with small anisotropy. However, among the generalised equations of state which are available today, BACKONE equations still yield the best description of the thermodynamic properties of fluids with nonpolar and polar molecules with moderate anisotropy, such as the halogenated ethane derivates discussed by Calero et al. (1998). Besides the actual advantages and disadvantages of the published equations, the strongest point of the BACKONE concept is its additive set-up which considers contributions from di- and quadrupole interactions explicitly for the first time. This concept avoids intercorrelations between the parameters φ, μ^{*2}, and Q^{*2} as they are frequently encountered for multiparameter extended corresponding states approaches, to a large extent and guarantees a high degree of numerical stability. Improvements with regard to the shortcomings of BACKONE equations seem to be possible and different approaches may adapt their general set-up as well.

7.2 Generalised Empirical Equations of State

As already mentioned, empirical approaches for a generalised description of thermodynamic properties of fluids are usually based on more or less simple multipararmeter equations of state which belong to the group of technical equations of state, group 3 in Chap. 5. Even if the generalisation succeeds without loss of quality, the resulting models cannot yield better results than the corresponding substance specific equations of state. Thus, most of the available models cannot satisfy the demands implied by advanced technical applications, since the underlying substance specific equations do not satisfy them, see the comparisons shown in Chap. 5. Based on this assessment, this section discusses just two generalised models, which are based on sufficiently accurate multiparameter equations of state and which therefore have the potential to match advanced demands.

7.2.1 The Approach by Platzer and Maurer

Based on the critical temperature, T_c, the critical density, ρ_c, the acentric factor ω, and the "*polar factor*", χ,[6] Platzer and Maurer (1989) used a four parameter corresponding states approach to generalise the coefficients of the equation of state established by Bender (1970). Written in terms of reduced properties the Bender-type equation of state reads

$$\frac{p}{\rho RT} = Z(\tau,\delta) = \sum_{i=1}^{14} n_i \tau^{t_i} \delta^{d_i} + \sum_{i=14}^{19} n_i \tau^{t_i} \delta^{d_i} \exp(-\delta^2), \qquad (7.18)$$

with $\tau = T_c/T$ and $\delta = \rho/\rho_c$; for the corresponding exponents t_i and d_i see Platzer and Maurer (1989). Based on this general form the coefficients were rewritten as

$$n_i = g_{4,i} + g_{1,i}\,\omega + g_{2,i}\,\chi + g^*\,g_{3,i}\,\omega\chi + g_{5,i}\,\chi^2. \qquad (7.19)$$

In this way, the 19 coefficients n_i of the Bender-type equation of state are expressed in terms of a system of simple correlations, which involve a total of 95 coefficients $g_{j,i}$. The 19 coefficients $g_{4,i}$ correspond to the coefficients n_i of a hypothetical simple fluid with $\omega = \chi = 0$, while the other $g_{j,i}$ describe the deviating behaviour of real fluids. The parameter g^* was chosen to be equal to 1 in general and equal to 10 for perfluorinated (completely fluorinated) substances. The values of the 95 coefficients $g_{j,i}$ were determined by a simultaneous fit to $p\rho T$ data and linearised Maxwell data (see Sect. 4.3.2) for 26 nonpolar and polar fluids. Since this fit involved only linear data and since the acentric and polar factors were treated as substance specific constants and not as adjustable parameters, the $g_{j,i}$ could be determined in a linear fit without starting values and with an unequivocal solution. For the resulting $g_{j,i}$ see Platzer and Maurer (1989).

When used in the way described above, the approach of Platzer and Maurer (1989) is completely predictive, since the coefficients of Bender-type equations can be determined based only on values of the acentric factor ω and the polar factor χ – no fits are involved in the application of Eqs. 7.18 and 7.19 to pure fluids. The necessary values for ω (see Chap. 6, footnote 3) and χ (see footnote 6) can be calculated from accurate vapour pressure equations which are available for a broad variety of substances today, see e.g. Reid et al. (1987). Besides ω and χ, only T_c and ρ_c are required to reduce the temperature and the density.[7]

---

[6] The polar factor was originally introduced by Halm and Stiel (1967) and describes reduced vapour pressures just like the acentric factor ω, see footnote 3 in Chap. 6. The polar factor is defined as $\chi = \log(p_s(0.6\,T_c)/p_c) + 1.7\,\omega + 1.552$.

[7] This completely predictive set-up differs clearly from the other models discussed in this chapter which involve three to five adjustable parameters. However, this set-up relies on accurate values for the critical temperature and density which are available only for substances with reasonably extensive data sets (see also Sect. 6.2.1.2). For these substances, three to five adjustable parameters can be determined as well, at least if the used model is numerically sufficiently stable.

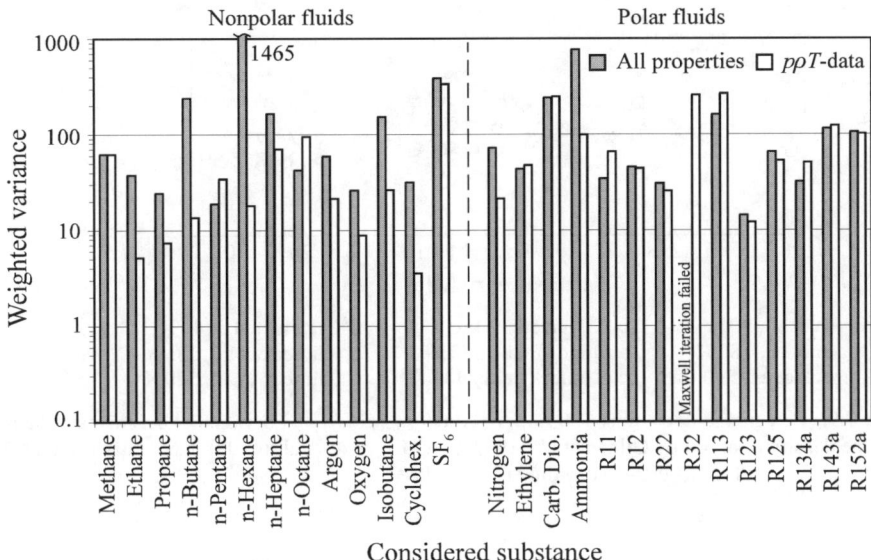

Fig. 7.8. Weighted variances which result from comparisons of data sets for 13 nonpolar and 14 polar fluids with values calculated from the model of Platzer and Maurer (1989).

7.2.1.1 Some Results

To assess the performance of the approach of Platzer and Maurer (1989) results calculated from the generalised Bender-type equation defined by Eqs. 7.18 and 7.19 were compared with the data sets described in Sects. 6.2.2 and 6.2.3. The experimental data were again weighted with the demanded uncertainties summarised in Table 6.1. Equation 7.18 was used in an integrated form as formulation for the residual part of the reduced Helmholtz energy, α^r, see Sect. 3.1.2.1. The required formulations for the ideal part of the reduced Helmholtz energy, α^o, were established based on the correlations for the isobaric heat capacity which were cited in Tables 6.3 and 6.5, respectively. Figure 7.8 shows the weighted variances which result from these comparisons. The comparisons to "all properties" include available data for all kinds of thermal and caloric properties, while comparisons to "$p\rho T$ data" involve only the data sets which are available for the $p\rho T$ relation of the corresponding substance.

From Fig. 7.8 it becomes obvious at once, that the model by Platzer and Maurer (1989) is not able to satisfy the formulated demands on the accuracy of calculated data, see footnote 3. The weighted variances resulting from comparisons to "all properties" exceed $\sigma_{wt}^2 = 10$ for all substances, and the variances which result from comparisons to "$p\rho T$ data" do so for most substances. On average, the results are worse than those obtained with BACKONE equations of state, see Fig. 7.1, but the model by Platzer and Maurer can be used for a broader group of substances. To

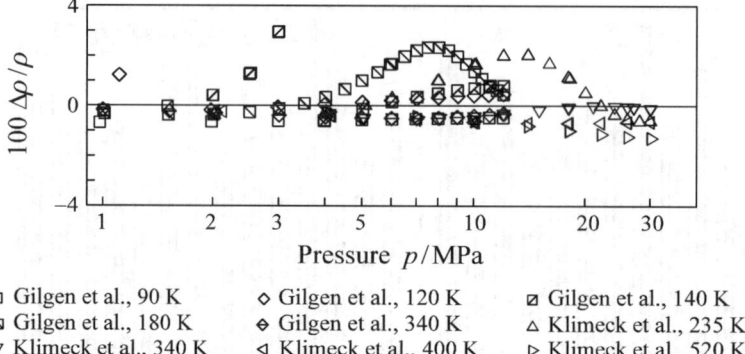

□ Gilgen et al., 90 K	◇ Gilgen et al., 120 K	◨ Gilgen et al., 140 K
◪ Gilgen et al., 180 K	◆ Gilgen et al., 340 K	△ Klimeck et al., 235 K
▽ Klimeck et al., 340 K	◁ Klimeck et al., 400 K	▷ Klimeck et al., 520 K

Fig. 7.9. Percentage deviations $100\, \Delta\rho/\rho = 100\,(\rho_{\mathrm{exp}} - \rho_{\mathrm{calc}})/\rho_{\mathrm{exp}}$ between experimental results for the density of argon by Gilgen et al. (1997a) and Klimeck et al. (1998) and values calculated from Eq. 7.18 with coefficients according to Eq. 7.19.

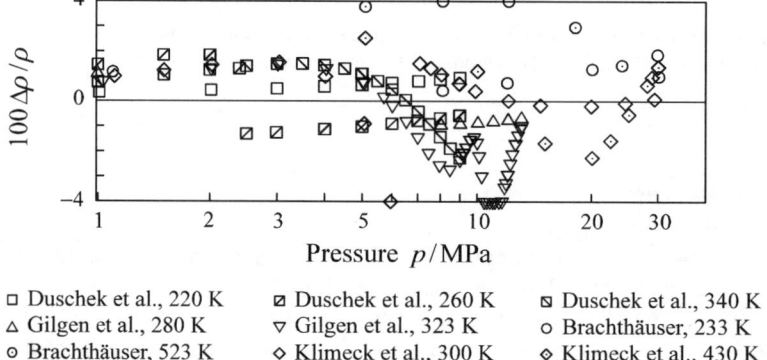

□ Duschek et al., 220 K	◨ Duschek et al., 260 K	◪ Duschek et al., 340 K
△ Gilgen et al., 280 K	▽ Gilgen et al., 323 K	○ Brachthäuser, 233 K
○ Brachthäuser, 523 K	◇ Klimeck et al., 300 K	◆ Klimeck et al., 430 K

Fig. 7.10. Percentage deviations $100\, \Delta\rho/\rho = 100\,(\rho_{\mathrm{exp}} - \rho_{\mathrm{calc}})/\rho_{\mathrm{exp}}$ between experimental results for the density of carbon dioxide by Duschek et al. (1990), Gilgen et al. (1992), Brachthäuser (1993), and Klimeck et al. (2000) and values calculated from Eq. 7.18 with coefficients according to Eq. 7.19.

illustrate the accuracy of the most sophisticated generalised multiparameter equations of state and to explain the foundation of the approach described in the next section, these results will be discussed in some more detail.

Figure 7.9 shows deviations between highly accurate experimental $p\rho T$ data for argon and values calculated from Eq. 7.18; for comparison see Figs. 5.4, 6.22, and 7.3. The experimental data are mostly represented within $\Delta\rho/\rho \leq \pm 1\,\%$. Significantly enlarged deviations are to be observed for gaseous states when approaching the phase boundary and for supercritical states in the range of medium densities. For the highest considered temperature, $T = 520$ K, the deviations at high pressures increase as well.

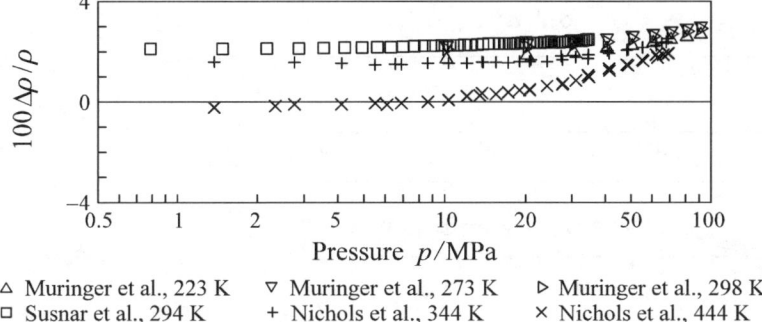

△ Muringer et al., 223 K   ▽ Muringer et al., 273 K   ▷ Muringer et al., 298 K
□ Susnar et al., 294 K   + Nichols et al., 344 K   × Nichols et al., 444 K

Fig. 7.11. Percentage deviations $100 \, \Delta\rho/\rho = 100 \, (\rho_{exp} - \rho_{calc})/\rho_{exp}$ between experimental data for the density of n-heptane measured by Muringer et al. (1985), Susnar et al. (1992), and Nichols et al. (1955) and values calculated from Eq. 7.18 with coefficients according to Eq. 7.19.

For less simple substances, larger deviations are to be observed, but the general tendencies remain the same. The deviations between selected $p\rho T$ data for carbon dioxide and values calculated from Eq. 7.18 remain within $\Delta\rho/\rho \le \pm 2\%$ in most cases, see Fig. 7.10. Enlarged deviations occur in the extended critical region, at supercritical temperatures for states in the range of medium densities, and for the highest considered temperatures. The same set of data was used for comparisons in Figs. 5.5, 5.11, and 6.41.

For higher alkanes, experimental data are available almost only in the liquid phase. Figure 7.11 shows deviations between accurate $p\rho T$ data in the liquid phase of n-heptane and values calculated from Eq. 7.18; for comparison see Fig. 6.21. The observed deviations exceed $\Delta\rho/\rho \approx \pm 2\%$ slightly, especially at low temperatures. At low temperatures, the deviations are almost independent of pressure, but at $T = 444$ K ($T/T_c \approx 0.82$) the pressure dependency becomes relevant. Based on comparisons with the available experimental data, the general tendencies again seem to be the same as described above. Thus, densities calculated from the model of Platzer and Maurer (1989) can be assumed to be accurate to $\Delta\rho/\rho \le \pm 2\%$ in general, but enlarged uncertainties are to be assumed in the regions described above.[8]

Typical uncertainties for thermal properties on the vapour-liquid phase boundary are of the same order as those for $p\rho T$ data. For isobutane, Fig. 7.12 shows deviations between experimental results for the vapour pressure and the saturated liquid density and values calculated from the model of Platzer and Maurer (1989); see also Fig. 5.33. Representative comparisons for saturated vapour densities are difficult, since accurate data for this property are available only for few substances. However, since the uncertainty of calculated saturated vapour densities is determined by the uncertainty of calculated vapour pressures and densities in the

---

[8] Significantly larger deviations may result from uncertain values of T_c, ρ_c, ω, and χ, see the discussion below.

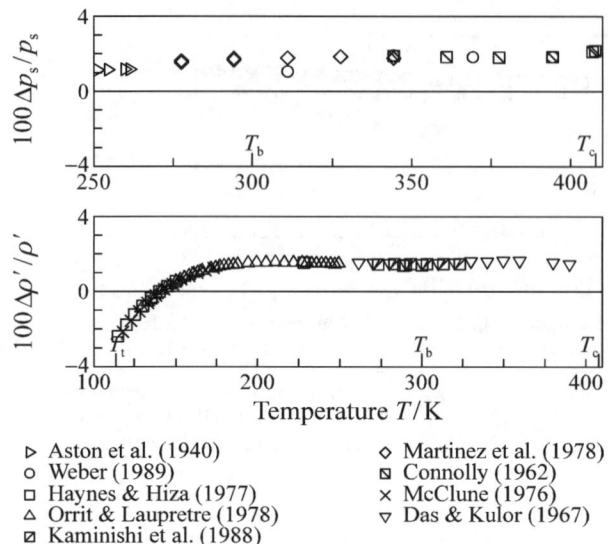

▷ Aston et al. (1940)   ◇ Martinez et al. (1978)
○ Weber (1989)   ⬛ Connolly (1962)
□ Haynes & Hiza (1977)   × McClune (1976)
△ Orrit & Laupretre (1978) ▽ Das & Kulor (1967)
⬛ Kaminishi et al. (1988)

Fig. 7.12. Percentage deviations $100 \Delta y / y = 100 (y_{exp} - y_{calc}) / y_{exp}$ with $y = p_s, \rho'$ between selected experimental results for isobutane and values calculated from Eq. 7.18 with coefficients according to Eq. 7.19.

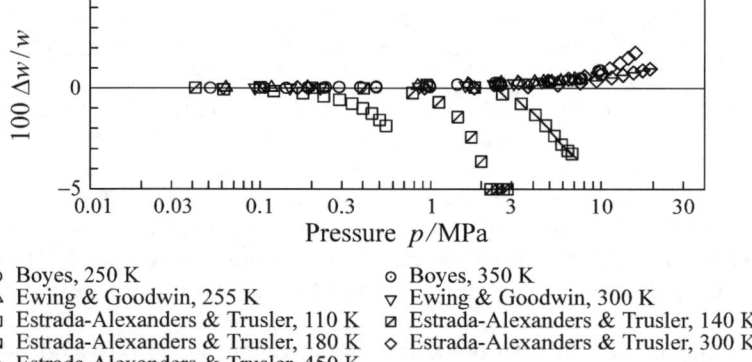

○ Boyes, 250 K   ○ Boyes, 350 K
△ Ewing & Goodwin, 255 K   ▽ Ewing & Goodwin, 300 K
□ Estrada-Alexanders & Trusler, 110 K   ⬛ Estrada-Alexanders & Trusler, 140 K
⬛ Estrada-Alexanders & Trusler, 180 K   ◇ Estrada-Alexanders & Trusler, 300 K
◆ Estrada-Alexanders & Trusler, 450 K

Fig. 7.13. Percentage deviations $100 \Delta w / w = 100 (w_{exp} - w_{calc}) / w_{exp}$ between experimental data for the speed of sound in argon by Ewing and Goodwin (1992), Boyes (1992), and Estrada-Alexanders and Trusler (1995) and values calculated from Eq. 7.18 with coefficients according to Eq. 7.19 and an independent formulation for α^o.

gas phase close to the phase boundary, enlarged uncertainties are to be expected for this property. Where comparisons are possible, they support this conclusion.

For caloric properties a more detailed examination is necessary. Figure 7.13 shows deviations between experimental results for the speed of sound in argon at

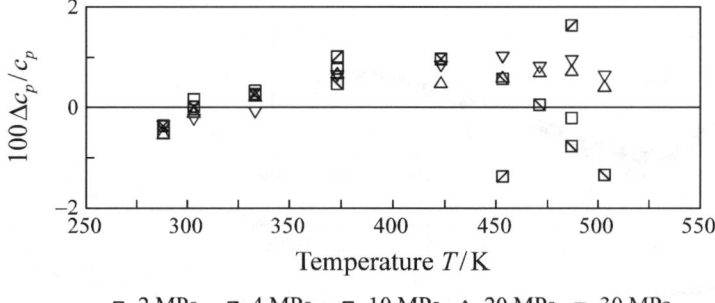

Fig. 7.14. Percentage deviations $100 \, \Delta c_p / c_p = 100 \, (c_{p,\text{exp}} - c_{p,\text{calc}}) / c_{p,\text{exp}}$ between experimental data for the isobaric heat capacity of R11 measured by Wirbser et al. (1992) and values calculated from Eq. 7.18 with coefficients according to Eq. 7.19 and an independent formulation for α°.

gaseous and gas-like supercritical states and values calculated from the model of Platzer and Maurer (1989). The same data set was used in Fig. 5.17.

Keeping in mind the general discussion on the sensitivity of such data, see Sect. 5.2, the observed deviations clearly support the thesis that the model of Platzer and Maurer (1989) shows significant shortcomings at gaseous states. Deviations up to $\Delta w / w \approx -12\,\%$ in the gas phase exceed results found for simple technical equations of state by more than one order of magnitude, see e.g. Fig. 5.21. At far supercritical gas-like states, the accuracy of speeds of sound calculated from Eq. 7.18 in combination with a suitable formulation for α° becomes satisfactory for common technical applications. The representation of speeds of sound at liquid states is unsatisfactory on the whole.

Different results were found for heat capacities in many cases. Figure 7.14 shows deviations between experimental results for the isobaric heat capacity of refrigerant R11 and values calculated from Eq. 7.18 in combination with the corresponding formulation for α°. The selected data cover supercritical states and liquid states at reduced temperatures $T / T_c > \approx 0.6$. The generalised equation describes these data almost as accurately as the corresponding simultaneously optimised equation of state, Eq. 6.13, see Fig. 6.31. However, deviations up to $\Delta c_v / c_v \approx \pm 30\,\%$ were found e.g. for corresponding states of refrigerant R143a.

Aside from these differences, one feature was common for all considered substances: the uncertainty of calculated caloric properties in the liquid phase increases at low temperatures. For refrigerant R134a and for n-butane, Fig. 7.15 shows deviations between measured heat capacities of the saturated liquid, c_σ, and values calculated from the model of Platzer and Maurer (1989). While the observed deviations stay within $\Delta c_\sigma / c_\sigma \approx \pm 10\,\%$ for R134a, Eq. 7.18 even predicts negative heat capacities for liquid n-butane. The large weighted variance obtained for the comparison to "all properties" for n-butane, see Fig. 7.8, results mainly from these unreasonable results for heat capacities at liquid states. In a similar way, the large weighted variance for n-hexane results from completely unreasonable results for speeds of sound at liquid states.

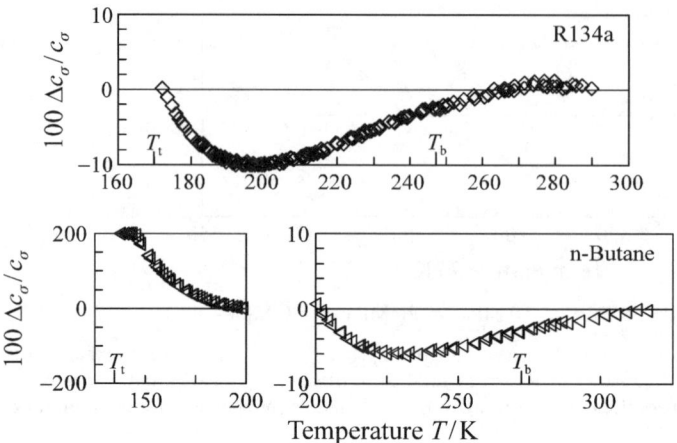

Fig. 7.15. Percentage deviations $100\,\Delta c_\sigma/c_\sigma = 100\,(c_{\sigma,\mathrm{exp}} - c_{\sigma,\mathrm{calc}})/c_{\sigma,\mathrm{exp}}$ between experimental results for R134a by Magee (1992) and for n-butane by Magee and Lüddecke (1998) and values calculated from Eq. 7.18 with coefficients according to Eq. 7.19 and an independent formulation for α^{o}.

These extreme differences, especially in the representation of caloric properties, explain the large difference in the results found for the model of Platzer and Maurer (1989) in comparisons e.g. by Türk et al. (1997) and by Karkaris et al. (1992); the obtained results depend completely on the considered property and substance. Just like other existing generalised multiparameter equations of state, Eqs. 7.18 and 7.19 should not be used for the calculation of caloric properties, unless a sufficient accuracy can be verified by comparison with experimental data for the considered fluid.

Independent of these comparisons, approaches like the one used by Platzer and Maurer (1989) can be criticised for two more general reasons. First of all, the accuracy of calculated properties depends directly on the accuracy of the values used for the critical temperature and the critical density. A shift in ρ_c directly results in a corresponding relative shift of calculated densities in the homogeneous regions. The way in which the representation of derived properties is distorted depends on the considered states and properties. Erroneous values of T_c result in more complex distortions, since they exercise an influence both on $\tau = T_c/T$ and on ω and χ, which depend on the values assumed for the critical temperature and the critical pressure. Thus, models which use corresponding states approaches in a predictive way can never be more accurate than the critical parameters used, even if the assumed similarity of fluids were an exact one. Keeping in mind the typical uncertainty of predicted critical parameters, see Table 6.2, such approaches cannot result in accuracies which satisfy advanced technical demands, except for substances with well known critical parameters. However, for these substances extensive data sets are available in general, allowing the development of more accurate substance specific equations of state as well.

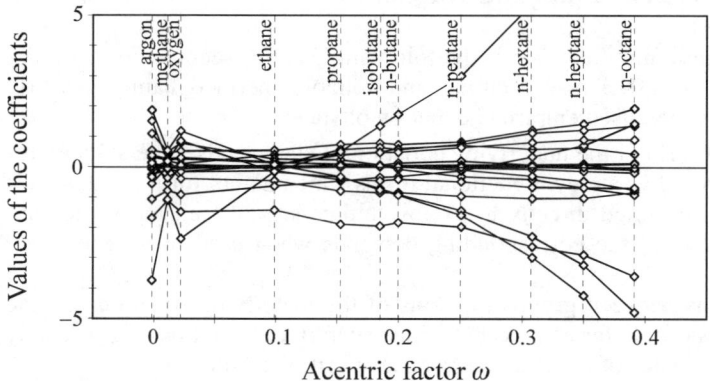

Fig. 7.16. Values of the coefficients of the integrated form of Eq. 7.18 which result from Eq. 7.19 for 11 nonpolar fluids.

The second point is related to the set-up of Eq. 7.19 in combination with the underlying functional form of Eq. 7.18. Figure 7.16 shows the values of the 19 coefficients n_i of the integrated form of Eq. 7.18 plotted over the acentric factor ω for 11 nonpolar substances. Compared to the coefficients of substance specific Bender-type equations, see Fig. 6.3, the resulting plots are of course rather systematic, but on average the coefficients are still large with alternating signs, indicating significant intercorrelations between groups of terms. And even for these substances, which should be described well by a three parameter corresponding states approach depending only on T_c, ρ_c, and ω, the coefficients which result from Eq. 7.19 show obvious discontinuities when plotted over ω. The set-up of Eq. 7.19 implies intercorrelations between the used values for ω and χ and yields "unphysical" oscillations for the coefficients n_i. Amplified by intercorrelations between the groups of terms in Eq. 7.18, these oscillations result in the unpredictable behaviour, especially of derived caloric properties, which was discussed above. Models which are based on simpler multiparameter equations of state such as BWR-type (Benedict et al., 1940) or Starling-type (Starling, 1973) equations minimise these problems, since intercorrelations between terms are less pronounced in these equations, see e.g. Fig. 6.3 for BWR-type equations. However, these simple equations of state are not accurate enough to fulfil advanced technical demands even when used with substance specific coefficients, see Chap. 5.

Türk et al. (1997) proposed the use of a readjusted reducing density instead of ρ_c to improve the results of the model by Platzer and Maurer (1989). This approach can be useful to improve the representation of certain properties, but it cannot solve the general problems discussed above. To use all of the parameters T_c, ρ_c, ω, and χ as adjustable parameters could improve the results of the model for substances with extensive data sets significantly, but when fitted to small data sets the intercorrelations described above cause unpredictable results especially for derived properties. Numerically more stable models are required to use the involved corresponding states parameters as adjustable parameters.

7.2.2 The Approach by Span and Wagner

Based on the discussion in Sect. 7.2.1, the following features seem to be necessary preconditions for a sufficiently accurate and reliable thermodynamic property model based on a generalised empirical equation of state:

- The functional form of the underlying equation of state needs to be suitable for the aspired level of accuracy. Functional forms which do not fulfil the demands on accuracy when fitted directly to extensive data sets of well measured substances cannot fulfil the corresponding demands when used in a generalised model.
- Intercorrelations between groups of terms of the underlying equation of state need to be reduced as far as possible, since such intecorrelations significantly increase the influence of uncertain substance specific parameters.
- Generalised coefficients should be formulated in terms of a single substance specific parameter to avoid intercorrelations which make results of the generalisation less reliable.
- Although the model needs to be based on an extended corresponding states approach in principle, the parameters of this approach must be adjustable to compensate for uncertainties of the critical parameters and for the limited accuracy of the extended corresponding states approach itself.

Starting from these fundamentals, Span and Wagner have developed a generalised empirical equation of state which is able to satisfy advanced technical demands on the accuracy of thermodynamic properties for the first time. The results of this ongoing project are summarised in this section; a corresponding article is being prepared (Span and Wagner, 2000d).

When expressing coefficients of a generalised empirical equation of state as a function of a single substance dependent parameter, the resulting model is necessarily restricted to nonpolar fluids, since a single parameter cannot properly account for both the influence of molecular anisotropy and polar interactions. However, the reduction of intercorrelations is the predominant task when working on a generalised model which is reliable at a high level of accuracy. Thus, the restriction to nonpolar fluids has to be accepted in a first step. The influence of polar interactions needs to be considered in a second step, preferably using an additive set-up like the one of the BACKONE equations of state, see Sect. 7.1. Since this development is not complete yet, this section is restricted to results for non- and weakly polar fluids.

The foundation of generalised empirical models is clearly the underlying equation of state. The simultaneous optimisation algorithm described in Sect. 6.1 resulted in equations of state which are both sufficiently accurate to fulfil the formulated demands on the accuracy of technical equations of state, see Table 6.1, and numerically stable enough to be fitted to small data sets. These equations build a significantly improved basis for a generalised empirical model. In Sect. 6.2.1 it was shown that simultaneously optimised equations of state with 12 terms are numerically stable enough to be fitted to small data sets. Due to their inferior accuracy, numerically more stable equations with only 10 terms (see Fig. 6.9) were rejected at that point; these equations failed to strictly fulfil the formulated de-

Table 7.1. Parameters of the simultaneously optimised functional form used in the generalised equation of state by Span and Wagner (2000d), Eq. 7.20.

i	Form	t_i	d_i	γ_i	p_i
1	Polynomial	0.125	1	0	0
2	Polynomial	1.125	1	0	0
3	Polynomial	1.250	2	0	0
4	Polynomial	0.250	3	0	0
5	Polynomial	0.750	8	0	0
6	Exponential	0.625	2	1	1
7	Exponential	2.000	3	1	1
8	Exponential	4.125	1	1	2
9	Exponential	4.125	4	1	2
10	Exponential	17.000	3	1	3

mands. However, when working on generalised equations, it became apparent that generalisations of the simultaneously optimised equation for nonpolar fluids with 10 terms yield better results than those of the corresponding equation with 12 terms, Eq. 6.12. The comparatively small, but already obvious intercorrelations between the terms in Eq. 6.12 make the results of a generalisation less accurate and less reliable. Of course, the shortcomings of the functional form with 10 terms cannot be overcome by using it in a generalised model, but in this case slightly enlarged uncertainties can easily be accepted, see also footnote 3. Simultaneously optimised equations of state with only 9 terms are numerically even more stable, but their performance becomes significantly worse, both on the basis of substance specific fits and on the basis of generalised use.

Finally, Span and Wagner (2000d) formulated the following equation of state:

$$\frac{a(T,\rho)}{RT} = \alpha^\circ(T,\rho) + \alpha^r(\tau,\delta)$$

$$= \alpha^\circ(T,\rho) + \sum_{i=1}^{10}\left[c_{1,i} + c_{2,i}\,w + c_{3,i}\,w^4\right]\tau^{t_i}\,\delta^{d_i}\,\exp(-\gamma_i\,\delta^{p_i}), \quad (7.20)$$

with $\tau = T_r/T$ and $\delta = \rho/\rho_r$. In Eq. 7.20, T_r, ρ_r, and w are the substance dependent adjustable parameters. The parameters t_i, d_i, p_i, and γ_i were determined by simultaneous optimisation of the functional form considering data sets for 13 non- and weakly polar substances, see Sect. 6.2.2. Table 7.1 summarises the corresponding results. The generalised coefficients $c_{j,i}$ in Eq. 7.20 were determined by a fit to data sets for argon, methane, oxygen, ethane, propane, isobutane, n-butane, n-pentane, n-hexane, n-heptane, and n-octane.[9] Nonlinear algorithms were used to be able to consider available data for all kinds of thermal and caloric properties and to fit the $c_{j,i}$ simultaneously with the substance specific parameters. The used data sets were

---

[9] For the formulations which were used to describe the contribution of the hypothetical ideal gas, α°, see Table 6.3.

Table 7.2. Substance independent coefficients of the generalised equation of state by Span and Wagner (2000d), Eq. 7.20.

i	$c_{1,i}$	$c_{2,i}$	$c_{3,i}$
1	$0.636479524 \cdot 10^{+0}$	$0.822473420 \cdot 10^{+0}$	$-0.186193063 \cdot 10^{+1}$
2	$-0.174667493 \cdot 10^{+1}$	$-0.954932692 \cdot 10^{+0}$	$0.105083555 \cdot 10^{+2}$
3	$-0.144442644 \cdot 10^{-1}$	$-0.745462328 \cdot 10^{+0}$	$0.164032330 \cdot 10^{+1}$
4	$0.679973100 \cdot 10^{-1}$	$0.182685593 \cdot 10^{+0}$	$-0.613747797 \cdot 10^{+0}$
5	$0.767320032 \cdot 10^{-4}$	$0.547120142 \cdot 10^{-4}$	$-0.693188290 \cdot 10^{-3}$
6	$0.218194143 \cdot 10^{+0}$	$0.761697913 \cdot 10^{+0}$	$-0.705727791 \cdot 10^{+1}$
7	$0.810318494 \cdot 10^{-1}$	$0.415691324 \cdot 10^{+0}$	$-0.290006245 \cdot 10^{+1}$
8	$-0.907368899 \cdot 10^{-1}$	$-0.825206373 \cdot 10^{+0}$	$-0.232497527 \cdot 10^{+0}$
9	$0.253122250 \cdot 10^{-1}$	$-0.240558288 \cdot 10^{+0}$	$-0.282346515 \cdot 10^{+0}$
10	$-0.209937023 \cdot 10^{-1}$	$-0.643818403 \cdot 10^{-1}$	$0.254250643 \cdot 10^{+1}$

identical with those described in Sect. 6.2.2 and were also weighted with the demanded uncertainties summarised in Table 6.1. The only difference was that data for the saturated vapour and liquid density, ρ'' and ρ', were used only for reduced temperatures $T/T_c \leq 0.995$. In this way, an unreasonably large influence of a few points in the critical regions was avoided. With the generalised parameters $c_{j,i}$ which are given in Table 7.2, Eq. 7.20 becomes an *empirical three parameter equation of state* for nonpolar fluids.

If the substance specific parameters T_r, ρ_r, and w are known, Eq. 7.20 can be used just like a usual substance specific equation of state with

$$n_i = c_{1,i} + c_{2,i}\, w + c_{3,i}\, w^4. \tag{7.21}$$

In general, the dominant contribution to the n_i results from the linear part in

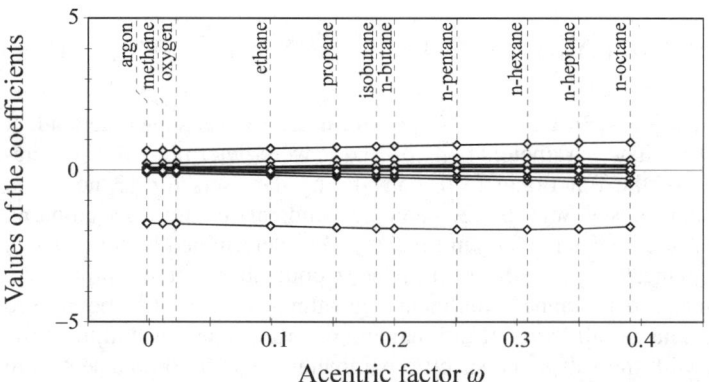

Fig. 7.17. Values of the coefficients n_i which result from Eq. 7.21 for 11 nonpolar fluids.

Eq. 7.21, namely from the $c_{1,i}$ and $c_{2,i}$. For most substances the term $c_{3,i} \cdot w^4$ yields only small corrections which are, however, necessary for an accurate description of the higher alkanes. To illustrate the numerical stability of Eq. 7.20, Fig. 7.17 shows the plot of the n_i resulting from Eq. 7.21 for the 11 nonpolar substances which were used to establish the model. When comparing Fig. 7.17 with Fig. 7.16, the advantages of the three parameter approach and of the numerically very stable simultaneously optimised functional form become obvious.

In some detail, results of the approach by Span and Wagner (2000d) for non- and weakly polar fluids will be discussed in Sect. 7.2.2.2. The numerical stability of Eq. 7.20 will be discussed in Sect. 7.2.2.3. The next section focuses on the techniques which are required to fit the substance specific parameters of Eq. 7.20.

7.2.2.1 Fitting the Substance Specific Parameters

While Eq. 7.20 can be used like common multiparameter equations of state in form of the reduced Helmholtz energy, the determination of the substance specific parameters T_r, ρ_r, and w requires different algorithms. The first fundamental difference is that the T_r, ρ_r, and w can be fitted only with nonlinear algorithms, since the

Table 7.3. Derivatives of the residual part α^r in Eq. 7.20 with respect to T and ρ

Derivative, abbreviation and formulation in $T_r, \rho_r, \tau = T_r/T$, and $\delta = \rho/\rho_r$

$$\alpha^r = \sum_{i=1}^{10} \left[c_{1,i} + c_{2,i} w + c_{3,i} w^4 \right] \delta^{d_i} \tau^{t_i} \exp(-\gamma_i \delta^{p_i})$$

$$\left(\frac{\partial \alpha^r}{\partial \rho}\right)_T = \alpha^r_\rho = \sum_{i=1}^{10} \left[c_{1,i} + c_{2,i} w + c_{3,i} w^4 \right] \frac{\delta^{d_i-1}}{\rho_r} (d_i - \gamma_i p_i \delta^{p_i}) \tau^{t_i} \exp(-\gamma_i \delta^{p_i})$$

$$\left(\frac{\partial^2 \alpha^r}{\partial \rho^2}\right)_T = \alpha^r_{\rho\rho} = \sum_{i=1}^{10} \left[c_{1,i} + c_{2,i} w + c_{3,i} w^4 \right] \frac{\delta^{d_i-2}}{\rho_r^2}$$
$$\cdot \left((d_i - \gamma_i p_i \delta^{p_i})(d_i - 1 - \gamma_i p_i \delta^{p_i}) - \gamma_i^2 p_i^2 \delta^{p_i}\right) \tau^{t_i} \exp(-\gamma_i \delta^{p_i})$$

$$\left(\frac{\partial \alpha^r}{\partial T}\right)_\rho = \alpha^r_T = \sum_{i=1}^{10} -\left[c_{1,i} + c_{2,i} w + c_{3,i} w^4 \right] \frac{t_i}{T_r} \delta^{d_i} \tau^{t_i+1} \exp(-\gamma_i \delta^{p_i})$$

$$\left(\frac{\partial^2 \alpha^r}{\partial T^2}\right)_\rho = \alpha^r_{TT} = \sum_{i=1}^{10} \left[c_{1,i} + c_{2,i} w + c_{3,i} w^4 \right] \frac{t_i(t_i+1)}{T_r^2} \delta^{d_i} \tau^{t_i+2} \exp(-\gamma_i \delta^{p_i})$$

$$\left(\frac{\partial^2 \alpha^r}{\partial \rho \partial T}\right) = \alpha^r_{\rho T} = \sum_{i=1}^{10} -\left[c_{1,i} + c_{2,i} w + c_{3,i} w^4 \right] \frac{\delta^{d_i-1}}{\rho_r} (d_i - \gamma_i p_i \delta^{p_i}) \frac{t_i}{T_r} \tau^{t_i+1} \exp(-\gamma_i \delta^{p_i})$$

$$\left(\frac{\partial^3 \alpha^r}{\partial \rho \partial T^2}\right) = \alpha^r_{\rho TT} = \sum_{i=1}^{10} \left[c_{1,i} + c_{2,i} w + c_{3,i} w^4 \right] \frac{\delta^{d_i-1}}{\rho_r} (d_i - \gamma_i p_i \delta^{p_i}) \frac{t_i(t_i+1)}{T_r^2} \tau^{t_i+2} \exp(-\gamma_i \delta^{p_i})$$

set-up of Eq. 7.20 is nonlinear with respect to these parameters; for details see Sect. 4.1. When fitting the parameters of Eq. 7.20, the critical temperature T_c, the critical density ρ_c, and the acentric factor ω can be used as starting solutions for T_r, ρ_r, and w, respectively.

Just as discussed for BACKONE equations of state in Sect. 7.1.1, the use of reducing parameters as adjustable parameters makes it advantageous to reformulate the residua given in Sects. 4.3.2 and 4.3.3 in terms of derivatives with respect to T and ρ instead of τ and δ. This step can easily be managed using the relations given in Eqs. 3.44–3.49; for an example see Eqs. 7.9 and 7.12. However, to evaluate the rewritten residua, derivatives of the residual part α^r with respect to T and ρ become necessary as well. These derivatives are summarised in Table 7.3. For reasons of simplification, the relations given in this section do not distinguish between polynomial and exponential terms; polynomial terms can simply be treated like exponential terms with $\gamma_i = 0$.

Besides the residua and the derivatives with respect to T and ρ, fitting routines need derivatives of the residua with respect to the adjustable parameters, see Sect. 4.1. These derivatives can be composed from the definition of the residua

Table 7.4. Derivatives of the residual part α^r and its derivatives with respect to T_r

Derivative and formulation in T_r, ρ_r, $\tau = T_r/T$, and $\delta = \rho/\rho_r$

$$\left(\frac{\partial \alpha^r}{\partial T_r}\right)_{w,\rho_r} = \sum_{i=1}^{10}\left[c_{1,i} + c_{2,i}\,w + c_{3,i}\,w^4\right]\delta^{d_i}\frac{t_i}{T_r}\tau^{t_i}\exp\!\left(-\gamma_i\delta^{p_i}\right)$$

$$\left(\frac{\partial \alpha^r_\rho}{\partial T_r}\right)_{w,\rho_r} = \sum_{i=1}^{10}\left[c_{1,i} + c_{2,i}\,w + c_{3,i}\,w^4\right]\frac{\delta^{d_i-1}}{\rho_r}\left(d_i - \gamma_i p_i \delta^{p_i}\right)\frac{t_i}{T_r}\tau^{t_i}\exp\!\left(-\gamma_i\delta^{p_i}\right)$$

$$\left(\frac{\partial \alpha^r_{\rho\rho}}{\partial T_r}\right)_{w,\rho_r} = \sum_{i=1}^{10}\left[c_{1,i} + c_{2,i}\,w + c_{3,i}\,w^4\right]\frac{\delta^{d_i-2}}{\rho_r^2}$$
$$\cdot\left((d_i - \gamma_i p_i \delta^{p_i})(d_i - 1 - \gamma_i p_i \delta^{p_i}) - \gamma_i^2 p_i^2 \delta^{p_i}\right)\frac{t_i}{T_r}\tau^{t_i}\exp\!\left(-\gamma_i\delta^{p_i}\right)$$

$$\left(\frac{\partial \alpha^r_T}{\partial T_r}\right)_{w,\rho_r} = \sum_{i=1}^{10}-\left[c_{1,i} + c_{2,i}\,w + c_{3,i}\,w^4\right]\frac{t_i(t_i+1)}{T_r^2}\delta^{d_i}\tau^{t_i+1}\exp\!\left(-\gamma_i\delta^{p_i}\right)$$

$$\left(\frac{\partial \alpha^r_{TT}}{\partial T_r}\right)_{w,\rho_r} = \sum_{i=1}^{10}\left[c_{1,i} + c_{2,i}\,w + c_{3,i}\,w^4\right]\frac{t_i(t_i+1)(t_i+2)}{T_r^3}\delta^{d_i}\tau^{t_i+2}\exp\!\left(-\gamma_i\delta^{p_i}\right)$$

$$\left(\frac{\partial \alpha^r_{\rho T}}{\partial T_r}\right)_{w,\rho_r} = \sum_{i=1}^{10}-\left[c_{1,i} + c_{2,i}\,w + c_{3,i}\,w^4\right]\frac{\delta^{d_i-1}}{\rho_r}\left(d_i - \gamma_i p_i \delta^{p_i}\right)\frac{t_i(t_i+1)}{T_r^2}\tau^{t_i+1}\exp\!\left(-\gamma_i\delta^{p_i}\right)$$

$$\left(\frac{\partial \alpha^r_{\rho TT}}{\partial T_r}\right)_{w,\rho_r} = \sum_{i=1}^{10}\left[c_{1,i} + c_{2,i}\,w + c_{3,i}\,w^4\right]\frac{\delta^{d_i-1}}{\rho_r}\left(d_i - \gamma_i p_i \delta^{p_i}\right)\frac{t_i(t_i+1)(t_i+2)}{T_r^3}\tau^{t_i+2}\exp\!\left(-\gamma_i\delta^{p_i}\right)$$

Table 7.5. Derivatives of the residual part α^r and its derivatives with respect to ρ_r

Derivative and formulation in T_r, ρ_r, $\tau = T_r/T$, and $\delta = \rho/\rho_r$

$$\left(\frac{\partial \alpha^r}{\partial \rho_r}\right)_{T_r,w} = \sum_{i=1}^{10} -\left[c_{1,i} + c_{2,i}\, w + c_{3,i}\, w^4\right]\left(d_i - \gamma_i p_i \delta^{p_i}\right)\frac{\delta^{d_i}}{\rho_r}\tau^{t_i}\exp\left(-\gamma_i \delta^{p_i}\right)$$

$$\left(\frac{\partial \alpha^r_\rho}{\partial \rho_r}\right)_{T_r,w} = \sum_{i=1}^{10}\left[c_{1,i} + c_{2,i}\, w + c_{3,i}\, w^4\right]\frac{\delta^{d_i-1}}{\rho_r^2}$$
$$\cdot\left(-d_i(d_i-1) + \gamma_i p_i \delta^{p_i}\left(2d_i - 1 + p_i - \gamma_i p_i \delta^{p_i}\right)\right)\tau^{t_i}\exp\left(-\gamma_i \delta^{p_i}\right)$$

$$\left(\frac{\partial \alpha^r_{\rho\rho}}{\partial \rho_r}\right)_{T_r,w} = \sum_{i=1}^{10}\left[c_{1,i} + c_{2,i}\, w + c_{3,i}\, w^4\right]\frac{\delta^{d_i-2}}{\rho_r^3}$$
$$\cdot\Big[-(d_i-2)\left(\left(d_i - \gamma_i p_i \delta^{p_i}\right)\left(d_i - 1 - \gamma_i p_i \delta^{p_i}\right) - \gamma_i^2 p_i^2 \delta^{p_i}\right)$$
$$+\left(\gamma_i p_i^2 \delta^{p_i}\left(2d_i - 1 - 2\gamma_i p_i \delta^{p_i}\right) + \gamma_i^2 p_i^3 \delta^{p_i}\right)$$
$$+\left(\left(d_i - \gamma_i p_i \delta^{p_i}\right)\left(d_i - 1 - \gamma_i p_i \delta^{p_i}\right) - \gamma_i^2 p_i^2 \delta^{p_i}\right)\left(\gamma_i p_i \delta^{p_i}\right)\Big]\tau^{t_i}\exp\left(-\gamma_i \delta^{p_i}\right)$$

$$\left(\frac{\partial \alpha^r_T}{\partial \rho_r}\right)_{T_r,w} = \sum_{i=1}^{10}\left[c_{1,i} + c_{2,i}\, w + c_{3,i}\, w^4\right]\left(d_i - \gamma_i p_i \delta^{p_i}\right)\frac{\delta^{d_i}}{\rho_r}\frac{t_i}{T_r}\tau^{t_i+1}\exp\left(-\gamma_i \delta^{p_i}\right)$$

$$\left(\frac{\partial \alpha^r_{TT}}{\partial \rho_r}\right)_{T_r,w} = \sum_{i=1}^{10} -\left[c_{1,i} + c_{2,i}\, w + c_{3,i}\, w^4\right]\left(d_i - \gamma_i p_i \delta^{p_i}\right)\frac{\delta^{d_i}}{\rho_r}\frac{t_i(t_i+1)}{T_r^2}\tau^{t_i+2}\exp\left(-\gamma_i \delta^{p_i}\right)$$

$$\left(\frac{\partial \alpha^r_{\rho T}}{\partial \rho_r}\right)_{T_r,w} = \sum_{i=1}^{10} -\left[c_{1,i} + c_{2,i}\, w + c_{3,i}\, w^4\right]\frac{\delta^{d_i-1}}{\rho_r^2}$$
$$\cdot\left(-d_i(d_i-1) + \gamma_i p_i \delta^{p_i}\left(2d_i - 1 + p_i - \gamma_i p_i \delta^{p_i}\right)\right)\frac{t_i}{T_r}\tau^{t_i+1}\exp\left(-\gamma_i \delta^{p_i}\right)$$

$$\left(\frac{\partial \alpha^r_{\rho TT}}{\partial \rho_r}\right)_{T_r,w} = \sum_{i=1}^{10} -\left[c_{1,i} + c_{2,i}\, w + c_{3,i}\, w^4\right]\frac{\delta^{d_i-1}}{\rho_r^2}$$
$$\cdot\left(-d_i(d_i-1) + \gamma_i p_i \delta^{p_i}\left(2d_i - 1 + p_i - \gamma_i p_i \delta^{p_i}\right)\right)\frac{t_i(t_i+1)}{T_r^2}\tau^{t_i+2}\exp\left(-\gamma_i \delta^{p_i}\right)$$

and from the derivatives of α^r and its derivatives with respect to the adjustable parameters, see Sect. 4.3.3. Derivatives with respect to w are almost as simple as for the linear coefficients n_i in typical multiparameter equations of state. Using the first density derivative of the reduced Helmholtz energy as an example, the derivative with respect to w becomes

$$\left(\frac{\partial \alpha_\rho^r}{\partial w}\right)_{T_r,\rho_r} = \sum_{i=1}^{10}\left(\frac{\partial[c_{1,i}+c_{2,i}w+c_{3,i}w^4]}{\partial w}\right)_{c_{j,i}}\frac{\delta^{d_i-1}}{\rho_r}(d_i-\gamma_i p_i \delta^{p_i})\tau^{t_i}\exp(-\gamma_i \delta^{p_i})$$

$$= \sum_{i=1}^{10}\left[c_{2,i}+4c_{3,i}w^3\right]\frac{\delta^{d_i-1}}{\rho_r}(d_i-\gamma_i p_i \delta^{p_i})\tau^{t_i}\exp(-\gamma_i \delta^{p_i}). \tag{7.22}$$

In the same way, the necessary relations can be derived from the other derivatives summarised in Table 7.3. However, for T_r and ρ_r the corresponding derivatives become more complex. Table 7.4 summarises the required derivatives with respect to T_r and Table 7.5 summarises the derivatives with respect to ρ_r. When using these relations in iterative nonlinear fits, the values for τ and δ have to be calculated with the current values for T_r and ρ_r in each step.

7.2.2.2 Results for Non- and Weakly Polar Fluids

To demonstrate both the capabilities and the limitations of Eq. 7.20, this section shows representative comparisons between experimental and calculated data for 15 non- and weakly polar fluids. The considered data sets correspond to those discussed in Sect. 6.2.2. Table 7.6 summarises the substance specific parameters as they were obtained from nonlinear fits to the corresponding data sets. For comparison, see the critical parameters T_c and ρ_c and the acentric factors ω which were

Table 7.6. Substance specific parameters of Eq. 7.20 for the 15 non- and weakly polar fluids considered in this section (to be used with $R = 8.314510$ J mol^{-1} K^{-1}).

Substance	T_r	ρ_r	w
n-Alkanes			
Methane	186.659809	163.413536	0.010528102
Ethane	295.159630	207.557649	0.095234716
Propane	354.964211	221.906745	0.149041513
n-Butane	406.785141	230.384826	0.194240287
n-Pentane	449.271155	233.873368	0.247058753
n-Hexane	487.762087	235.700888	0.298052404
n-Heptane	525.389862	235.977855	0.350780196
n-Octane	565.427917	234.605116	0.402698435
Other non- or weakly polar substances			
Argon	147.707801	540.014968	0.000305675
Oxygen	150.875090	439.519141	0.023479051
Nitrogen	122.520245	316.134310	0.043553140
Ethylene	273.316763	216.108926	0.085703183
Isobutane	390.355535	228.302484	0.178714317
Cyclohexane	526.231121	274.647526	0.221837522
SF$_6$	304.013497	747.815849	0.181815238

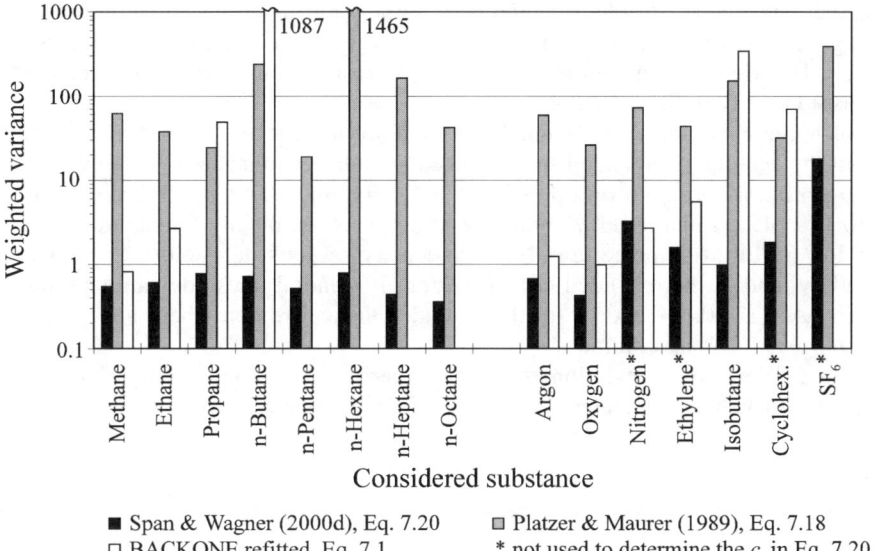

Fig. 7.18. Weighted variances which result from comparisons of experimental data with results obtained from different generalised empirical equations of state for 15 non- and weakly polar fluids.

summarised in Table 6.3. The significant differences between T_r and T_c which are observed for some of the substances can be interpreted as an additional proportionality factor in the effective substance specific coefficients, see Eq. 6.11. The ρ_r and w differ from ρ_c and ω only moderately.

In terms of the weighted variance, Fig. 7.18 compares the results obtained from Eq. 7.20 with those obtained from BACKONE equations of state which were fitted to the same data sets and from comparisons with the predictive model of Platzer and Maurer (1989). The data sets were weighted with the demanded uncertainties summarised in Table 6.1. For alkanes and other common nonpolar fluids, Eq. 7.20 results in weighted variances $\sigma_{wt}^2 \approx 0.4 - 1.0$. Thus, for nonpolar fluids Eq. 7.20 already satisfies the demands initially formulated for the variance of "more accurate future models," see footnote 3. The new formulation is clearly superior to existing generalised equations of state, except for simple nonpolar fluids with small anisotropy where BACKONE equations yield comparable results. However, comparisons which are based only on a single quality criterion cannot describe the performance of an equation of state completely. Representative comparisons with reliable experimental data are given below.

Although designed for nonpolar substances, Eq. 7.20 still yields results which are as good as those obtained with four parameter QUBACKONE equations for fluids with weakly polar molecules like nitrogen and ethylene. Significantly worse results were found only for sulfurhexafluoride. Even though the molecules of sulfurhexafluoride are almost perfectly anisotropic and no polarity is expected, the results found for the thermodynamic properties do not agree with those for other nonpolar substances. Problems which were already discussed for equations of state with generalised functional form, see Sect. 6.2.2, become more pronounced when using a less flexible generalised equation. The results found for cyclohexane are discussed briefly in Sect. 7.2.2.3.

For nonpolar fluids, experimental $p\rho T$ data are mostly represented by Eq. 7.20 within the accuracy demanded for advanced technical applications, namely within $\Delta\rho/\rho \leq \pm 0.2\%$ for pressures up to 30 MPa and within $\Delta\rho/\rho \leq \pm 0.5\%$ at pressures up to 100 MPa. Enlarged deviations are to be observed mainly for liquid states at very low (reduced) temperatures, for high density gaseous states close to the phase boundary, and in the extended critical region. In general, these deviations remain within $\Delta\rho/\rho \leq \pm 0.5\%$ except for the extended critical region where uncertainties of $\Delta p/p \approx \pm 0.3\%$ are common.

Figure 7.19 shows deviations between experimental results for the density of propane and values calculated from Eq. 7.20. Most of the data at moderate pres-

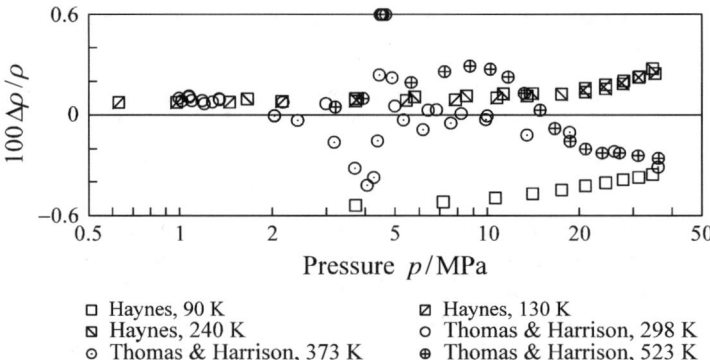

Fig. 7.19. Percentage deviations $100\,\Delta\rho/\rho = 100\,(\rho_{exp} - \rho_{calc})/\rho_{exp}$ between experimental results for the density of propane by Haynes (1983) and Thomas and Harrison (1982) and values calculated from Eq. 7.20.

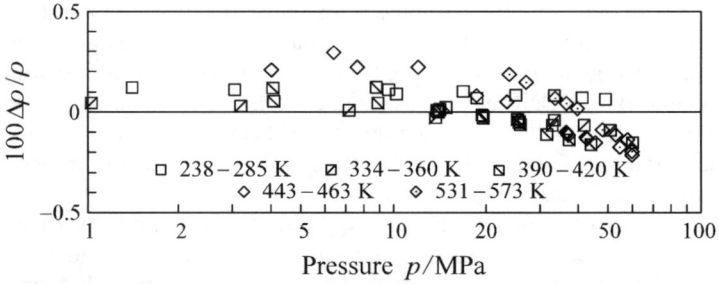

Fig. 7.20. Percentage deviations $100\,\Delta\rho/\rho = 100\,(\rho_{exp} - \rho_{calc})/\rho_{exp}$ between experimental data for the density of n-pentane measured by Kratzke et al. (1985) and values calculated from Eq. 7.20.

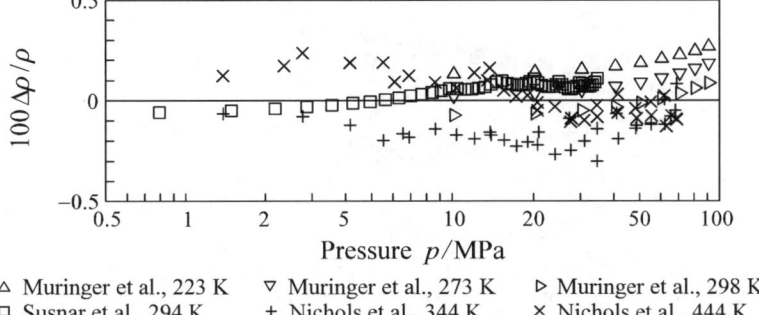

△ Muringer et al., 223 K   ▽ Muringer et al., 273 K   ▷ Muringer et al., 298 K
□ Susnar et al., 294 K   + Nichols et al., 344 K   × Nichols et al., 444 K

Fig. 7.21. Percentage deviations $100\,\Delta\rho/\rho = 100\,(\rho_{\text{exp}}-\rho_{\text{calc}})/\rho_{\text{exp}}$ between experimental data for the density of n-heptane measured by Muringer et al. (1985), Susnar et al. (1992), and Nichols et al. (1955) and values calculated from Eq. 7.20.

sures are represented within $\Delta\rho/\rho \leq \pm 0.2\,\%$. Significantly enlarged deviations become obvious for liquid states at $T = 90$ K ($T/T_c \approx 0.24$) and in the critical region ($T = 373$ K, $T/T_c \approx 1.01$) including the high density gas phase close to the critical temperature. Data for liquid states at $T = 130$ K ($T/T_c \approx 0.35$) are again represented within the expected uncertainties. Comparisons with the same set of data were shown for a SIMBACKONE equation of state in Fig. 7.4.

Figure 7.20 illustrates deviations between measured liquid densities for n-pentane and values calculated from Eq. 7.20. With only a few exceptions, the data are represented by the generalised equation of state within $\Delta\rho/\rho \leq \pm 0.2\,\%$ up to high pressures. For comparison, see also Fig. 6.20 where the same data set is compared with results obtained from the corresponding simultaneously optimised equation of state with 12 terms, Eq. 6.12.

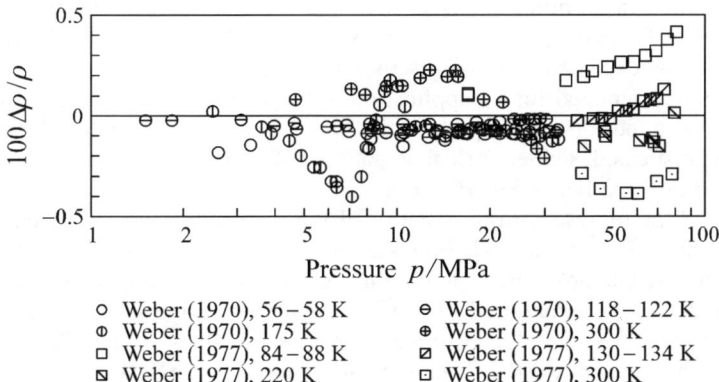

○ Weber (1970), 56–58 K   ⊖ Weber (1970), 118–122 K
⊕ Weber (1970), 175 K   ⊕ Weber (1970), 300 K
□ Weber (1977), 84–88 K   ⊠ Weber (1977), 130–134 K
⊠ Weber (1977), 220 K   □ Weber (1977), 300 K

Fig. 7.22. Percentage deviations $100\,\Delta\rho/\rho = 100\,(\rho_{\text{exp}}-\rho_{\text{calc}})/\rho_{\text{exp}}$ between experimental data for the density of oxygen and values calculated from Eq. 7.20.

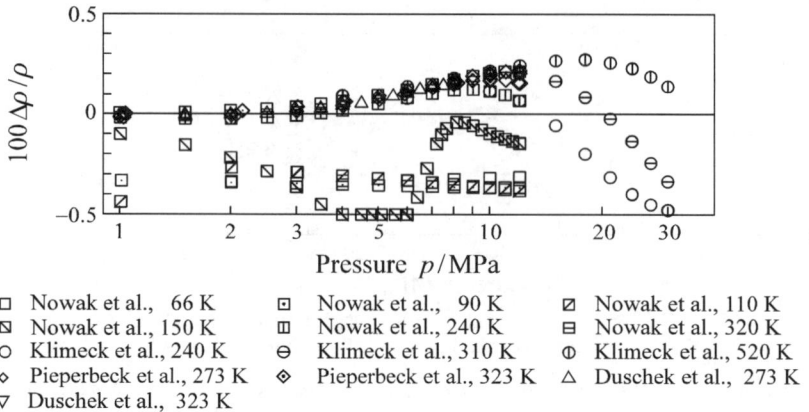

Fig. 7.23. Percentage deviations $100\,\Delta\rho/\rho = 100\,(\rho_{\mathrm{exp}} - \rho_{\mathrm{calc}})/\rho_{\mathrm{exp}}$ between experimental data for the density of nitrogen by Nowak et al. (1997a), Klimeck et al. (1998), Pieperbeck et al. (1991), and Duschek et al. (1988) and values calculated from Eq. 7.20.

For n-heptane, measured liquid densities are compared with results obtained from Eq. 7.20 in Fig. 7.21. Again, the experimental data are mostly represented within $\Delta\rho/\rho \leq \pm 0.2\,\%$ up to high pressures. For comparison see also Fig. 6.21.

Figure 7.22 shows results for the density of oxygen at gaseous, liquid, and supercritical states. Except for a few points on the isotherm $T = 300$ K the experimental data are represented within the limits defined by Table 6.1. The data at $T = 175$ K ($T/T_c \approx 1.13$) for which enlarged density deviations become obvious in Fig. 7.22 are represented well within $\Delta p/p \leq \pm 0.2\,\%$.

For fluids with weakly polar molecules, enlarged deviations are to be observed. This fact is illustrated in Fig. 7.23 where results from Eq. 7.20 are compared with selected $p\rho T$ data for nitrogen; for comparison see also Figs. 5.3, 5.10, and 7.5. For substances like nitrogen and ethylene, typical uncertainties of $p\rho T$ data calculated from Eq. 7.20 are in an order of $\Delta\rho/\rho \approx \pm 0.3\,\% - \pm 0.5\,\%$. On the isotherm $T = 150$ K ($T/T_c \approx 1.19$) the observed deviations increase up to $\Delta\rho/\rho \approx -0.6\,\%$. In general, this result was expected, see the discussion at the beginning of Chap. 5. Equation 7.20 was not intended for an application to fluids with polar molecules. However, the results found for weakly polar fluids are still comparable with results obtained from sophisticated semiempirical equations of state which take polar interactions into account explicitly, see Fig. 7.5.

Results found for the representation of saturated liquid densities are consistent with the results which were discussed for $p\rho T$ data in the homogeneous liquid region above. Accurate data are represented within $\Delta\rho'/\rho' \leq \pm 0.2\,\%$ on the whole. Enlarged uncertainties are encountered only when approaching the critical temperature and in some cases at reduced temperatures below about $T/T_c \approx 0.35$.[10] Slightly larger deviations are found for vapour pressures in general. The typical

---

[10] For most fluids, this shortcoming is not relevant since reduced temperatures $T/T_c < 0.35$ correspond to states below the triple point temperature, see Table 6.3.

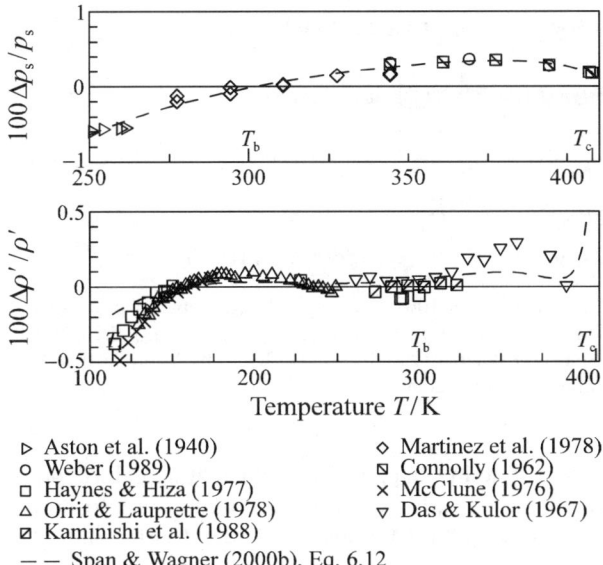

- ▷ Aston et al. (1940)
- ○ Weber (1989)
- □ Haynes & Hiza (1977)
- △ Orrit & Laupretre (1978)
- ⊠ Kaminishi et al. (1988)
- ◇ Martinez et al. (1978)
- ⊠ Connolly (1962)
- × McClune (1976)
- ▽ Das & Kulor (1967)
- — — Span & Wagner (2000b), Eq. 6.12

Fig. 7.24. Percentage deviations $100\,\Delta y/y = 100\,(y_{exp} - y_{calc})/y_{exp}$ with $y = p_s, \rho'$ between selected experimental results for isobutane and values calculated from Eq. 7.20. Values calculated from the corresponding simultaneously optimised equation of state with 12 terms, Eq. 6.12, are plotted as dashed lines for comparison.

uncertainty of calculated vapour pressures is $\Delta p_s/p_s \leq \pm 0.4\,\%$; as usual, increased relative deviations are to be observed for very low vapour pressures. Correspondingly enlarged deviations can also be found for saturated vapour densities which depend directly on the accuracy of calculated vapour pressures. Figure 7.24 illustrates these features using selected experimental results for the vapour pressure and the saturated liquid density of isobutane as an example. For comparison, results of the corresponding simultaneously optimised equation of state with 12 terms, Eq. 6.12, are plotted as dashed lines. The same set of data was used in comparisons with an MBWR-type multiparameter equation of state (Fig. 5.33) and with the generalised model by Platzer and Maurer (Fig. 7.12).

Caloric properties on the vapour-liquid phase boundary are represented generally within the accuracy expected for advanced technical equations of state. For example, Fig. 7.25 shows deviations between experimental results for the heat capacity c_σ of saturated liquid n-butane and values calculated from Eq. 7.12, from the simultaneously optimised equation of state by Span and Wagner (2000b), Eq. 6.12, and from the MBWR-type equation by Younglove and Ely (1987). Results from the generalised equation by Platzer and Maurer (1989) were compared to the same data set in Fig. 7.15. Equations 7.20 and 6.12 yield results which agree with the data within $\Delta c_\sigma/c_\sigma \approx \pm 1\,\% - \pm 2\,\%$. The 32 term equation by Younglove and Ely (1987) yields results of similar accuracy as well, but close to the triple point temperature significantly larger deviations are observed, see also the discus-

7 Generalised Equations of State

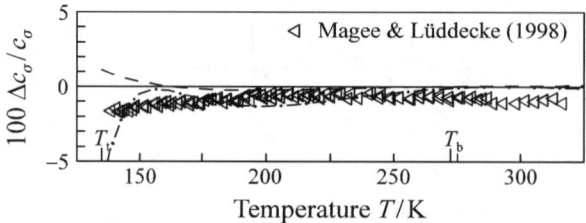

– – – Span & Wagner (2000b), Eq. 6.12   – · – · – Younglove & Ely (1987)

Fig. 7.25. Percentage deviations $100\,\Delta c_\sigma/c_\sigma = 100\,(c_{\sigma,\mathrm{exp}} - c_{\sigma,\mathrm{calc}})/c_{\sigma,\mathrm{exp}}$ between experimental results for n-butane by Magee and Lüddecke (1998) and values calculated from Eq. 7.20. Values calculated with the equations by Span and Wagner (2000b), Eq. 6.12, and by Younglove and Ely (1987) are plotted as dash-dotted lines for comparison.

sion on MBWR-type equations of state in Chap. 5. When calculating caloric properties on the phase boundary from Eq. 7.20, increased uncertainties are to be expected only at very low reduced temperatures for speeds of sound in the saturated liquid.

With respect to the representation of thermal properties on the phase boundary, Eq. 7.20 is clearly inferior to the simultaneously optimised equations of state by Span and Wagner (2000b), Eq. 6.12. However, the accuracy of Eq. 7.20 is still sufficient for most technical applications and Eq. 7.20 is far superior to other generalised models. With regard to the representation of caloric properties on the phase boundary, Eq. 7.20 comes close to the performance of Eq. 6.12.

Regarding the representation of caloric properties at homogeneous states, Eq. 7.20 is only slightly inferior to the corresponding simultaneously optimised equations of state, Eq. 6.12. Significantly enlarged deviations are observed only

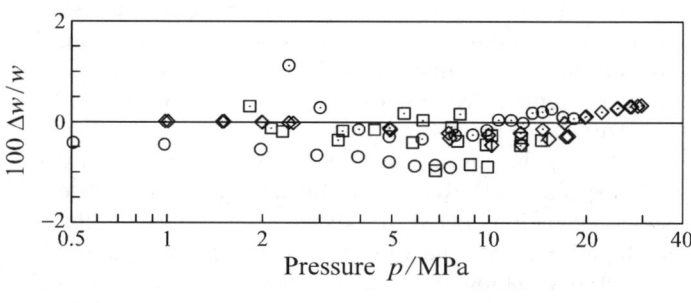

□ Straty, 210 K   ▫ Straty, 240 K
◇ Trusler & Costa Gomes, 250 K   ◆ Trusler & Costa Gomes, 350 K
○ Van Itterbeek et al., 111 K   ● Van Itterbeek et al., 170 K

Fig. 7.26. Percentage deviations $100\,\Delta w/w = 100\,(w_{\mathrm{exp}} - w_{\mathrm{calc}})/w_{\mathrm{exp}}$ between experimental data for the speed of sound in methane by Straty (1975), Trusler and Costa Gomes (1996), and van Itterbeek et al. (1967) and values calculated from Eq. 7.20.

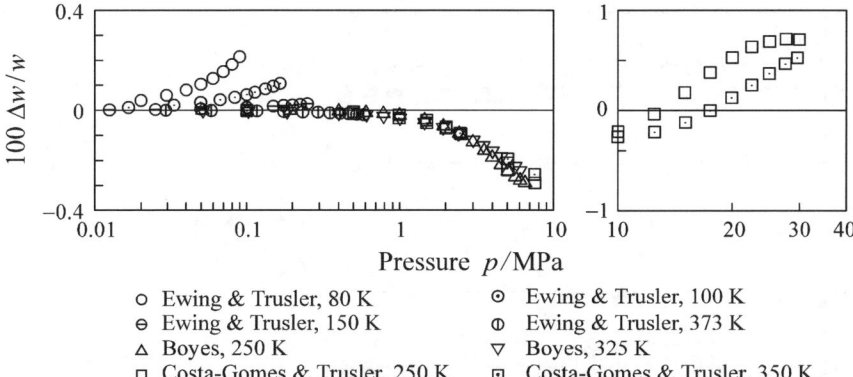

Fig. 7.27. Percentage deviations $100 \, \Delta w/w = 100 \, (w_{\text{exp}} - w_{\text{calc}})/w_{\text{exp}}$ between experimental data for the speed of sound in nitrogen by Ewing and Trusler (1992), Boyes (1992), and Costa Gomes and Trusler (1998a) and values calculated from Eq. 7.20.

for speeds of sound at low temperature $(T/T_c < \approx 0.35)$ liquid states. As an example for the representation of speeds of sound at typical liquid and supercritical states, Fig. 7.26 shows the representation of experimental results for methane. Data at liquid states are usually represented within $\Delta w/w \leq \pm 2\%$ and data at supercritical states are represented within $\Delta w/w \leq \pm 1\%$. In general, the demands formulated in Table 6.1 are satisfied.

For weakly polar fluids slightly enlarged uncertainties are encountered for caloric properties as well. However, at gaseous and supercritical states the enlarged uncertainties hardly become obvious. Figure 7.27 shows deviations between highly accurate speeds of sound for nitrogen and values calculated from Eq. 7.20. The same set of data was used for comparisons to different equations of state in Figs. 5.16, 5.18, 5.21, and 6.24. At gaseous and gas-like supercritical states, Eq. 7.20 is clearly inferior to the simultaneous optimised equation of state, Eq. 6.12. At higher pressures, both equations yield similar results. The observed deviations stay within the demanded uncertainty of $\Delta w/w \leq \pm 1\%$ and they are significantly smaller than those found for common technical equations of state, see Fig. 5.21.

In Fig. 7.28, a comparison to the recent data from Abdulagatov (1998) is used to assess the accuracy of isochoric heat capacities calculated from Eq. 7.20 for higher alkanes. The representation of these data is similar to the predictive results obtained from the simultaneously optimised equation, Eq. 6.12, see Fig. 6.15. Significantly enlarged deviations can again be found in the extended critical region, where simple equations of state cannot be expected to represent isochoric heat capacities accurately, see Sects. 4.5.1 and 4.5.2. Keeping in mind that relative deviations for isochoric heat capacities are usually larger than those for isobaric heat capacities, the representation of these data satisfies typical technical demands. The performance of Eq. 7.20 is clearly superior to other generalised and technical equations of state, especially at liquid states.

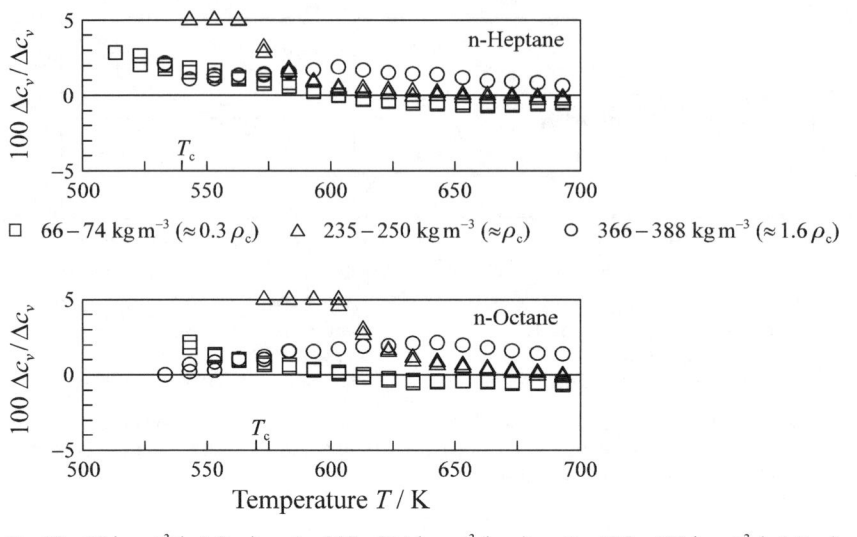

Fig. 7.28. Percentage deviations $100\,(c_{v,\text{calc}} - c_{v,\text{exp}})/c_{v,\text{exp}}$ between experimental data for the isochoric heat capacity of n-heptane and n-octane by Abdulagatov (1998) and values calculated from Eq. 7.20.

For a broad variety of nonpolar fluids which ranges from fluids with spherical molecules such as argon or methane to molecules with a pronounced elongation such as n-heptane or n-octane, Eq. 7.20 describes thermodynamic properties with an accuracy which comes close to the accuracy of simultaneously optimised equations of state with 12 terms, see Sect. 6.2.2. Compared to simultaneously optimised equations, disadvantages are to be observed mainly for thermal properties on the phase boundary, for liquid states at low (reduced) temperatures, and for high density gaseous states close to the phase boundary. With only three substance specific parameters, the performance of Eq. 7.20 is comparable with that of Bender-type equations of state, see Sect. 5.1, and satisfies the demands which were initially formulated for advanced generalised models. For nonpolar substances, for which only very restricted data sets are available, the use of generalised equations of state in the form of Eq. 7.20 is a promising alternative to simultaneously optimised equations of state which were discussed in Sect. 6.2. In this case, the superior numerical stability of the three parameter equation of state may be more important than the superior accuracy of the 12 term equation – however, this superior numerical stability still needs to be proved.

7.2.2.3 Numerical Stability

In Sect. 6.2.1.1 it was shown that equations of state with simultaneously optimised functional form are numerically much more stable than common technical equa-

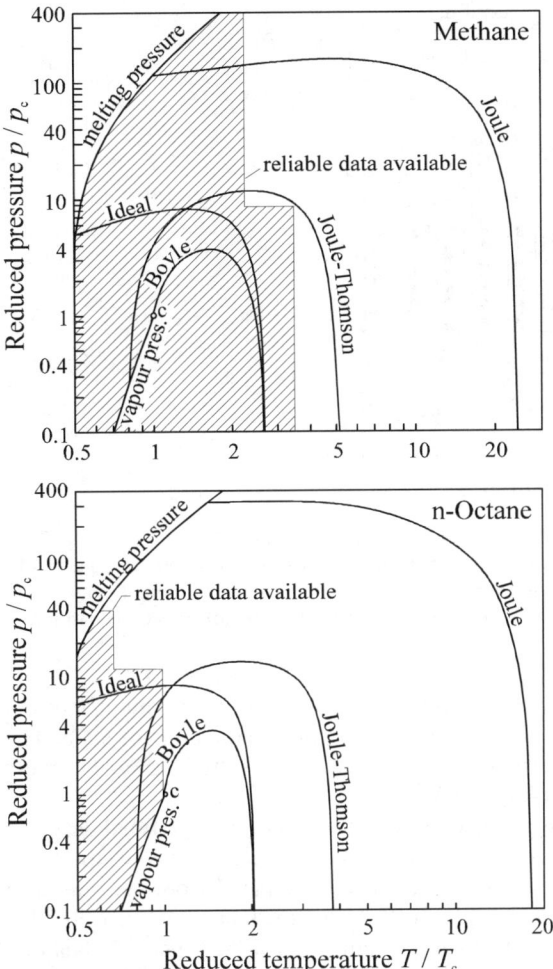

Fig. 7.29. Ideal curves of the compression factor calculated from the generalised equation of state by Span and Wagner (2000d), Eq. 7.20, for methane and n-octane. For a definition of the ideal curves see Sect. 4.6.3. The hatched areas correspond to the regions, where reliable data are available for thermodynamic properties of the corresponding substance.

tions of state. Thus, comparisons between the numerical stability of Eq. 6.12 and Eq. 7.20 necessarily have to be comparisons on a very high level.

As a first indication for the numerical stability of the generalised equation of state by Span and Wagner (2000d), Eq. 7.20, Fig. 7.29 shows plots of the Ideal curve, the Boyle curve, the Joule-Thomson inversion curve, and the Joule inversion curve (see also Sect. 4.6.3) obtained for methane and n-octane. Up to the Joule curve, the new three parameter equation of state is able to predict proper

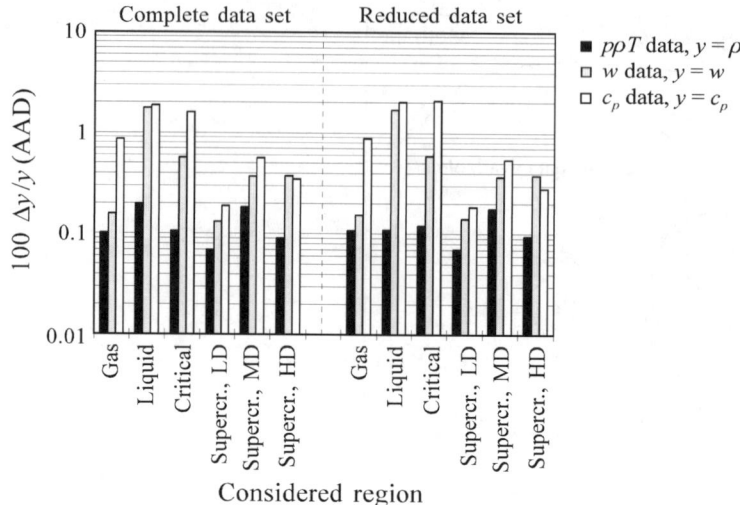

Fig. 7.30. Percentage average absolute deviations between densities, speeds of sound, and isobaric heat capacities calculated from Eq. 7.20 for methane and from the reference equation by Setzmann and Wagner (1991). For an explanation of the different supercritical regions see Sect. 6.2.1.1. In the critical region, pressure deviations were used instead of density deviations.

ideal curves even when fitted to very restricted data sets. However, this is still not an advantage when compared to simultaneously optimised equations of state. The numerically stable functional form of these equations resulted in qualitatively similar results, see Figs. 6.13, 6.14, and 6.35.

More systematic comparisons can be carried out based on well defined data sets calculated from highly accurate reference equations of state. Therefore, the comparisons carried out in Sect. 6.2.1.1 for the data set calculated from the reference equation of state by Setzmann and Wagner (1991) for methane were repeated for the generalised equation of state by Span and Wagner, Eq. 7.20. Under the heading "*complete data set*", Fig. 7.30 shows results of a fit of Eq. 7.20 to all 511 $p\rho T$, 511 c_p, 511 w, 20 p_s, 20 ρ', and 20 ρ'' data which were calculated from the equation by Setzmann and Wagner (1991). As expected, the results for Eq. 7.20 are in general inferior to the corresponding results of Eq. 6.12, see Fig. 6.11 for comparison. However, for all properties and in all regions, the average absolute deviations (see Eq. 6.10) found for Eq. 7.20 are smaller than the uncertainties demanded in Table 6.1. For a generalised equation of state, this result is absolutely satisfactory and it is consistent with the comparisons shown in the preceding section.

Under the heading "*reduced data set*", Fig. 7.30 shows results which were obtained from fitting Eq. 7.20 to just *8 data points*: 4 $p\rho T$ data at $T = 230$ K and $p = 5$ MPa, 10 MPa, 15 MPa, and 20 MPa, 2 saturated liquid densities at $T = 130$ K and 170 K and 2 vapour pressures at the same temperatures. T_c, ρ_c, and ω were used as starting values for the substance specific parameters of Eq. 7.20. The comparisons consider only those data from the complete data set, which were not used

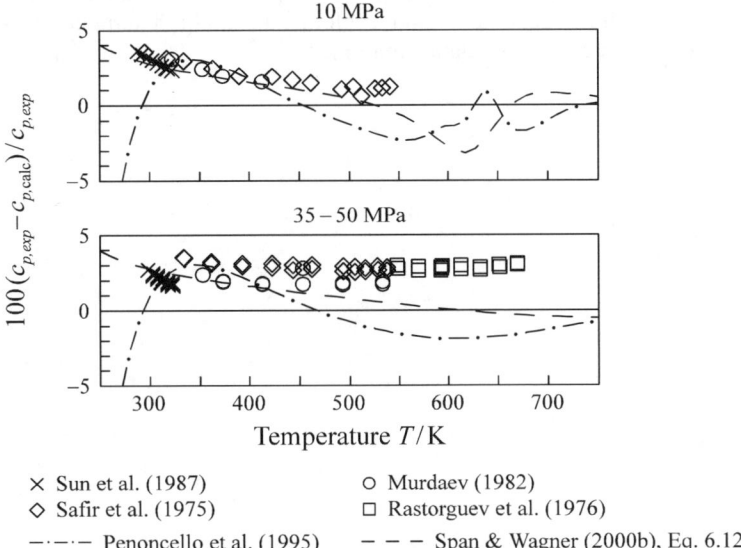

Fig. 7.31. Percentage deviations between experimental results for the isobaric heat capacity of cyclohexane and values calculated from Eq. 7.20. For comparison, values calculated from the equation of Penoncello et al. (1995) and from the simultaneously optimised equation of state, Eq. 6.12, are plotted as dash-dotted lines.

to fit the equation. *No significant differences become obvious between the equation fitted to all 1593 data and the equation which was fitted to just 8 data.* Obviously, a very small number of accurate experimental results is sufficient to adjust the substance specific parameters of Eq. 7.20. Any attempt to fit the 12 coefficients of Eq. 6.12 to the same data set is bond to fail due to numerical limitations – the number of data has to be larger than the number of fitted parameters when using the algorithms described in Sect. 4.1.

The numerical stability of Eq. 7.20 can be advantageous for investigations of quite large, but inconsistent data sets as well. Based on results of simultaneously optimised equations of state, the data set for cyclohexane was discussed in Sect. 6.2.2. When fitted to the selected data set for cyclohexane, Eq. 7.20 results in an enlarged weighted variance, see Fig. 7.18. However, this enlarged weighted variance does not result primarily from data at supercritical temperatures, but from enlarged deviations to old data for the saturated liquid density at high temperatures. These data are represented well (and may be too well) by Eq. 6.12 and by the equation of Penoncello et al. (1995). Isobaric heat capacities at high temperatures are represented by Eq. 7.20 as well as by Eq. 6.12 and even better than by the equation by Penoncello et al. (1995), see Fig. 7.31. The remaining systematic offset may very well be caused by Eq. 7.20 and seems at any rate more reasonable than the trends predicted by the numerically less stable equations of state. From these results, it seems likely that the observed problems are caused by inconsistencies among the experimental results for thermal properties. New data for the density of saturated and homogeneous liquid cyclohexane at high temperatures may already be sufficient to clarify the data situation.

8 Describing Mixtures with Multiparameter Equations of State

When discussing the application of certain thermodynamic property models to mixtures, problems result frequently from the fact that demands on such models are very different depending on the assumed application. Thus, it is important to characterise the type of intended application *before* discussing certain models. To do so, one may distinguish between two major groups of applications which result in completely different demands on thermodynamic property models.[1]

The first group of applications deals mainly with the prediction of vapour-liquid, liquid-liquid, and vapour-vapour equilibria, with the description of very complex systems such as mixtures which contain polyatomic molecules or a large amount of electrolytes, with strongly associating mixtures, or even with systems where mixture effects have to be described in combination with chemical reactions. Such problems are typical, e.g., for the development of new products, processes, plants and apparatuses in chemical engineering. Thermodynamic property models which are applied are usually G^E models and cubic equations of state in combination with group contribution methods (see for instance Gmehling et al., 1993; Gmehling, 1995), models for polymer solutions (see for instance Sadowski et al., 1997; Sadowski and Arlt, 1995), models for electrolyte solutions (see for instance Kuranov et al., 1997), and models which take into account chemical reactions (see for instance Maurer, 1996). Compared to empirical multiparameter equations of state, such models yield only qualitatively correct results – the scientific challenge lies in the complexity or variety of the described systems. And even results which are only qualitatively correct are quite often difficult to obtain for the mentioned tasks in the chemical and petrochemical industry, but in other areas as well.

The second group of applications deals with the prediction of phase equilibria as well, but the focus lies mainly on different thermodynamic properties such as densities and caloric properties on the phase boundaries and at homogeneous states, often up to quite high pressures. Such problems are typical for engineering tasks in power, pipelining, and refrigeration techniques, air and gas liquefaction plants, for hydrocarbon processing, natural gas and oil production, for the design and optimisation of apparatuses and machines consuming large amounts of energy,

---

[1] To distinguish between only two groups of applications is of course a gross simplification which needs further refinement when discussing the applicability of a certain model for a given task. However, this attempt is helpful in explaining the general concept of using multiparameter equations of state to describe mixtures.

and last but not least for custody transfer based on measured fluid flows. Thermodynamic models which are applied are usually empirical mixture approaches like those of Bender (1973) and Starling (1973), extended corresponding states models like the one by Plöcker et al. (1978), shape factor concepts (see for instance Ely, 1990), accurate equations of state for pure fluids (see Table 5.1) or for well defined mixtures (see for instance Starling and Savidge, 1992; Lemmon, 1997), but still simple cubic equations of state as well. The described systems are less complicated than those discussed above – the scientific challenge lies in the accuracy of the description. Accurate results are frequently needed to develop competitive techniques or to improve the performance of known processes, apparatuses, and machines.

In the foreseeable future, the development of more accurate mixture models which are based on empirical multiparameter equations of state will address the second group of applications. The models which are in use today are frequently obsolete and the potential for improvements is as large as the variety of applications. Tasks like the development of an accurate thermodynamic property model for all processes which deal with natural gas are major scientific challenges (Span, 1996; Jaeschke, 1998). However, these challenges do not interfere with those which result from the first group of applications – in fact developments in both areas supplement each other.

8.1 Using Composition Dependent Sets of Coefficients

For a pure substance, the most general way to write a simple empirical equation of state in form of the Helmholtz energy is

$$\frac{a(T,\rho)}{RT} = \alpha^{\circ}(T,\rho) + \sum_{i=1}^{I} n_i \left(\frac{T_r}{T}\right)^{t_i} \left(\frac{\rho}{\rho_r}\right)^{d_i} \exp\left(-\gamma_i \left(\frac{\rho}{\rho_r}\right)^{p_i}\right), \tag{8.1}$$

see also Sect. 3.1.1. In general, T_r and ρ_r correspond to the critical temperature and density. The parameter γ_i is equal to 0 for polynomial terms and usually equal to 1 for exponential terms.

With regard to the ideal gas term, it is easy to rewrite Eq. 8.1 for a mixture. From thermodynamics it is known that the relation

$$\frac{a_M^\circ(T,\rho,\mathbf{x})}{RT} = \alpha_M^\circ(T,\rho,\mathbf{x}) = \sum_{j=1}^{C} x_j\, \alpha_j^\circ(T,\rho) + \sum_{j=1}^{C} x_j \ln(x_j) \tag{8.2}$$

holds for the reduced Helmholtz energy of a mixture of ideal gases with C components. In Eq. 8.2, R is the molar gas constant, ρ the molar density, and the vector $\mathbf{x}$ and the x_j denote the molar composition of the mixture.

However, no thermodynamically exact mixing rules are known for the residual part of the Helmholtz energy. Following the common set-up, e.g., for the application of cubic equations of state to mixtures (see Reid et al., 1987), one may rewrite the residual part of the reduced Helmholtz energy as

$$\frac{a_M^r(T,\rho,\mathbf{x})}{RT} = \alpha_M^r(T,\rho,\mathbf{x}) = \sum_{i=1}^{I} m_i(\mathbf{x}) \left(\frac{T_{r,M}}{T}\right)^{t_i} \left(\frac{\rho}{\rho_{r,M}}\right)^{d_i} \exp\left(-\gamma_i \left(\frac{\rho}{\rho_{r,M}}\right)^{p_i}\right) \quad (8.3)$$

with
$$m_i(\mathbf{x}) = \sum_{j=1}^{C} x_j\, n_{j,i}. \quad (8.4)$$

However, this very simple *mixing rule* results in a number of problems. First of all, Eqs. 8.3 and 8.4 assume that equations of state with identical structure, thus with identical values for I, t_i, d_i, γ_i, and p_i are available for all C components of the mixture. This has not been a problem when corresponding approaches in terms of pressure were discussed in combination with modified BWR-type equations in the 70's, since equations with identical functional form were available for a variety of substances. But such approaches could not be applied to state-of-the-art reference equations with substance specific optimised functional form.

The second problem results from the reducing parameters $T_{r,M}$ and $\rho_{r,M}$ of the mixture. Unlike the corresponding parameters for the pure fluids, these parameters do not correspond to the critical parameters of the mixture. They are pseudocritical parameters which have to be determined from the pure fluid parameters by some kind of mixing and combination rules. Common formulations are, e.g., the *van der Waals mixing rules*

$$\rho_{r,M}^{-1} = \sum_{j=1}^{C}\sum_{k=1}^{C} x_j x_k\, \rho_{c,j,k}^{-1} \quad \text{and} \quad T_{r,M}\, \rho_{r,M}^{-1} = \sum_{j=1}^{C}\sum_{k=1}^{C} x_j x_k\, T_{c,j,k}\, \rho_{c,j,k}^{-1} \quad (8.5)$$

with $\rho_{c,j,j} = \rho_{c,j}$, $T_{c,j,j} = T_{c,j}$, and with the so called *Lorentz-Berthelot combination rule*

$$\rho_{c,j,k}^{-1} = \frac{1}{8}\left(\rho_{c,j}^{-1/3} + \rho_{c,k}^{-1/3}\right)^3 \quad \text{and} \quad T_{c,j,k} = \sqrt{T_{c,j}\, T_{c,k}}. \quad (8.6)$$

The van der Waals mixing rules can be justified by molecular considerations. However, less complicated linear or quadratic mixing rules such as

$$\rho_{r,M}^{-1} = \sum_{j=1}^{C} x_j\, \rho_{c,j}^{-1}, \quad T_{r,M} = \sum_{j=1}^{C} x_j\, T_{c,j} \quad (8.7)$$

and
$$\rho_{r,M}^{-1} = \sum_{j=1}^{C}\sum_{k=1}^{C} x_j x_k\, \rho_{c,j,k}^{-1}, \quad T_{r,M} = \sum_{j=1}^{C}\sum_{k=1}^{C} x_j x_k\, T_{c,j,k} \quad (8.8)$$

may be used as well, where Eqs. 8.5, 8.7, and 8.8 can be used with a variety of combination rules.[2] Significantly better results can be obtained when using mixing and combining rules which introduce *binary interaction parameters*, empirical

---

[2] For a discussion of different mixing and combination rules see for instance Plöcker (1977).

parameters which exercise an influence on $\rho_{r,M}$ and $T_{r,M}$. A typical formulation for an adjustable combination rule is

$$\rho_{c,j,k}^{-1} = \frac{1}{8} b_{j,k}^{\rho} \left(\rho_{c,j}^{-1/3} + \rho_{c,k}^{-1/3} \right)^3 \quad \text{and} \quad T_{c,j,k} = b_{j,k}^{T} \sqrt{T_{c,j} T_{c,k}}, \tag{8.9}$$

where the parameter $b_{j,k}^{\rho}$ and $b_{j,k}^{T}$ have to be adjusted to data for each binary subsystem.[3] Recently, the influence of such adjustable parameters has been discussed in some detail by Tillner-Roth (1998). Finally, the decision for certain mixing and combining rules is a purely empirical one. The only constraint which is strictly valid is that Eq. 8.3 needs to connect smoothly to the pure fluid limits.

The third problem involved by Eq. 8.3 is related to the mixing rules for the coefficients m_i. In the way Eq. 8.4 is formulated, it is assumed that the mixture behaves ideally, or that any non-ideality can be compensated by suitable expressions for $\rho_{r,M}$ and $T_{r,M}$. A more rigorous way to determine the coefficients of Eq. 8.3 would be

$$m_i = \sum_{j=1}^{C} \sum_{k=1}^{C} x_j x_k n_{j,k,i} \tag{8.10}$$

with $n_{j,j,i} = n_{j,i}$. The I parameters $n_{j,k,i}$ for the binary subsystem j, k could be fitted to corresponding data theoretically. However, to do so would require a large number of data for each binary subsystem. Thus, in practice empirical combination rules with or without adjustable interaction parameters again have to be formulated.

If the necessary parameters are determined, Eqs. 8.2 and 8.3 can be used to calculate properties in the homogeneous phases just like pure component equations of state, see Sect. 3.2. The corresponding relations do not involve derivatives of α with respect to composition, and thus derivatives of Eqs. 8.2 and 8.3 can be used as given in Tables 3.1–3.3. The mixture is described as a single pseudo-pure fluid; the discussed mixture approach belongs to the group of *one-fluid models*. The only exceptions are calculations of phase equilibria where different strategies are required, since the compositions of vapour and liquid phase are not identical and depend on temperature and pressure. Algorithms which are suitable to calculate phase equilibria for mixtures were discussed in detail by Heidemann (1983), Michelsen (1982a/b, 1984), and Elhassan et al. (1996).[4]

However, when approaches based on a mixing of coefficients were actually developed in the 70's, the set-up of these models was less systematic than the one discussed above. This is especially true in the case of corresponding states where the results were not considered properly. For mixtures of nitrogen, argon, and

---

[3] In multicomponent mixtures, $b_{j,k}^{\rho} = b_{j,k}^{T} = 1$ can be used frequently for the binary interaction parameters of subsystems involving only minor components.

[4] Since the problems which have to be solved for mixtures are much more complicated, realised algorithms are still less reliable than those for pure fluids. Further work on this area will be necessary to apply multiparameter equations of state to increasingly complex mixtures. This statement holds for the mixture approaches discussed in the following sections as well. See also the discussion on multiple Maxwell loops, Sect. 3.3.5.1.

oxygen, Bender (1973) wrote his pressure explicit equation of state in a form which corresponds to

$$\frac{p}{\rho RT} = 1 + \left(n_1 - \frac{n_2}{T} - \overline{B}\right)\rho + C\rho^2 + D\rho^3 + E\rho^4 + F\rho^5 +$$
$$+ \left(G\rho^2 + H\rho^4\right)\exp\left(-n_{20}\rho^2\right), \tag{8.11}$$

where density and gas constant are molar properties. The $\overline{B}$, C, D, E, F, G, and H in Eq. 8.11 are temperature dependent functions which summarise polynomials in T each with 1−3 coefficients n_i. Simple mixing rules without underlying combination rules or adjustable binary parameters were formulated for n_1, n_2, n_{20}, $\overline{B}$, C, D, E, F, G, and H.

The set-up of Eq. 8.11 becomes questionable mainly due to the use of dimensional values for temperature and density. For an equation of state which is formulated in this way, the coefficients n_i become very different for similar substances with different values of T_c and ρ_c, see the relations given for conversion of Eq. 3.25. For the critical temperature, this effect is compensated for by using mixing rules for groups of terms which already include the temperature dependence. But the mixing rules cannot compensate for different values of ρ_c. Thus, Eq. 8.11 could be applied successfully to mixtures of argon, nitrogen, and oxygen, since these substances have rather similar critical properties, but it yields unsatisfactory results for mixtures of fluids with very different (molar) critical densities. The lack of binary interaction parameters makes an adjustment to mixtures with significantly different interactions between molecules of the same kind and of different kinds impossible.

To apply mixing rules to groups of terms rather than to single coefficients is a reasonable way to minimise effects which are caused by the numerical instability of Bender-type equations of state, see Fig. 6.3. In this way, intercorrelations between single terms are partly compensated within the corresponding group of terms. However, such mixing rules still fail if substances like higher alkanes are involved for which unreasonable sets of pure component coefficients were determined.

Starling (1973) addressed the problem of lacking numerical stability even more rigorously. He wrote his modification of the BWR-type equation of state with dimensional coefficients in terms of molar density and temperature as well and formulated mixing rules for every single coefficient. The mixing rules for the coefficients which contribute to the second virial coefficient contain an adjustable binary interaction parameter which was introduced to improve the description of phase equilibria in particular. For coefficients which contribute to higher virial coefficients, cubic mixing rules are used. The set-up of these mixing rules implies a combination rule without adjustable parameters. However, the substance specific coefficients are not determined by fits to data sets for individual substances but in a generalised way based on linear relations in terms of the acentric factor ω. Thus, an unreasonable scatter of the coefficients becomes impossible but the accuracy in the description of the pure components is sacrificed especially for polar substances which cannot be described accurately with three parameter corresponding states approaches. The general critique on mixture approaches which use dimensional variables is valid for the approach of Starling (1973) as well.

The introduction of binary interaction parameters and reasonable sets of coefficients for substances with restricted data sets led to a reliable description of typical light hydrocarbon mixtures. Although the obtained results are reasonably accurate only at gaseous and gas-like supercritical states,[5] the model of Starling (1973) is still in use especially in the petrochemical industry.

Frequently, it has been argued that mixture approaches like those of Bender (1973) and Starling (1973) cannot yield satisfactory results, since mixing rules have to be formulated for parameters which do not have a physical meaning. Based on the discussion above and on the findings in Chaps. 5–7, this conclusion seems questionable. Actually, the observed problems seem to depend mainly on an inappropriate consideration of corresponding states principles, on missing or not sufficiently effective binary interaction parameters, and on lacking numerical stability and accuracy of the underlying pure component equations of state. Approaches which are based on mixture rules for coefficients cannot be applied to recent reference equations of state with substance specific functional forms. However, the new class of simultaneously optimised (see Sect. 6.2) and generalised (see Sect. 7.2.2) equations of state for advanced technical applications could form a new basis for such approaches. At least for mixtures of nonpolar fluids, more accurate formulations seem possible even though different approaches may have a higher potential. But the application of equations of state with mixture dependent coefficients is numerically more convenient than the application of the models described in Sects. 8.2 and 8.3, especially if no phase equilibrium calculations are involved.

8.1.1 The AGA8-DC92 Equation of State for Natural Gases

In 1981, the Gas Research Institute in Chicago initiated a research project which was aimed to improve the capabilities for accurate computation of compressibility factors of natural gases beyond the temperature, pressure, and composition ranges which were addressed by earlier equations of state. The whole project was carried out in cooperation with the Transmission Measurement Committee of the American Gas Association (AGA) involving contributions from the Groupe Européen de Recherches Gazières (GERG) as well. The theoretical work was carried out mainly by the groups around *K. E. Starling*, University of Oklahoma, and *R. T Jacobsen*, Center for Applied Thermodynamic Studies, University of Idaho. In 1992, the American Gas Association presented the final result of this project, the so called AGA8-DC92 equation of state by Starling and Savidge (1992). Written in terms of the compression factor, this equation reads

---

[5] See also Chap. 5 for a discussion on the performance of Starling-type equations of state for pure fluids.

8.1 Using Composition Dependent Sets of Coefficients

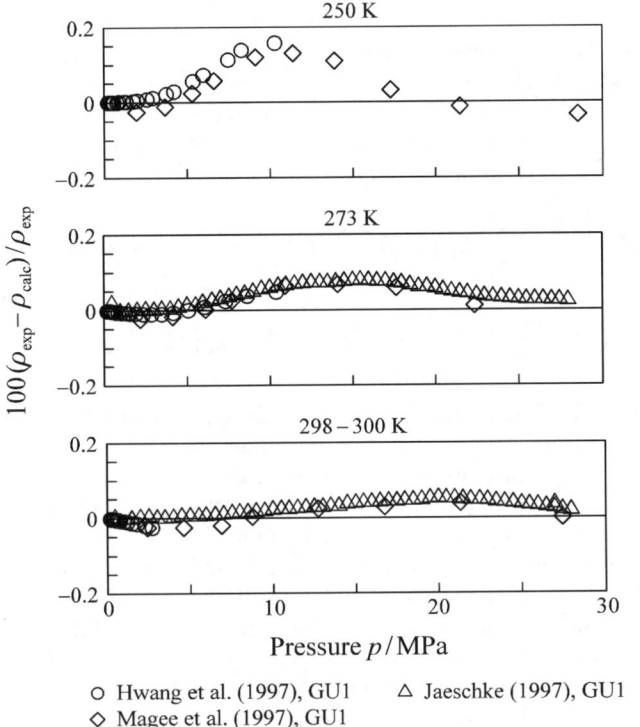

○ Hwang et al. (1997), GU1   △ Jaeschke (1997), GU1
◇ Magee et al. (1997), GU1

Fig. 8.1. Percentage deviations between accurate experimental results for the density of a natural gas like mixture considered in a round robin test which was initiated by the Gas Research Institute and values calculated from the AGA8-DC92 equation of state. For the composition of the investigated mixture (code GU1), see the cited references.

$$\frac{p(T,\rho,\mathbf{x})}{\rho RT} = Z(T,\rho,\mathbf{x}) = 1 + \left(\frac{B}{K^3} - \sum_{n=13}^{18} C_n^* T^{-u_n}\right)\delta$$

$$+ \sum_{n=13}^{58} C_n^* T^{-u_n} \left(b_n - c_n k_n \delta^{k_n}\right) \delta^{b_n} \exp\left(-c_n \delta^{k_n}\right), \quad (8.12)$$

with $\delta = \rho K^3$ and density and gas constant in molar units. Independent, composition dependent formulations are given for the mixture size parameter K, the second virial coefficient B, and the coefficients C_n^*. These relations involve a number of further parameters which will not be discussed here; for details on the very complex set-up of the AGA8-DC92 equation of state, see Starling and Savidge (1992) or Jaeschke and Schley (1996a). Although the set-up of the used coefficients is a semiempirical one, the functional form of the equation of state was established with empirical means and Eq. 8.12 can be used like an empirical multiparameter equation for a pseudo-pure fluid, if the composition dependent coefficients are determined.[6]

Since it was known that Eq. 8.12 was developed as an equation using the form of the Helmholtz energy (see Starling et al., 1991), it is consistent to re-integrate the published formulation in the form of the compression factor; different groups did so, see Span and Wagner (1994) and Jaeschke et al. (1998). The resulting equation for the reduced residual Helmholtz energy reads

$$\frac{a^r(T,\rho,\mathbf{x})}{\rho RT} = \alpha^r(T,\rho,\mathbf{x}) = \left(\frac{B}{K^3} - \sum_{n=13}^{18} C_n^* \tau^{u_n}\right)\delta + \sum_{n=13}^{58} C_n^* \tau^{u_n} \delta^{b_n} \exp(-c_n \delta^{k_n})$$

$$= \left(\sum_{n=1}^{18} a_n \tau^{u_n} f(\mathbf{x}) - \sum_{n=13}^{18} C_n^* \tau^{u_n}\right)\delta + \sum_{n=13}^{58} C_n^* \tau^{u_n} \delta^{b_n} \exp(-c_n \delta^{k_n}). \quad (8.13)$$

The inverse temperature $\tau = T_r/T$ with $T_r = 1\,\mathrm{K}$ was introduced to achieve formal consistency with the formulations discussed in the preceding chapters. The function $f(\mathbf{x})$ depends only on the given composition and describes a mixing rule for the polynomial terms which contribute to the second virial coefficient. None of the parameters changed due to the integration. In combination with an appropriate formulation for the ideal gas contribution to the reduced Helmholtz energy (see Sect. 3.1.1.1), Eq. 8.13 can be used for the calculation of thermal and caloric properties of natural gases when adapting the prescriptions given in Sect. 3.2.1.

The capabilities and the restrictions of the AGA8-DC92 equation of state have been discussed in detail in a number of articles and reports (Jaeschke and Schley, 1996a/b; Klimeck et al., 1996; Jaeschke et al., 1998). In the range $290\,\mathrm{K} \leq T \leq 350\,\mathrm{K}$ and $p \leq 30\,\mathrm{MPa}$, Eqs. 8.12 and 8.13 describe $p\rho T$ data of typical natural gases within an uncertainty of $\Delta\rho/\rho \leq \pm 0.1\,\%$. In the same range, accurate speed of sound data are represented within $\Delta w/w \leq \pm 0.2\,\%$ and other caloric properties like heat capacities and enthalpy differences within $\Delta y/y \leq \pm 1\,\%$, if Eq. 8.13 is combined with a formulation for α^o which is based on the equations by Jaeschke and Schley (1995). With these uncertainties, the AGA8-DC92 equation is one of the most accurate equations of state which have been published for multi-component mixtures to date.[7] For the calculation of compressibility factors of natural gases based on a detailed gas analysis the equation has been declared an international standard (ISO, 1994).

However, the performance of the AGA8-DC92 equation decreases for conditions and compositions which are untypical for natural gas pipelining applications. Increased uncertainties are observed, e.g., for certain natural gas compositions and generally at temperatures $T < \approx 270\,\mathrm{K}$; for details see Jaeschke and Schley (1996b). Typical examples for the performance of the AGA8-DC92 equation of state are shown in Fig. 8.1 for the representation of $p\rho T$ data and in Fig. 8.2 for the representation of speed of sound data. Accurate results can be found for binary mixtures of the main components of natural gases, unless the considered states are

---

[6] The composition dependent parameters of the AGA8-DC92 equation are formulated as functions of the molar composition considering 21 typical natural gas components.

[7] See also Sect. 8.3.2 describing the model by Lemmon and Jacobsen (1999a).

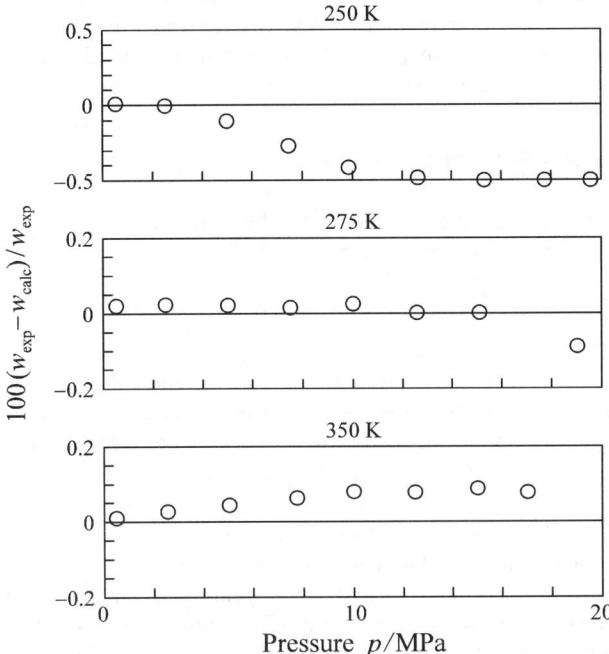

Fig. 8.2. Percentage deviations between accurate experimental results for the speed of sound in a natural gas like quintic mixture by Costa Gomes and Trusler (1998b) and values calculated from the AGA8-DC92 equation of state in combination with ideal gas properties according to Jaeschke and Schley (1995). For the composition of the investigated mixture, see Costa Gomes and Trusler (1998b).

located close to the dew point curve or in the extended critical region of the mixture. The AGA8-DC92 equation is strictly not valid for the calculation of vapour-liquid phase equilibria and for the calculation of thermodynamic properties at liquid states.

8.2 Extended Corresponding States Approaches

When Pitzer et al. (1955) extended the classical corresponding states approach by introducing an additional parameter, the acentric factor ω (see Chap. 6, footnote 3), they considered only pure component data to establish their reference tables. However, based on approaches which define effective critical parameters, the corresponding relations were soon used for mixtures as well. The major disadvantages of the algorithm proposed by Pitzer and co-workers were that it was based on tables rather than on equations of state and that it considered only states at reduced temperatures $T/T_c \geq 0.8$. During the 70's new models were developed which used the basic idea of the extended corresponding states approach for convenient computational use. At the same time, both the range of validity was extended and the accuracy which could be achieved with extended corresponding states models was improved. Two of these approaches which are still relevant in technical and in scientific applications today will be discussed briefly in the following sections.

8.2.1 Interpolation Between Reference Fluids

In accordance with the approach by Pitzer et al. (1955), the compression factor Z of a non- or weakly polar fluid can be written as

$$Z(\theta,\pi) = Z^{(0)}(\theta,\pi) + \frac{\omega}{\omega^{(r)}}\left(Z^{(r)}(\theta,\pi) - Z^{(0)}(\theta,\pi)\right), \tag{8.14}$$

where $\theta = T/T_c$ is the reduced temperature and $\pi = p/p_c$ is the reduced pressure. Both properties are reduced with the critical parameters of the fluid under investigation. $Z^{(0)}$ corresponds to the compression factor of a hypothetical simple fluid with spherical molecules and $\omega^{(0)} = 0$. $Z^{(r)}$ is the compression factor of a reference fluid with anisotropic molecules and $\omega^{(r)} \neq 0$. To facilitate the evaluation of Eq. 8.14, Lee and Kessler (1975) developed modified BWR-type equations of state with 11 terms which represent $Z^{(0)}$ and $Z^{(r)}$. The formulation for $Z^{(0)}$ was developed using reduced data for argon, krypton, and methane. To guarantee a reasonable extrapolation to reduced temperatures below the triple point temperature of common simple fluids, low temperature data from other substances were included in the fit using Eq. 8.14 for a suitable transformation of the data. Octane was chosen as reference fluid since it is the fluid with the most pronounced anisotropy among the fluids with reasonably large data sets. However, the coefficients of the final equation for $Z^{(r)}$ were readjusted to fit data for a variety of substances as well as possible. The equations for $Z^{(0)}$ and $Z^{(r)}$ are formulated using a reduced specific volume which is defined as

$$\frac{v^{(0/r)}(\theta,\pi)}{v_r} = \frac{p_c v^{(0/r)}}{RT_c}. \tag{8.15}$$

Thus, two reduced volumes are determined from the equations for $Z^{(0)}$ and $Z^{(r)}$. The reducing parameter v_r depends on the critical parameters of the investigated fluid, while the specific volume v results from the equation for the simple fluid ($v^{(0)}$) in the first step and from the equation for the reference fluid ($v^{(r)}$) in the second step. The equations for $Z^{(0)}$ and $Z^{(r)}$ which depend on θ and v/v_r are evaluated at two different states to avoid additional iterations which would be necessary to evaluate the equations at the value of $v(\theta,\pi)/v_r$ which corresponds to the fluid under investigation. Relations which are necessary to calculate common derived properties are given in the article by Lee and Kessler (1975).

Up to this point, the described model is a generalised equation of state for pure fluids rather than a mixture model. According to comparisons given by the authors, the average absolute error of densities at liquid and liquid-like supercritical states calculated with this approach is $\Delta\rho/\rho$ (AAD) $\approx 1\% - 4\%$; for a comparison with recent generalised equations of state see Sect. 7.2.

To describe mixtures, Lee and Kessler (1975) introduced van der Waals like mixing rules for the pseudocritical volume and the pseudocritical temperature of the mixture, see Eqs. 8.5 and 8.6. Together with a linear mixing rule for the acentric factor, these relations yield the required input parameters for mixture calculations which are again based on the *"one-fluid concept"*. However, without any adjustable parameters, this approach yields reliable results only for rather simple mixtures.

By introducing new mixing rules, Plöcker et al. (1978) extended the approach by Lee and Kessler (1975) to asymmetric mixtures. Based on the modified Lorentz-Berthelot combining rules

$$\rho_{c,j,k}^{-1} = \frac{1}{8}\left(\rho_{c,j}^{-1/3} + \rho_{c,k}^{-1/3}\right)^3 \quad \text{and} \quad T_{c,j,k} = b_{j,k}^T \cdot \sqrt{T_{c,j} T_{c,k}} \tag{8.16}$$

the mixing rules by Plöcker et al. (1978),

$$v_{r,M} = \sum_{j=1}^{C}\sum_{k=1}^{C} x_j x_k v_{c,j,k}, \quad T_{r,M} = v_{r,M}^{-\eta} \sum_{j=1}^{C}\sum_{k=1}^{C} x_j x_k T_{c,j,k} v_{c,j,k}^{\eta},$$

$$\text{and} \quad \omega_M = \sum_{j=1}^{C} x_j \omega_j, \tag{8.17}$$

introduce a pseudochritical temperature $T_{r,M}$ and a pseudocritical specific volume $v_{r,M}$. The pseudocritical pressure of the mixture is determined according to

$$p_{r,M} = (0.2905 - 0.085\omega_M)\frac{RT_{r,M}}{v_{r,M}}. \tag{8.18}$$

The exponent η in the mixing rule for $T_{r,M}$ is a substance independent constant with $\eta = 0.25$. The binary interaction parameter $b_{j,k}^T$ in the combining rule for $T_{r,M}$ can be adjusted most effectively to vapour-liquid equilibrium data for the corresponding binary mixture. For symmetric mixtures $b_{j,k}^T \approx 1$ can be assumed.

In the form modified by Plöcker et al. (1978), the approach by Lee and Kessler (1975) is frequently used in chemical engineering until to date. Improvements seem to be possible especially with regard to the underlying modified MBWR-type equations for $Z^{(0)}$ and $Z^{(r)}$. A corresponding attempt was made by Soave (1995), but still no state-of-the-art techniques have been used to establish the involved empirical formulations. However, even further improvements with regard to the used equations of state could not overcome the principle shortcomings of the simple extended corresponding states approach, see the discussion in Sect. 7.2.2. Thus, the chances to adapt these kinds of approaches to advanced technical needs on accuracy seem to be rather limited.

8.2.2 The Shape Factor Concept

For the residual Helmholtz energy and the compression factor, the simple corresponding states theorem can be written as

$$a_j^{\mathrm{r}}(T,\rho) = a_o^{\mathrm{r}}\bigl(T\cdot(T_{c,o}/T_{c,j}), \rho\cdot(\rho_{c,o}/\rho_{c,j})\bigr) \tag{8.19}$$

and
$$z_j(T,\rho) = z_o\bigl(T\cdot(T_{c,o}/T_{c,j}), \rho\cdot(\rho_{c,o}/\rho_{c,j})\bigr). \tag{8.20}$$

In Eqs. 8.19 and 8.20, the index j refers to the fluid of interest and the index o to an arbitrary reference fluid. With the so called "equivalent substance reducing parameters" $f_j = T_{c,j}/T_{c,o}$ and $h_j = \rho_{c,o}/\rho_{c,j}$, Eqs. 8.19 and 8.20 can be rewritten as

$$a_j^{\mathrm{r}}(T,\rho) = a_o^{\mathrm{r}}\bigl(T/f_j, \rho\cdot h_j\bigr) \tag{8.21}$$

and
$$z_j(T,\rho) = z_o\bigl(T/f_j, \rho\cdot h_j\bigr). \tag{8.22}$$

However, it is well known that this simple corresponding states analogy holds only for nonpolar fluids with spherical molecules – and even for such fluids the achieved accuracies cannot satisfy advanced technical demands.

To overcome this limitation, Leland and co-workers (see Leach et al., 1968; Fisher and Leland, 1970) introduced so called "*shape factors*", which are temperature and density dependent corrections of the equivalent substance reducing parameters.[8] With the shape factors θ and ϕ the substance reducing parameters in Eqs. 8.21 and 8.22 become

$$f_j = \frac{T_{c,j}}{T_{c,o}}\cdot\theta_j(\rho,T) \tag{8.23}$$

and
$$h_j = \frac{\rho_{c,o}}{\rho_{c,j}}\cdot\phi_j(\rho,T). \tag{8.24}$$

The relations between Eqs. 8.21 or 8.22 and derived thermodynamic properties become rather complex when using temperature and density dependent shape factors according to Eqs. 8.23 and 8.24 and accurate generalised formulations for the shape functions θ and ϕ were not achieved through the 70's and 80's.[9]

However, mixture approaches can be applied to shape functions very effectively. When used for shape functions, a frequently used modified van der Waals one-fluid approach becomes

---

[8] Theoretically, shape factors were introduced to account for different shapes of intermolecular potentials.

[9] More recently, Huber and Ely (1994) presented generalised shape functions $\theta(T)$ and $\phi(T)$ which map the vapour pressure and the saturated liquid density of the considered fluid to the corresponding properties of a reference fluid. Applied to refrigerants, this approach resulted in density deviations which are reasonably small for a predictive model which is based only on T_c, ρ_c, and ω. However, comparisons for more sensitive properties such as caloric properties in the liquid were not presented.
Based on recent molecular theory, Estela-Uribe and Trusler (1998) again took up the shape factor concept as a semiempirical approach and fitted shape functions directly to experimental data in a limited range of states.

$$h_{\mathbf{x}} = \sum_{j=1}^{C}\sum_{k=1}^{C} x_j\, x_k\, h_{j,k} \quad \text{and} \quad f_{\mathbf{x}} = h_{\mathbf{x}}^{-1}\sum_{j=1}^{C}\sum_{k=1}^{C} x_j\, x_k\, f_{j,k}\, h_{j,k} \qquad (8.25)$$

with the combining rules

$$h_{j,k} = \left(h_j^{1/3} + h_k^{1/3}\right)^3 \cdot \frac{1 - b_{j,k}^h}{8} \quad \text{and} \quad f_{j,k} = \sqrt{f_j\, f_k}\cdot\left(1 - b_{j,k}^f\right). \qquad (8.26)$$

The $b_{j,k}^h$ and $b_{j,k}^f$ in Eq. 8.26 are adjustable parameters which can be fitted to data of the corresponding binary mixture. For simple symmetric systems, both parameters can be assumed to be 0. Thus, with $f_{\mathbf{x}}$ and $h_{\mathbf{x}}$ the shape factor concept can be applied to mixtures as well; for details on the required complex dependencies see Mollerup (1980) and Ely (1990).

To be able to make use of these promising mixing rules, Ely and co-workers developed the so called *"exact shape factor concept,"* see Ely (1990). Assuming that accurate multiparameter equations of state are known for all components of a mixture and for a reference fluid, the shape factors for the pure components can be computed from Eqs. 8.21 and 8.22 exactly (at least within the comparatively small uncertainty of the equations of state). To do so, a two dimensional Newton-Raphson iteration is applied for each of the components to determine the values for f_j and h_j which fulfil Eqs. 8.21 and 8.22 simultaneously for the given values of T and ρ. Based on f_j and h_j, the corresponding shape factors can be calculated easily from Eqs. 8.23 and 8.24 if they are needed for further calculations. The f_j and h_j obtained in this way satisfy the assumed corresponding states similarity, Eqs. 8.21 and 8.22, exactly but they are valid only for the investigated combination of T and ρ; the temperature and density dependence of the shape factors is still unknown. Thus, the exact shape factor concept has no predictive capabilities – it is just an algorithm to map known thermodynamic surfaces onto the thermodynamic surface of a reference fluid.

However, the values obtained for the f_j and h_j of the pure components can now be used in a mixture approach according to Eqs. 8.25 and 8.26. In this way, equivalent substance reducing parameters $h_{\mathbf{x}}$ and $f_{\mathbf{x}}$ are obtained which can be used to calculate thermodynamic properties of the corresponding mixture using Eqs. 8.21 and 8.22. The approach is not limited to binary mixtures; multicomponent mixtures can be considered in the same way, but for each of the components in the mixture, an iterative solution is required to determine the corresponding values of f_j and h_j. Relations which are necessary to determine derived properties were given by Ely (1990).

Compared to the algorithms described before, the exact shape factor concept is numerically rather inconvenient since two dimensional iterations are involved for each of the components in the mixture and since the relations for derived properties become quite complex. The major advantage of the shape factor concept is that accurate equations of state can be used for the pure components. These equations may have different functional forms and they may even be explicit in different properties. Accurate equations of state for the main components of a mixture can be combined with simple equations for the minor components – restrictions result only from the realised program codes, not from the underlying model.

The exact shape factor concept has been used mainly by scientists at the National Institute of Standards and Technology (NIST) based on pressure explicit MBWR-type equations of state for the involved pure substances (Ely, 1990). The well known computer program DDMIX (for a recent version see Friend, 1992) which is frequently used in scientific applications is based on this approach.[10] Clarke et al. (1993) reported an application of the exact shape factor concept to mixtures of nitrogen, argon, and oxygen, where the shape factor calculations are based on reference equations of state explicit in the Helmholtz energy.

With regard to the accuracy of calculated mixture properties, the exact shape factor concept is superior to other extended corresponding states approaches at least at homogeneous states. However, its complex form and its numerical inconvenience were always considered as disadvantages. Nowadays, the use of exact shape factors in mixture calculations is superseded by more accurate (see Lemmon, 1997) and less complex multi-fluid approaches.

8.3 Helmholtz Models with Departure Functions

In terms of the Helmholtz energy, a simple one-fluid corresponding states approach can be written as

$$\frac{a_M(T,\rho,\mathbf{x})}{RT} = \alpha_M(T,\rho,\mathbf{x})$$

$$= \alpha_M^o(T,\rho,\mathbf{x}) + \alpha_o^r\big(T\cdot(T_{c,o}/T_{r,M}(\mathbf{x})), \rho\cdot(\rho_{c,o}/\rho_{r,M}(\mathbf{x}))\big). \quad (8.27)$$

The reduced Helmholtz energy of the mixture can be determined exactly, see Eq. 8.2. For the residual part, it is assumed that the reduced Helmholtz energy of the mixture behaves like the residual Helmholtz energy of a reference fluid at the same reduced temperature and density, whereby a suitable mixing rule has to be used for the composition dependent pseudocritical parameters $T_{r,M}$ and $\rho_{r,M}$. The extended corresponding states approaches presented in Sect. 8.2 account for differences between the mixture and the reference fluid by using either an adapted reference fluid (see Sect. 8.2.1) or adapted reducing parameters (see Sect. 8.2.2).

However, a more rigorous way to account for differences between the compounds of a mixture is to rewrite Eq. 8.27 as

$$\frac{a_M(T,\rho,\mathbf{x})}{RT} = \alpha_M(T,\rho,\mathbf{x}) = \alpha_M^o(T,\rho,\mathbf{x}) + \sum_{j=1}^{C} x_j\, \alpha_j^r\big(T/T_{r,M}(\mathbf{x}), \rho/\rho_{r,M}(\mathbf{x})\big). \quad (8.28)$$

In this way, differences between the reduced Helmholtz surfaces of the components are directly taken into account. Eq. 8.28 is referred to as the "*multi-fluid approximation*," since equations of state for the C components of the mixture are directly involved in the calculation of mixture properties.

---

[10] DDMIX uses the exact shape factor concept to calculate properties at homogeneous states. Vapour-liquid equilibria are calculated using cubic equations of state.

The theoretical background of the multi-fluid approach was recently outlined by Tillner-Roth (1998), who discussed different kinds of mixing rules for the pseudo-critical parameters $T_{r,M}$ and $\rho_{r,M}$ as well. To summarise the corresponding results very briefly, one may conclude that van der Waals mixing and Lorentz-Berthelot combination rules (see Eqs. 8.5 and 8.6) are on average advantageous for mixtures of nonpolar substances, while linear mixing rules (see Eq. 8.7) without specific combination rules can be advantageous for mixtures of polar substances. Considerable improvements can be achieved by combination rules which involve adjustable binary interaction parameters. As a modification of the Lorentz-Berthelot combination rules, the adjustable parameters $b_{j,k}^T$ and $b_{j,k}^\rho$ were introduced in Eq. 8.9. In the same way, adjustable parameters can be introduced into the simple linear combination rules

$$\rho_{c,j,k}^{-1} = \frac{b_{j,k}^\rho}{2}\left(\rho_{c,j}^{-1} + \rho_{c,k}^{-1}\right) \quad \text{and} \quad T_{c,j,k} = \frac{b_{j,k}^T}{2}\left(T_{c,j} + T_{c,k}\right) \tag{8.29}$$

which can be used, e.g., with the quadratic mixing rules given in Eq. 8.8. In general, the representation of vapour-liquid equilibria data is very sensitive with regard to changes of the binary interaction parameter in the combination rule for the pseudocritical temperature, $b_{j,k}^T$. Thus, $b_{j,k}^T$ should be adjusted to vapour-liquid equilibrium data if possible. The binary interaction parameter $b_{j,k}^\rho$ is related mainly to properties in the homogeneous region and should only be adjusted if reliable data for such properties are available.

However, the general idea behind Eq. 8.28 is not new. If all of the equations of state which are used to represent the α_j^r in Eq. 8.28 have the same functional form, Eq. 8.28 can be rewritten in the form of Eqs. 8.3 and 8.4. These equations were introduced as a generalised formulation for methods which are based on composition dependent sets of coefficients. Compared to models based on this concept (see Sect. 8.1), the application of Eq. 8.28 is advantageous due to two reasons:

- State-of-the-art reference equations of state with an optimised functional form can be used to represent the pure component contributions α_j^r. This is very important for accurately describing rather simple mixtures, where the uncertainty of the equations of state which are used to describe the pure components dictates the accuracy of the mixture model. The same is true if one component is overly dominant in a mixture. In this case, the equation of state for the main component needs to be as accurate as possible while the equations for the minor components can be rather simple.
- The use of temperatures and densities which are reduced with pseudocritical properties of the mixture leads to advantages for systems where the critical temperatures and densities of the components are significantly different. However, this idea has been used by simple corresponding states approaches before and recent reference equations of state are already formulated in reduced properties anyway.

Beside these advantages, the multi-fluid approach has one disadvantage as well:
- The same mixing rule is applied to the whole equation of state. The concept of composition dependent coefficients made it possible to use special mixing and

combination rules for single coefficients or groups of coefficients. In this way, theoretically appropriate mixing rules have been used mostly for terms which contribute to the second and third virial coefficient while purely empirical mixing rules were used for all other terms.

Thus, Eq. 8.28 is a consistent advancement of known mixture approaches rather than a completely new concept. Certain advantages were to be expected, but no real breakthrough with regard to the accuracy which can be achieved in mixture calculations occurred. This breakthrough came with the introduction of "*departure functions*," which were introduced by two groups independently of each other (see Tillner-Roth, 1993; Lemmon, 1996).

As an extension of Eq. 8.28, the Helmholtz energy of a mixture can be rewritten as

$$\frac{a_M(T,\rho,\mathbf{x})}{RT} = \alpha_M(T,\rho,\mathbf{x})$$

$$= \alpha_M^o(T,\rho,\mathbf{x}) + \sum_{j=1}^{C} x_j \, \alpha_j^r(\tau_M,\delta_M) + \Delta\alpha_M^r(\tau_M,\delta_M,\mathbf{x}), \quad (8.30)$$

where $\tau_M(\mathbf{x}) = T_{r,M}/T$ and $\delta_M(\mathbf{x}) = \rho/\rho_{r,M}$. The departure function $\Delta\alpha_M^r(\tau_M,\delta_M,\mathbf{x})$ accounts for non-idealities of mixtures, which cannot be described accurately enough with the simple multi-fluid approach, Eq. 8.28. Although the general set-up of Eq. 8.30 is similar to the set-up of common G^E models, $\Delta\alpha_M^r$ is not an excess property for two reasons: The contribution to the Helmholtz energy is defined at given values of T and ρ rather than at given values of T and p, as it is common for excess properties. And the first two terms in Eq. 8.30 which are identical with Eq. 8.28, correspond to a quite accurate multi-fluid mixture approach rather than to an ideal mixture of real fluids. Thus, the contribution of $\Delta\alpha_M^r$ is in general much smaller than the contribution of formulations for $g^E(T,p)$; for details see Tillner-Roth (1998).

Just as for multiparameter equations of state for pure fluids, little information on suitable functional forms for $\Delta\alpha_M^r(\tau_M,\delta_M,\mathbf{x})$ is available from theory. A general set-up which is based on experiences with multiparameter equations of state for pure substances can be written as

$$\Delta\alpha_M^r(\tau_M,\delta_M,\mathbf{x}) = \sum_{j=1}^{C}\sum_{k=1}^{C} f(x_j,x_k)\left[\sum_{i=1}^{I} n_i \, \delta_M^{d_i} \, \tau_M^{t_i} \, \exp(-\gamma_i \delta_M^{p_i})\right]_{j,k}. \quad (8.31)$$

With regard to the temperature and density exponents d_i, p_i, and t_i, the general guidelines explained in Sect. 4.4.1 may be adapted, but strict rules cannot be formulated yet. However, to avoid problems in the limit of low densities where Eq. 8.30 has to approach the mixture of ideal gases which is represented by $\alpha_M^o(T,\rho)$, it is important to use only integer density exponents with $d_i \geq 1$, $p_i \geq 1$.

The composition dependent function $f(x_j,x_k)$ has to guarantee that the contribution of $\Delta\alpha_M^r$ vanishes for mixtures with $x_j \to 1$ and $x_{i\neq j} \to 0$ where j may correspond to any component in the mixture. Thus, $f(x_j,x_k) = 0$ has to hold for $j = k$ by definition. A suitable general approach for $f(x_j,x_k)$ is

$$f(x_j, x_k)_{j \neq k} = x_j x_k^{\beta_{jk}} \quad \text{and} \quad f(x_j, x_k)_{j = k} = 0. \tag{8.32}$$

Binary interaction parameters similar to β_{jk} have yet been used only to account for the asymmetry of binary mixtures (see Tillner-Roth, 1993/1998).[11] To describe the strongly non-ideal binary mixture ammonia/water, Tillner-Roth and Friend (1998) introduced additional composition dependencies into single terms of the density and temperature dependent part of Eq. 8.31.

Although the set-up of Eqs. 8.30 – 8.32 seems quite complex, the application of formulations of this form is rather convenient. For a given composition $\mathbf{x}$, $T_{r,M}$ and $\rho_{r,M}$ can be calculated from the corresponding mixing and combination rules using the critical parameters of the pure substances and, where necessary, binary interaction parameters. With $T_{r,M}$ and $\rho_{r,M}$ the reduced properties τ_M and δ_M can be calculated from T and ρ, and with T, ρ, $\mathbf{x}$, τ_M, and δ_M all contributions in Eq. 8.30 can be evaluated in straight forward calculations. Since $T_{r,M}$ and $\rho_{r,M}$ depend only on composition and not on temperature and density, the relations which are given for the calculation of thermodynamic properties in Sect. 3.2.1 can easily be adapted to the set-up of Eq. 8.30.[12] Mixture specific algorithms are required for the calculation of phase equilibria, see also Sect. 8.1.

The use of specific departure functions for each binary subsystem in a multi-component mixture, which is implied by the way in which Eq. 8.31 is written, is in most cases not realistic. Reasonable specific departure functions can only be established for well measured binary systems, and such systems are rare. Thus, departure functions have been used in two different forms yet which will be described in the following sections.

8.3.1 Mixture Specific Departure Functions

Multi-fluid models with mixture specific departure functions have been published for mixtures of HFC and HCFC refrigerants and for the system ammonia/water to date (see Tillner-Roth, 1993/1998; Tillner-Roth et al., 1998; Tillner-Roth and Friend, 1998). Tillner-Roth et al. (1998) apply these models to ternary systems as well, but essentially the models were developed for binary mixtures. Written for a binary mixture, Eq. 8.31 becomes

$$\Delta \alpha_M^r (\tau_M, \delta_M, x_1) = f(x_1) \sum_{i=1}^{I} n_i \delta_M^{d_i} \tau_M^{t_i} \exp(-\gamma_i \delta_M^{p_i}), \tag{8.33}$$

where x_2 is replaced by $x_2 = 1 - x_1$.

To establish a departure function which describes a certain binary mixture, the simple multi-fluid model is adjusted to the available data in a first step. To do so,

---

[11] Exponents like β_{jk} are effective binary interaction parameters, but they need to be used with care. In Eq. 8.32, $\beta_{jk} \geq 1$ has to hold to avoid infinite slopes in the pure component limit.

[12] α^o and its derivatives can be replaced by Eq. 8.2 and its derivatives and α^r and its derivatives can be replaced by $\Sigma x_j \alpha_j^r + \Delta \alpha_M^r$ and the corresponding derivatives. Compared to shape factor concepts, this is a major practical advantage.

suitable mixture and combination rules have to be selected and the binary interaction parameters in the reducing functions have to be determined by nonlinear fits to appropriate data, see above. If this is accomplished, the parameters d_i, t_i, and p_i can be determined using linear optimisation algorithms, see Sect. 4.4. Linear data at homogeneous states can be used directly by adapting the algorithms given in Sect. 4.3.2 to the requirements of a functional form according to Eq. 8.30. Implicit linear and nonlinear data need to be precorrelated, see Sect. 4.3.2 as well. The necessary linearisation of vapour-liquid phase equilibrium data becomes more complex for mixtures, see Tillner-Roth and Friend (1998). The coefficients n_i in Eq. 8.33 can be determined in a direct nonlinear fit to the original data. To ensure a proper consideration of implicit linear and nonlinear data, the cyclic process of precorrelation, linear optimisation, and nonlinear fit has to be repeated until no significant difference remains between linear and nonlinear results.[13] Adjustable interaction parameters in $f(x_1)$, see Eq. 8.32, cannot be considered in linear optimisation algorithms, since they are nonlinearly related to $\Delta\alpha_M^r$ and to the coefficients n_i. Thus, such interaction parameters hinder the convergence of the described cyclic process.

Based on multi-fluid models in the form of Eq. 8.30 which incorporate departure functions with up to $I = 8$ terms, binary mixtures of HFC and HCFC refrigerants have been described within an accuracy which comes close to the accuracy of the corresponding reference equations for the pure fluids. Enlarged uncertainties in certain regions can be attributed mainly to unsolved problems in the extended critical region of mixtures or to enlarged uncertainties of experimental results for mixtures.[14] With a departure function which contains 14 terms and a more sophisticated composition dependence, accurate results were obtained for the system ammonia/water as well. However, the application of multi-fluid models with mixture specific departure functions will always be restricted to selected mixtures where extensive data sets are available for the considered binary mixture or for all binary subsystems of a multicomponent mixture.

8.3.2 Generalised Departure Functions

To overcome the restrictions which result from the application of mixture specific departure functions, Lemmon (1996) supplemented his simple multi-fluid approach with a generalised departure function. Using the nomenclature introduced above, the approach by Lemmon (1996) reads

$$\frac{a_M(T,\rho,\mathbf{x})}{RT} = \alpha_M(T,\rho,\mathbf{x})$$

$$= \alpha_M^o(T,\rho,\mathbf{x}) + \sum_{j=1}^{C} x_j \, \alpha_j^r(\tau_M,\delta_M) + \Delta\alpha_M^r(\tau_M,\delta_M,\mathbf{x}), \qquad (8.34)$$

---

[13] Nonlinear optimisation algorithms (see Sect. 4.4.6) have not yet been applied to the development of departure functions.

[14] This is especially true when comparing the representation of vapour-liquid equilibrium data for pure fluids and for mixtures.

with $$\Delta \alpha_{\text{M}}^{\text{r}}(\tau_{\text{M}}, \delta_{\text{M}}, \mathbf{x}) = \left(\sum_{j=1}^{C} \sum_{k=1}^{C} x_j \, x_k \, F_{j,k} \right) \cdot \sum_{i=1}^{I} n_i \, \delta_{\text{M}}^{d_i} \, \tau_{\text{M}}^{t_i} \, \exp\!\left(-\gamma_i \delta_{\text{M}}^{p_i}\right). \quad (8.35)$$

Aside from differences in the used reducing functions, $T_{\text{r,M}}(\mathbf{x})$ and $\rho_{\text{r,M}}(\mathbf{x})$, the main difference between the models by Lemmon (1996) and Tillner-Roth (1993) lies in the way in which the departure function is formulated. The formulation which is given in Eq. 8.35 assumes that there is a single function[15] with $I = 10$ terms which is valid for the departure of all binary subsystems. The only parameter which depends on the considered system is the factor $F_{j,k}$ which scales the influence of the universal formulation.

From a theoretical point of view, there is no reason to believe that this approach is valid. The way in which the departure function is used by Tillner-Roth (1993/1998), see Sect. 8.3.1, accounts for the residuum contribution which results from limitations of the simple multi-fluid approach and of the underlying reducing functions. These residua may be very different, even if the deviations from the simple "ideal mixture of real fluids" model are rather similar for two systems.

However, Lemmon (1996) uses the departure function in a slightly different way. The scaling factor $F_{j,k}$ is adjusted *simultaneously* with the three binary interaction parameters[16] in the reducing functions, $T_{\text{r,M}}(\mathbf{x})$ and $\rho_{\text{r,M}}(\mathbf{x})$, in a nonlinear fit to data of the corresponding binary mixture. In this way, the flexibility of the reducing functions is used to find a solution which describes the given mixture adequately, using a departure function which is flexible only with regard to its amplitude, but not with regard to its shape in terms of τ_{M}, δ_{M}, and $\mathbf{x}$.

The development of a generalised departure function is numerically more complex than the development of a specific departure function, since different binary mixtures have to be considered simultaneously and since both the functional form of the departure function and the interaction parameters in the reducing functions need to be determined. To do so, Lemmon (1996) used nonlinear fits of the interaction parameters for all considered binary mixtures and linear optimisation runs to determine the functional form of the generalised departure function and its coefficients. Both steps were repeated in a cyclic process. Data for 28 binary mixtures of methane, ethane, propane, isobutane, n-butane, nitrogen, carbon dioxide, argon, neon, R32, R125, R134a, and R152a were used to establish the generalised departure function.

Results of the generalised multi-fluid model are reported for a variety of mixtures (Lemmon and Jacobsen, 1999a/2000; Lemmon and Tillner-Roth, 1999; Lemmon et al., 1999). For binary refrigerant mixtures, comparisons with multi-fluid models with mixture specific departure functions show that the generalised model is slightly inferior with regard to accuracy. However, the results obtained with the generalised multi-fluid model are still significantly better than those ob-

---

[15] In addition, the universal function is supplemented by additional terms for the system R32/R125 and by one additional term for the system R32/R134a.

[16] For binary systems with poor data situation, only a subset of the available interaction parameters needs to be fitted. For the remaining parameters, simple symmetric solutions are assumed in this case.

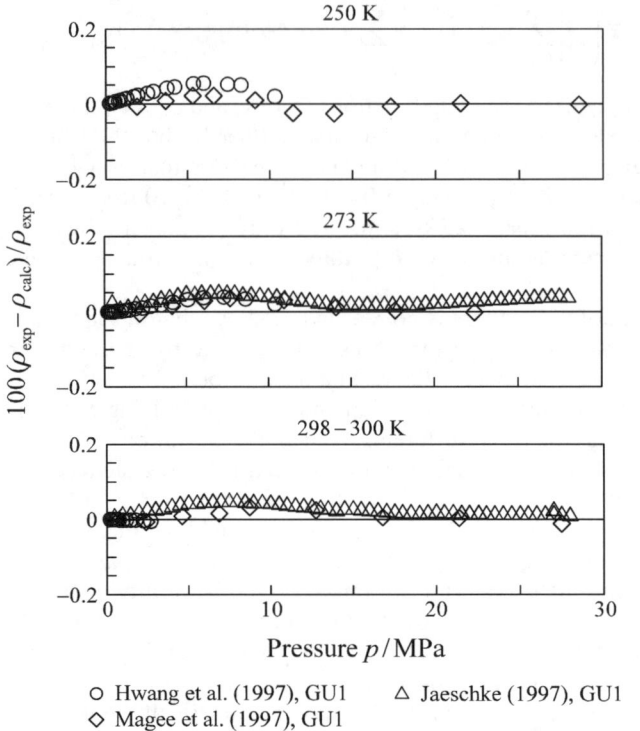

○ Hwang et al. (1997), GU1   △ Jaeschke (1997), GU1
◇ Magee et al. (1997), GU1

Fig. 8.3. Percentage deviations between accurate experimental results for the density of a natural gas like mixture and values calculated from the multi-fluid model by Lemmon and Jacobsen (1999a). For the composition of the investigated mixture (code GU1) see the cited references.

tained with the exact shape factor concept, see Sect. 8.2.2. Tillner-Roth (1998) questioned that it will be possible to describe strongly non-ideal systems like ammonia/water with generalised departure functions − most likely it will not be possible, at least not until generalised departure functions have been developed for characteristic types of non ideal mixtures.

However, the potential of the generalised model becomes obvious when working on multicomponent mixtures. Figures 8.3 and 8.4 show deviations between accurate experimental data for the density and the speed of sound of a natural gas like mixture and values calculated from the formulation by Lemmon and Jacobsen (1999a). The same data sets have been used for comparisons with the AGA8-DC92 equation of state in Figs. 8.1 and 8.2. At temperatures $T \geq 270$ K which are typical for natural gas pipelining applications, advantages and disadvantages of both equations of state are arbitrary in most cases. For typical natural gas compositions the AGA8-DC92 equation of state is in general slightly superior, but examples which show the opposite can be found easily as well. At temperatures $T < 270$ K the accuracy of the multi-fluid model decreases due to the lack of accu-

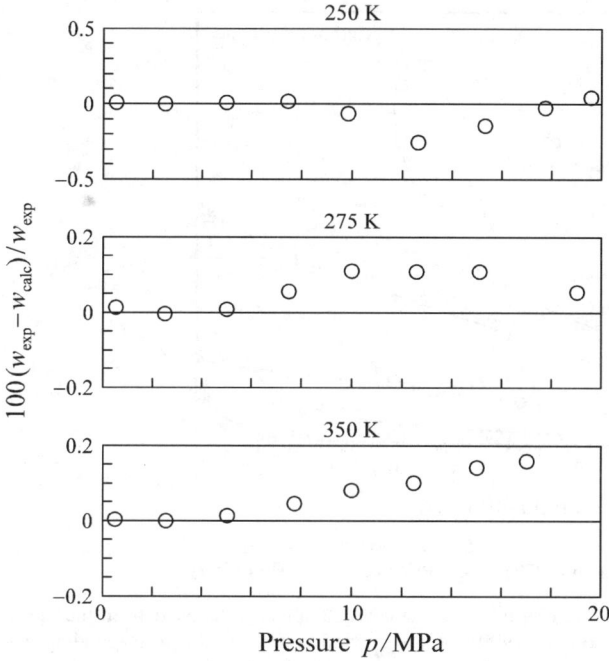

Fig. 8.4. Percentage deviations between accurate experimental results for the speed of sound in a natural gas like quintic mixture by Costa Gomes and Trusler (1998b) and values calculated from the multi-fluid model by Lemmon and Jacobsen (1999a). For the composition of the investigated mixture see Costa Gomes and Trusler (1998b).

rate data for the involved binary mixtures, but still it meets the data which are available for natural gas like mixtures at these states better than the AGA8-DC92 equation of state. And while the AGA8-DC92 equation of states fails close to phase boundaries of binary and multicomponent mixtures, the multi-fluid model yields reasonably accurate results at such states and even predicts vapour-liquid equilibria, see Fig. 8.5, and properties at liquid states.

Based on multi-fluid models, the work on a future property model for all applications of natural gases continues internationally. Current topics of this work are the consideration of compounds which have not yet been included in the corresponding models and further improvements of the accuracy in regions where extremely high demands are formulated by the natural gas industry. It seems clear that the resulting model will contain a generalised departure function in the end, since only few of the binary subsystems which are relevant for natural gases have been investigated to a degree that allows the development of mixture specific departure functions. However, it is still questionable whether the necessary accuracy can be achieved based only on a generalised departure function. The use of specific departure functions may become mandatory for the well measured binary

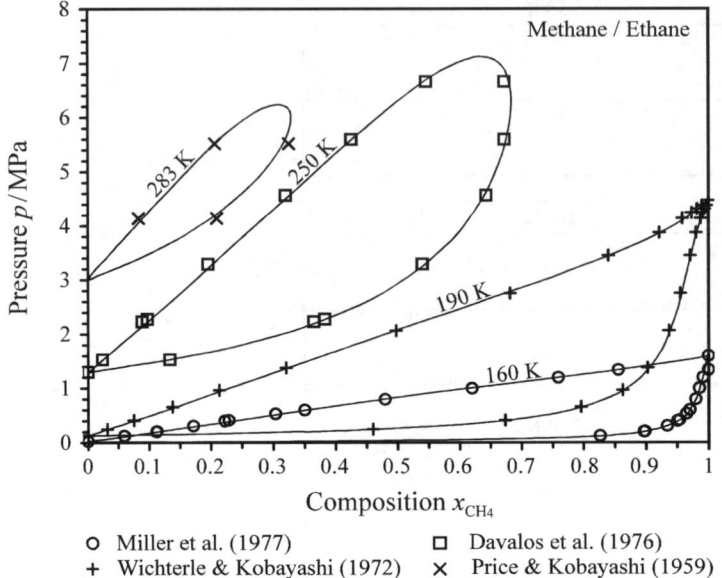

Fig. 8.5. Phase equilibria of the binary mixture methane/ethane as calculated from the multi-fluid model of Lemmon and Jacobsen (1999a). Experimental results on the corresponding isotherms are given for comparison.

systems which involve only the main components of natural gases. Thus, the use of a *"hybrid departure function"* which reads

$$\Delta \alpha_M^r(\tau_M, \delta_M, \mathbf{x}) = \sum_{j=1}^{C} \sum_{k=1}^{C} f(x_j, x_k) F_{j,k} \left[\sum_{i=1}^{I} n_i \delta_M^{d_i} \tau_M^{t_i} \exp(-\gamma_i \delta_M^{p_i}) \right]_{j,k} \quad (8.36)$$

seems to be a promising alternative for the multicomponent mixture (Klimeck et al., 1999b). In Eq. 8.36, the scaling factor $F_{j,k}$ becomes 1 for binary systems for which specific departure functions can be developed. In this case, system specific values are used for the parameters I, n_i, d_i, t_i, γ_i, and p_i. For less well measured binary subsystems which are described with a generalised set of parameters I, n_i, d_i, t_i, γ_i, and p_i, the scaling factor $F_{j,k}$ is used as the only adjustable parameter of the departure function. A multi-fluid model which uses a departure function according to Eq. 8.36 is extremely flexible and, even more important, it can later be extended to further binary subsystems.

References

Abdulagatov IM (1998) Experimental results for the isochoric heat capacity of n-heptane and n-octane. Private communication.

Abdulagatov IM, Polikhronidi NG, Batyrova RG (1994) Measurement of the isochoric heat capacities C_V of carbon dioxide in the critical region. J. Chem. Thermodynamics 26: 1031–1045.

Ahrendts J, Baehr HD (1979a) Die direkte Verwendung von Meßwerten beliebiger thermodynamischer Zustandsgrößen zur Bestimmung kanonischer Zustandsgleichungen (in German). Forsch. Ing.-Wes. 45: 1–10.

Ahrendts J, Baehr HD (1979b) Die Anwendung nichtlinearer Regressionsverfahren bei der Aufstellung thermodynamischer Zustandsgleichungen (in German). Forsch. Ing.-Wes. 45: 51–56.

Ahrendts J, Baehr HD (1981a) Direct application of experimental values for any thermodynamic variables of state in establishing canonical equations of state. Int. Chem. Eng. 21: 557–571.

Ahrendts J, Baehr HD (1981b) The use of nonlinear regression processes in establishing thermodynamic equations of state. Int. Chem. Eng. 21: 572–579.

Albright PC, Chen ZY, Sengers JV (1987b) Crossover from singular to regular thermodynamic behavior of fluids in the critical region. Phys. Rev. B 36: 877–880.

Albright PC, Edwards TJ, Chen ZY, Sengers JV (1987a) A scaled fundamental equation for the thermodynamic properties of carbon dioxide in the critical region. J. Chem. Phys. 87: 1717–1725.

Albright PC, Sengers JV, Nicoll JF, Ley-Koo M (1986) A crossover description for the thermodynamic properties of fluids in the critical region. Int. J. Thermophys. 7: 75–85.

Altunin VV, Gadetskii OG (1971) Equation of state and thermodynamic properties of liquid and gaseous carbon dioxide. Thermal Engineering 18 (3): 120–125 (English translation).

Altunin VV, Geller VZ, Kremenevskaya EA, Perelshtein II, Petrov EK (1987b) Thermodynamic properties of freons, part 2. Hemisphere, Washington New York London.

Altunin VV, Geller VZ, Petrov EK, Rasskazov DC, Spiridonov GA (1987a) Thermodynamic properties of freons, part 1. Hemisphere, Washington New York London.

Aly FA, Lee LL (1981) Self-consistent equations for calculating the ideal gas heat capacity, enthalpy and entropy. Fluid Phase Equilibria 6: 169–179.

Angus S (1983) Guide for the preparation of thermodynamic tables and correlations of the fluid state. CODATA Bull. 51, Pergamon, Oxford New York Toronto.

Angus S, Armstrong B (1971) Thermodynamic tables of the fluid state – 1. argon. Butterworths, London.

Angus S, Armstrong B, de Reuck KM (1974) Thermodynamic tables of the fluid state – 2. ethylene, 1972. Butterworths, London.

Angus S, Armstrong B, de Reuck KM (1976) Thermodynamic tables of the fluid state – 3. carbon dioxide. Pergamon, Oxford.

Angus S, Armstrong B, de Reuck KM (1978) Thermodynamic tables of the fluid state – 5. methane. Pergamon, Oxford.

Angus S, Armstrong B, de Reuck KM (1980) International thermodynamic tables of the fluid state – 7. propylene (propene). Pergamon, Oxford.

Angus S, Armstrong B, de Reuck KM (1985) International thermodynamic tables of the fluid state – 8. chlorine. Pergamon, Oxford.
Angus S, de Reuck KM (1977) Thermodynamic tables of the fluid state – 4. helium-4. Pergamon, Oxford.
Angus S, de Reuck KM, Armstrong B (1979) International thermodynamic tables of the fluid state – 6. nitrogen. Pergamon, Oxford.
Aston JG, Kennedy RM, Schumann SC (1940) The heat capacity and entropy, heats of fusion and vaporization and the vapor pressure of isobutane. J. Am. Chem. Soc. 62: 2059–2063.
Barber CR (1969) The International Temperature Scale of 1968. Metrologia 5: 35–44.
Baehr HD (1963) Das Verhalten der spezifischen Wärmekapazität c_v und der Entropie am kritischen Punkt des Wassers (in German). BWK 15: 514–522.
Baehr HD (1992) Thermodynamik, 8th edn. (in German). Springer, Berlin Heidelberg New York.
Baehr HD, Diedrichsen C (1988) Berechnungsgleichungen für Enthalpie und Entropie der Komponenten von Luft und Verbrennungsgasen (in German). BWK 40: 30–33.
Baehr HD, Tillner-Roth R (1991) Measurement and correlation of the vapour pressures of 1,1,1,2-tetrafluoroethane (R 134a) and of 1,1-difluoroethane (R 152a). J. Chem. Thermodynamics 23: 1063–1068.
Baehr HD, Tillner-Roth, R (1995) Thermodynamic properties of environmentally acceptable refrigerants. Springer, Berlin Heidelberg New York.
Barreiros SF, Calado JCG, Clancy P, Nunes da Ponte M, Street WB (1982) Thermodynamic properties of liquid mixtures of argon + krypton. J. Phys. Chem. 86: 1722–1729.
Bender E (1970) Equations of state exactly representing the phase behavior of pure substances. In: Bonila CF (ed.) Proc. 5th Symp. Thermophys. Prop., pp. 227–235. ASME, New York.
Bender E (1973) The calculation of phase equilibria from a thermal equation of state applied to the pure fluids argon, nitrogen, oxygen and their mixtures. C. F. Müller, Karlsruhe.
Bender R (1982) Untersuchungen zur zwischenmolekularen Wechselwirkung in binären Gasgemischen niedriger Dichte (in German). Dissertation, TH-Karlsruhe, Karlsruhe.
Bender R, Bier K, Maurer G (1981) Messung der spezifischen Wärme und des Joule-Thomson-Koeffizienten von Kohlendioxid bei Temperaturen von –40 °C bis 200 °C und Drücken bis 15 bar (in German). Ber. Bunsenges. Phys. Chem. 85: 778–784.
Benedict M, Webb GB, Rubin LC (1940) An empirical equation for thermodynamic properties of light hydrocarbons and their mixtures. J. Chem. Physics 8: 334–345.
Benedict M, Webb GB, Rubin LC (1942) An empirical equation for thermodynamic properties of light hydrocarbons and their mixtures II. Mixtures of methane, ethane, propane, and n-butane. J. Chem. Physics 10: 747–758.
Benoit SZ (1924) Sur une méthode de résolution des équations normales etc. (procédé du commandant Cholesky). Bull. Géodésique 2.
Bischoff M (1988) Die Berechnung der thermischen Zustandsgleichung für Methanol unter Verwendung der Wagnerschen Regressionsanalyse (in German). Diploma thesis at the Institute for Thermodynamic, University Hannover, Hannover.
Biswas SN, Trappeniers NJ, Kortbeek PJ, ten Seldam CA (1988) Apparatus for the measurement of compressibility isotherms of gases up to 10 kbar: Experimental data for argon at 298.15 K. Rev. Sci. Instrum. 59: 470–476.
Blanke W, Weiß R (1990) Isochoric (pvT) measurements on CCl2F2 (R12) in the liquid state from 120 K to 470 K and at pressures up to 300 bar. Int. Inst. of Refrig., Commission B1, Herzlia (Israel).
Boublik T (1975) Hard convex body equations of state. J. Chem. Phys. 63: 4084.
Boublik T, Nezbeda I (1977) Equation of state for hard dumbells. Chem. Phys. Letters 46: 315–316.
Boyes SJ (1992) The speed of sound in gases with application to equations of state and sonic nozzles. Dissertation, University of London, London.

Boyes SJ, Weber LA (1995) Vapour pressures and gas-phase (p, ρ_n, T) values for CF_3CHF_2 (R125). J. Chem. Thermodynamics 27: 163–174.
Brachthäuser K (1993) Entwicklung eines neuen Dichtemeßverfahrens für große Temperatur- und Druckbereiche und Aufbau einer Apparatur zur Messung der Dichte fluider Stoffe im Temperaturbereich von −40 °C bis +250 °C bei Drücken bis zu 300 bar (in German). Dissertation, Ruhr-Universität Bochum, Bochum.
Brule MR, Lin CT, Lee LL, Starling KE (1982) Multiparameter corresponding-states correlation of coal-fluid thermodynamic properties. AIChE J. 28: 616–625.
Brown EH (1960) On the thermodynamic properties of fluids. Bul. Int. Inst. Refrig. 1: 169–178.
Calero S, Wendland M, Fischer J (1998) Description of alternative refrigerants with Backone equations. Fluid Phase equilibria, 152: 1–22.
Carnahan NF, Starling KE (1969) Equation of state for nonattracting rigid spheres. J. Chem. Phys. 51: 635–636.
Carnahan NF, Starling KE (1972) Intermolecular repulsions and the equation of state for fluids. AIChE J. 18: 1184–1189.
Cerný V (1985) Thermodynamical approach to the traveling salesman problem: An efficient simulation algorithm. J. Optim. Theo. Appl. 45: 41–51.
Chao J, Wilhoit RC, Zwolinski BJ (1973) Ideal gas thermodynamic properties of ethane and propane. J. Phys. Chem. Ref. Data 2: 427–437.
Chapela GA, Rowlinson JS (1974) Accurate representation of thermodynamic properties near the critical point. J. Chem. Soc., Faraday (I) 70: 584–593.
Chase MW Jr., Davies CA, Downey JR Jr., Frurip DJ, McDonald RA, Syverud AN (1985) JANAF Thermochemical tables, 3rd edn. J. Phys. Chem. Ref. Data: 14, Suppl. 1, Part I–II.
Chen SS, Kreglewski A (1977) Application of the augmented van der Waals theory for fluids I. Pure fluids. Ber. Bunsenges. Phys. Chem. 81: 1048–1052.
Chen SS, Wilhoit RC, Zwolinski BJ (1975) Ideal gas thermodynamic properties and isomerization of n-butane and isobutane. J. Phys. Chem. Ref. Data 4: 859–868.
Chen ZY, Albright PC, Sengers JV (1990a) Crossover from singular critical to regular classical thermodynamic behavior of fluids. Phys. Review A 41: 3161–3177.
Chen ZY, Abbaci A, Tang S, Sengers JV (1990b) Global thermodynamic behavior of fluids in the critical region. Phys. Review A 42: 4470–4484.
Cibulka L, Hnedkovsky L (1996) Liquid densities at elevated pressure of n-alkanes from C5 to C16: A critical evaluation of experimental data. J. Chem. Eng. Data 41: 657–668.
Clarke WP, Jacobsen RT, Beyerlein SW, Penoncello SG (1993) An extended corresponding states model for predicting thermodynamic properties of the N_2-Ar-O_2 mixture. Fluid Phase Equilibria 88: 13–24.
Claus P, Kleinrahm R, Wagner W (2000) Measurement of the (p,ρ,T) relation of ethylene in the temperature range from 240 K to 370 K at pressures up to 30 MPa. To be submitted to J. Chem. Thermodynamics.
Cohen RE, Taylor BN (1986) The 1986 adjustment of the fundamental physical constants; CODATA Bul. 63. Pergamon, Oxford.
Cole WA, de Reuck KM (1989) An interim analytic equation of state for sulfurhexafluoride. Int. J. Thermophys. 11: 189–199.
Collmann H.-J., Span R, Wagner W (1996) Introduction of a pairwise exchange of terms into the stepwise regression analysis. Unpublished results.
Conolly JF (1962) Ideality of n-butane : isobutane solutions. J. Phys. Chem. 66: 1082–1086.
Coplen TB (1997) Atomic weights of the elements 1995. J. Phys. Chem. Ref. Data 26: 1239–1253.
Costa Gomes MF, Trusler JPM (1998a) The speed of sound in nitrogen at temperatures between $T = 250$ K and $T = 350$ K and at pressures up to 30 MPa. J. Chem. Thermodynamics 30: 527–534.

Costa Gomes MF, Trusler JPM (1998b) The speed of sound in two methane-rich mixtures at temperatures between 250 K and 350 K and at pressures up to 20 MPa. J. Chem. Thermodynamics 30: 1121–1129.

Das TR, Kuloor NR (1967) Thermodynamic properties of hydrocarbons: Part II – isobutane. Indian J. Technol. 5: 40–45.

Davalos J, Anderson WR, Phelps RE, Kidnay AJ (1976) Liquid-vapor equilibria at 250.00 K for systems containing methane, ethane, and carbon dioxide. J. Chem. Eng. Data 21: 81–84.

Debenedetti PG (1996) Metastable liquids - Concepts and principles. Princeton Univ. Press, Princeton.

Defibaugh DR, Morrison G (1992) Compressed liquid densities and saturation densities of pentafluoroethane (R125). Fluid Phase Equilibria 80: 157–166.

Deiters UK (1981a) A new semiempirical equation of state for fluids. Part I: Derivation. Chem. Eng. Sci. 36: 1139–1146.

Deiters UK (1981b) A new semiempirical equation of state for fluids. Part II: Application to pure substances. Chem. Eng. Sci. 36: 1147–1151.

Deiters UK (1982) A new semiempirical equation of state for fluids. Part III: Application to phase equilibria in binary mixtures. Chem. Eng. Sci. 37: 855–861.

Dobbs ER, Finegold L (1960) Measurement of the velocity of sound in liquid argon and liquid nitrogen at high pressures. J. Acoust. Soc. Am. 32: 1215–1220.

Douslin DR, Harrison RH (1973) Pressure, volume, temperature relations of ethane. J. Chem. Thermodynamics 5: 491–512.

Douslin DR, Harrison RH (1976) Pressure, volume, temperature relations of ethylene. J. Chem. Thermodynamics 8: 301–330.

Dressner M, Bier K (1993) Thermische Mischungseffekte in binären Gasmischungen mit neuen Kältemitteln (in German). VDI Fortschritt-Berichte 3, No. 332, VDI, Düsseldorf.

Duschek W, Kleinrahm R, Wagner W (1988) Measurement and correlation of the (pressure, density, temperature) relation of nitrogen in the temperature range from 273.15 to 323.15 K at pressures up to 8 MPa. J. Chem. Thermodynamics 20: 1069–1077.

Duschek W, Kleinrahm R, Wagner W (1990) Measurement and correlation of the (pressure, density, temperature) relation of carbon dioxide. I. The homogeneous gas and liquid regions in the temperature range from 217 K to 340 K at pressures up to 9 MPa. J. Chem. Thermodynamics 22: 827–840.

Eder FX (1981) Arbeitsmethoden der Thermodynamik. Band I: Temperaturmessung (in German). Springer, Berlin Heidelberg New York.

Eder FX (1983) Arbeitsmethoden der Thermodynamik. Band II: Thermische und kalorische Stoffeigenschaften (in German). Springer, Berlin Heidelberg New York.

Edison TA, Anisimov MA, Sengers JV (1998) Critical scaling laws and an excess Gibbs energy model. Fluid Phase Equilibria 151: 429–438.

Edwards TJ (1984) Specific heat measurements near the critical point of carbon dioxide. Ph.D. thesis, Univ. Western Australia, Western Australia.

Elhassan AE, Craven RJB, de Reuck KM (1997) The area method for pure fluids and an analysis of the two phase region. Fluid Phase Equilibria 130: 167–187.

Elhassan AE, Lopez AA, Craven RJB (1996) Solution of the multiphase equilibrium problem for pure component, binary and ternary systems using the area method. J. Chem. Soc., Faraday Trans. 92: 4419–4433.

Ely JF (1986) An equation of state model for pure CO_2 and CO_2 rich mixtures. In: Proc. 65th annual convention of the Gas Processor Association. San Antonio, TX.

Ely JF (1990) A predictive, exact shape factor extended corresponding states model for mixtures. Adv. in Cryog. Eng. 35: 1511–1520.

Ely JF, Magee JW, Bain BC (1989) Isochoric (p,v,T) measurements on CO_2 and on (0.982 CO_2 + 0.018 N_2) from 250 to 330 K at pressures to 35 MPa. J. Chem. Thermodynamics 21: 879–894.

Erickson DD, Leland TW (1986) Application of critical-region scaling to pure component equations of state. Int. J. Thermophysics 7: 911–922.
Erickson DD, Leland TW, Ely JF (1987) A method for improving equations of state near the critical point. Fluid Phase Equilibria 37: 185–205.
Ernst G, Hochberg UE (1989) Flow calorimetric results for the heat capacity c_p of CO_2, of C_2H_6, and of (0.5 CO_2 + 0.5 C_2H_6) at high pressures. J. Chem. Thermodynamics 21: 407–414.
Ernst G, Maurer G, Wiederuh E (1989) Flow calorimeter for the accurate determination of the isobaric heat capacity at high pressures; results for carbon dioxide. J. Chem. Thermodynamics 21: 53–65.
Estela-Uribe JF, Trusler JPM (1998) Shape factors for light hydrocarbons. Fluid Phase Equilibria 151: 225–234.
Estrada-Alexanders AF, Trusler JPM (1995) The speed of sound in gaseous argon at temperatures between 110 K and 450 K and at pressures up to 19 MPa. J. Chem. Thermodynamics 27: 1075–1089.
Estrada-Alexanders AF, Trusler JPM (1996) Thermodynamic properties of gaseous argon at temperatures between 110 K and 450 K and densities up to 6.8 mol/dm^3 determined from the speed of sound. Int. J. Thermophysics 17: 1325–1347.
Estrada-Alexanders AF, Trusler JPM (1997) The speed of sound and derived thermodynamic properties of ethane at temperatures between 220 K and 450 K and pressures up to 10.5 MPa. J. Chem. Thermodynamics 29: 991–1015.
Eubank PT, Elhassan AE, Barrufet MA, Whiting WB (1992) Area method for prediction of fluid-phase equilibria. Ind. Eng. Chem. Res. 31: 942–949.
Ewers J, Wagner W (1982a) Eine Methode zur Optimierung der Struktur von Zustandsgleichungen und ihre Anwendung zur Aufstellung einer Fundamentalgleichung für Sauerstoff (in German). VDI-Forsch.-Heft 609: 27–34, VDI, Düsseldorf.
Ewers J, Wagner W (1982b) A method for optimizing the structure of equations of state and its application to an equation of state for oxygen. In: Sengers JV (ed.) Proceedings of the eighth symposium on thermophysical properties. Vol. 1: Thermophysical properties of fluids. 78–87, ASME, New York.
Ewing MB, Goodwin ARH (1992) An apparatus based on a spherical resonator for measuring the speed of sound in gases at high pressures. Results for argon at temperatures between 255 K and 300 K and at pressures up to 7 MPa. J. Chem. Thermodyn. 24: 531–547.
Ewing MB, Goodwin ARH, McGlashan ML, Trusler JPM (1988) Thermophysical properties of alkanes from speeds of sound determined using a spherical resonator 2. n-butane. J. Chem. Thermodynamics 20: 243–256.
Ewing MB, Goodwin ARH, Trusler JPM (1989) Thermophysical properties of alkanes from speeds of sound determined using a spherical resonator 3. n-pentane. J. Chem. Thermodynamics 21: 867–877.
Ewing MB, Trusler JPM (1992) Second acoustic virial coefficients of nitrogen between 80 and 373 K. Physica A 184: 415–436.
Fischer J (1999) Molecular dynamic simulations for a two-center Lennard Jones fluid with dipole; results from M. Lisal (Prague) and J. Fischer (Vienna). Private communication.
Fisher GD, Leland TW (1970) Corresponding states principle using shape factors. Ind. Eng. Fundam. 9: 537–544.
Fisher RA (1934) Two new properties of mathematical likelihood. Proc. Roy. Soc. London A144: 285–307.
Friend DG, Huber ML (1994) Thermophysical property standard reference data from NIST. Int. J. Thermophysics 15: 1279–1288.
Friend DG, Ingham H, Ely JF (1991) Thermophysical properties of ethane. J. Phys. Chem. Ref. Data 20: 275–347.
Fletcher R (1971) A modified marquardt subroutine for non-linear least squares. United Kingdom atomic energy authority research group report AERE-R 6799, Harwell.

Fox JR (1983) Method for construction of nonclassical equations of state. Fluid Phase Equilibria 14: 45–53.
Friend DG (1992) NIST standard reference database 14: NIST mixture property Program, Version 9.08. National Institute of Standards and Technology, Standard Reference Data Program, Gaithersburg.
Fujii K (1994) Accurate measurements of the speed of sound velocity in pure water under high pressure. Proceedings 12th Symp. on Thermophys. Prop., NIST, Boulder CO.
Fujii K, Masui R (1993) Accurate measurements of the sound velocity in pure water by combining a coherent phase-detection technique and a varaiable path-length interferometer. J. Acoust. Soc. Am. 93: 276–282.
Funke M, Kleinrahm R, Wagner W (2000a) Measurement and correlation of the (p,ρ,T) relation of sulphur hexafluoride (SF_6). I. The homogeneous gas and liquid regions in the temperature range from 224 K to 340 K at pressures up to 12 MPa. Submitted to J. Chem. Thermodynamics.
Funke M, Kleinrahm R, Wagner W (2000b) Measurement and correlation of the (p,ρ,T) relation of sulphur hexafluoride (SF_6). II. Saturated-liquid and saturated-vapour densities and vapour pressures along the entire coexistence curve. Submitted to J. Chem. Thermodynamics.
Garipis D, Stamatoudis M (1992) Comparison of generalized equations of state to predict gas-phase heat capacity. AIChE J. 38: 302–307.
Garnjost H (1974) Druck–Volumen–Temperaturmessungen mit Ammoniak und Wasser (in German). Ph.D. thesis, Ruhr-Universität Bochum, Bochum.
Garvin D, Parker VB, White HJ Jr. (1987) CODATA thermodynamic tables. Springer, Berlin, Heidelberg, New York.
Gauß CF (1809) Theoria motus corporum coelestium in sctionobus coicis solem ambientum. Perthes and Besser, Hamburg.
Gilgen R, Kleinrahm R, Wagner W (1992) Supplementary measurements of the (pressure, density, temperature) relation of carbon dioxide in the homogeneous region at temperatures from 220 K to 360 K and pressures up to 13 MPa. J. Chem. Thermodynamics 24: 1243–1250.
Gilgen R, Kleinrahm R, Wagner W (1994) Measurement and correlation of the (pressure, density, temperature) relation of argon. I. The homogeneous gas and liquid regions in the temperature range from 90 K to 340 K at pressures up to 12 MPa. J. Chem. Thermodynamics 26: 383–398.
Gillis KA (1997) Thermodynamic properties of seven gaseous halogenated hydrocarbons from acoustic measurements: $CHClFCF_3$, CHF_2CF_3, CF_3CH_3, CHF_2CH_3, $CF_3CHFCHF_2$, $CF_3CH_2CF_3$, and $CHF_2CF_2CH_2F$. Int. J. Thermophysics 18: 73–135.
Gmehling J (1995) From UNIFAC to modified UNIFAC to PSRK with the help of DDB. Fluid Phase Equilibria 107: 1–29.
Gmehling J, Fischer K, Li J, Schiller M (1993) Status and results of group contribution methods. Pure and Appl. Chem. 55: 919–926.
Golovskii EA, Zagoruchenko VA, Tsymarnii VA (1973) Experimental study of the compressibility of ethylene up to a pressure of 2000 bar. Izvestiya Vysshikh Uchebuykh Zavedenii, Neft i Gaz 16: 73–76.
Goodwin ARH, Defibaugh DR, Weber LA (1992) The vapour pressure of 1,1,1,2-tetrafluoroethane (R134a) and chlorodifluoromethane (R22). Int. J. Thermophysics 13: 837–854.
Goodwin ARH, Marsh KN, Wakeham WA (2000) IUPAC Experimental thermodynamics, Vol. VI: measurements of the thermodynamic properties of single phases. In preparation, Blackwell Science, Oxford.
Goodwin RD, Haynes WM (1982a) Thermophysical properties of propane from 85 to 700 K at pressures to 70 MPa. Nat. Bur. Stand. Monograph 170, Boulder CO.
Goodwin RD, Haynes WM (1982b) Thermophysical properties of normal butane from 135 to 700 K at pressures to 70 MPa. Nat. Bur. Stand. Monograph 169, Boulder CO.

Goodwin RD, Haynes WM (1982c) Thermophysical properties of isobutane from 114 to 700 K at pressures to 70 MPa. Nat. Bur. Stand. Tech. Note 1051, Boulder CO.

Grigorev MA, Muradev RM, Rastorguev YL (1975) PVT dependence of cyclohexane at temperatures and pressures (in Russian). Izv. Vyssh. Uchebn. Zaved. Neft Gaz 18: 61.

Grini PG, Owren GA (1997) Enthalpy increment measurements on nitrogen between the temperatures 160 K and 260 K, at the pressures (0.3, 3.0, 5.0, and 15.0) MPa. J. Chem. Thermodynamics 29: 37–42.

Gunn RD, Chueh PL, Prausnitz JM (1966) Inversion temperatures and pressures for cryogenic gases and their mixtures. Cryogenics 6: 324–329.

Gupta D, Eubank PT (1997) Density and virial coefficients of gaseous butane from 265 to 450 K at pressures to 3.3 MPa. J. Chem. Eng. Data 42: 961–970.

Gurvich LV, Veyts IV, Alcock CB (1989): Thermodynamic properties of individual substances, Volume 1, 4th edn. Hemisphere, New York Washington Philadelphia London.

Haar L, Gallagher JS, Kell GS (1982) The anatomy of the thermodynamic surface of water: the formulation and comparison with data. In: Sengers JV (ed.) Proc. 8th Symp. Thermophys. Prop., pp. 298–302, ASME, New York.

Haar L, Shenker SH (1971) Equations of state for dense gases. J. Chem. Phys. 55: 4951–4958.

Halm RL, Stiel LI (1967) A fourth parameter for the vapor pressure and entropy of vaporization of polar fluids. AIChE J. 13: 351–355.

Händel G, Kleinrahm R, Wagner W (1992) Measurement of the (pressure, density, temperature) relation of dichlorodifluoromethane (R12) and chlorodifluoromethane (R22) in parts of the homogeneous gas and liquid region and of the saturation curve. J. Chem. Thermodynamics 24: 697–713.

Haselton HT Jr., Sharp WE, Newton RC (1978) CO_2 fugacities at high temperatures and pressures from experimental decarbonation reactions. Geophys. Res. Letters 5: 753–757.

Haynes WM (1983) Measurements of densities and dielectric constants of liquid propane from 90 to 300 K at pressures to 35 MPa. J. Chem. Thermodynamics 15: 419–424.

Haynes WM, Hiza MJ (1977) Measurements of the orthobaric liquid densities of methane, ethane, propane, isobutane, and normal butane. J. Chem. Thermodynamics 9: 179–187.

Hayward ATJ (1971) Negative pressure in liquids: can it be harnessed to serve man? American Scientist 59: 434–443.

Heidemann RA (1983) Coputation of high pressure phase equilibria. Fluid Phase Equilibria 14: 55–78.

de Hemptinne J-C, Ungerer P (1995) Accuracy of the volumetric predictions of some important equations of state for hydrocarbons, including a modified version of the Lee-Kesler method. Fluid Phase Equilibria 106: 81–109.

Herzberg G (1950) Molecular spectra and molecular structure, I. spectra of diatomic molecules, 2nd edn., cor. reprint 1989. Van Nostrand Reinhold, New York.

Herzberg G (1966) Molecular spectra and molecular structure, III. electronic spectra and electronic structure of polyatomic molecules, 2nd edn., cor. reprint 1991. Van Nostrand Reinhold, New York.

Herzberg G, Herzberg L (1972) Constants of polyatomic molecules. In: Gray DE (ed.) American Institute of Physics handbook, 3rd edn., McGraw-Hill, New York.

Hill PG (1990) A unified fundamental equation for the thermodynamic properties of H_2O. J. Phys. Chem. Ref. Data 19: 1233–1274.

Huber ML, Ely JF (1994) A predictive extended corresponding states model for pure and mixed refrigerants including an equation of state for R134a. Int. J. Refrig. 17: 18–31.

Huang FH, Li MH, Lee LL, Starling KE, Chung FTH (1985) An accurate equation of state for carbon dioxide. J. Chem. Eng. Jap. 18: 490–496.

Hust JG, McCarty RD (1967) Curve-fitting techniques and applications to thermodynamics. Cryogenics 7: 200–206.

Hwang C-A, Simon PP, Hou H, Hall KR, Holste JC, Marsh KN (1997) Burnett and pycnometric (p, V_m, T) measurements for natural gas mixtures. J. Chem. Thermodynamics 29: 1455–1472.

ISO (1992) ISO 31, Quantities and units – Part 4: Heat. International Organization for Standardization, Genève.

ISO (1994) ISO 12213 Natural gas – calculation of compression factor. Draft International Standard. International Organization for Standardization, Genève.

Van Itterbeek A, van Dael W (1961) The velocity of sound in liquid argon and liquid nitrogen at high pressures. Cryogenics 1: 226–228.

Van Itterbeek A, van Dael W (1962) Velocity of sound in liquid oxygen and liquid nitrogen as a function of temperature and pressure. Physica 28: 861–870.

Van Itterbeek A, Thoen J, Cops A, van Dael W (1967) Sound velocity measurements in liquid methane as a function of pressure. Physica 35: 162–166.

IUPAC (1993) Quantities, Units and Symbols in Physical Chemistry, Recommendations 1993; Reprinted with corrections 1995. Blackwell, Oxford.

IUPAC (1996) Größen, Einheiten und Symbole in der Physikalischen Chemie (in German). Hausmann (ed.), VCH, Weinheim.

Jacobsen RT, Jahangiri M, Stewart RB, McCarty RD, Levelt Sengers JMH, White J Jr., Sengers JV, Olchowsky GA, de Reuck KM, Angus S, Cole RJ, Craven B, Wakeham, WA (1988) International thermodynamic tables of the fluid state – 10 – Ethylene. Blackwell, Oxford.

Jacobsen RT, Penoncello SG, Lemmon RT (1992) A fundamental equation for dichlorodifluoromethane (R11). Fluid Phase Equilibria 80: 45–56.

Jacobsen RT, Penoncello SG, Lemmon RT (1997) Thermodynamic properties of cryogenic fluids. Plenum Press, New York.

Jacobsen RT, Stewart RB (1973) Thermodynamic properties of nitrogen including liquid and vapor phases from 63 to 2000 °K with pressures up to 10,000 MPa. J. Phys. Chem. Ref. Data 2: 757–922.

Jacobsen RT, Stewart RB, Jahangiri M (1986a) Thermodynamic properties of nitrogen from the freezing line to 2000 K at pressures to 1000 MPa. J. Phys. Chem. Ref. Data 15: 735–909.

Jacobsen RT, Stewart RB, Jahangiri M (1986b) A thermodynamic property formulation for nitrogen from the freezing line to 2000 K at pressures to 1000 MPa. Int. J. Thermophysics 7: 503–511.

Jacobsen RT, Stewart RB, Jahangiri M, Penoncello SG (1986c) A new fundamental equation for thermodynamic property correlations. Adv. in Cryog. Eng. 31: 1161–1168.

Jaeschke M (1997) Results of refractive index measurements on natural gas like mixtures. Private communication, Ruhrgas AG, Dorsten.

Jaeschke M (1998) Zustandsgleichungen für eine sichere Erdgasversorgung (in German). Lecture held at the VDI GVC/GET session on thermodynamics, Leipzig.

Jaeschke M, Humphreys AE (1990) The GERG Databank of High Accuracy Compressibility Measurements. GERG Tech. Monograph 4, VDI, Düsseldorf.

Jaeschke M, Schley P (1995) Ideal-gas thermodynamic properties for natural gas applications. Int. J. Thermophys. 6: 1381–1392.

Jaeschke M, Schley P (1996a) Berechnung des Realgasfaktors von Erdgasen mit der AGA8-DC92 Zustandsgleichung. I. Aufbau der Gleichung (in German). GWF Gas Erdgas 137: 339–345.

Jaeschke M, Schley P (1996b) Berechnung des Realgasfaktors von Erdgasen mit der AGA8-DC92 Zustandsgleichung. II. Vergleich mit Meßwerten (in German). GWF Gas Erdgas 137: 420–426.

Jaeschke M, Schley P, Busch C (1998) Berechnung kalorischer Zustandsgrößen von Erdgasen mit der AGA8-DC92-Zustandsgleichung (in German). GWF Gas Erdgas 139: 714–719.

Jahangiri M, Jacobsen RT, Stewart RB, McCarty RD (1986) Thermodynamic properties of ethylene from the freezing line to 450 K at pressures to 260 MPa. J. Phys. Chem. Ref. Data 15: 593–734.

Ji WR, Lempe DA (1997) Density improvement of the SRK equation of state. Fluid Phase Equilibria 130: 49–63.

Ji WR, Lempe DA (1998) A systematic study of cubic three-parameter equations of state for deriving a structurally optimized PVT relation. Fluid Phase Equilibria 147: 85–103.

Jin GX, Tang S, Sengers JV (1992) Thermodynamic properties of methane in the critical region. Int. J. Thermophysics 13: 671–684.

Jing Z, Fam AT (1987) An algorithm for computing continous Chebyshev approximations. Mathm. of Computations 48: 691–710.

Kamei A, Beyerlein SW, Jacobsen RT (1995) Application of nonlinear regression in the development of a wide range formulation for HFC-22. Int. J. Thermophys. 16: 1155–1164.

Kaminishi GI, Yokoyama C, Takahashi S (1988) Saturated liquid densities of n-butane - isobutane, n-butane - propane, isobutane - propane, and n-butane - isobutane - propane mixtures. Sekiyu Gakkaishi 31: 433–438.

Karkaris A, Kalfopoulus T, Stamatoudis M (1992) Comparison of generalised methods to predict gas-phase heat capacity. Ind. Eng. Chem. Res. 31: 1830–1833.

Katti RS, Jacobsen RT, Stewart RB, Jahangiri M (1986) Thermodynamic properties for neon for temperatures from the triple point to 700 K at pressures to 700 MPa. Adv. Cryo. Eng. 31: 1189–1197.

Keenan JH, Keyes FG, Hill PG, Moore JG (1969) Steam tables - thermodynamic properties of water including vapor, liquid, and solid phases. John Wiley & Sons, New York London Sidney Toronto.

Kell GS, Whalley E (1975) Reanalysis of the density of liquid water in the range $0-150\,°C$ and $0-1$ kbar. J. Chem. Phys. 48: 3496–3503.

Kelly RL (1987) Atomic and ionic spectrum lines below 2000 Angstroms: hydrogen through krypton. J. Phys. Chem. Ref. Data 16, Suppl. 1, Part I–III.

Kerimov AM, Apaev TA (1974) Experimental investigation of the P-V-T relationship of cyclohexane over a wide range of temperature and pressure. Fluid-Mech. Sov. Res. 3(4): 100–104.

Kerns WJ, Anthony RG, Eubank PT (1974) Volumetric properties of cyclohexane vapor. Am. Inst. Chem. Eng. 70 (140): 14–21.

Killner J, Craven RJB (1996) Comparisons of equations of state with experimental data for R32 and R125. Final report to the IEA-Annex 18. IUPAC Thermodynamic Tables Project Centre, Imperial College, London.

Kirkpatrick S, Gelatt CD, Vecchi MP Jr. (1983) Optimization by simulated annealing. Science 220: 671–680.

Kiselev SB (1998) Cubic crossover equation of state. Fluid Phase Equilibria 147: 7–23.

Kiselev SB, Friend DG (1999) Revision of a multiparameter equation of state to improve the representation in the critical region: application to water. Fluid Phase Equilibria 155: 33–55.

Kiselev SB, Kostyukova IG, Povodyrev AA (1991) Universal crossover behavior of fluids and fluid mixtures in the critical region. Int. J. Thermophysics 12: 877–895.

Kiselev SB, Rainwater JC, Huber ML (1998) Binary mixtures in and beyond the critical region: thermodynamic properties. Fluid Phase Equilibria 151: 469–478.

Kiselev SB, Sengers JV (1993) An improved parametric crossover model for the thermodynamic properties of fluids in the critical region. Int. J. Thermophysics 14: 1–32.

Kleinrahm R, Wagner W (1986) Measurement and correlation of the equilibrium liquid and vapour densities and the vapour pressure along the coexistence curve of methane. J. Chem. Thermodynamics 18: 739–760.

Klimeck J, Kleinrahm R, Wagner W (1998) An accurate single-sinker densimeter and measurements of the (p, ρ, T) relation of argon and nitrogen in the temperature range from (235 to 520) K at pressures up to 30 MPa. J. Chem. Thermodynamics 30: 1571–1588.

Klimeck J, Kleinrahm R, Wagner W (2000) Measurements of the (p, ρ, T) relation of methane and carbon dioxide in the temperature range from (235 to 520) K at pressures up to 30 MPa using a new accurate single-sinker densimeter. Submitted to J. Chem. Thermodynamics.

Klimeck R, Span R, Kleinrahm R, Wagner W (1996) Fundamental equations for caloric properties of natural gases – collecting of data and test of existing equations. Final Report to the GERG working group 1.3, Ruhr-Universität Bochum, Bochum.

Klimeck R, Span R, Wagner W (1999a) Development of a reference equation for thermal and caloric properties of natural gases − theortical results of phase one. Report to N.V. Nederlandse Gasunie, Snam SpA, Enagas, Gaz de France, Statoil AS, and DVGW e.V.; Ruhr-Universität Bochum, Bochum.

Klimeck R, Span R, Wagner W (1999b) Development of a reference equation of state for thermal and caloric properties of natural gases − first results on mixtures. Lecture held at a meeting of working group 1.34 of the Groupe Européen de Recherches Gazières, Trondheim.

Klomfar J, Hruby J, Šifner O (1993) Measurements of the p-v-T behaviour of the refrigerant R134a in the liquid phase. Int. J. Thermophysics 14: 727−738.

Kohlen R (1987) Das fluide Zustandsgebiet von R22. VDI Fortschritt-Berichte 19, No. 14, VDI, Düsseldorf.

Kohler F, Haar L (1981) A new representation for thermodynamic properties of a fluid. J. Chem. Phys. 75: 388−394.

Kortbeek PJ, Trappeniers NJ, Biswas SN (1988) Compressibility and sound velocity measurements on N_2 up to 1 GPa. Int. J. Thermophysics 9: 103−116.

Kraft K, Leipertz A (1994) Thermal diffusivity and ultrasonic velocity of saturated R125. Int. J. Thermophysics 15: 387−399.

Kratzke H, Müller S, Bohn M, Kohlen R (1985) Thermodynamic properties of saturated and compressed liquid n-pentane. J. Chem. Thermodynamics 17: 283−294.

Kriebel C, Mecke M, Winkelmann J, Vrabec J, Fischer J (1998) An equation of state for dipolar two-center Lennard-Jones molecules and its application to refrigerants. Fluid Phase Equilibria 142: 15−32.

Kruse A (1997) Neue Zustandsgleichungen für industrielle Anwendungen im technisch relevanten Zustandsgebiet von Wasser (in German). Dissertation, Ruhr-Universität Bochum, Bochum.

Kuranov G, Rumpf B, Maurer G, Smirnova N (1997) VLE modelling for aqueous systems containing methyldiethanolamine, carbon dioxide and hydrogen. Fluid Phase Equilibria 136: 147−162.

Kurzeja N, Tielkes Th, Wagner W (1999) The nearly classical behavior of a pure fluid on the critical isochore very near the critical point under the influence of gravity. Int. J. Thermophys. 20: 531−561.

Kurzeja N, Tielkes Th, Wagner W (2000) The nearly classical behavior of a pure fluid on the phase boundary very near the critical point under gravity influence. To be submitted to Int. J. Thermophys.

Kurzeja N, Wagner W (2000) The curvature of the pressure-temperature relationship of a pure fluid on the critical isochore very near the critical point under gravity influence. To be submitted to Int. J. Thermophys.

Lacam A (1956) Etude experimentale de la propagation des ultrasons dans les fluides en fonction de la pression (1200 atmospheres) et de la temperature (200 °C) (in French). J. Recherches du CNRS 34: 25−56.

Lallemand M, Vidal D (1977) Variation of the polarizability of noble gases with density. J. Chem. Phys. 66: 4776−4780.

Leach JW, Chappelear PS, Leland TW (1968) Use of molecular shape factors in vapor-liquid equilibrium calculations with the corresponding states principle. AIChE J. 14: 568−576.

Lee BI, Kessler MG (1975) A generalised thermodynamic correlation based on three-parameter corresponding states. AIChE J. 21: 510−527.

Lemmon EW (1996) A generalised model for the prediction of the thermodynamic properties of mixtures including vapor-liquid equilibrium. Ph. D. thesis, Univ. of Idaho, Moscow.

Lemmon EW (1997) Evaluation of thermodynamic property models for mixtures of R-32, R-125, and R-134a. Report to IEA Annex 18, Nat. Inst. Stand. and Techn., Boulder CO.

Lemmon EW, Jacobsen RT (1998) A multiple regression procedure for thermodynamic equations of state introducing direct nonlinear fits of temperature and density exponents. Private communication, National Institute of Standards and Technology, Boulder CO.

Lemmon EW, Jacobsen RT (1999a) A generalised model for the description of the thermodynamic properties of mixtures. Int. J. Thermophysics 20: 825–835.
Lemmon EW, Jacobsen RT (1999b) An international standard formulation for the thermodynamic properties of 1,1,1-trifluoroethane (HFC-143a) for temperatures from 161 to 500 K and Pressures to 60 MPa. Submitted to J. Phys. Chem. Ref. Data.
Lemmon EW, Jacobsen RT (2000) Thermodynamic properties of mixtures of refrigerants R-32, R-125, R-134a, and R-152a. Int. J. Thermophysics, in Press.
Lemmon EW, Jacobsen RT, Beyerlein SW (1992) The prediction of the vapor-liquid equilibrium thermodynamic properties of N2-Ar-O2 mixtures including air. Adv. Cryog. Eng. 37B: 1107–1113.
Lemmon EW, Jacobsen RT, Penoncello SG, Friend DG (1999) Thermodynamic properties of air from 60 to 2000 K at pressures to 2000 MPa and of mixtures of nitrogen, argon and oxygen. Submitted to J. Phys. Chem. Ref. Data.
Lemmon EW, Tillner-Roth R (1999) A Helmholtz energy equation of state for calculating the thermodynamic properties of fluid mixtures. Fluid Phase Equilibria 165: 1–21.
Li J, Tillner-Roth R, Sato H, Watanabe K (1997) An equation of state for 1,1,1-trifluoroethane (R-143a). In: Haynes WM (Ed.) Proc. 13th Symp. Thermophys. Prop., NIST, Boulder Co.
Lucas K (1991) Applied statistical thermodynamics. Springer, Berlin Heidelberg New York.
Lucas K, Delfs U, Buss V, Speis M (1993) Ideal-gas properties of new refrigerants from quantum mechanical abinitio calculations. Int. J. Thermophys. 14: 993–1006.
Lüddecke TO (1991) Vergleich verschiedener Strukturoptimierungsverfahren zur Aufstellung thermodynamischer Gleichungen (in German). Student thesis at the Institute for Thermodynamic, University Hannover, Hannover.
Lüddecke TO, Magee JW (1996) Molar heat capacity at constant volume of difluoromethane (R32) and pentafluoroethane (R125) from the triple-point temperature to 345 K at pressures to 35 MPa. Int. J. Thermophysics 17: 823–849.
Luettmer-Starthmann J, Tang S, Sengers JV (1992) A parametric model for the global thermodynamic behavior of fluids in the critical region. J. Chem. Phys. 97: 2705–2717.
Magee JW (1991) Molar heat capacity (C_v) for saturated and compressed liquid and vapor nitrogen from 65 to 300 K at pressures to 35 MPa. J. Res. Nat. Bur. Stand. Techn. 96: 725–740.
Magee JW (1992) Measurements of molar heat capacities at constant volume (C_V) for 1,1,1,2-tetrafluoroethane (R134a). Int. J. Refrig. 15: 372–380.
Magee JW (1996) Isochoric p-ρ-T measurements on difluoromethane (R32) from 142 to 396 K and pentafluoroethane (R125) from 178 to 398 K at pressures to 35 MPa. Int. J. Thermophysics 17: 803–822.
Magee JW (1998) Molar heat capacity at constant volume of 1,1-difluoroethane (R152a) and 1,1,1-trifluoroethane (R143a) from the triple-point temperature to 345 K at pressures to 35 MPa. Int. J. Thermophysics 19: 1397–1420.
Magee JW, Ely JF (1986) Specific heats (c_v) of saturated and compressed liquid and vapor carbon dioxide. Int. J. Thermophysics 7: 1163–1181.
Magee JW, Haynes WM, Hiza MJ (1997) Isochoric (p,ρ,T) measurements for five natural gas mixtures from $T = (225$ to $350)$ K at pressures to 35 MPa. J. Chem. Thermodynamics 29: 1439–1454.
Magee JW, Howley JB (1992) Vapour pressure measurements on 1,1,1,2-tetrafluoroethane (R134a) from 180 to 350 K. Int. J. Refrig. 15: 362–364.
Magee JW, Lüddecke TOD (1998) Molar heat capacity at constant volume of n-butane at temperatures from 141 to 342 K and at pressures to 33 MPa. Int. J. Thermophysics 19: 129–144.
Marquardt DW (1963) An algorithm for least-squares estimation of nonlinear parameters. J. Soc. Indust. Appl. Math. 11: 431–441.
Martinez-Ortiz JA, Manley DB (1978) Vapor pressures for the system isobutane–isobutylene–n-butane. J. Chem. Eng. Data 23: 165–167.

Marx V, Pruß A, Wagner W (1992) Neue Zustandsgleichungen für R12, R22, R11 und R113, Beschreibung des thermodynamischen Zustandsverhaltens bei Temperaturen bis 525 K und Drücken bis 200 MPa (in German). VDI Fortschritt-Berichte 6, No. 57, VDI, Düsseldorf.

Mason EA, Spurling TH (1969) The virial equation of state. In: The Int. Enc. of Phys. Chem. and Chem. Phys., Pergamon, Oxford.

Maurer G (1996) Phase equilibria in chemical reactive fluid mixtures. Fluid Phase Equilibria 116: 39–51.

McBride BJ, Gordon S, Reno MA (1993) Coefficients for calculating thermodynamic and transport properties of individual species. NASA Technical Memorandum 4513.

McCarty RD (1970) Provisional thermodynamic functions for helium 4 for temperatures from 2 to 1500 K with pressures to 100 MN/m^2 (1000 atmospheres). Nat. Bur. Stand. Report 9762, Boulder CO.

McCarty RD, Arp VD (1990) A new wide range equation of state for helium. Adv. Cryog. Eng. 35: 1465–1475.

McClune CR (1976) Measurement of the densities of liquefied hydrocarbons from 93 to 170 K. Cryogenics 16: 289–295.

McCullough JP, Scott DW (1968) Experimental thermodynamics. Volume I: Calorimetry of non-reacting systems. Butterworth, London.

Mecke M, Müller A, Winkelmann J, Fischer J (1997) An equation of state for two-center Lennard-Jones fluids. Int. J. Thermophysics 18: 683–698.

Mecke M, Müller A, Winkelmann J, Vrabec J, Fischer J, Span R, Wagner W (1996) An accurate Van der Waals type equation of state for the Lennard-Jones fluid. Int. J. Thermophysics 17: 391–404.

Meyer-Pittroff R, Grigull U (1973) Eine kanonische Zustandsgleichung für Kohlendioxid (in German). Wärme- und Stoffübertragung 3: 134–145.

Michels A, Wijker H, Wijker HK (1949) Isotherms of argon between 0 °C and 150 °C and pressures up to 2900 atmospheres. Physica 15: 627–633.

Michelsen ML (1982a) The isothermal flash problem. Part I. Stability. Fluid Phase Equilibria 9: 1–19.

Michelsen ML (1982b) The isothermal flash problem. Part II. Phase-split calculation. Fluid Phase Equilibria 9: 21–40.

Michelsen ML (1984) Calculation of critical points and phase boundaries in the critical region. Fluid Phase Equilibria 16: 57–76.

Miller DG (1970) Joule-Thomson inversion curve, corresponding states, and simpler equations of state. Ind. Eng. Chem. Fundam. 9: 585–589.

Miller RC, Kidnay AJ, Hiza MJ (1977) Liquid + vapor equilibria in methane + ethene and in methane + ethane from 150.00 to 190.00 K. J. Chem. Thermodynamics 9: 167–178.

Mills RL, Liebenberg DH, Bronson JC (1975) Sound velocity and the equation of state of N_2 to 22 kbar. J. Chem. Phys. 63: 1198–1204.

Mills MB, Wills MJ, Bhirud VL (1980) The calculation of density by the BWRS equation of state in process simulation contexts. AIChE J. 26: 902–910.

Mohr PJ, Taylor BN (2000) CODATA Recommended values of the fundamental physical constants: 1998. Accepted for publication in J. Phys. Chem. Ref. Data.

Moldover MR, Trusler JPM, Edwards TJ, Mehl JB, Davis RS (1988) Measurement of the universal gas constant R using a spherical acoustic resonator. J. Res. Nat. Bur. Stand. 93: 85–144.

Mollerup J (1980) Thermodynamic properties from corresponding states theory. Fluid Phase Equilibria 4: 11–34.

Moore CE (1971) Atomic energy levels. Nat. Stand. Ref. Data Ser., No. 35, Nat. Bur. Stand. (U.S.), Washington.

Moore MJ, Sieverding CH (1976) Two-phase steam flow in turbines and separators. Hemisphere, Washington London.

Morris EC (1984) Improved and extended high-pressure *PVT* measurements for argon. J. Chem. Phys. 81: 581–582.

Morris EC, Wylie RG (1980) Accurate method for high pressure *PVT* measurements and results for argon for $T = -20$ to $+35$ °C and p in the range 200–480 MPa. J. Chem. Phys. 73: 1359–1367.

Morrison G, Ward DK (1991) Thermodynamic properties of two alternative refrigerants: 1,1-dichloro-2,2,2-trifluoroethane (R123) and 1,1,1,2-tetrafluoroethane (R134a). Fluid Phase Equilibria 62: 65–86.

Morsy TE (1963) Zum thermischen und kalorischen Verhalten realer fluider Stoffe (in German). Ph.D. thesis, University Karlsruhe, Karlsruhe.

Müller A, Winkelmann J, Fischer J (1996) Backone family of equations of state: 1. Nonpolar and polar pure fluids. AIChE J. 42: 1116–1126.

Muradev RM (1982) Creation of a data bank of an automatic system for users of thermophysical properties. Thermal equation of state of liquid cyclohexane. Elek. Avu. Ob'ekt. Neft Prom. Groznyi 217–220.

Muringer MJP, Trappeniers NJ, Biswas SN (1985) The effect of pressure on the sound velocity and density of toluene and n-heptane up to 2600 bar. Phys. Chem. Liq. 14: 273–296.

Nagel M, Bier K (1998) Saturation densities of new refrigerant mixtures. Int. J. Refrig. 21: 556–566.

Nellis WJ, Mitchell AC (1980) Shock compression of liquid argon, nitrogen, and oxygen to 90 GPa (900 kbar). J. Chem. Phys. 73: 6137–6145.

Nichols WB, Reamer HH, Sage BH (1955) Phase equilibria in hydrocarbon systems. Volumetric behavior of n-heptane. Ind. Eng. Chem. 10: 2219–2221.

Niepmann R (1984) Thermodynamic properties of propane and n-butane - 2. Speeds of sound in the liquid up to 60 MPa. J. Chem. Thermodynamics 16: 851–860.

Nieuwoudt JC, Neindre BL, Tufeu R, Sengers JV (1987) Transport properties of isobutane. J. Chem. Eng. Data 32: 1–8.

Novikov II, Trelin YS (1960) Velocity of sound in vapour-liquid equilibrium phases; velocity of sound propagation in saturated vapours of CO_2 (in Russian). Zurnal Prikladnoj mechaniki i Technickeskoj Fiziki 2: 112–115.

Nowak P, Kleinrahm R, Wagner W (1996a) Measurement and correlation of the (p,ρ,T) relation of ethylene I. The homogeneous gaseous and liquid regions in the temperature range from 105 K to 340 K at pressures up to 12 MPa. J. Chem. Thermodynamics 28: 1423–1439.

Nowak P, Kleinrahm R, Wagner W (1996b) Measurement and correlation of the (p,ρ,T) relation of ethylene II. Saturated-liquid and saturated-vapour densities and vapour pressures along the entire coexistance curve. J. Chem. Thermodynamics 28: 1441–1460.

Nowak P, Kleinrahm R, Wagner W (1997a) Measurement and correlation of the (p,ρ,T) relation of nitrogen I. The homogeneous gas and liquid regions in the temperature range from 66 K to 340 K at pressures up to 12 MPa. J. Chem. Thermodynamics 29: 1137–1156.

Nowak P, Kleinrahm R, Wagner W (1997b) Measurement and correlation of the (p,ρ,T) relation of nitrogen II. Saturated-liquid and saturated-vapour densities and vapour pressures along the entire coexistence curve. J. Chem. Thermodynamics 29: 1157–1174.

Le Neindre B, Vodar B (1975) Experimental thermodynamics. Volume II: Experimental thermodynamics of non-reacting fluids. Butterworth, London.

Orrit JE, Laupretre JM (1978) Density of liquified natural gas components. Adv. Cryog. Eng. 23: 573–579.

Outcalt SL, McLinden MO (1995) Equations of state for the thermodynamic properties of R32 (Difluoromethane) and R125 (Pentafluoroethane). Int. J. Thermophysics 16: 79–89.

Outcalt SL, McLinden MO (1996) A modified Benedict-Webb-Rubin equation of state for the thermodynamic properties of R152a (1,1-difluoroethane). J. Phys. Chem. Ref. Data 25: 605–636.

Panasiti MD, Lemmon EW, Penoncello SG, Jacobsen RT, Friend DG (1999) Thermodynamic properties of air from 60 to 2000 K at pressures up to 2000 MPa. Int. J. Thermophys. 20: 217–228.

Peng D, Robinson D (1976) A new two constant equation of state. Ind. Eng. Chem. Fundamentals 15: 59–64.

Pennington RE, Kobe KA (1954) Contributions of vibrational anharmonicity and rotation-vibration interaction to thermodynamic functions. J. Chem. Phys. 22: 1442–1447.

Penoncello SG, Jacobsen RT, Goodwin ARH (1995) A thermodynamic property formulation for cyclohexane. Int. J. Thermophysics 16: 519–531.

Penoncello SG, Jacobsen RT, Lemmon EW (1992) A fundamental equation for dichlorodifluoromethane (R-12). Fluid Phase Equilibria 80: 57–70.

Perel'shtein II (1970) Experimental investigation of compressibility of Freon-12. Thermodyn. Prop. of Matter and Substance: 225–234.

Piao CC, Noguchi M (1998) An international standard equation of state for the thermodynamic properties of HFC-125 (pentafluoroethane). J. Phys. Chem. Ref. Data 27: 775–806.

Pieperbeck N, Kleinrahm R, Wagner W (1991) Results of (pressure, density, temperature) measurements on methane and on nitrogen in the temperature range from 273.15 K to 323.15 K at pressures up to 12 MPa using a new apparatus for accurate gas-density measurements. J. Chem. Thermodynamics 23: 175–194.

Pitzer KS, Lippmann DZ, Curl RF, Huggins CM Jr., Petersen DE (1955) The volumetric and thermodynamic properties of fluids. II. Compressibility factor, vapor pressure and entropy of vaporization. J. Am. Chem. Soc. 77: 3433–3440.

Pitzer KS, Schreiber DR (1988) Improving equation-of-state accuracy in the critical region: Equations for carbon dioxide and neopentane as examples. Fluid Phase Equilibria 41: 1–17.

Pitzer KS, Sterner SM (1995a) Equations of state valid contiuously from zero to extreme pressures with H_2O and CO_2 as examples. Int. J. Thermophysics 16: 511–518.

Pitzer KS, Sterner SM (1995b) Equations of state valid contiuously from zero to extreme pressures for H_2O and CO_2. J. Chem. Phys. 101: 3111–3116.

Platzer B (1990) Eine Generalisierung der Zustandsgleichung von Bender zur Berechnung von Stoffeigenschaften unpolarer und polarer Fluide und deren Gemische (in German). Dissertation, Universität Kaiserslautern, Kaiserslautern.

Platzer B, Maurer G (1989) A generalised equation of state for pure polar and nonpolar fluids. Fluid Phase Equilibria 51: 223–236.

Plöcker U (1977) Berechnung von Hochdruck-Phasengleichgewichten mit einer Korrespondenzmethode unter besonderer Berücksichtigung asymmetrischer Gemische (in German). Dissertation, TU Berlin, Berlin.

Plöcker U, Kamp H, Prausnitz J (1978) Calculation of high-pressure vapor-liquid equilibria from a corresponding-states correlation with emphasis on asymmetric mixtures. I&EC Proc. Des. & Dev. 17: 324–332.

Pollak R (1974) Die thermodynamischen Eigenschaften von Wasser - dargestellt durch eine kanonische Zustandsgleichung für die fluiden homogenen und heterogenen Zustände bis 1200 Kelvin und 3000 bar (in German). Dissertation, Ruhr-Univ. Bochum, Bochum.

Pollak R (1975) Eine neue Fundamentalgleichung zur konsistenten Darstellung der thermodynamischen Eigenschaften von Wasser (in German). BWK 27: 210–213.

Polt A (1987) Zur Beschreibung thermodynamischer Eigenschaften reiner Fluide mit „Erweiterten BWR-Gleichungen" (in German). Dissertation, Univ. Kaiserslautern, Kaiserslautern.

Polt A, Maurer G (1992) The Bender equation of state for describing properties of krypton, neon, fluorine, sulfur dioxide and water over a wide range of states. Fluid Phase Equilibria 73: 27–38.

Polt A, Platzer B, Maurer G (1992) Parameter der thermischen Zustandsgleichung von Bender für 14 mehratomige Stoffe (in German). Chem. Technik 22: 216–224.

Povodyrev AA, Jin GX, Kiselev SB, Sengers JV (1996) Crossover equation of state for the thermodynamic properties of mixtures of methane and ethane in the critical region. Int. J. Thermophysics 17: 909–944.

Preston-Thomas H (1976) The International Practical Temperature Scale of 1968; amended edition of 1975. Metrologia 12: 7–17.

Preston-Thomas H (1990) The International Temperature Scale of 1990. Metrologia 27: 3–10.

Price RA, Kobayashi R (1959) Low temperature vapor-liquid equilibrium in light hydrocarbon mixtures: Methane - ethane - propane system. J. Chem. Eng. Data 4: 40–52.

Pruß A, Wagner W (1995) Eine neue Fundamentalgleichung für das fluide Zustandsgebiet von Wasser für Temperaturen von der Schmelzlinie bis zu 1273 K bei Drücken bis zu 1000 MPa (in German). VDI Fortschritt-Berichte, 6, No. 320, VDI, Düsseldorf.

Rabinovich VA, Vassermann AA, Nedostup VI, Veksler LS (1988) Thermodynamic properties of neon, argon, krypton, and xenon. Hemisphere, Washington New York London.

Rastorguev Y, Grigorev BA, Safir LI (1975) PVT dependence of cyclohexane in the liquid phase (in Russian). Izv. Vyssh. Uchebh. Zaved. Fiz., Neft Gaz 18 (1): 66–70.

Rastorguev Y, Grigorev BA, Safir LI (1976) Experimental study of the isobaric heat capacity of cyclohexane in the range 275 – 400 °C at critical pressures (in Russian). Izv. Sev.-Kauk. Nauch. Tsen. Vys. 4 (2): 107–108.

Rechenberg I (1973) Evolutionsstrategie (in German). Problemata, Frommann-Holzboog, Stuttgart.

Redlich O, Kwong JNS (1949) On the thermodynamics of solutions. V: An equation of state. Fugacities of gaseous solutions. Chem. Reviews 44: 233–244.

Reid CR, Prausnitz JM, Poling BE (1987) The properties of gases and liquids, 4th edn. McGraw-Hill, New York.

de Reuck (1979) SEEQ - a computer program for a method of correlation using a search procedure based on a step-wise least-squares technique. Report PC/R33, IUPAC Thermodynamic Tables Research Centre, Imperial College, London.

de Reuck (1991) Extrapolation of accurate equations of state outside the range of the experimental data, 1st draft. Personal communication to the participants of the 5th International Workshop on Equations of State, Imperial College, London.

de Reuck KM (1990) Thermodynamic tables of the fluid state – 11. fluorine. Pergamon, Oxford.

de Reuck KM, Armstrong B (1979) A method of correlation using a search procedure based on a stepwise least-squares technique and its application to an equation of state for propylene. Cryogenics 19: 505–512.

de Reuck KM, Craven RJB (1993) International thermodynamic tables of the fluid state – 12. methanol. Blackwell Scientific, London.

Riedel L (1963) Berechnung der kritischen Daten der verzweigten Paraffin-Kohlenwasserstoffe (in German). Chem. Eng. Tech. 35: 433–439.

Robertson SL, Babb SE Jr., Scott GJ (1969) Isotherms of argon to 10000 bars and 400°. J. Chem. Phys. 50: 2160–2166.

Rowlinson JS (1965) The equation of state of dense systems. Rep. Prog. Phys. 28: 169–199.

Rowlinson JS, Swinton FL (1982) Liquids and liquid mixtures, 3rd edn. Butterworths, Great Britain.

Rusby RL (1990) The conversion of thermal reference values to the ITS-90. J. Chem. Thermodynamics 23: 1153–1161.

Rusby RL, Hudson RP, Durieux M (1994) Revised values for $(t_{90}-t_{68})$ from 630° C to 1064° C. Metrologia 31: 149–153.

Saager B, Fischer J, Neumann M (1991) Reaction field simulations of monoatomic and diatomic dipolar fluids. Mol. Simulation 8: 27–49.

Saager B, Fischer J (1992) Construction and application of physically based equations of state – Part II. The dipolar and quadrupolar contributions to the Helmholtz energy. Fluid Phase Equilibria 72: 67–88.

Saager B, Hennenberg R, Fischer J (1992) Construction and application of physically based equations of state – Part I. Modification of the BACK equation. Fluid Phase Equilibria 72: 41–66.

Sachs L (1973) Statistische Auswertemethoden (in German). Springer, Berlin Heidelberg New York.

Sadowski G, Arlt W (1995) High pressure phase equilibrium in multicomponent hydrocarbon systems using the MFLG model. J. Supercrit. Fluids 8: 273–281.

Sadowski G, Mokrushina LV, Arlt W (1997) Finite and infinite dilution activity coefficients in polycarbonate systems. Fluid Phase Equilibria 139: 391–403.

Safir LI, Gerasimov AA, Grigorev BA (1975) Experimental study of cyclohexane specific heat (in Russian). Izv. Vyssh. Uchebn. Zaved., Neft Gaz 11 (11): 61–65.

Sato H, Watanabe K, Levelt Sengers JMH, Gallagher JS, Hill PG, Starub J, Wagner W (1991) Sixteen thousand evaluated thermodynamic property data for water and steam. J. Phys. Chem. Ref. Data 20: 1023–1044.

Sauermann P, Holzapfel K, Oprzynski J, Kohler F, Poot W, de Loos TW (1995) The pvT properties of ethanol and hexane. Fluid Phase Equilibria 112: 249–272.

Saul A, Wagner W (1989) A fundamental equation for water covering the range from the melting line to 1273 K at pressures up to 25000 MPa. J. Phys. Chem. Ref. Data 18: 1537–1564.

Savidge JL, Shen JJS (1989) Sound speed of natural gas. In: Proc. 4 th IGRC, pp. 511–519, Tokyo, Government Institutes Inc., Rockville, MD.

Savidge KE, Starling JL (1992) Compressibility factors of natural gas and other related hydrocarbon gases. American Gas Association Transmission Measurement Committee Report No. 8, Arlington.

Schaber A (1965) Zum thermischen Verhalten fluider Stoffe (in German). Ph.D. thesis, University Karlsruhe, Karlsruhe.

Schäfer K (1960) Statistische Theorie der Materie, Band 1, allgemeine Grundlagen und Anwendungen auf Gase (in German). Vandenhock & Ruprecht, Göttingen.

Schley P (1994) Entwicklung von Korrelationsgleichungen für die isobare Wärmekapazität im Zustand des idealen Gases und Kombination der neuen Gleichungen mit der AGA8-DC92 Zustandsgleichung zur Berechnung kalorischer Zustandsgrößen von Erdgasen (in German). Diploma thesis, Ruhr-Universität Bochum, Bochum and Ruhrgas AG, Dorsten.

Schlünder E-U, et al. (1997) VDI-Wärmeatlas, 8th edn. (in German). Springer, Berlin Heidelberg New York.

Schmidt R, Wagner W (1985) A new form of the equation of state for pure substances and its application to oxygen. Fluid Phase Equilibria 19: 175–200.

Schofield P (1969) Parametric representation of the equation of state near a critical point. Phys. Rev. Letters 22: 606–608.

Schofield P, Lister JD, Ho JT (1969) Correlation between critical coefficients and critical exponents. Phys. Rev. Letters 23: 1098–1102.

Schreiner K (1986) Beschreibung des thermischen Verhaltens reiner Fluide mit druckexpliziten kubischen Zustandsgleichungen. VDI Fortschritt-Berichte 3, No. 125, VDI, Düsseldorf.

Schwartzentruber J, Renon H, Watanasiri S (1989) Development of a new cubic equation of state for phase equilibrium calculations. Fluid Phase Equilibria 52: 127–134.

Sengers JV (1994) Effects of critical fluctuations on the thermodynamic and transport properties of supercritical fluids. In: Kiran E and Levelt Sengers JMH (ed.) Supercritical fluids. Kluwer, Dordrecht.

Sengers JV, Levelt Sengers JMH (1986) Thermodynamic behavior of fluids near the critical point. Ann. Rev. Phys. Chem. 37: 189–222.

Setzmann U, Span R, Wagner W (1990) Two methods for optimizing the structure and length of correlation equations. Documentation of a software development for the National Engineering Laboratory, Ruhr-University Bochum, Bochum.

Setzmann U, Wagner W (1989a) A new method for optimising the structure of thermodynamic correlation equations. Int. J. Thermophys. 10: 1103–1126.

Setzmann U, Wagner W (1989b) Zwei Verfahren zur Optimierung der Struktur und der Länge von Korrelationsgleichungen (in German). Documentation of a software development for the Linde AG, Ruhr-University Bochum, Bochum.

Setzmann U, Wagner W (1991) A new equation of state and tables of thermodynamic properties for methane covering the range from the melting line to 625 K at pressures up to 1000 MPa. J. Phys. Chem. Ref. Data 20: 1061–1155.

Sharif MAR, Groves TK (1989) Apparatus for the measurement of decompression wave front velocity based sound speeds and of associated densities and isothermal compressibility coefficients in moderately dense gases. Chem. Eng. Comm. 86: 199–223.

Sharif MAR, Groves TK (1990) A critical evaluation of three equations of state for dense gases against observed data. Chem. Eng. Com. 98: 89–111.

Shimanouchi T (1972) Tables of molecular vibrational frequencies, consolidated volume. Nat. Stand. Ref. Data Ser., No. 39, Nat. Bur. Stand. (U.S.), Washington.

Shimanouchi T (1977) Tables of molecular vibrational frequencies, consolidated volume II. J. Phys. Chem. Ref. Data 6: 993–1102.

Shimanouchi T, Matsuura H, Ogawa Y, Harada I (1978) Tables of molecular vibrational frequencies, part 9. J. Phys. Chem. Ref. Data 7: 1323–1443.

Shimanouchi T, Matsuura H, Ogawa Y, Harada I (1980) Tables of molecular vibrational frequencies, part 10. J. Phys. Chem. Ref. Data 9: 1149–1254.

Shmonov VM, Shmulovich KI (1974) Molar volumes and equations of state of carbon dioxide in the 100 - 1000° and 2000 - 10000 bar range. Akad. Wouk. SSSR 217: 935–938.

Shubert KB, Ely JF (1995) Application of a new selection algorithm to the development of a wide-range equation of state for refrigerant R134a. Int. J. Thermophys. 16: 101–110.

Šifner O, Klomfar J (1994) Thermodynamic properties of xenon from the triple point to 800 K with pressures up to 350 MPa. J. Phys. Chem. Ref. Data 23: 63–152.

Sirota AM, Khromykh YA, Gol'dshtein II (1979) An experimental investigation of the specific heat of sulphur hexafluoride. Thermal Eng. 26: 733–738.

Smukala J, Span R, Wagner W (1999) A new equation of state for ethylene covering the fluid region for temperatures from the melting line to 450 K at pressures up to 300 MPa. Submitted to J. Phys. Chem. Ref. Data.

Soave GS (1972) Equilibrium constants from a modified Redlich-Kwong equation of state. Chem. Eng. Sci. 27: 1197–1203.

Soave GS (1995) A noncubic equation of state for the treatment of hydrocarbon fluids at reservoir conditions. I&EC Research 34: 3981–3994.

Somayajulu GR (1989) Estimation procedures for critical constants. J. Chem. Eng. Data 34: 106–120.

Span R (1993) Eine neue Fundamentalgleichung für das fluide Zustandsgebiet von Kohlendioxid bei Temperaturen bis zu 1100 K und Drücken bis zu 800 MPa (in German). VDI Fortschritt-Berichte 6, No. 285, VDI, Düsseldorf.

Span R (1996) Equations of state for natural gases – current status and future developments. Lecture held at a meeting of working group 1.3 of the Groupe Européen de Recherches Gazières, Dorsten.

Span R (1998a) Simple equations of state for technical applications with an improved behaviour in the meta- and instable regions. Unpublished results.

Span R (1998b) BACKONE-type equations of state for 14 nonpolar and polar fluids. Unpublished results.

Span R, Collmann HJ, Wagner W (1998a) Simultaneous optimization as a method to establish generalized functional forms for empirical equations of state. Int. J. Thermophysics 19: 491–500.

Span R, Kleinrahm R, Wagner W (1995a) Ein Programmpaket zur interaktiven Berechnung der Zustandsgrößen von Ethylen (in German). Software developed for ARG, EC, and DSM, Ruhr-Universität Bochum, Bochum.

Span R, Kleinrahm R, Wagner W, Konopka G, Jaeschke M, Bitkow P (1995b) Vereinfachte Zustandsgleichungen für Erdgase (in German). gwf – Gas / Erdgas 136: 644–648.
Span R, Lemmon EW, Jacobsen RT, Wagner W (1998b) A reference quality equation of state for nitrogen. Int. J. Thermophys. 19: 1121–1132.
Span R, Lemmon EW, Jacobsen RT, Wagner W, Yokozeki A (1999) A reference equation of state for the thermodynamic properties of nitrogen for temperatures from 63.151 to 1000 K and pressures to 2200 MPa. Submitted to J. Phys. Chem. Ref. Data.
Span R, Wagner W (1994) Analytic integration of the AGA8-DC92 equation of state. Private communication to members of the working group 1.3 of the Groupe Européen de Recherches Gazières, Ruhr-Universität Bochum, Bochum.
Span R, Wagner W (1996) A new equation of state for carbon dioxide covering the fluid region from the triple-point temperature to 1100 K at pressures up to 800 MPa. J. Phys. Chem. Ref. Data 25: 1509–1596.
Span R, Wagner W (1997) On the extrapolation behavior of empirical equations of state. Int. J. Thermophys. 18: 1415–1443.
Span R, Wagner W (2000a) Equations of state for technical applications. I. Simultaneously optimised functional forms for nonpolar and polar fluids. To be submitted to Int. J. Thermophysics.
Span R, Wagner W (2000b) Equations of state for technical applications. II. Results for nonpolar fluids. To be submitted to Int. J. Thermophysics.
Span R, Wagner W (2000c) Equations of state for technical applications. III. Results for polar fluids. To be submitted to Int. J. Thermophysics.
Span R, Wagner W (2000d) An accurate empirical three parameter equation of state for nonpolar fluids. To be submitted to Fluid Phase Equilibria.
Starling KE (1973) Fluid thermodynamic properties for light petroleum systems. Gulf publishing, Houston.
Starling KE, Fitz CW, Chen YC, Rondon E, Jacobsen RT, Beyerlein SW, Clarke WP, Lemmon EW, Savidge JL (1991) GRI high accuracy natural gas equation of state for gas measurement applications. Annual report to the Gas Research Institute, University of Oklahoma, Tulsa.
Starling KE, Savidge JL (1992) Compressibility factors of natural gases using the AGA-DC92 equation. AGA Transmission Measurement Com. Rep. No. 8, sec. Ed., Arlington, Virginia.
Sterner SM, Pitzer KS (1994) An equation of state for carbon dioxide valid from zero to extreme pressures. Contrib. Mineral. Petrol. 117: 362–374.
Stewart RB, Jacobsen RT (1989) Thermodynamic properties of argon from the triple point to 1200 K with pressures to 1000 MPa. J. Phys. Chem. Ref. Data 18: 639–798.
Straty GC (1975) Hypersonic velocities in saturated and compressed fluid methane. Cryogenics 15: 715–731.
Straty GC (1980) (p,V,T) of compressed fluid ethene. J. Chem. Thermodynamics 12: 709–716.
Straty GC, Tsumara R (1976) PVT and vapor pressure measurements on ethane. J. Res. Nat. Bur. Stand. 80A: 35–39.
Straty GC, Younglove BA (1973) Velocity of sound in saturated and compressed fluid oxygen. J. Chem. Thermodynamics 5: 305–312.
Straub D (1964) Zur Theorie eines allgemeinen Korrespondenzprinzips der thermischen Eigenschaften fluider Stoffe (in German). Ph.D. thesis, University Karlsruhe, Karlsruhe.
Straub J (1967) Optische Bestimmung von Dichteschichtungen im kritischen Zustand (in German). Chem. Ing. Techn. 39: 291–296.
Straub J (1972) Das nicht klassische Verhalten fluider Stoffe im kritischen Zustand (in German). Wärme und Stoffübertragung 5: 56–63.
Strobridge TR (1962) The thermodynamic properties of nitrogen from 64 to 300° K between 0.1 and 200 atmospheres. National Bureau of Standards, Technical Note 129, Washington.
Sun TF, Kortbeek PJ, Trappeniers NJ, Biswas SN (1987) Acoustic and transport properties of benzene and cyclohexane as function of pressure and temperature. Phys. Chem. Liq. 16: 163–178.

Sunaga H, Tillner-Roth R, Sato H, Watanabe K (1998) A thermodynamic equation of state for pentafluoroethane (R-125). Int. J. Thermophysics 19: 1623–1635.
Susnar SS, Budziak CJ, Hamza HA, Neumann AW (1992) Pressure dependence of the density of n-alkanes. Int. J. Thermophysics 13: 443–452.
Sychev VV, Vassermann AA, Golovsky EA, Kozlov AD, Spiridonov GA, Tsymarny VA (1987g) Thermodynamic properties of ethylene. Hemisphere, Washington New York London.
Sychev VV, Vassermann AA, Kozlov AD, Spiridonov GA, Tsymarny VA (1984) Thermodynamic properties of helium. Gosstandard, Moscow.
Sychev VV, Vassermann AA, Kozlov AD, Spiridonov GA, Tsymarny VA (1987a) Thermodynamic properties of helium. Hemisphere, Washington New York London.
Sychev VV, Vassermann AA, Kozlov AD, Spiridonov GA, Tsymarny VA (1987b) Thermodynamic properties of nitrogen. Hemisphere, Washington New York London.
Sychev VV, Vassermann AA, Kozlov AD, Spiridonov GA, Tsymarny VA (1987e) Thermodynamic properties of oxygen. Hemisphere, Washington New York London.
Sychev VV, Vassermann AA, Kozlov AD, Spiridonov GA, Tsymarny VA (1987f) Thermodynamic properties of air. Hemisphere, Washington New York London.
Sychev VV, Vassermann AA, Kozlov AD, Tsymarny VA (1991) Thermodynamic properties of propane. Hemisphere, Washington New York London.
Sychev VV, Vassermann AA, Kozlov AD, Tsymarny VA (1995) Thermodynamic properties of butane. Begell House, New York.
Sychev VV, Vassermann AA, Kozlov AD, Zagoruchenko VA, Spiridonov GA, Tsymarny VA (1987d) Thermodynamic properties of ethane. Hemisphere, Washington New York London.
Sychev VV, Vassermann AA, Zagoruchenko VA, Kozlov AD, Spiridonov GA, Tsymarny VA (1987c) Thermodynamic properties of methane. Hemisphere, Washington New York London.
Tegeler C, Span R, Wagner W (1997) Eine neue Fundamentalgleichung für das fluide Zustandsgebiet von Argon für Temperaturen von der Schmelzlinie bis 700 K und Drücke bis 1000 MPa (in German). VDI Fortschritt-Berichte 3, No. 480, VDI, Düsseldorf.
Tegeler C, Span R, Wagner W (1999) A new equation of state for argon covering the fluid region for temperatures from the melting line to 700 K at pressures up to 1000 MPa. J. Phys. Chem. Ref. Data 28: 779–850.
Teja AS, Lee RJ, Rosenthal D, Anselme M (1990) Correlation of the critical properties of alkanes and alkanols. Fluid Phase Equilibria 56: 153–169.
Tiesinga BW, Sakonidou E, van den Berg HR, Luettmer-Strathmann J, Sengers JV (1994) The thermal conductivity of argon in the critical region. J. Chem. Phys. 101: 6944–6963.
Thomas RHP, Harrison RH (1982) Pressure-volume-temperature relations for propane. J. Chem. Eng. Data 27: 1–11.
Thomas W, Zander M (1966) Thermische Zustandsgrößen von Äthylen (in German). Zeitschr. angew. Phys. 20: 417–419.
Tillner-Roth R (1993) Die thermodynamischen Eigenschaften von R152a, R134a und ihren Gemischen (in German). Forschungsberichte des DKV, No. 41, DKV, Stuttgart.
Tillner-Roth R (1995) A fundamental equation of state for 1,1-difluoroethane (HFC-152a). Int. J. Thermophysics 16: 91–100.
Tillner-Roth R (1996) A nonlinear regression analysis for estimating low-temperature vapor pressures and enthalpies of vaporization applied to refrigerants. Int. J. Thermophys. 17: 1365–1385.
Tillner-Roth R (1998) Fundamental equations of state. Shaker, Aachen.
Tillner-Roth R, Baehr HD (1992) Burnett measurements and correlation of gas-phase (p,ρ,T) of 1,1,1,2-tetrafluoroethane (R134a) and of 1,1,-difluoroethane (R152a). J. Chem. Thermodynamics 24: 413–424.
Tillner-Roth R, Baehr HD (1993) Measurements of liquid, near critical, and supercritical (p,ρ,T) of 1,1,1,2-tetrafluoroethane (R134a) and of 1,1,-difluoroethane (R152a). J. Chem. Thermodynamics 25: 277–292.

Tillner-Roth R, Baehr HD (1994) An international standard formulation for the thermodynamic properties of 1,1,1,2-tetrafluoroethane (HFC-134a) for temperatures from 170 K to 455 K and pressures up to 70 MPa. J. Phys. Chem. Ref. Data 23: 657–729.

Tillner-Roth R, Friend DG (1998) A Helmholtz free energy formulation of the thermodynamic properties of the mixture {water + ammonia}. J. Phys. Chem. Ref. Data 27: 63–96.

Tillner-Roth R, Harms-Watzenberg F, Baehr HD (1993) Eine neue Fundamentalgleichung für Ammoniak (in German). DKV-Tagungsbericht 20, II: 167–181, DKV, Stuttgart.

Tillner-Roth R, Li J, Yokozeki A, Sato H, Watanabe K (1998) Thermodynamic properties of HFC refrigerants. JAR, Tokyo.

Tillner-Roth R, Yokozeki A (1997) An international standard equation of state for difluoromethane (R-32) for temperatures from the triple point at 136.34 K to 435 K at pressures up to 70 MPa. J. Phys. Chem. Ref. Data 26: 1273–1328.

Toledo PG, Reich R (1988) A comparison of enthalpy prediction methods for nonpolar and polar fluids and their mixtures. Ind. Eng. Chem. Res. 27: 1004–1010.

Trappeniers NJ, Wassenaar T, Wolkers GJ (1976) Isotherms and thermodynamic properties of ethylene at temperatures between 0 and 150 °C and at densities up to 500 amagat. Physica 82A: 305–311.

Trebble MA, Bishnoi PR (1987) Development of a new four-parameter cubic equation of state. Fluid Phase Equilibria 35: 1–18.

TRC (1972-1993) Thermodynamic Tables. Thermodynamic Research Centre, Texas A&M University, College Station.

Trusler JPM (1991) Physical acoustics and metrology of fluids. Adam Hilger, Bristol Philadelphia New York.

Trusler JPM, Costa Gomes MF (1996) The speed of sound in methane and in two methane-rich gas mixtures. Final Report to GERG WG 1.3, Imperial College, London.

Trusler JPM, Zarari MP (1996) The speed of sound in gaseous propane at temperatures between 225 K and 375 K and at pressures up to 0.8 MPa. J. Chem. Thermodynamics 28: 329–335.

Türk M, Crone M, Bier K (1997) Mixing behaviour of equal amounts of substance of 1,1,1,2-terafluoroethane and 1,1-difluoroethane II. Representation of thermal properties by equations of state. J. Chem. Thermodynamics 29: 369–383.

Vacek V, Zollweg JA (1993) Speed of sound in compressed sulfur hexafluoride. Fluid Phase Equilibria 88: 219–226.

Verschaffelt JE (1896) Measurements on capillary ascension of liquefied carbonic acid near the critical temperature. Com. Phys. Lab. Univ. Leiden 28: 1–15.

Vetere A (1995) Methods to predict the critical constants of organic compounds. Fluid Phase Equilibria 109: 17–27.

de Vries B (1997) Experimental results for the density of R125. Private Communication, Univ. Hannover.

de Vries B, Tillner-Roth R, Baehr HD (1995) The thermodynamic properties of HFC-124. Proc. 19th Int. Congr. of Refrig., 582–589, Den Haag.

Vukalovich MP, Altunin VV (1962) Experimental investigation of the specific volume of CO_2 at temperatures 200 - 750 °C and pressures to 600 kg / cm^2. Teploenergetica 9 (5): 56–62.

van der Waals JD (1873) Over de Continuiteit van den Gas en Vloistof Toestand. Dissertation, Univ. Leiden, Leiden.

Wagner W (1970) Eine thermische Zustandsgleichung zur Berechnung der Phasengleichgewichte flüssig-gasförmig für Stickstoff (in German). Dissertation, TU Braunschweig, Braunschweig.

Wagner W (1972) A method to establish equations of state exactly representing all saturated state variables applied to nitrogen. Cryogenics 12: 214–221.

Wagner W (1974) Eine mathematisch statistische Methode zum Aufstellen thermodynamischer Gleichungen - gezeigt am Beispiel der Dampfdruckkurve reiner fluider Stoffe (in German). Fortschr.-Ber. VDI-Z., 3, 39, VDI-Verlag, Düsseldorf.

Wagner W (1977) A new correlation method for thermodynamic data applied to the vapour-pressure curve of argon, nitrogen and water. Report PC/T15, IUPAC Thermodynamic Tables Research Centre, Imperial College, London.

Wagner W, Cooper JR, Dittmann A, Kijima J, Kretzschmar H-J, Kruse A, Mareš R, Oguchi K, Sato H, Stöcker I, Šifner O, Takaishi Y, Tanishita I, Trübenbach J, Willkommen Th (2000a) The IAPWS industrial formulation 1997 for the thermodynamic properties of water and steam. Submitted to ASME J. Eng. for Gas Turbines and Power 122: 150–182.

Wagner W, Ewers J, Schmidt R (1982) An equation for the ideal-gas heat capacity of molecular oxygen for temperatures from 30 K to 3000 K. Ber. Bunsenges. Phys. Chem. 86: 538–540.

Wagner W, Kruse A (1998) Properties of water and steam. Springer, Berlin Heidelberg New York.

Wagner W, Kurzeja N, Pieperbeck B (1992) The thermal behaviour of pure fluid substances in the critical region – experiences from recent prT measurements on SF6 with a multi-cell apparatus. Fluid Phase Equilibria 79: 151–174.

Wagner W, Marx V, Pruß A (1993) A new equation of state for chlorodifluoromethane (R22) covering the entires fluid region from 116 K to 550 K at pressures up to 200 MPa. Rev. Int. Froid 16: 373–389.

Wagner W, Pollak R (1974) A simple, but very accurate vapor pressure equation for water. In: Proc. 8th Int. Conf. on the Prop. of Water and Steam, Giens, 1974, Edition Européennes Thermique et Industries, Paris, Vol. 2, 787–796.

Wagner W, Pruß A (1997) Die neue internationale Standard-Zustandsgleichung für Wasser für den allgemeinen und wissenschaftlichen Gebrauch (in German with the English IAPWS release as appendix). In: Jahrbuch 97, VDI-GVC, pp. 134–156, VDI, Düsseldorf.

Wagner W, Pruß A (2000) The IAPWS formulation 1995 for the thermodynamic properties of ordinary water substance for general and scientific use. Submitted to J. Phys. Chem. Ref. Data.

Wagner W, de Reuck KM (1987) International thermodynamic tables of the fluid state – 9 – Oxygen. Blackwell, Oxford.

Wagner W, de Reuck KM (1996) International thermodynamic tables of the fluid state – 13 – Methane. Blackwell, Oxford.

Wagner W, Saul A, Pruß A (1994) International equations for the pressure along the melting and along the sublimation curve of ordinary water substance. J. Phys. Chem. Ref. Data 23: 515–527.

Wagner W, Span R (1993) Special equations of state for methane, argon and nitrogen for the temperature range from 270 to 350 K at pressures up to 30 MPa. Int. J. Thermophys. 14: 699–725.

Wagner W, Span R, Bonsen C (2000b) Wasser und Wasserdampf (in German). Springer electronic media, Berlin Heidelberg New York.

Waxman M (1980) Private communication to the NBS. In: Goodwin RD, Haynes WM (1982c) Thermophysical properties of isobutane from 114 to 700 K at pressures to 70 MPa. Nat. Bur. Stand. Tech. Note 1051, Boulder CO.

Weber LA (1970) P-V-T thermodynamic and related properties of oxygen from the triple point to 300 K at pressures to 33 MN/m^2. J. Res. Nat. Bur. Stand. 74A: 93–129.

Weber LA (1977) Thermodynamic and related properties of oxygen from the triple point to 300 K at pressures to 1000 bar. NASA Ref. Publ. 1011, NBSIR 77–865.

Weber LA (1981) Measurements of the heat capacities C_v of dense gaseous and liquid nitrogen and nitrogen trifluoride. J. Chem. Thermodynamics 13: 389–403.

Weber LA (1989a) Vapor pressures and gas-phase pvT data for 1,1,1,2-tetrafluoroethane. Int. J. Thermophysics 10: 617–627.

Weber LA (1989b) Simple apparatus for vapor-liquid equilibrium measurements with data for the binary systems of carbon dioxide with n-butane and isobutane. J. Chem. Eng. Data 34: 171–175.

Wegner FJ (1972) Corrections to scaling laws. Phys. Review B 5: 4529–4536.

Wichterle I, Kobayashi R (1972) Vapor-liquid equilibrium of methane-ethane system at low temperatures and high pressures. J. Chem. Eng. Data 17: 9–12.
Wilson KG (1974) Renormalization group and critical phenomena. Phys. Review 4: 3174–3184.
Wilson KG (1979) Problems in physics with many sacles of length. Scientific American 241 (2): 140–157.
Wilson KG (1986) Die Renormierungsgruppe (in German). In: Teilchen, Felder und Symmetrien. Spektrum der Wissenschaft, Heidelberg.
Wirbser H, Bräuning G, Ernst G (1992) Flow calorimetric specific heat capacities of the refrigerants CFCl3 (R11) and CF2ClCFCl2 (R113) at pressures between 0.6 MPa and 30 MPa and temperatures between 288.15 K and 503.15 K. J. Chem. Thermodynamics 24: 783–784.
Wong YL, Cheng SC, Groeneveld DC (1990) Generalized thermodynamic and transport properties evaluation for nonpolar fluids. Heat Transfer Eng. 11: 60–72.
Wooley HW (1983) A switch function applied to the thermodynamic properties of steam near and not near the critical point. Int. J. Thermophysics 4: 51–95.
Yamada T (1973) An improved generalized equation of state. AIChE J. 19: 286–291.
Ye S, Lagourette B, Alliez J, Saint-Guirons H, Montel F (1992) Comparison with experimental data of ultrasonic velocity in pure hydrocarbons calculated from equations of state. Fluid Phase Equilibria 74: 157–175.
Yokoyama C, Takahashi S (1991) Saturated liquid densities of 2,2-dichloro-1,1,1-trifluoroethane (HCFC-123), 1,2-dichloro-1,2,2-trifluoroethane (HCFC-123a), 1,1,1,2-tetrafluoroethane (HFC-134a) and 1,1,1-trifluoroethane (HFC-143a). Fluid Phase Equilibria 67: 227–240.
Younglove BA (1981) Velocity of sound in liquid propane. J. Res. Nat. Bur. Stand. 86 (2): 165–170.
Younglove BA, Ely JF (1987) Thermophysical properties of fluids. II. Methane, ethane, propane, isobutane, and normal butane. J. Phys. Chem. Ref. Data 16: 577–798.
Younglove BA, McLinden MO (1994) An international standard equation of state for the thermodynamic properties of refrigerant 123 (2,2-dichloro-1,1,1-trifluoroethane). J. Phys. Chem. Ref. Data 23: 731–779.
Zhang H-L, Sato H, Watanabe K (1997) A new three-parameter cubic equation of state for refrigeration engineering calculations. Int. J. Refrig. 20: 421–440.
Zubarev VN, Telegin GS (1962) The impact compressibility of liquid nitrogen and solid carbon dioxide. Sov. Phys. Doklady 7: 34–36.
Zurmühl R (1964) Matrizen (in German). Springer, Berlin Göttingen Heidelberg.
Zurmühl R, Falk S (1992) Matrizen und ihre Anwendungen, Bd. 1 (in German). Springer, Berlin Heidelberg New York.

Index

acentric factor 222, 292
AGA8-DC92 129, 324
air, reference equation for 174
ammonia
– characteristic parameters of 263
– data set for 275
– reference equation for 174
– technical equation for 265
ancillary equations 47, 78
anisotropy parameter 279
argon
– characteristic parameters of 247
– data set for 256
– reference equation for 174
– technical equation for 248
atomic energy level 69

BACKONE equations of state 29, 279
– fitting to data 281
– results of 285
bank of terms 7, 91
– defining a 91
Bender-type equation of state 6, 27, 220
– generalised form 292
– numerical stability of 222
binary interaction parameters 321, 333
Boyle curve 168
BWR-type equation of state 5
– numerical stability of 222

caloric properties, comparisons with 191
– at vapour-liquid phase equilibrium, comparisons with 207
carbon dioxide
– characteristic parameters of 263
– data set for 274
– reference equation for 174
– technical equation for 265
Center for Applied Thermodynamic Studies (CATS) 10
challenges 12
chlorine, reference equation for 174

combination rules 321
computer simulations 18
constraints 64, 112, 128
corresponding states 278, 300, 327
critical
– amplitudes 135
– exponents 135, 144
– exponents from analytic equations of state 135
– parameters ⇨ critical parameters
– point 133
– point universality 134
– region 133
– region, capabilities of empirical multiparameter equations of state 141
– region, new experimental results 138
– region terms 31, 150
– scaling ⇨ scaled equations of state
critical parameters
– estimation of 243
– influence of uncertain 242
– typical uncertainties of 243
cubic equations of state 5, 277
– starting values from 48, 52
cyclohexane
– characteristic parameters of 247
– data set for 250, 317
– reference equation for 174
– technical equation for 249

degrees of freedom 66
density
– typical uncertainties of 178
– uncertainty in the critical region 187
departure functions 334
– generalised 336
– hybrid 339
– mixture specific 335
dipole moment 280
dryness fraction ⇨ vapour fraction

electronic state 69

EOM 9, 106
equations of state
– explicit in the Helmholtz energy 16
– explicit in pressure 25
– integration of pressure explicit 26
– reference ⇨ reference equations of state
– simultaneously optimised ⇨ simultaneously optimised equations
– technical ⇨ technical equations of state
– typical performance of 173
enthalpy 37, 40
– of evaporation 43
– typical uncertainty of 203
entropy 37, 40
error propagation, law of 86
– limitations of 86
ethane
– characteristic parameters of 247
– data set for 252
– reference equation for 174
– technical equation for 248
ethene ⇨ ethylene
ethylene
– characteristic parameters of 247
– data set for 258
– reference equation for 174
– technical equation for 248
evolutionary optimisation method ⇨ EOM
experimental data
– assessment of 61
– assigning weights to 85
– at high temperatures and pressures 162
– beyond the range of primary data 163
– corrections of 89
exponential terms 93
extrapolation behaviour 161
– influence of the functional form 166

Fisher F statistic 101
fitting empirical equations to data
– fundamentals 62
– linear problems 64
– nonlinear problems 65
fugacity coefficient 40
fundamental equations of state 9

gas constant 17
Gaussian bell shaped terms 31, 150
generalised coefficients 292, 301
– fitting of 303
generalised empirical equations of state 278, 291
– based on the Bender equation 292

– based on a simultaneously optimised equation 300
– necessary features of 300
– numerical stability 302, 314
– results of 293, 306
Gibbs energy 40
group contribution methods 278

hard sphere terms 29, 93
– simplified 229
helium, reference equation for 174
Helmholtz energy
– of the ideal gas 18, 66
– residual part 23, 74
– thermodynamic properties from 35
Hugoniot curve 164

IAPWS-95 174
IAPWS-IF97 3
Ideal curve 168
ideal curves 168
– characteristic points of 170
– deformations of 171
IEA-Annex 18 10
intercorrelations 102, 222, 235, 300
internal energy 36, 40
International Association for the Properties of Water and Steam (IAPWS) 12
IPTS-68 16
isentropic compressibility 41
isentropic expansion coefficient 41
isobaric heat capacity 37, 40
– experimental results for the ideal gas 72
– fitting equations for the ideal gas 71
– of the ideal gas 19, 66
– typical uncertainty of 202
isobutane
– characteristic parameters of 247
– data set for 259
– reference equation for 174
– technical equation for 249
isochoric heat capacity 40
– typical uncertainty of 205
isothermal compressibility 41
isothermal expansion coefficient 41
isothermal throttling coefficient 41
iterative calculations 46
– based on pressure and density 51
– based on pressure and enthalpy 52
– based on pressure and entropy 54
– based on temperature and pressure 48
– of phase equilibria 54
ITS-90 16

Index

IUPAC Thermodynamic Tables Project Centre   12

Joule inversion curve   168
Joule-Thomson coefficient   40
Joule-Thomson inversion curve   168
journals, relevant   11

Lagrangian multipliers   64
linear data   77
linear preselection   120
liquid fraction   45

Massieu function   16
maximum likelihood method   62
Maxwell loops   55
– multiple   56
MBWR-type equation of state   6, 28
metastable states   45, 57
methane
– characteristic parameters of   247
– data set for   250
– reference equation for   174
– technical equation for   248
methanol, reference equation for   174
mixing rules   321
mixture models
– based on shape factors   330
– composition dependent coefficients   320
– extended corresponding states   327
– general types of   319
– multi fluid   332
– one fluid   322, 329
– with departure functions   ⇨ departure functions
molecular energy levels   70
molar masses   17
monographs, relevant series of   11
multiproperty fitting   6, 74

n-butane
– characteristic parameters of   247
– data set for   253
– reference equation for   174
– technical equation for   248
n-heptane
– characteristic parameters of   247
– data set for   254
– technical equation for   248
n-hexane
– characteristic parameters of   247
– data set for   254
– technical equation for   248

n-octane
– characteristic parameters of   247
– data set for   255
– technical equation for   248
n-pentane
– characteristic parameters of   247
– data set for   253
– technical equation for   248
National Institute of Standards and Technology (NIST)   11
natural gases   129, 324, 338
neon, reference equation for   174
Newton-Raphson algorithm   47
nitrogen
– characteristic parameters of   247
– data set for   257
– reference equation for   174
– technical equation for   249
NLOPT   125
NLREG   121
nonanalytic terms   33, 152
nonlinear data   83
nonlinear optimisation algorithms   117
– speeding up   126

OPTIM   107
– control parameters   112
– nonlinear form   ⇨ NLOPT
– preselected terms   113
– sublimits for functional forms   115
optimisation algorithms   8, 90
– automated   128
– future perspectives   131
– nonlinear   117
– quality criterion   97, 118, 225
– simultaneous   223
oxygen
– characteristic parameters of   247
– data set for   256
– reference equation for   174
– technical equation for   249

Planck-Einstein function   19, 68
3-/5-point contact   135
polar factor   292
polynomial equations   35
polynomial terms   92
power laws   134
precorrelation
– of the density   77
– factor   79
pressure   40
– derivatives of   40

propane
- characteristic parameters of 247
- data set for 252
- reference equation for 174
- technical equation for 248
propene ⇨ propylene
property tables 1
propylene, reference equation for 174
pseudocritical parameters 321, 332
pseudo-stable states 57

quadrupol moment 280
quality ⇨ vapour fraction

R11 (trichlorofluoromethane)
- characteristic parameters of 263
- data set for 265
- reference equation for 174
- technical equation for 264
R12 (dichlorodifluoromethane)
- characteristic parameters of 263
- data set for 266
- reference equation for 174
- technical equation for 264
R22 (chlorodifluoromethane)
- characteristic parameters of 263
- data set for 267
- reference equation for 174
- technical equation for 264
R32 (trifluoromethane)
- characteristic parameters of 263
- data set for 268
- reference equation for 174
- technical equation for 264
R113 (1,1,2-trichlorotrifluoroethane)
- characteristic parameters of 263
- data set for 268
- reference equation for 174
- technical equation for 264
R123 (2,2-dichloro-1,1,1-trifluoroethane)
- characteristic parameters of 263
- data set for 270
- reference equation for 174
- technical equation for 264
R124 (1-chloro-1,2,2,2-tetrafluoroethane), reference equation for 174
R125 (pentafluoroethane)
- characteristic parameters of 263
- data set for 270
- reference equation for 174
- technical equation for 264
R134a (1,1,1,2-tetrafluoroethane)
- characteristic parameters of 263

- data set for 272
- reference equation for 174
- technical equation for 264
R143a (1,1,1-trifluoroethane)
- characteristic parameters of 263
- data set for 272
- reference equation for 174
- technical equation for 265
R152a (1,1-difluoroethane)
- characteristic parameters of 263
- data set for 273
- reference equation for 174
- technical equation for 265
reader profiles 3
reference equations of state 1, 6
- group one 173
- group two 175
- highly accurate 11, 173
- selection of 174
- simultaneously optimised 275
reference state 21
regression matrix 97
regula falsi algorithm 47
- modified 48
renormalisation group theory 134
residua 62, 75
- explicit linear 75
- for linear algorithms 77
- for nonlinear algorithms 80
- implicit linear 76
- implicit nonlinear 76
rigid rotator, harmonic oscillator 68
- corrections to 70
rotational modes
- characteristic temperature 67
- external 67
- internal 70

saturated liquid heat capacity, c_σ 43, 80
- typical uncertainty of 212
scaled equations of state 138
- crossover 139
- kernel term 159
- linear model 138
- revised and extended 139
- simple 138
- switching 157
- transformation 139, 158
SEEQ 98
semiempiric equations of state 277
set-up of this book 3
SF_6 ⇨ sulphur hexafluoride
shape factor models 278, 329

– exact 331
shock-wave measurements 164
SIMOPT 224
simulated annealing 132
simultaneous optimisation 223
simultaneously optimised equations
– for nonpolar substances 248
– for polar substances 262
– for technical applications 227
– numerical stability of 235
– reference quality 275
– required data sets 245
– with 10 and 12 terms 232
software 1, 15
speed of sound 39, 40
– at ideal gas state 73
– typical uncertainty of 193
square-well terms 229
stability criteria 57
stepwise regression analysis 7, 98
– modified form in OPTIM 109
– nonlinear form ⇨ NLREG
– pairwise exchange of terms 103, 116
Starling-type equation of state 6
Student t test 100
sulphur hexafluoride
– characteristic parameters of 247
– data set for 262
– technical equation for 249
sum of squares 62, 75
– in matrix donation 97
– reduced 225
switching functions 157

technical equations of state 6, 176, 227
– bank of terms for 229
– demands on accuracy 228
temperature scales 16
thermal properties, comparisons with 177
– at vapour-liquid phase equilibrium, comparisons with 207
transformation 158
translational modes 67

uncertainty of calculated properties
⇨ *property*, typical uncertainties of
– some general assessments 214
unstable states 57

vapour fraction 45
– equivalent for metastable states 46
vapour-liquid equilibria
– consideration of 6

– iterative calculation of 54
– linearised residua for 77
– nonlinear residua for 83
– properties of the coexisiting phases 42
– properties in the two phase region 45
– comparisons with 206
vibrational modes 67
– characteristic temperature 68
– frequency 68
– wavenumber 68
virial coefficients
– second acoustic 40
– second thermal 40
– third thermal 40
– typical representation of 190
virial equation of state 5
volume expansivity 41

water, reference equation for 174
weighted variance 231
weighting factor 88
weighting of experimental data 85
– for technical equations of state 90, 228

xenon, reference equation for 174

Printing: Mercedes-Druck, Berlin
Binding: Buchbinderei Lüderitz & Bauer, Berlin